EXTENDED DEFECTS IN SEMICONDUCTORS

This book surveys the properties, effects, roles and characterization of structurally extended defects in semiconductors. The basic properties of extended defects (dislocations, stacking faults, grain boundaries and precipitates) are outlined, and their effect on the electronic properties of semiconductors, their role in semiconductor devices and techniques for their characterization are discussed. These topics are among the central issues in the investigation and applications of semiconductors and in the operation of semiconductor devices. Elucidation of the effects of extended defects on electronic properties of materials is especially important in view of the current advances in electronic device development that involve defect control and engineering at the nanometre level. The authors preface their treatment with an introduction to semiconductor materials and the book concludes with a chapter on point defect maldistributions.

This text is suitable for advanced undergraduate and graduate students in materials science and engineering, and for those studying semiconductor physics. The book may also serve as a reference for researchers in a wide variety of fields in the physical and engineering sciences.

D. B. Holt is an emeritus professor in the Department of Materials at Imperial College, London. His research has focused on the science and technology of semiconducting materials and devices, and developing and applying techniques for their chracterization in the scanning electron microscope.

B. G. Yacobi is an adjunct professor in the Department of Materials Science and Engineering at the University of Toronto. He has worked in several areas of solid-state physics, including the synthesis, applications and characterization of various electronic and photonic materials and devices.

EXTENDED DEFECTS IN SEMICONDUCTORS

Electronic Properties, Device Effects and Structures

D. B. Holt

Imperial College of Science, Technology and Medicine, London

B. G. Yacobi

University of Toronto

CAMBRIDGE
UNIVERSITY PRESS

CAMBRIDGE
UNIVERSITY PRESS

University Printing House, Cambridge CB2 8BS, United Kingdom

Cambridge University Press is part of the University of Cambridge.

It furthers the University's mission by disseminating knowledge in the pursuit of education, learning and research at the highest international levels of excellence.

www.cambridge.org
Information on this title: www.cambridge.org/9781107424142

First published 2007
First paperback edition 2014

A catalogue record for this publication is available from the British Library

Library of Congress Cataloguing in Publication data

Holt, D. B.
Extended defects in semiconductors : electronic properties, device effects and structures / D. B. Holt, B. G. Yacobi.
p. cm.
Includes bibliographical references.
ISBN-13: 978-0-521-81934-3 (hardback)
ISBN-10: 0-521-81934-2 (hardback)
1. Semiconductors--Defects. I. Yacobi, B. G. II. Title.
TK7871.852.E98 2007
621.3815'2--dc22   2006037298

ISBN 978-0-521-81934-3 Hardback
ISBN 978-1-107-42414-2 Paperback

Dedication and History

The idea for this book first arose during the (Northern-Hemisphere) Academic year (1969–70), which one of the authors, DBH, spent in the Department of Applied Physics of the University of New South Wales, in Sydney, Australia, working with Professor Dan Haneman (DH). DBH was then concerned with the use of the concept of dangling bonds in relation to defects while DH used the concept in relation to the then-new and unexpected ordered structures on atomically clean semiconductor free surfaces recently discovered by the then-new technique of LEED (low-energy electron diffraction) of which he had been one of the earliest users. It occurred to us (DH and DBH) that it would be useful to produce a critical overview of the whole field of the defects and surfaces of semiconductors from a common point of view. Work was begun on the writing of such a book; however, the field expanded explosively, changing and advancing so rapidly and giving rise to so large a literature on the topics concerned at such a rate that new information kept appearing more rapidly than it could be read, understood and reviewed. Consequently, after 10 or 15 years, the attempt had to be reluctantly abandoned as a failure. Much later, when the field had matured considerably and progress and change slowed somewhat, the present authors (DBH and BGY) took up the challenge again and when successful completion began to appear possible DBH tried via the Internet to obtain information to enable him to contact DH to inform him that the book was likely to appear at long last and to hopefully obtain his agreement and approval. Unfortunately, what was first obtained from the Internet was the obituary of DH. So Dan: Here is our book, at last, DBH. I am very sorry that it only appeared posthumously, in your case. However, if as we hope, it proves useful to folks in the field and students trying to enter it, this will be, in considerable part, to your credit.

Contents

Preface

One of the central issues in the investigation of semiconductors (and solid-state materials in general) is related to the study of various defects and their effects on materials' properties and the operation of electronic devices.

The topics of electronic properties of extended defects (i.e., dislocations, stacking faults, grain boundaries, and precipitates) in semiconductors and the influence of these defects on various electronic devices have been of great importance and interest for several decades. During this period of intensive research and development of semiconductor materials and devices, the majority of the defects and the mechanisms of their formation were elucidated. This was accompanied with concurrent efforts in eliminating the unwanted defects. For controlling properties of semiconductors through defect engineering, it is essential to understand the interactions between various defects and their effect on semiconductor and device characteristics.

With the development of various microscopy techniques, including scanning probe techniques, the fundamental properties of various defects have been better understood and many details have been further clarified.

The main objective of this book is to outline the basic properties of extended defects, their effect on electronic properties of semiconductors, their role in devices, and the characterization techniques for such defects. We hope that this book will be useful to both undergraduate and graduate students and researchers in a wide variety of fields in physical and engineering sciences.

D. B. Holt
London
B. G. Yacobi
Toronto

1

Semiconducting materials

1.1 Materials development and crystal growth techniques

This chapter outlines the nature and importance of semiconductors. The industrially important semiconductors are tetrahedrally coordinated, diamond and related structure IVB, III-V and related materials. The sp^3 tetrahedral covalent bonding is stiff and brittle, unlike the metallic bond, which merely requires closest packing to minimize the energy. The atomic core structures of extended defects in semiconductors depend on this stiff, brittle bonding and in turn give rise to the electrical and optical properties of defects.

The semiconductors' closely related adamantine (diamond-like) crystal structures and energy band diagrams are outlined. There are a large number of families of such semiconducting compounds and alloys, some of which are non-crystalline. However, only a few have been developed to the highest levels of purity and perfection so that single crystal wafers are available. Instead, with modern epitaxial growth techniques, thin films, quantum wells, wires and dots and artificial superlattices can be produced. This can be done with many semiconductor materials, including alloys of continuously variable composition, with the necessary quality on one of the few available types of wafer. These epitaxial materials have 'engineered' energy band structures and hence electronic and optoelectronic properties and can be designed for incorporation into devices to meet new needs. It is largely to this field that materials development has moved, except for the occasional development of an additional material like GaN.

The chapter closes with a brief account of the way that competitive materials development, responding to economic demand, determines which materials enter production. Cases of materials competition covered include those of Si vs. Ge, of III-Vs vs. Si, and of GaN versus other electroluminescent III-V compounds and alloys.

Semiconductors (half conductors) originally meant materials with conductivities between those typical of metals and of insulators. However, this definition would,

1

Table 1.1. *Typical conductivity (at room-temperature) ranges
attainable by doping particular cyberconductors*

	Conductivity $(\Omega\text{-cm})^{-1}$	
Material	Heavily doped n^+ and p^+	Intrinsic or compensated
Si	$\sim10^{+3}$ n and p	$\sim10^{-4}$
Ge	$\sim2\times10^{+3}$ n and p	$\sim2\times10^{-2}$
GaAs	$\sim10^{+3}$ p and 10^{+4} n type	$\sim10^{-8}$ (semi-insulating)

strictly, include both electronic and 'fast ionic' (superionic) conductors. Moreover, the conductivity of any semiconductor can be varied in sign and magnitude over a wide range (see Table 1.1). They have high resistivities at room temperature when pure (intrinsic) but can be given large conductivities by impurity doping them n- or p-type. This makes the production of microelectronic devices and integrated circuits possible. Also the conductivity can be controlled during device operation by injecting additional carriers through a contact such as that to the base of a bipolar transistor or by shining light on a photoconductive or photovoltaic detector. Thus the essential feature of semiconductors is not that they are 'semi' (half) conductors but that they are 'cyber' (controllable, changeable) conductors. (Cyber is from a Greek root word, meaning steersman, and is used in the same way as in 'cybernetics' (the science of feedback and control.)) This behaviour has historically come to be identified with a particular group of materials, including diamond, which, except when doped, is the best insulator known.

1.1.1 Microelectronics and microphotonics

Cyberconductivity is exploited in discrete solid state electronic devices and in microelectronic integrated circuits that are basic to electronics: the technology that manipulates electrons.

There is a second, rapidly growing, field of application of these magical materials: photonics, i.e. the technology that manipulates photons. In this field their important properties are their ability to emit, detect and controllably transmit light, i.e. they are not only 'cyber' conductors but also 'cyber' emitters.

1.1.2 Cyberconductors

Research on germanium led to the invention of transistors and the rise of solid state electronics. Silicon became dominant with the invention of planar technology for integrated circuits. The III-V compounds and alloys dominate in optoelectronics including lightwave communications. This story of materials development and competition is recounted in Section 1.8. The crystal structures, chemical bonding and electronic energy band structures of these materials are all diamond-like. This book is essentially concerned with the deviations from crystalline perfection in real crystals of these dominant practical cyberconductors, the adamantine materials.

Defects generally strongly affect semiconductor properties and the operation of semiconductor devices. This is because defects in a semiconductor may introduce energy levels (in the forbidden energy gap) affecting the electronic properties of the material. They may also result in device degradation and failure during operation.

The most important application of semiconductors is that for the manufacture of integrated circuits for microelectronics. This uses Si in the form of large crystals of the highest purity and perfection of any material. Semiconductors, however, are not always employed in this form. Devices that must be large in area at a low cost do not employ single crystals of adamantine semiconductors. Examples are found in electrophotographic reproduction (reprographics or xerography including laser printers), lighting and displays, and terrestrial solar photovoltaic power generation. Reprography usually employs layers of non-crystalline ('amorphous') Se or of polycrystalline or powdered ZnO or polymer coatings. Lighting and displays use powder phosphor materials of many kinds in fluorescent lights, CRT and TV screens and electroluminescent display panels. Terrestrial photovoltaic devices employ poly- and non-crystalline Si or compound heterojunctions, as will be discussed in Section 1.7.5 and Section 1.8. So defects like grain boundaries and powder particle surfaces can sometimes be tolerated.

Recent developments have drawn attention to the fact that the cyberconductors are also cyberoptical materials. That is, many of these materials are also intermediate between metals (opaque and metallically shiny) and insulators (transparent or translucent). Moreover, they can be made to emit or detect light. Also, while transmitting light, their electrooptical properties can be exploited to process signals in the form of light.

To measure the intrinsic properties of cyberconductors and gain an understanding of semiconductor physics it was first necessary to grow sufficiently pure and perfect crystals. Defect microstructure could then be resolved and its effects on physical properties studied. The impressive improvement of the Czochralski growth technique and the invention of zone refining and methods of controlled doping were other necessary preconditions for both semiconductor science and technology.

Zone refining (Pfann 1952), developed for Ge in the late 1940s, also worked for Si. Narrow molten zones were run along a bar of the material pulled through a furnace with narrow hot zones. The electrically important impurities in Ge and Si fortunately had higher solubilities in the liquid zones than in the solid. Hence they were swept along and accumulated in the end of the rod which was cut off. Much higher purities than ever before were first reached so the properties of intrinsic Ge (and later Si) could be observed. It also made it possible to investigate the effects of controlled doping on these properties. Now, however, the necessary purification takes place in the gaseous phases occurring in the extraction of these elements from their ores.

Many techniques of *crystal growth* were developed to produce the relatively large and perfect single crystals needed to determine the basic properties of the new semiconductors that were being discovered in the early post-Second World War period (see e.g. the books of Pamplin 1980 and Brice 1986). Such methods produce controlled crystal growth and can be classified according to whether this is from liquid or gaseous material. For semiconductors, the most important methods are the Czochralski (Fig. 1.1) and Bridgman methods of growth from melt, and the liquid, vapour phase and molecular beam methods of epitaxial growth of thin films. Bridgman growth employs a long container or 'boat' drawn through a furnace (horizontally or vertically), so a single crystal grows from a nucleus at the end where freezing starts.

Czochralski growth from melts (named for the Polish worker who originated it) proved best able to be scaled up and it is now the method in industrial use for producing large silicon crystals to be sliced into wafers. Czochralski crystal 'pulling' is shown schematically in Fig. 1.1. Automated Czochralski growth, using feedback

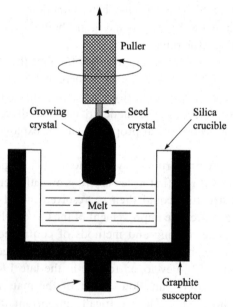

Figure 1.1 Schematic illustration of a Czochralski crystal growth mehtod. A seed crystal gripped by a vertical rod is dipped into the melt and slowly raised while being rotated. Molten material freezes at the seed interface as heat is extracted by conduction up the rod and by radiation outward from the rising crystal. During growth the crucible is raised to maintain the melt surface at a constant height in the furnace. (The growth chamber and the RF inductive heating coils are not shown.)

control to keep the cylindrical 'boules' constant in diameter is used for growing crystals to be sawn and polished into wafers of Si. Annual world production of Czochralski silicon crystals already by the mid 1980s was estimated to be 4000 to 5000 tonnes per annum (e.g. Brice, 1986). The boules are sawn and polished into slices or wafers, totalling about half this mass. This final 1986 total of ~2000 tonnes of Si wafers, each ~500 μm thick thus covered a total area of 1 km^2! Obviously the world output of single crystal silicon for the production of integrated circuits (ICs) and other devices is now much greater than these earlier figures.

Si and Ge can be pulled in vacuum or under inert or reducing atmospheres but at their melting points III-V compounds present dangerously high pressures of their toxic constituent gases. Floating a layer of relatively insoluble boric oxide on the surface of the molten compound was found to be a practical method to reduce this volatility. Crystals pulled through this layer are said to be 'encapsulated' in boric oxide. Most of the slices of III-V compounds available are of such 'liquid encapsulated Czochralski' (LEC) material. The only semiconductors of which large enough crystals to provide slices are available today are Si, Ge, GaAs, GaP, GaSb and InP (and, in the past, InSb). To give an idea of the 'state of the art': at the time of writing the Si IC industry is re-equipping to use 300 mm (12 inch) diameter slices. The most developed III-V compound GaAs is used in industry as 100 or (mainly) 150 mm (6 inch) wafers while 200 mm (8 inch) slices are becoming available. The reason more semiconductors have not been developed to this level is the cost. The late Prof. C. H. L. Goodman used to say that such research and development efforts are so great that the natural unit involved is the man-millenium (thousand man years)!

Recent progress and some of the present issues in materials development and crystal growth techniques were outlined in reviews by Hurle and Rudolph (2004) and Mullin (2004) (see Section 6.5 and Section 6.6). For scaling-up crystal diameter, the former note, among other issues (see Section 6.6), a need for improved furnace design to reduce the dislocation density in the larger crystals. In a review on the melt growth of III-V compounds, Mullin (2004) described the advances, advantages and practical limitations of the liquid encapsulation, vertical gradient freeze, vapour pressure controlled Czochralski and hot wall pulling growth techniques.

The range of materials and properties available to the semiconductor device industry, however, is greatly increased by two things. These are epitaxial growth and the quantum confined structures made possible by it, as we shall now see.

1.1.3 Epitaxy

Prior to the invention of planar technology, integrated circuits and the rise of the silicon solid state electronics industry, epitaxy was of minor academic interest. Si integrated circuits are made by processing carried out only on one face of flat, circular wafers. This planar technology is now universally used for semiconductor microelectronics and optoelectronics device fabrication. The epitaxial growth of a differently doped Si layer on a Si slice was found to make possible valuable new 'abrupt' *p-n* junctions (i.e. junctions with very rapid, short-range changes from *n-* to *p*-type doping). These could not be obtained by the alternative in-diffusion or ion implantation methods.

The only III-V compound wafers available are of LEC material. These, it was found, cannot be used to make devices due to the presence of unknown and uncontrolled point and other defects. Fortunately, it was found that III-V layers of device quality could be produced by epitaxy. LEC slices are, therefore, used only as substrates on which to grow such epitaxial layers. III-V devices are exclusively fabricated in such material so epitaxy has become a vital semiconductor production technique.

Epitaxy is oriented overgrowth, i.e. the growth of one crystal on another in a single, well-defined, related orientation. (Si on Si growth is thus not, strictly speaking, epitaxy at all. It has been dubbed 'homoepitaxy'.) Using a single crystal substrate and the correct epitaxial growth conditions, therefore, a monocrystalline deposit can be obtained. Two cubic materials, like any pair of the most important semiconductors, usually grow epitaxially with the simple parallel orientation relationship. For general accounts see Matthews (1975), Stradling and Klipstein (1990), Pashley (1991), Davies and Williams (1994), Bachmann (1995), and Mahajan and Sree Harsha (1999).

In addition to the use of epitaxial deposition to produce abrupt Si 'homoepitaxial' *p-n* junctions, and to grow good III-V material on unusable LEC wafers it is also used to produce heterojunctions (interfaces between two different semiconducting materials). The deposited material can be an alloy of two or more III-V compounds, of continuously variable composition. Repeated epitaxial growth of thin layers of different semiconductors is possible. It produces a 'single crystal', i.e. material with a single crystallographic orientation consisting of an epitaxial layered structure. This can be used to produce band gap engineered device structures and quantum confined structures like quantum wells as will be discussed later. This additional vast range of new possibilities removed any continuing necessity to develop new semiconducting compounds to obtain new properties. Consequently, except for GaN, no new compound has been developed in recent decades. Thin films can be deposited epitaxially from the liquid or vapour phase or by molecular beam methods.

Vapour phase epitaxy (VPE) is produced by chemical vapour deposition (CVD); that is by reactions of gases on heated wafers in equipment like that in Fig. 1.2. This has been essentially replaced by a later, superior variant, MOVPE (metal organic VPE, sometimes known as MOCVD, i.e. MO chemical vapour deposition), in which the elements required are provided by the reactions of metal organic precursor gases. For further information see Stringfellow (1989). This can grow more complex III-V alloys and quantum well structures, etc. Doping impurity gases can be added, computer-operated valves can control flow to grow any numbers and thicknesses of

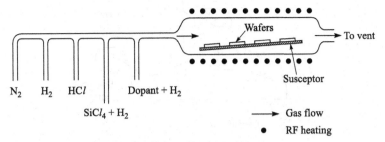

Figure 1.2   Schematic diagram of a reactor for the epitaxial growth of silicon by the hydrogen reduction of silicon tetrachloride. (After Sze 1985; *Semiconductor Devices: Physics and Technology*; Copyright 1985. Reprinted with permission of John Wiley & Sons, Inc.) Other chemical vapour deposition techniques are modified according to the gases involved.

successive thin films. Finally, the thermodynamics of the reactions is understood, so the process can be optimized.

In *liquid phase epitaxy (LPE)* of III-V compounds, substrates held in a carbon 'slider' are pushed under a carbon boat containing small volumes of supersaturated solutions, e.g. of GaP in Ga, as shown in Fig. 1.3. The solvent element (Ga or In) suppresses vacancies on sites of that element in the deposit. Lower growth temperatures can be obtained than for simple melts and growth rates can be high, but rough surfaces and interfaces can result and contamination by carry over of material from one melt to the next is possible. This technique worked surprisingly well in some cases but has largely fallen out of use now.

Molecular beam epitaxy (MBE) is a sophisticated form of evaporation in ultra high vacuum (see Fig. 1.43). Its relatively low growth temperatures make abrupt interfaces between different materials possible as little or no interdiffusion occurs during growth. It also allows continuous monitoring by methods taken from surface physics, such as reflection high-energy electron diffraction (RHEED), and good process control (for further details see Parker 1985, Herman and Sitter 1989 and Cho 1995).

Growth techniques, such as MBE and MOCVD, can produce structures in which layers of different materials and/or thicknesses (as small as a few monolayers) alternate. Such structures can have the characteristics of two-dimensional (2D) or one-dimensional (1D) quantum confined systems with novel properties, to be discussed below. The equipment used for MBE growth of multiquantum well (MQW) structures (for which purpose MBE was developed) is described in Section 1.7.3.

1.1.4 Crystal perfection

To minimize grown-in defects, crystals must be produced under controlled conditions to give, e.g. flat melt-crystal interfaces. Dash (1958) developed a Czochralski method of 'necking down' the seed crystal (Fig. 1.1), i.e. reducing the

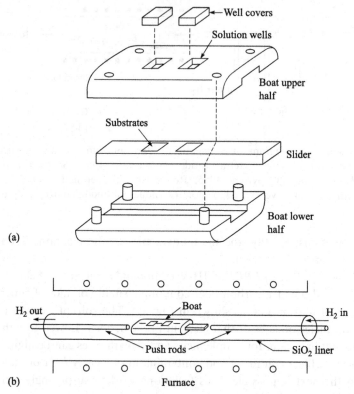

Figure 1.3   Liquid phase epitaxy equipment: (a) Construction of the boat and slider which are machined from reactor-grade graphite blocks. (b) The arrangement of the boat in the furnace. The substrates, held in the slider, are pushed under the melts, one after the other, to grow successive epitaxial layers. (After Sze 1985; *Semiconductor Devices: Physics and Technology*; Copyright 1985. Reprinted with permission of John Wiley & Sons, Inc.)

diameter by fast pulling. During necking down of the initial seed crystal the dislocations grew out the sides. Large, entirely dislocation-free crystals of Si, Ge and GaAs can be grown from such seeds. Structural defects reappear with each new material, and generally each new material at first exhibits some new type of defect in profusion. This requires research to identify the defects characteristic of the new material, their mechanism of generation and means of passivating (rendering electrically harmless) or eliminating them. At the time of writing this type of development is in progress in the case of GaN. New processing-induced defects, introduced into initially good material during device fabrication, including epitaxial growth, also keep appearing with each new device structure and production process.

   To grow high perfection material by epitaxy the film and substrate materials must closely match in symmetry and lattice parameter at the interface. This ensures that the unit meshes in the two materials at the original growth surface match in shape and size, respectively. This, in turn, allows low-energy bonding to occur across the interface. Symmetry mismatch opens the way for the formation of domains like the antiphase domains that can occur in a sphalerite-structure material such as GaAs grown on an (001)-oriented diamond-structure substrate such as Si. Lattice parameter matching minimizes the interface elastic strain, which favours misfit dislocation formation. However, for relatively small misfits, high-quality hetero-junctions can be produced in pseudomorphic or strained-layer structures provided the epitaxially deposited layers are sufficiently thin. Then no misfit dislocations are generated at the interface and the lattice mismatch is accommodated by long-range elastic strain. Fig. 1.4(a) illustrates lattice-mismatched sphalerite-structure materials

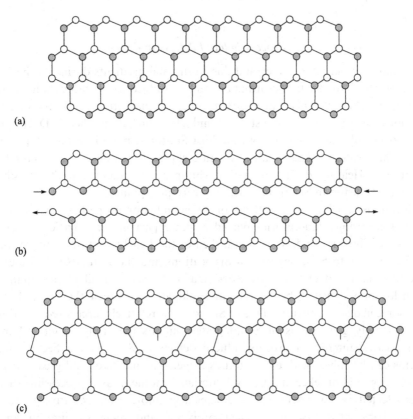

(a)

(b)

(c)

Figure 1.4   Misfit dislocations in a (111) heterojunction seen in projection on the $(01\bar{1})$ plane for the case in which the repeat distance $p = 5b = (5 + 1)a$. Here a and b are the lattice constants of the film (top) and substrate (bottom) respectively. (a) Perfect crystal of intermediate lattice parameter $\lambda = \lambda_1/2 + \lambda_2/2$. This is the reference structure. (b) The structure is cut along the (111) plane and expanded to λ_2 on one side and contracted to λ_1 on the other. (c) When the two crystals are bonded together there is a Vernier of period p. A dangling bond occurring at a spacing p marks the positions of successive misfit dislocations. (After Holt 1966.)

(envisaged as derived from an intermediate material by increasing the lattice constant of the substrate and decreasing that of the material of the film). The epitaxial heterostructures are shown with (Fig. 1.4b) the mismatch accommodated by long-range strain, and with (Fig. 1.4c) the mismatch accommodated by the localized strain at misfit dislocations. This will be discussed more fully in Section 4.6.

The historical advances in understanding the major mechanisms of formation of defects in semiconductor crystals were outlined by Hurle and Rudolph (2004) and Mahajan (2004). The former point out that dislocations can be eliminated in Si, but not in III-V and II-VI semiconductor compounds, in which the lower thermal conductivity and yield stresses prohibit sufficient reduction of the thermal stresses to prevent multiplication of dislocations (Hurle and Rudolph 2004).

1.1.5 Process-induced defects in devices

Differential contraction in the crystal due to thermal gradients during cooling after growth can lead to plastic deformation and introduce local dislocation densities greater than $10^6 \, \text{cm}^{-2}$. This is at least two orders of magnitude above the threshold for serious device effects in most semiconductors including silicon. Diffusion of dopant atoms of size different from the host Si atoms, notably B and P, produces strains. The resultant stresses can also cause slip and introduce 'diffusion-induced dislocations'. Heating during oxidation can produce stacking faults in silicon. Heating or diffusion in Si can produce impurity precipitates.

Dislocations can have large effects on semiconductor properties, so workers in industrial electronics laboratories played a major part in early, basic dislocation, research. W. Shockley and W. T. Read, then of the Bell Laboratories, for example, gave their names to Shockley partial dislocations and the Frank-Read dislocation source. As one result of this early work, material became available economically, with dislocation densities below the threshold level at which they ceased to be troublesome in discrete devices of Ge or Si and in the relatively large-sized constituent devices in early generations of Si integrated circuits (generally about $10^4 \, \text{cm}^{-2}$). Now industrial interest in process-induced defects in Si is limited because many problems can be solved with existing knowledge, trial and error and greater care in processing ('good housekeeping'). New materials, new methods of processing, smaller devices affected by fewer defects, and devices using more defect-sensitive properties, however, constantly raise new defect problems and re-arouse interest among industrial workers.

1.2 Electron energy levels and energy bands

The properties of semiconductors and defects depend on the way the wave functions and energies of electrons are influenced by the periodic potential in the perfect crystal and by the surroundings of the defect. More detailed accounts are given, for example, in Ziman (1972) and Kittel (1996) and in Section 1.5 following the necessary crystallography.

1.2.1 The broadening of energy levels into bands

The energies of the electrons surrounding the nucleus of an isolated atom take discrete values, indicated schematically at the right in Fig. 1.5. When the atoms are brought together to form a solid, the electric fields due to the other atoms acting on a particular atom modify its electron energy levels. That is, the potential term in Schrödinger's equation is altered, leading to different solutions. The energies of the inner-core electrons are changed only slightly (less than about 0.1 eV) because they are well shielded by the outer electrons, but the energies of the exposed outer electrons may be altered by up to several eV. Hence instead of a few sharp energy levels, there are now whole ranges of energies as shown at the left in the figure.

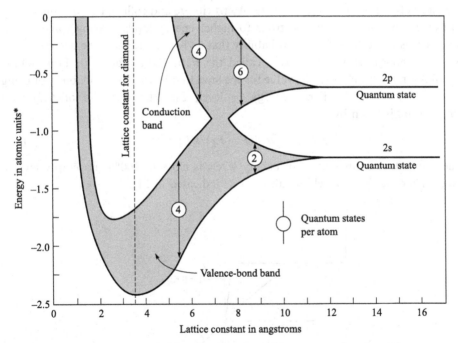

Figure 1.5   The outer energy levels of carbon spread into energy bands as the atoms are brought together to form a diamond. The 2s and 2p levels mix to form the filled valence and empty conduction bands. The energy is measured in Rydbergs = 13.5 electron volts and the interatomic distance in units of one tenth of a nm. The negative energy of the valence band states corresponds to these being 'bonding states' and the positive energy of the conduction band to these states being 'antibonding'. (After Shockley 1950.)

The electrons do not all occupy the lowest of these energy levels, due in part to the Pauli exclusion principle. In simple language this states that no more than two electrons (and these must have opposite spin) can have the same (spatial) wave function within any system in thermodynamic equilibrium. (Strictly, we should only speak of the wave function for the entire assembly of electrons, which must be antisymmetric with respect to the interchange of any two electrons. This many-body description, while more accurate, is not as helpful in gaining an elementary picture of a solid (see Section 1.5).

Since there are usually only one or a few one-electron wave functions per energy level, the electrons are distributed over a number of energy levels. These are derived from the original atomic levels, perturbed by the multi-atom environment, into energy bands (Fig. 1.5). The gaps between them consist of ranges of energies that cannot be possessed by electrons in the crystal just as certain energies are not possible for electrons in an atom.

In an atom the electrons are usually in the lowest energy levels consistent with the Pauli principle, and need to be given energy to excite them to the higher levels. Likewise, in the solid at the absolute zero of temperature, pairs of electrons fill up the lowest levels. At room temperature the atoms vibrate and collide, transferring energy to excite some of the outer electrons to higher energy levels, leaving some vacant lower levels called 'holes'. The probability that a given energy level be occupied by a pair of (opposite spin) electrons is called the Fermi-Dirac factor or Fermi-Dirac distribution function $f(E)$. The value for the level that has a 50% probability of being occupied is called the Fermi energy or level, E_F. The variation of $f(E)$ with temperature is given by

$$f(E) = [\exp\{(E - E_F)/kT\} + 1]^{-1} \qquad (1.1)$$

The occupation probability f of energy levels is close to 100% (no empty levels), except near the Fermi level and above it as indicated in Fig. 1.6.

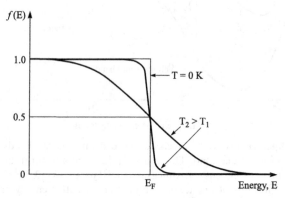

Figure 1.6   A plot of the Fermi function $f(E)$ as a function of energy at different temperatures.

1.2.2 The dependence of electrical conductivity on energy band structure

If the Fermi level E_F occurs well within an energy band, the material is a metal. If it occurs between energy bands, with no near-by energy levels, the material is an insulator, but if there are energy levels within a few kT (see Fig. 1.6), the material is a semiconductor at this temperature. This follows since an electric field increases the proportion of electrons that have a velocity in the direction of the field provided that there are empty energy states into which electrons may transfer.

In metals, since the probability that the levels around the Fermi energy are occupied is about 50%, there are many empty levels near occupied levels. Thus conduction can occur. The effects of electron scattering by imperfections and by quanta of thermal vibrational energy called phonons set limits on this.

In insulators, the Fermi level lies between, and far from, the bands. The energy levels in the valence band (below the Fermi level) have so high a probability of being occupied that there are virtually no vacant states in this band. The electric field can therefore cause little or no redistribution of electron velocities, and negligible 'hole' current in this band. The conduction band lies well above the Fermi level so there are very few occupied levels and thus little 'free electron' conductivity either.

Semiconductors are useful because the electrical conductivity can readily be changed. This can be done by adding small concentrations of donor or acceptor atoms to introduce free charge carriers. Donors have electrons in levels at an energy E_D just below the lowest conduction band state of energy E_C. If the Fermi level lies $3kT$ or more below the donor levels, they will be empty (Fig. 1.6). That is, the electrons will be thermally activated from these shallow levels (donated) into conduction band states to produce 'n-type' (negative charge carrier) conduction and the carrier density will be equal to the donor doping density. Acceptors have an empty energy level just above the highest valence band state at E_V. They all accept electrons giving p-type (positive, hole carrier) conductivity, provided the Fermi level is $3kT$ or more above the acceptor level at E_A. Conductivity can also be altered by injecting carriers across a contact or junction, or by irradiation as in solar photovoltaic power devices and charged particle detectors.

The electrical conductivity of a semiconductor is given by

$$\sigma = nq\mu_e + pq\mu_h \tag{1.2}$$

or, when the concentration of one type of dopant is many orders of magnitude greater than that of the other, by $\sigma = nq\mu_e$ or $\sigma = pq\mu_h$ depending on whether the material is n-type, or p-type, respectively, where n (p) is the density of free electrons (holes), q the charge on the electron and μ_e (μ_h) is the electron (hole) mobility, i.e. the velocity of the carriers per unit electric field. This mobility is affected by scattering by imperfections and phonons. It also depends on the form of the electron energy bands. The latter dependence is often expressed in the form of an 'effective mass' for the carriers. It can be shown that if the band in the vicinity of the Fermi level is narrow, the mobility is small, by simple reasoning of a type that is useful in considering new and unknown situations. If the band is narrow, the original atomic energy levels are not much spread by the interactions with their neighbours in the solid. Since they are not strongly influenced by their neighbours' fields they will also not be strongly influenced by an external field and the mobility of such electrons will be low. This is frequently the case in organic materials to such an extent that the conduction is considerably reduced. Conversely if the band is wide, one can argue that the mobility should be high. This is basically true but detailed effects of the interatomic potentials can modify this result in certain cases.

1.2.3 Band theory and chemical bonding

Crystals should be treated by deriving the wave functions of the electrons directly from the periodic potential in which they move. This is often difficult, e.g. for complex crystals such as multi-element compounds and alloys and near surfaces or defects. Here one is forced to use the atomic bond approach. This viewpoint is most applicable if the bonds of neighbouring atoms do not overlap much so the bonding electrons are not much influenced by anything except nearest neighbour atoms. The electron wave functions for isolated atoms, modified by any translational and other symmetry considerations are used for the atoms near the surface or defect, etc. The well-known LCAO (linear combinations of atomic orbitals) method uses such sums of wave functions with undetermined multipliers and fixes the latter by comparison with the results of other methods of calculation. The value of the simple bond approach is that it gives a qualitative idea of the nature of the forces between the atoms.

In the diamond structure (Fig. 1.7), each atom is surrounded by four others at the corners of a regular tetrahedron. The bonds to the four neighbours are separated by the tetrahedral angle 109°28'. Isolated atoms of C, Si and Ge have four valence electrons outside the last closed shell (which gives a spherical charge distribution), two in s states and two in mutually perpendicular p states, with the forms shown in Fig. 1.8. An appropriate mixture of these states produces the sp^3 'hybrid' orbital that has tetrahedral symmetry. The energy of the four electrons in sp^3 states and in the two s and two p states is little different. In III-V compounds each atom is again tetrahedrally coordinated so an appreciable degree of sp^3 hybridization is present, although there is also some ionic bonding.

In the NaCl structure (Fig. 1.9), each atom is surrounded by six atoms of the other element at the corners of an octahedron. The bonds to these six neighbours are orthogonal and can be made by overlapping the p_x, p_y and p_z orbitals of the neighbouring atoms. This is p^3 bonding and is important in PbS, PbSe and PbTe which are semiconducting materials that have had some practical importance and that have the NaCl structure.

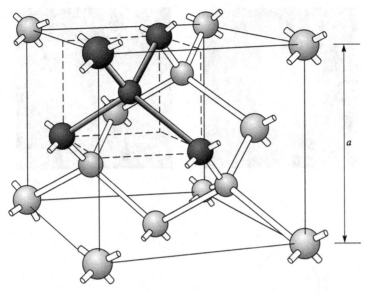

Figure 1.7   The unit cell of the diamond cubic structure. The broken lines delineate a cube containing an atom surrounded by its four nearest neighbours arranged at the corners of a regular tetrahedron. These five atoms and the covalent bonds between them are shaded more heavily than the rest. (After Shockley 1950.)

1.3 Bonding and the crystal chemistry of semiconducting materials

Silicon and germanium are elements of column IVB of the Periodic Table (Fig. 1.10). They crystallize, like carbon from the same column, in the diamond structure (Fig. 1.7) with tetrahedral coordination. These elements also form a semiconducting compound, SiC, and the Si-Ge semiconducting alloys. SiC crystallizes in over 150 'polytype' structures, all tetrahedrally coordinated semiconductors with different forbidden energy gaps (Verma and Krishna 1966). Much work was done on SiC in industrial laboratories and some SiC devices have appeared on the market. Si and Ge, in all proportions, form solid solutions with the diamond structure (Wang and Alexander 1955, Dismukes and Ekstrom 1965).

   The chemical bonding of C, Si, Ge and, in one crystalline form, Sn, is thus largely covalent. That is, the lowering of energy on bonding is due to pairing electrons with opposite spins in hybridized sp^3 orbitals. The tetrahedral coordination of the atoms in diamond structure materials can be regarded as due to the directional character of these orbitals.

1.3.1 Adamantine semiconductors as electron compounds

The Periodic Table can be divided into crystal-chemical classes of elements as in Fig. 1.10 (Hume-Rothery and Raynor 1954). Class III elements, from groups IVB

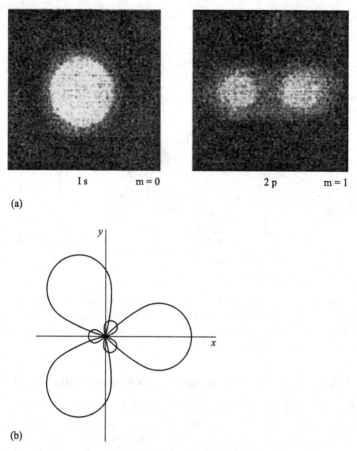

(a)

(b)

Figure 1.8 Atomic wave functions: (a) The 1s and the 2p wave functions plotted as probability density clouds (from Finkelnburg 1950 after H. E. White), (b) An sp^3 hybrid orbital plotted as an equal probability contour (s and p represent quantum numbers whose value affects the wave function shape).

to VIIB, crystallize with covalent bonds, to which each of the neighbouring atoms contributes one electron so each atom gets a share in a complete octet or quantum shell of eight bonding electrons. Hence Hume-Rothery's '8-N Rule' states that the crystal structures of the elements of this class will be characterized by coordination numbers (numbers of nearest neighbours) = 8-N where N is the number of valence electrons, i.e. the column (group) number of the element in the periodic table. Thus C (diamond), Si and Ge each have four valence electrons and in the crystal each atom has four neighbours with which it shares in an octet of electrons. The elements of class II are influenced by the 8-N rule too.

Intermetallic compounds, with particular crystal structures, occur in several alloy systems (different mixtures of elements) at compositions corresponding to specific

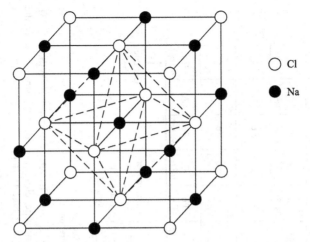

○ Cl

● Na

Figure 1.9   Unit cell of the sodium chloride structure. The Na ions form a face centred cubic array as do the Cl ions. Each ion is surrounded by six nearest neighbour ions of the other kind arranged at the corners of a regular octahedron. The broken lines are the edges of the octahedron of ions coordinating the one at the centre of the unit cell. PbS, PbSe and PbTe are semiconductors that have this structure.

electron/atom ratios. That is, the sum of the valence electrons divided by the number of atoms in a formula unit is a constant for the particular structure. These structures occur due to a low-energy fit of the Fermi surface, which depends on the electron/atom ratio, in the Brillouin zone characteristic of the structure (Hume-Rothery and Raynor 1954). Four electrons per atom just fill the Brillouin zone of the diamond structure, so the adamantine materials are electron/atom ratio 4 structures. Brillouin zones will be discussed in Section 1.5.5.

1.3.2 III-V and other families of adamantine semiconductors

Such ideas led H. Welker and N. A. Goryunova in 1952 to suggest that compounds of equal numbers of atoms of elements from the IIIB and VB columns of the periodic table should be very similar to the IVB elements. The bonds should be largely covalent since they come from Classes II and III of Fig. 1.10 and the electron/atom ratio $= (III + V)/2 = 4$, as for the IVB elements. A few III-V compounds were known to form and have the sphalerite (zincblende) structure (Fig. 1.11) (Welker and Weiss 1956, Goryunova 1965). The atom sites are just as in the diamond structure but half the sites are occupied by the column III atoms (e.g. Ga) and the rest by the column V atoms (e.g. As). It was soon established that most of the III-V compounds do crystallize in the sphalerite structure and have optical and electronic properties closely related to those of the IVB elements, see Table 1.2 and Madelung (1964).

I	IIa	IIIa	IVa	Va	VIa	VIIa	VIII			Ib	IIb	IIIb	IVb	Vb	VIb	VIIb	VIII
1 H 1s																	2 He 1s²
3 Li 2s	4 Be 2s²											5 B 2s²2p	6 C 2s²2p²	7 N 2s²2p³	8 O 2s²2p⁴	9 F 2s²2p⁵	10 Ne 2s²2p⁶
11 Na 3s	12 Mg 3s²											13 Al 3s²3p	14 Si 3s²3p²	15 P 3s²3p³	16 S 3s²3p⁴	17 Cl 3s²3p⁵	18 Ar 3s²3p⁶
19 K 4s	20 Ca 4s²	21 Sc 3d4s²	22 Ti 3d²4s²	23 V 3d³4s²	24 Cr 3d⁴4s²	25 Mn 3d⁵4s²	26 Fe 3d⁶4s²	27 Co 3d⁷4s²	28 Ni 3d⁸4s²	29 Cu 3d¹⁰4s	30 Zn 3d¹⁰4s²	31 Ga 4s²4p	32 Ge 4s²4p²	33 As 4s²4p³	34 Se 4s²4p⁴	35 Br 4s²4p⁵	36 Kr 4s²4p⁶
37 Rb 5s	38 Sr 5s²	39 Yt 4d5s²	40 Zr 4d²5s²	41 Nb 4d⁴5s	42 Mo 4d⁵5s	43 Tc 4d⁶5s	44 Ru 4d⁷5s	45 Rh 4d⁸5s	46 Pd 4d¹⁰	47 Ag 4d¹⁰5s	48 Cd 4d¹⁰5s²	49 In 5s²5p	50 Sn 5s²5p²	51 Sb 5s²5p³	52 Te 5s²5p⁴	53 I 5s²5p⁵	54 Xe 5s²5p⁶
55 Cs 6s	56 Ba 6s²	Rare earths	72 Hf 5d²6s²	73 Ta 5d³6s²	74 W 5d⁴6s²	75 Re 5d⁵6s²	76 Os 5d⁶6s²	77 Ir 5d⁹	78 Pt 5d⁹6s	79 Au 5d¹⁰6s	80 Hg 5d¹⁰6s²	81 Tl 6s²6p	82 Pb 6s²6p²	83 Bi 6s²6p³	84 Po 6s²6p⁴	85 At 6s²6p⁵	86 Rn 6s²6p⁶
87 Fr 7s	88 Ra 7s²	Actinide series															

← Complete 'shells'

Key

Atomic number → 79 Au 5d¹⁰ 6s ← Chemical symbol

Outer electron configuration

Figure 1.10 The Periodic Table of the elements: (i) An abbreviated form showing the A and B sub-groups (after Wilkes 1973), (ii) Hume-Rothery's classes of elements: class I elements generally crystallize with the metallic close-packed structures or with the body-centred cubic structure. Class III elements crystallize with 8–N coordination structures. Class II elements have crystal structures influenced by 8–N coordination requirements (after Hume-Rothery and Raynor 1954, *The Structure of Metals and Alloys*, 3rd edn. London: Institute of Metals, 1954 © Maney Publishing).

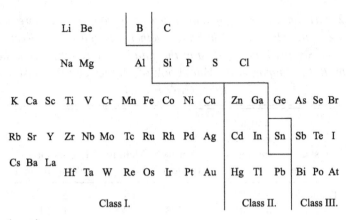

Li	Be				B	C										
Na	Mg				Al	Si	P	S	Cl							
K	Ca	Sc	Ti	V	Cr	Mn	Fe	Co	Ni	Cu	Zn	Ga	Ge	As	Se	Br
Rb	Sr	Y	Zr	Nb	Mo	Tc	Ru	Rh	Pd	Ag	Cd	In	Sn	Sb	Te	I
Cs	Ba	La	Hf	Ta	W	Re	Os	Ir	Pt	Au	Hg	Tl	Pb	Bi	Po	At

Class I.   Class II.   Class III.

Figure 1.10. (cont.)

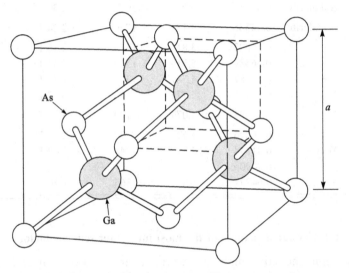

As

a

Ga

Figure 1.11 The sphalerite or zincblende structure. The atom sites form an array identical with that in the diamond structure. However, one atom of the basis unit, see Figure 1.13, belongs to the one element A, represented by the smaller, lighter spheres, and the other atom belongs to element B, represented by the larger, stippled spheres. Hence, the structure is sometimes said to consist of one f.c.c. sublattice of B (As) atoms (occupying the sites of the cubic unit cell shown) and the A (Ga) atoms occupy another offset by a quarter of a structure cell body diagonal, $\frac{1}{4}\langle 111 \rangle$, from the first.

Several materials, already used in optoelectronics, like ZnS and CdS, were then recognized as II-VI compounds of equal numbers of atoms of IIB and VIB elements with, again, the electron/atom ratio $4 = \{(II + VI)/2\}$ (Aven and Prener 1967). They occur with both the sphalerite structure and a second tetrahedrally coordinated structure: wurtzite (Fig. 1.12 and Section 1.4.4).

Table 1.2. *Structure, energy gap and properties of some of the more important semiconducting materials; I and D indicate whether it is indirect or direct energy gap, respectively (see Section 1.5.6). (Most of the data here were taken from the extensive tabulation by B. R. Pamplin in the* Handbook of Physics and Chemistry, *62nd edn, 1981, CRC Press.)*

Material	Crystal Structure	Latttice Parameters (nm) at room T	Density (gm/cc)	Melting Point K	Energy Gap (eV) at room T	Electron Mobility (cm^2/Vs) at room T
IVB: C	Diamond	0.35597	3.51	4300	5.4 I	1800
" Si	"	0.54307	2.3283	1685	1.12 I	1900
" Ge	"	0.56575	5.3234	1231	0.67 I	3800
IV-IV: SiC	Sphalerite	0.4348	3.21	3070	2.30 I	1000
III-V: GaP	Sphalerite	0.54505	4.13	1750	2.24 I	300
" AlAs	"	0.56622	3.81	1870	2.16 I	1200
" GaAs	"	0.56531	5.316	1510	1.42 D	8800
" InP	"	0.58687	4.787	1330	1.35 D	4600
" InSb	"	0.64787	5.775	798	0.17 D	78,000
II-VI: ZnS	"	0.54093	4.079	2100	3.68 D	180
" ZnSe	"	0.56676	5.42	1790	2.70 D	540
" CdS	Wurtzite	0.41348 a 6.7490 c	4.82	1748	2.42 D	400
" CdTe	Sphalerite	0.6477	5.86	1365	1.56 D	1200

1.4 Crystal structures of the most important semiconductors

Because the semiconducting materials described above, and many others, related more complex materials (see Section 1.6), have similar bonding and electronic structures it is not surprising that they have related crystal structures.

1.4.1 Space lattices, crystal structures, unit cells and symmetry elements

A space (Bravais) lattice is defined as an array of points in three dimensions in which the surroundings of all points are the same. It can be proved mathematically that there are only 14 lattices that satisfy this restrictive condition. That the surroundings are the same means that any point can be taken as origin and the set of lattice translation vectors constituting the translation group will generate all the other points. The lattice translation vectors are of the form:

$$\mathbf{R}_m = M_1\mathbf{a}_1 + M_2\mathbf{a}_2 + M_3\mathbf{a}_3. \tag{1.3}$$

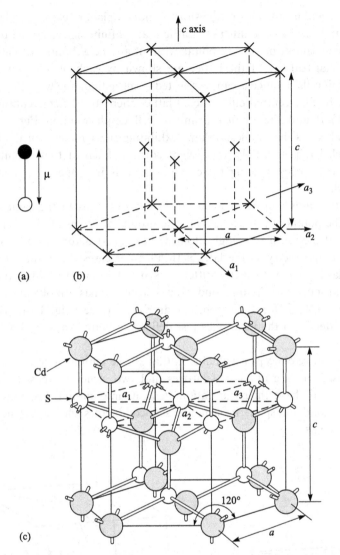

Figure 1.12 The wurtzite crystal structure (c) is produced by placing (a) a basis unit consisting of an A atom, represented by the lighter ('alabaster') sphere, and a B atom represented by the darker ('black') sphere at each of the sites marked x in (b) the hexagonal array. The space lattice of this structure is simple or primitive hexagonal. The atoms of the basis unit are separated by a distance μ along the c-axis. For the ideal wurtzite structure $\mu = 0.375\,c$ and $c/a = 1.633$. Most wurtzite-structure compounds are very near ideal, i.e. they have μ, c and a values satisfying these ratios.

Where a_1, a_2, and a_3 are a set of the shortest non-coplanar vectors (basic vectors) in the lattice and the M's are integers. Lattices are infinite since, by the definition of a lattice, there cannot be surface points without any neighbours on one side.

The essential features of the lattice are shown by each of its volume elements, known as unit cells. Unit cells can be selected in more than one way. For example, in the case of the face-centred cubic (f.c.c.) lattice there are the conventional unit cell of Fig. 1.13(a) and the smaller primitive cell emphasized in Fig. 1.13(b). The conventional cell is the smallest volume exhibiting the symmetry of the lattice. The primitive cell has points only at its eight corners. Each of these points is shared among the eight cells meeting there. Thus there is just one $= (8 \times 1/8)$ point per primitive cell.

Crystal structures consist of identical groups of atoms in the same orientation at each of the points of a space lattice. The groups of atoms are known as basis or motif units. If the motif is a single atom, the crystal structure is an array of atoms with the same geometry as the lattice. In all other cases the crystal structure and the space lattice have distinctly different symmetry so terms like 'the diamond lattice' are misnomers. The diamond crystal structure has a motif of two atoms at 0,0,0 and 1/4,1/4,1/4 for each point of an f.c.c. lattice (Fig. 1.14). Real crystal structures differ from the ideal ones just defined in many ways, each of which is a type of defect.

The space lattice concept is valuable in that all the innumerable crystal structures must be based on one of just 14 lattices and two structures based on the same space lattice must have their translation elements in common. For example, as we shall see, diamond and spaherite structure semiconductors and f.c.c. structure metals share slip systems.

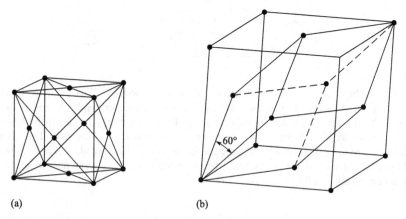

(a)   (b)

Figure 1.13   Unit cells of the face-centred cubic space lattice. (a) The structure unit cell showing the face-centred arrangement of the lattice points, (b) The smaller primitive unit cell, showing its relation to the larger structure cell.

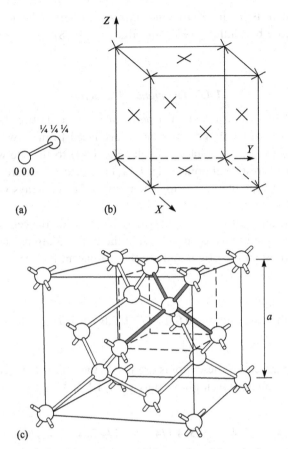

¼ ¼ ¼

0 0 0

(a)   (b)

(c)

Figure 1.14   The diamond cubic structure (c) is produced by placing a basis unit of two atoms (a), one on the lattice point 0,0,0 and the other displaced from it by ¼ of the cube diagonal from it, at each point of the f.c.c. space lattice (b). In the diamond structure both types of site 0,0,0 and 1/4,1/4,1/4 are occupied by the same kind of atom. In the sphalerite structure (see Figure 1.11) each type of site is occupied by a different element. The tetrahedral coordination of one atom in Figure 1.7 is marked by the heavily drawn bonds to its four nearest neighbour atoms, occupying half the corners of the small dashed cube.

Symmetry elements are operations that carry the crystal to a congruent (indistinguishable) situation. They include the lattice translations of Equation (1.3) and the elements of the point group of the crystal. The latter are the macroscopic symmetry elements that were found by studies of the external morphology of crystals prior to X-ray crystallography.

The valid combinations of the translation groups (the groups of lattice translations of Equation (1.3)) with the above types of symmetry elements, plus screw axes and glide planes constitute the 230 space groups and specify the symmetry of all possible crystal structures. A readable introductory account of crystallography with emphasis on crystalline defects is in Kelly and Groves (1970). Classic crystallography textbooks are those by Phillips (1971) and Buerger (1978).

1.4.2 The diamond structure

This has a motif of two atoms at each point of the f.c.c. lattice. One occupies the lattice site, e.g. that at 0,0,0 and the other is displaced from it by a quarter of the body diagonal of the conventional unit cell (Fig. 1.14) to the site with coordinates 1/4,1/4,1/4. Four nearest neighbours (Fig. 1.14c) tetrahedrally coordinate each atom. The atoms of each type of site occupy the points of f.c.c. arrays (sub-lattices) like those in Fig. 1.11.

There are eight unit cell corner atoms, each 1/8th inside the cell = 1 atom. There are six face-centre site atoms, each 1/2 inside the cell = 3 atoms. These four atoms can be taken as those at the origin and the three adjacent face centre sites, with unit cell coordinates:

first sub-lattice sites :   0,0,0   1/2,1/2,0   1/2,0,1/2   0,1/2,1/2.

The four motif partners of these atoms are inside the unit cell at second sub-lattice sites with the coordinates given above plus 1/4,1/4,1/4:

1/4,1/4,1/4   3/4,3/4,1/4   3/4,1/4,3/4   1/4,3/4,3/4.

Thus there are a total of eight atoms in the conventional unit cell of the diamond crystal structure.

The diamond structure can be pictured as consisting of {111} double-atom planes stacked in a particular order. The array of atoms in such planes is that of the well-known close packing of spheres, such as those marked '*a*' in Fig. 1.15a. The motif partners of these atoms are in sites vertically above, marked α in Fig. 1.15b. There are three sets of equivalent sites for stacking double-atom planes: '*a*', '*b*' (triangle apex down sites in Fig. 1.15a) and '*c*' (triangle apex up sites). Diamond structure stacking is the cyclic permutation of all three, i.e. $a\alpha$, $b\beta$, $c\gamma$, $a\alpha$, $b\beta$, $c\gamma$...or, omitting the Greek letters: *a, b, c, ...* The other adamantine crystal structures can be described in terms of similar stacking.

The translation symmetry of the diamond crystal structure is that of the f.c.c. lattice on which it is based. Its point group symmetry is $\bar{4}$3m, i.e. it includes a fourfold roto-inversion axis, symbol $\bar{4}$, a threefold rotation axis (3) and a mirror plane (m). (Roto-inversion axes produce congruence after n-fold rotation followed

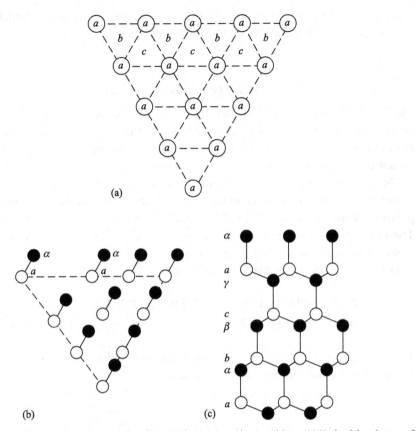

Figure 1.15   The construction of the diamond structure by stacking (111) double planes of atoms. (a) Close packed atom positions, *a*, in a (111) plane, (b) *a* and *α* positions of the atoms in a (111) double plane, (c) Projection of the diamond structure onto the (111) plane to show the *aαbβcγ*... stacking sequence.

by inversion through a centre lying on the axis.) A number of other symmetry elements must also be present 'by symmetry' as a consequence of the presence and relative orientations of the three listed in the International symbol $\bar{4}3m$. For details see Kelly and Groves (1970).

The space group symmetry is Fd3m. The 'F' indicates that the structure is based on a face-centred (cubic) lattice; d, 3 and m indicate the presence of a diamond glide plane, threefold rotation axis and a mirror plane. Glide planes carry points to other lattice points by mirror imaging plus translation parallel to the plane in a direction indicated by the letter. These elements again imply the presence of many others. Space group symbols and elements are fully explained in *The International Tables for X-ray Crystallography* (Kynoch Press: Birmingham 1952), and stereographic representations of the point groups together with unit cell representations of the space groups are given. The coordinates of all the points of special symmetry in the unit cell are listed together with their local symmetries.

1.4.3 The sphalerite structure

The sphalerite structure occurs only in binary AB compounds (e.g. III-V, II-VI or I-VII). The array of atomic sites is the same as in the diamond structure but the two atoms of the basis unit are of different chemical elements so one f.c.c. sublattice is occupied by A atoms and the other by B atoms.

The stacking sequence is as in diamond but the Latin letter planes are occupied by A atoms and the Greek letter planes by B atoms in the standard convention for polarity. A atoms are tetrahedrally coordinated by four B atoms and vice versa. The two types of atom give the basis unit polar directionality and the structure is non-centrosymmetric, i.e. has no centre of inversion.

The positions of the atoms in the unit cell are:

A sites: 0,0,0 1/2,1/2,0 1/2,0,1/2 0,1/2, 1/2
B sites: 1/4,1/4,1/4 3/4,3/4,1/4 3/4,1/4,3/4 1/4,3/4,3/4

The space group is $F\bar{4}3m$, i.e. it is f.c.c.-based with a fourfold roto-inversion axis, a threefold rotation axis and a mirror plane.

1.4.4 The wurtzite structure

The wurtzite structure can be produced by stacking double atom planes in cyclic permutation of two of the possible positions, e.g. $a\alpha$ $b\beta$ $a\alpha$ $b\beta$... The Latin letter planes are occupied by A atoms and the Greek letter planes by B atoms of the binary compound AB.

The wurtzite structure is related to the hexagonal close packed (h.c.p.) structure that results from stacking close packed planes of atoms in the sequence: *abab*... That is, the wurtzite structure has a basis of two atoms, one A and one B, placed at each point of an h.c.p. array as shown in Fig. 1.16. (There is only one hexagonal space lattice: the simple hexagonal lattice. Both the h.c.p. and wurtzite structures are based on it.) There are four atoms in the primitive hexagonal unit cell (Fig. 1.16):

A atoms at 0,0,0 and 2/3,1/3,1/2 and
B atoms at 0,0,μ and 2/3,1/3,1/2 + μ

In all semiconducting compounds with this structure, $\mu \cong 3/8c = 0.375c$, the value for exact tetrahedral coordination of the atoms. The c/a ratio is also always near the ideal value of $\sqrt{8/3} = 1.633$.

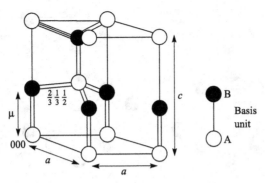

Figure 1.16   The primitive cell of the hexagonal space lattice of the wurtzite crystal structure. This is a fraction of the volume of the hexagonal structure unit cell shown in Figure 1.12.

The wurtzite structure, like the sphalerite structure, is non-centrosymmetric and exhibits polarity, i.e. the opposite directions [0001] and [000$\bar{1}$] are not equivalent. The space group symmetry of the wurtzite structure is P6$_3$mc, i.e. it is based on the primitive hexagonal lattice and has a 6$_3$ screw axis and mirror and c-glide planes. Screw axes carry lattice points to others by a rotation about the axis (sixfold in this case) plus a translation parallel to the axis (through 3/6 of the repeat distance).

1.4.5 The NaCl structure

The lead chalcogenide narrow-gap semiconducting compounds (see Section 4.1.6) crystallize with this structure.

The NaCl (rocksalt) structure occurs for a third of all AB compounds (Kelly and Groves 1970). The variously bonded (see Section 4.1.6) NaCl structure compounds include the semiconducting lead chalcogenides, PbS, PbSe and PbTe. The NaCl stucture has a face centred cubic space lattice with a basis of two atoms: a B atom at 0,0,0 and an A atom at 0,0,1/2 as shown in Fig. 1.17a.

This structure differs from the diamond cubic, sphalerite and wurtzite structures in that the atoms are not tetrahedrally bonded and thus tetrahedrally coordinated, i.e. in the NaCl structures the atoms do not have four nearest neighbours occupying sites at the corners of a tetrahedron. In the lead chalcogenides with the NaCl structure the atoms are octahedrally coordinated, i.e. have six nearest neighbours placed at the corners of a regular octahedron as shown in Fig. 1.17b.

1.5 Symmetry, Bloch waves and energy band theory

There are of the order of 10^{19} electrons and atoms per mm^3 in solids. Solution of Schrödinger's equation for so many particles is not possible. Severely simplifying assumptions have to be made. The simplest theories use the

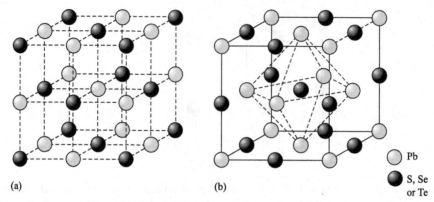

Figure 1.17   (a) The NaCl structure with an f.c.c. lattice and a basis of a B atom at an f.c.c. site and an A atom displaced from it by half the lattice parameter cubic cell edge (after Blakemore 1985), (b) The dashed lines are the edges of the octahedron of A (lighter) sites surrounding the B (black) site at the centre of the cubic unit cell.

one-electron approximation. This assumes that valid wave functions can be found for an electron moving in a constant, spatially periodic potential − as if the atoms and other electrons were stationary.

The chemical bond approach finds solutions by using atomic wave functions for the two 'bonded' atoms, to which the electron is assumed to be localized. The other form of one-electron theory in effect makes the opposite assumption and employs wave functions that spread throughout the entire macroscopic crystal. The latter are solutions of Schrödinger's equation of the form known to exist as the result of Bloch's theorem (also known as Floquet's theorem to mathematicians). These solutions only occur because of the translational symmetry (periodicity) of crystal structures.

1.5.1 Translational symmetry and Bloch's theorem

Translational symmetry in crystals means that the surroundings of atoms on periodically distributed sites are the same, assuming imperfections to be far away. This is less so as a surface or other defect is approached, so imperfections reduce the applicability of those results that rely on symmetry.

The values of the potential of an electron at equivalent points in the structure must be equal since the surroundings are the same. Therefore the probability of finding an electron at equivalent sites is expected to be the same. That is, $|\Psi_n(\mathbf{r})|^2$, (where $\Psi_n(\mathbf{r})$ is the wave function derived from an atomic state of quantum number n), which determines the probability of finding the electron at $\mathbf{r}$, must have the same periodicity as the potential function $V(\mathbf{r})$. This restriction on $|\Psi_n(\mathbf{r})|^2$ leads to a corresponding restriction on $\Psi_n(\mathbf{r})$. Bloch's theorem states that:

$$\Psi_n(\mathbf{r}) = u_n(\mathbf{k},\mathbf{r})\exp(i\mathbf{k}\cdot\mathbf{r}) \qquad (1.4)$$

where $u_n(\mathbf{k},\mathbf{r})$, often written as $u_k^n(\mathbf{r})$, is a function with the same periodicity as $V(\mathbf{r})$, and k is a quantized vector $= 2\pi m/Na$ where a is the spacing of equivalent atoms in the direction k, N is the total number of equivalent atom sites in the solid in this direction, and m is an integer. Clearly,

$$|\Psi_n(\mathbf{r})|^2 = u_n(\mathbf{k},\mathbf{r})^2 \tag{1.5}$$

which has the same periodicity as $u_k^n(\mathbf{r})$ and hence as $V(\mathbf{r})$, as required.

We can physically infer several features of the wave function. Near any nucleus, an electron is well shielded from interactions with neighbouring atoms so its wave function is like that of an electron in the free atom. Further away, interactions with neighbouring atoms become important and the function is more like that of a free electron. Since $\Psi(\mathbf{r})$ is the product of $u_k(\mathbf{r})$ (dropping the superscript n) and $\exp(i\mathbf{k}\cdot\mathbf{r})$, which varies slowly for small k, $u_k(\mathbf{r})$ itself must have the atomic form near atom sites particularly for small k as in the wave functions for sodium in Fig. 1.18.

Bringing many atoms together in the crystal converts the electron wave function from an atomic form $u(\mathbf{r})$ to a related form $u_k(\mathbf{r})$ (where k has a minor effect close to the nucleus) multiplied by $\exp(i\mathbf{k}\cdot\mathbf{r})$, a plane wave of wavelength $2\pi/k$. k is the wave vector and $\hbar k$ is the 'crystal momentum'. All electrons have a wave function of the Bloch form, extending throughout the solid. 'Tightly bound' (inner core) electrons and 'loosely bound' (outer, valence) electrons differ in $u_k(\mathbf{r})$. For inner core electrons $u_k(\mathbf{r})$ is large near atom sites and dies away quickly, but for outer electrons, $u_k(\mathbf{r})$ extends further out from the site. This enables us to understand modern methods for producing approximate Bloch functions in usable analytical forms. A more mathematical, but readable account of this topic appears in Ziman (1972).

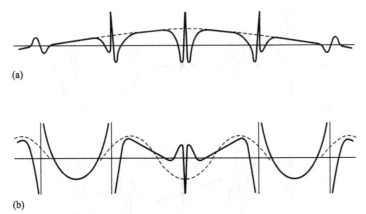

(a)

(b)

Figure 1.18   Bloch functions (solid lines) for electrons propagating through the periodic potential in a crystal. The dashed lines show the unmodulated plane waves to which the functions approximate between the ion cores. The top curve is for small k, i.e. a long wavelength and the lower curve is for large k, i.e. for a short wavelength (the dotted curve). (After Ziman 1972.)

Herring added plane wave terms to a sum of atomic orbitals in the method of orthogonalized plane waves (OPW). The atomic orbitals are included only inside the ion core 'cells' shown in Fig. 1.19. Simplifications in calculation arise from the pseudo-potential idea that the valence electrons in semiconductors are not strongly affected by the details of the ion core potential so it can be chosen as a simple smoothly varying function. In many simple solids, the potential is nearly constant between the ion cores so a muffin tin potential is used (Fig. 1.20). This is spherically symmetrical inside some fraction of the Wigner-Seitz cell (coordination polyhedron) for the atoms and constant between. The Wigner-Seitz cell around a site is formed by planes perpendicularly bisecting the lines to the nearest lattice points. The solutions of Schrödinger's equation inside the spherical regions are spherical harmonics and plane waves are the solutions in the flat potential regions. The two are matched on the surface of the spheres to give augmented plane wave (APW) solutions. Many variants of these approaches have been developed for particular purposes.

1.5.2 Electron energy bands

Any energy level of an electron in an isolated atom splits into many levels due to interactions with neighbouring atoms in the crystal (Fig. 1.5). In an atom, the energy

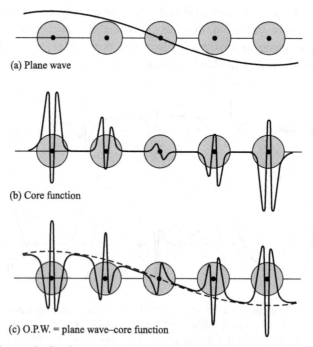

(a) Plane wave

(b) Core function

(c) O.P.W. = plane wave–core function

Figure 1.19   The synthesis of an orthogonalized plane wave by combination of a plane wave with core functions. (After Ziman 1972.)

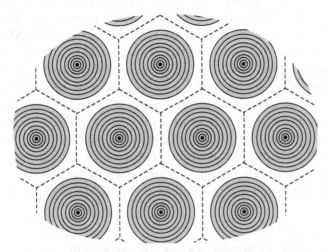

Figure 1.20   The form of 'muffin tin' potentials is spherically symmetrical throughout some fraction of the polyhedral Wigner-Seitz cell around each atom. (After Ziman 1972.)

corresponded to a wave function $u_{nlm}(\mathbf{r})$ while in the crystal the corresponding wave function is a related function, say $w_{nlm}(k \cdot r)$, multiplied by $\exp(ik \cdot r)$. There is now a set of extra wave functions, corresponding to all the values of k, each with its own energy level $E(k)$. The different energy levels each correspond to an electron having a different wave vector $\mathbf{k}$.

The number of these energy states is the number of values of k. $k = 2\pi m/Na$, as stated above, where $m = 1, 2 \ldots N$. Hence the number of values of k in a given direction is equal to the number N of equivalent atoms in this direction. The values of k are specified for a given structure and size of crystal (the values of a and N). To find the energy levels means solving the Schrödinger equation:

$$H\Psi = E\Psi \tag{1.6}$$

where H is the Hamiltonian energy operator, to find $E(\mathbf{k})$ for each $\Psi(\mathbf{k},\mathbf{r})$. This is not possible, as mentioned earlier, since there are of the order of 10^{19} particles per mm^3 and simplified methods like the one electron approximation must be used.

1.5.3 Effects of point symmetry

The wave functions of crystals have the Bloch form due to translational symmetry, and the point symmetry of particular crystals enables us to deduce further properties. Suppose, for example, that a certain plane has fourfold rotational symmetry. Then if the crystal is rotated by one or more quarter turns about a fourfold rotation axis perpendicular to the plane, the appearance from a fixed observation point is unaltered.

The simplest such case is that of rows of atoms intersecting at right angles (Fig. 1.21). The wave function of an electron attached to an atom in such a plane would have to be the same in the equivalent directions +y, +x. Otherwise the electron density would be different along these directions and therefore so would the bond strength and the atom spacing, violating the assumed condition of symmetry.

Other properties of wave functions can be deduced from the other symmetry elements of the surroundings of an atom. To deal with the full three dimensional symmetry requires mathematical group theory methods. In dealing with any new situation it is useful to find the symmetry as this gives initial clues about the electron spatial distributions and hence certain features of the energy. Since the wave function varies with direction, so does the energy. It is customary to plot E versus $\mathbf{k}$, the wave vector, i.e. the extra quantum number introduced by the translational symmetry.

1.5.4 Momentum-space and reciprocal lattices

k has the dimensions of reciprocal length and it is customary to refer to plots against k as in 'reciprocal space'. These are simply convenient graphical representations of energy band structures and diffraction phenomena. An E versus $\mathbf{k}$ graph represents the variation of the electron energy with the wave function $u_k(\mathbf{r})\exp(i\mathbf{k}\cdot\mathbf{r})$ and corresponding electron density distribution in the direction in real space that corresponds to the direction of k in reciprocal space.

Let the basic vectors of a lattice be $\mathbf{a}_1$, $\mathbf{a}_2$ and $\mathbf{a}_3$. The reciprocal unit cell is defined by the vectors $\mathbf{b}_1$, $\mathbf{b}_2$ and $\mathbf{b}_3$, where:

$$\mathbf{b}_1 = \frac{2\pi\mathbf{a}_2 \times \mathbf{a}_3}{(\mathbf{a}_1\cdot\mathbf{a}_2 \times \mathbf{a}_3)} = \frac{2\pi\mathbf{a}_2 \times \mathbf{a}_3}{V} \tag{1.7}$$

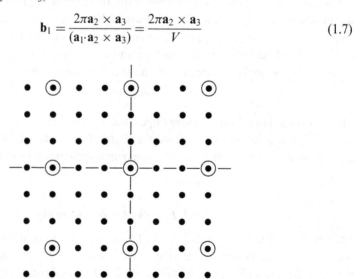

Figure 1.21   Two dimensional network of atoms illustrating fourfold rotational symmetry.

where V is the volume of the (direct lattice) unit cell, and $\mathbf{b}_2$ and $\mathbf{b}_3$ are given by similar expressions. Conversely:

$$\mathbf{a}_1 = \frac{2\pi\mathbf{b}_2 \times \mathbf{b}_3}{(\mathbf{b}_1 \cdot \mathbf{b}_2 \times \mathbf{b}_3)} \tag{1.8}$$

and $\mathbf{a}_2$ and $\mathbf{a}_3$ are given by similar expressions. Cubic system crystals have the basic lattice vectors, $\mathbf{a}$, all at right angles and it is easy to see that the reciprocal lattice vectors are simply the reciprocals of the direct lattice vectors, e.g. $\mathbf{b}_1 = 2\pi/\mathbf{a}_1$ and in the same directions. This is only true for cubic system crystals.

The product of translation vectors in the real and reciprocal lattices is a multiple of 2π and this is of advantage in the formalism as we now demonstrate. Let the direct and reciprocal lattice vectors be direct (recall Equation 1.3): $\mathbf{R}_m = M_1\mathbf{a}_1 + M_2\mathbf{a}_2 + M_3\mathbf{a}_3$ and reciprocal:

$$\mathbf{K}_n = N_1\mathbf{b}_1 + N_2\mathbf{b}_2 + N_3\mathbf{b}_3 \tag{1.9}$$

where the Ms and Ns are integers. Then:

$$\mathbf{R}_m \cdot \mathbf{K}_n = 2\pi(M_1N_1 + M_2N_2 + M_3N_3) \tag{1.10}$$

and the bracketed term is a pure number as stated above. The vectors $\mathbf{K}_n$ constitute the translation group of the reciprocal lattice (i.e. they connect any reciprocal lattice point to all the others).

The Bloch wave function (recall Equation 1.4) is $\Psi_n(\mathbf{r}) = u_n(\mathbf{k},\mathbf{r})\exp(i\mathbf{K}\cdot\mathbf{r})$,

$$\text{where} \quad u_k(\mathbf{r} + \mathbf{R}_j) = u_k(\mathbf{r}) \tag{1.11}$$

$u_k(\mathbf{r})$ has the periodicity of the lattice and may be expanded as a Fourier series of terms with this periodicity:

$$\begin{aligned} f_k(\mathbf{r} + \mathbf{R}_m) &= A_n(\mathbf{k})\exp(il\mathbf{K}_n\cdot\mathbf{r})\exp\{il\mathbf{K}_n\cdot\mathbf{R}_m\} \\ &= A_n(\mathbf{k})\exp(il\mathbf{K}_n\cdot\mathbf{r}) = f_k(\mathbf{r}) \end{aligned} \tag{1.12}$$

(where l is an integer) since $\exp(il\mathbf{K}_n\cdot\mathbf{R}_m) = \exp(il2\pi) = 1$. Hence $f_k(\mathbf{r})$ has the lattice periodicity, i.e. the general form:

$$u_k(\mathbf{r}) = \Sigma_n A_n(\mathbf{k})\exp\{i\,(l\mathbf{K}_n + \mathbf{k})\cdot\mathbf{r}\} \tag{1.13}$$

and

$$\Psi_k(\mathbf{r}) = \Sigma_n A_n(\mathbf{k})\exp\{i\,(l\mathbf{K}_n + \mathbf{k})\cdot\mathbf{r}\} \tag{1.14}$$

The validity of this expression does not depend on the value of l so k may be increased by any number of reciprocal lattice vectors without affecting the form of the wave function. The electron energy E is unchanged if k is increased by one or more reciprocal lattice vectors since in the Schrödinger equation $H\psi = E\psi$, the solution for E is unaffected by any multiplication constant in ψ, as it cancels out. Hence it is sufficient to plot the energy for values of k less than a reciprocal lattice vector, i.e. less than $\mathbf{b}_1$, $\mathbf{b}_2$, $\mathbf{b}_3$ the unit reciprocal lattice vectors. Thus the range of k values to be considered lies within a unit cell of the reciprocal lattice.

1.5.5 Brillouin zones and constant energy surfaces

The cell just described, called the first Brillouin zone (Bz), is the Wigner-Seitz cell of the reciprocal lattice, bounded by planes bisecting the vectors from the origin to the nearest reciprocal lattice points. In crystals with structures based on the f.c.c. lattice it is a cubo-octahedron (Fig. 1.22). This is the Bz of diamond and sphalerite structure materials and is marked with the standard Bouckaert, Smoluchowski and Wigner (BSW) notation: upper case Greek letters for the origin and directions of special symmetry and upper case Latin letters for points of special symmetry on the Bz boundary.

The energy band structure is represented by plots of E versus k in the special symmetry directions [together with plots of the density of states, $N(E)$ versus E]. As in Fig. 1.23 the right- and left-hand halves of the plots generally refer to different directions. Useful information is also represented by the locus of values of k for a given value of E. Figure 1.24 illustrates the geometrical shapes of the constant-energy surfaces for conduction electrons near the lowest energy state in the conduction band in germanium (Fig. 1.24a), shown in reciprocal space, and in silicon (Fig. 1.24b), shown relative to the cubic crystal axes. That for the Fermi energy, E_F, is called the Fermi surface.

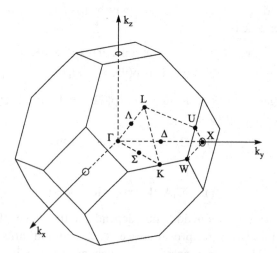

Figure 1.22   The first Brillouin zone for crystals based on the face centred cubic space lattice showing the standard notation used to label points and directions of special symmetry in the zone. This is the Brillouin zone for materials with the diamond and sphalerite crystal structures.

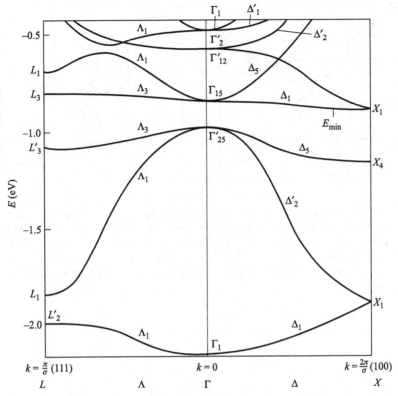

Figure 1.23   The energy bands in silicon labelled in accordance with the notation in Figure 1.22. Thus energy curves plotted versus k in the $<100>$ directions are labelled Δ, etc. The various branches for wave functions with $u_k(\mathbf{r})$ derived from different atomic levels are labelled Δ_1, Δ_2, Δ_5. [After Kleinman and Phillips 1960. Reprinted with permission from *Physical Review*, 118, 1153–67, 1960 (http://publish.aps.org/linkfaq.html). Copyright 1960 by the American Physical Society.]

If the Fermi surface occurs near the Brillouin zone boundary the wave functions satisfy the conditions for Bragg reflection. The Bloch functions have wavelengths $\lambda = 2\pi/k$ and for normal incidence on (hkl) planes will be reflected if $d_{hkl} = \lambda/2 = \pi/k$, i.e. π/a, the value at the Bz boundary in the direction a. The reflected and incident waves set up two standing waves, one with nodes on the atom planes and the other with nodes midway between them. These waves have the same k but different E so an energy gap occurs for waves with values of k lying on the zone boundaries (Fig. 1.25).

The energy gaps at the Bz boundary for different k directions may not coincide in value, so there is not necessarily an overall energy gap in the crystal. Semiconductors and insulators are those materials in which there is, and the Fermi surface passes approximately through the middle of the gap so the valence

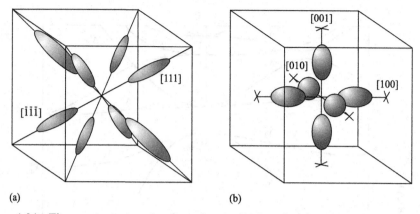

(a)   (b)

Figure 1.24 The constant energy surfaces for conduction electrons near the lowest energy state in the conduction band (a) in germanium, shown in reciprocal space, and (b) in silicon shown relative to the cubic crystal axes. The lowest conduction band minima in Ge lie at the points L and those in Si at points part of the way out in the Δ directions. [After Sze 1981 (Physics of Semiconductor Devices; Copyright 1981; Reprinted with permission of John Wiley & Sons, Inc.)]

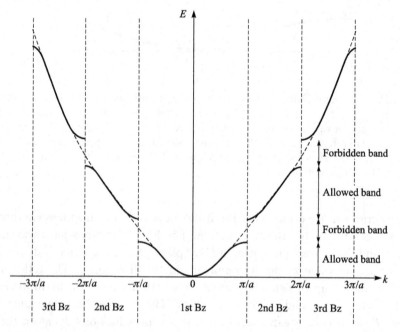

Figure 1.25 Energy versus wave vector for an electron in a field-free vacuum (broken line) and in a crystal (solid line); $k = \pm n\pi/a$ is the value of the wave vector at a Brillouin zone boundary where Bragg reflection of the electrons takes place. This is referred to as the extended zone representation of the nearly-free electron model, showing the modification of the parabolic $E(k)$ dependence for free electrons at the band edges corresponding to $k = n\pi/a$. The first three Brillouin zones (Bz) are indicated.

band states inside the zone are all occupied and separated by energy gaps from the conduction band states. The Bz of Fig. 1.22 contains states holding four electrons per atom site so diamond and sphalerite structure materials have electron/atom ratios = 4 and the Fermi surface coincides with the Brillouin zone boundary.

1.5.6 *The form of the forbidden energy gap*

The gap is said to be direct if the lowest state in the conduction band and the highest in the valence band are at the same point in k space and indirect if they are not (Fig. 1.26). In Ge the conduction band minima are at the points L (Fig. 1.24) but the valence band maximum is at Γ, the origin of k space, so the gap is indirect, as it is in Si also. Electrons and holes in such materials differ in crystal momentum (Fig. 1.26) so the radiative recombination probability is relatively low and such materials are not used in semiconductor lasers. The energy band structure of GaAs, however, is direct (Fig. 1.27), so the radiative recombination probability is relatively high and GaAs can be used in lasers.

The occurrence of two conduction band minima of the form shown in Fig. 1.28 is essential for the transferred electron (Gunn) effect so only materials with this form of energy band structure, like GaAs and InP can be used in transferred electron oscillators (TEOs). More detailed accounts of these device phenomena will be found in Sze (1985). Much semiconductor materials science now is 'energy band engineering', that is, the production of optimal energy band structures for particular

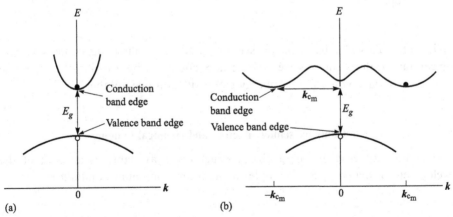

Figure 1.26 (a) Direct and (b) indirect energy band gaps. Electron-hole pair recombination across a direct gap can result in the emission of the energy as a photon. Such recombination across an indirect gap must result in the emission or absorption of at least one phonon to conserve the momentum $\hbar k_{c_m}$ where k_{c_m} is the value of k at the conduction band minimum.

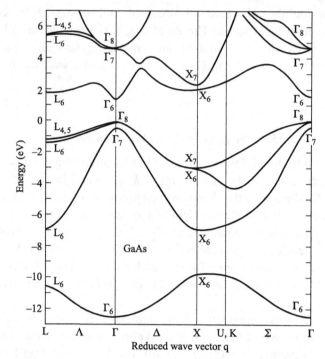

Figure 1.27   The electron energy band structure of GaAs (electron energy vs. the reduced wave vector for the four valence bands and the first several conduction bands) labelled with the notation of Figure 1.22. The top of the valence bands is zero on this scale (from Blakemore 1982 after Chelikowsky and Cohen, 1976). (Reprinted with permission from *Journal of Applied Physics*, Vol. **53**, pp. R123−R181. Copyright 1982, American Institute of Physics).

applications. This can be done by selecting and controlling the composition of semiconductor alloys and, as we will see in Section 1.7, by growing thin epitaxial layers of materials with different energy gap widths on one another.

1.6 Complex semiconductors and chemical bonding

A wide range of electron energy band structures is available as a result of the development of many types of complex adamantine compounds and alloys.

1.6.1 Cross substitution and ternary compounds

Goodman and Douglas (1954) first suggested that ternary (three element) adamantine materials could be obtained from known binary compounds by 'cross substitution', i.e. replacing one element by pairs from other groups of the periodic

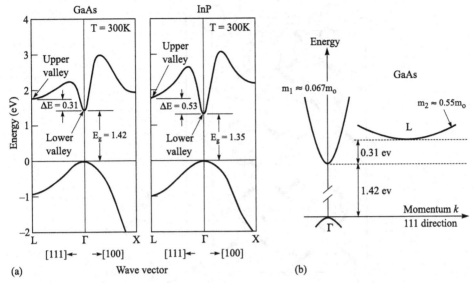

Figure 1.28 Electrons in the two lowest energy band minima in GaAs (or InP which is similar) will have different effective masses, m^*, since $1/m^* = (1/\hbar^2)(d^2E/dk^2)$, and hence mobilities. The densities of states in the higher minima are also much greater than those in the lower, direct-gap minima. Such energy band structures are essential for the occurrence of transferred electron effects. (After Rees and Gray 1976. *IEEE Journal of Solid State and Electron Devices*, Vol. **1**, pp. 1–8; © 2002 IEEE.)

table while keeping the valence electron to atom ratio = 4. Austin *et al.* (1956) produced I-III-VI$_2$ compounds by replacing the IIB element in the II-VI compounds by equal numbers of IB and IIIB atoms. Similarly the semiconducting II-IV-V$_2$ compounds, obtained by cross substituting for the III element in III-V materials, were predicted and confirmed (Goodman 1957).

Most of them have the sphalerite structure, at least at elevated temperatures, with the VB atoms on the sites of one f.c.c. sublattice and the IIB and IVB atoms randomly distributed over the sites of the other f.c.c. sublattice. Below a critical temperature many II-IV-V$_2$ compounds adopt the chalcopyrite structure (Fig. 1.29) with the IIB and IVB atoms in an ordered arrangement on one sublattice and the crystal distorts tetragonally. That is, one edge of the cubic unit cell shortens slightly and, to include a complete stacking sequence of the ordered atom planes, the unit cell is twice as high in that direction. Thus the chalcopyrite unit cell height, c, is slightly less, generally, than twice the edge of the base, a (Fig. 1.29). For reviews of these materials see Shay and Wernick (1975) and Miller *et al.* (1981).

Numerous additional families of semiconducting compounds were developed by successive cross substitution. The ideas employed and the many families of compounds produced were reviewed by Mooser and Pearson (1960) and Parthe (1966).

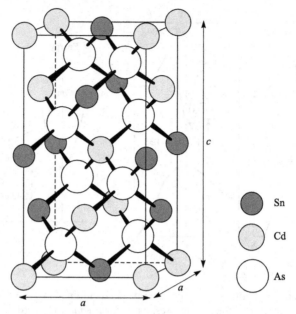

Figure 1.29   A unit cell of the chalcopyrite structure.

1.6.2 Semiconductor alloys

Many pairs of sphalerite-structure compounds can be combined to give a continuous range of substitutional solid solutions with the sphalerite structure (Woolley 1962). Such miscible systems, i.e. those with no intermediate change of structure (phase) at certain compositions, have the simple form of pseudo-binary phase diagram typified by AlSb-InSb (Fig. 1.30 from one of the first examples reported). In these alloys, for all compositions, the VB (Sb) atoms occupy one f.c.c. sublattice of the sphalerite crystal structure and the IIIB atoms (Al and In in this case) occupy the other sublattice. If the occupation is completely random, the material is an ideal solid solution.

Energy band gaps vary continuously in such systems (Fig. 1.31) and this helps make energy band engineering possible. For example, in GaAs-AlAs, the direct gap at Γ (the origin of k space) is the minimum one from GaAs to $Ga_{0.62}Al_{0.38}As$ (for detailed discussion, see e.g. Vurgaftman *et al.* 2001). For more Al, the indirect X gap is the smallest. This nearly ideal system was employed in first successful superlattice (multiquantum well) structures as well as in the epitaxial double heterostructure lasers of the first-generation fibre optic communication links. (These topics are discussed in more detail in Section 1.7). In GaAs-GaP alloys the gap is direct from GaAs to $GaAs_{0.65}P_{0.35}$ (Goryunova *et al.* 1968). The latter composition is used in light emitting diodes because the wider direct gap due to the addition of P means that the recombination radiation is in the visible. Many other

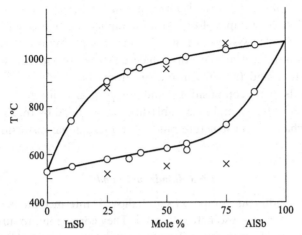

Figure 1.30 Phase diagram for the AlSb–InSb system. O: heating; X: cooling. (After Goryunova 1965; *The Chemistry of Diamond Like Semiconductors*. Reprinted with permission of the MIT Press.)

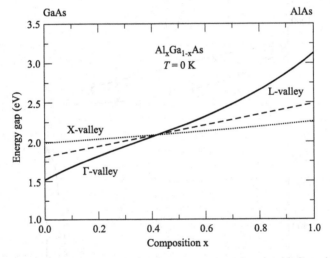

Figure 1.31 Bandgap energy versus aluminum concentration in $Al_xGa_{1-x}As$ alloys. The energy gap values between the valence band maximum at Γ (see Figure 1.22) and the conduction band minima at Γ, i.e. the direct gap, and at X, i.e. the indirect gap in Al rich alloys, and at L are plotted. (After Vurgaftman *et al.* 2001. Reprinted with permission from *Journal of Applied Physics*, Vol. **89**, pp. 5815–75. Copyright 2001, American Institute of Physics.)

III-V and II-VI alloys have been developed for optoelectronic applications. There is much current interest also in the relatively new materials GaN and $Ga_xIn_{1-x}N$ alloys for blue and green LEDs and lasers and for high brightness LEDs suitable for possible lighting applications.

More complex compounds can also be combined in miscible alloy systems but not all are ideal. A simple example of the sort of complication that can arise is shown by the InAs-ZnSnAs$_2$ alloy system of Fig. 1.32. For compositions and temperatures in the α phase region the crystals have the sphalerite structure with As atoms occupying one f.c.c. sublattice and the In, Zn and Sn atoms randomly distributed over the other. In crystals with compositions and temperatures in the β phase region, however, the sites of the second f.c.c. sublattice are occupied by In, Zn and Sn atoms in the ordered chalcopyrite structure but with no tetragonal distortion so $c/a = 2$.

1.6.3 Bonds and bands

Bonding in semiconducting compounds and alloys is intermediate between covalent and ionic. Both pictures are sometimes useful. They correspond to different types of interaction following the transfer of charges from the one atom to the other. Extreme ionic bonds in e.g. II-VI compounds would involve transferring two electrons from the IIB to the VIB atom to form II^{2+} and VI^{2-} ions which would result in coulombic

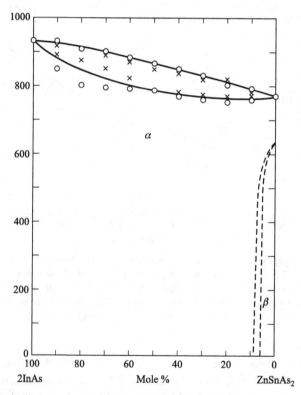

Figure 1.32　Phase diagram for the 2InAs−ZnSnAs$_2$ system. Open circles: cooling; crosses: heating. (After Borchers and Maier 1963.)

attraction. Extreme covalent bonds would involve transferring two electrons the opposite way to form II^{2-} and VI^{2+} ions each with four valence electrons that would then form four shared-electron sp^3 bonds. Neutral bonding is the intermediate case that involves no charge on the atoms. The most useful treatment of bonding by Phillips relates it to energy bands.

Phillips' (1973) approach to chemical bonding in the adamantine semiconductors is both empirical and band theoretical. He distinguished the bonding energy E_b from the forbidden gap energy E_g. On the band picture E_b, the energy separation of the bonding and antibonding orbitals which influences structural properties, is the difference between the mean energies of the valence and the conduction band states $\bar{E}_v$ and $\bar{E}_c$ (Fig. 1.33). The energy gap is that between the highest valence band level E_v and the lowest conduction band level E_c. E_b can be obtained from spectroscopic measurements of the difference in energy between bonding and antibonding states for corresponding molecules.

The bonding energy E_b consists of a covalent or homopolar contribution E_h and an ionic contribution C. According to Phillips:

$$E_b = E_h + iC \tag{1.15}$$

since, in band theory, the covalent and ionic potentials have to be combined via the crystal structure factor in a manner which multiplies them by factors that are 90° out of phase (Fig. 1.34).

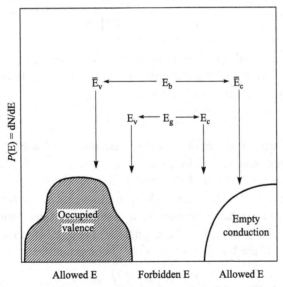

Figure 1.33   The filled and empty energy bands of a semiconductor. The bonding energy E_b is larger than the forbidden gap energy E_g. $N(E)$ is the density of states, i.e. the number of allowed states per unit energy, per unit volume. (After Phillips 1973. Reprinted from *Bonds and Bands in Semiconductors*. Copyright 1973, with permission from Elsevier.)

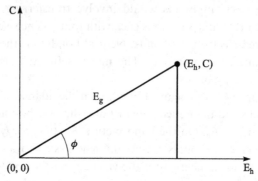

Figure 1.34   The phase angle ϕ in the E_b, C plane is a measure of the relative ionic and covalent contributions to the bonds in $A^N B^{8-N}$ compounds. (After Phillips 1973. Reprinted from *Bonds and Bands in Semiconductors*. Copyright 1973, with permission from Elsevier.)

The covalent contribution is of the empirical form:

$$E_h \propto l_{AB}^{-2.5} \tag{1.16}$$

where l_{AB} is the (nearest-neighbour) bond length. The constant of proportionality is obtained by plotting the bond energies of the IVB diamond structure elements against l_{AB}. Empirical values of C are derived from the observed dielectric constants, of $A^N B^{8-N}$ materials as follows. The dielectric constant, i.e. the square of the optical index of refraction in the limit of low frequencies, $\varepsilon(0)$, can be written as:

$$\varepsilon(0) = 1 + \left(\frac{\hbar \omega_p}{E_h}\right)^2 A \tag{1.17}$$

where $\hbar$ is $h/2\pi$ (h is Plank's constant), ω_p is the plasma frequency, i.e. the frequency of oscillation of disturbances in the positions of the valence electrons, and A is a constant approximately equal to 1. Now:

$$\omega_p = 4\pi n_o q^2 / m \tag{1.18}$$

where n_o is the average number of s-p valence electrons per atom, q is the electronic charge and m is the mass of an electron.

Substituting Eqs. (1.15) and (1.18) in Equation (1.17) gives C^2 in terms of the dielectric constant $\varepsilon(0)$ and the valence electron/atom ratio n_o so both E_h and C can be determined from optical properties in a manner consistent with band theory. Using Equation (1.15) and the phase angle ϕ (Fig. 1.34), Phillips defines the fractional ionicity of the bonding in an $A^N B^{8-N}$ compound as:

$$f_i = \sin^2\phi = C^2 / E_b^2 \tag{1.19}$$

and the fractional covalency is given by:

$$f_h = \cos^2\phi = E_h^2 / E_b^2 \tag{1.20}$$

The values of E_h and C for 68 $A^N B^{8-N}$ compounds are plotted in Fig. 1.35. The diamond, sphalerite and wurtzite (adamantine) structures occur for $f_i < 0.785$ (below the line), in keeping with the simple idea that directed sp^3 covalent bonding, corresponding to high values of f_h, produces tetrahedral coordination.

This approach to bonding has had considerable success in the calculation of point defect energies.

1.7 Energy band engineering

An alternative to developing ever more new semiconducting compounds and alloys, with novel properties required for particular device applications, has evolved in

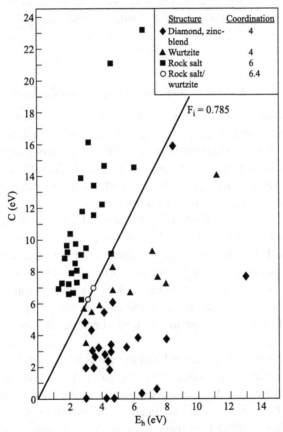

Figure 1.35 A plot of the values of E_b and C for $A^N B^{8-N}$ compounds on which the coordination numbers are indicated by the shapes of the points. The line $F_i = 0.785$ separates all the fourfold, tetrahedrally coordinated crystals from all the sixfold, octahedrally coordinated materials. (After Phillips 1973. Reprinted from *Bonds and Bands in Semiconductors*. Copyright 1973, with permission from Elsevier.)

recent decades. This uses energy band engineering employing epitaxial layer structures also known as heterostructures. These layers can be so thin (a few atomic layers, less than the carrier thermal de Broglie wavelength) that new quantum mechanical phenomena appear. These give such 'designer' structures new properties, different from, and often superior to, those of the materials of which the layers consist. They can be described as low-dimensional or quantum confined structures. They are referred to as quantum wells if they are of quantum confined size in one dimension only. Quantum wires are limited in two dimensions and quantum dots are limited in all three spatial dimensions.

Almost all research and development on semiconductor materials is now concentrated in this field. Even the work on a new material with outstanding properties for light emitting devices, GaN, soon came to concern the development of such structures involving also GaInN, etc. Prediction of the properties of quantum-confined heterostructures is quantitative and firmly based on quantum mechanics.

1.7.1 Heterojunctions and energy band alignment

The new possibilities first appeared in research on heterojunctions, their one-dimensional energy band diagrams and electronic and optoelectronic properties. A heterojunction is the interface between two different semiconducting materials and is a more complex analogue of a *p-n* junction between differently doped regions in a single material. Heterojunctions result from all cases of semiconductor hetero-epitaxy (growth of one semiconductor on another).

The energy band diagrams of *p-n* junctions were well understood through the work of Shockley and others in the earliest days of solid state electronics. A *p-n* junction can be represented by the conduction band minimum and the valence band maximum drawn normal to the junction as in Fig. 1.36. The Fermi level, which in equilibrium must be everywhere constant, is near the conduction band edge in the *n*-doped material and near the valence band edge in the *p*-doped material. Hence band bending and built-in fields occur as shown, because there is a potential change V_d across the junction.

To construct heterojunction energy band diagrams we must take into account the fact that the band gap, E_g, changes from one material (E_{g1}) to the other (E_{g2}). So band edge discontinuities, ΔE_V and ΔE_C, occur at the heterojunction and their magnitudes must be found for each particular combination of materials (Fig. 1.37). (For details see Jaros 1989.) It is these discontinuities that are used in 'band gap engineering'. For example, a discontinuity (step) in a band edge can be a potential barrier to carriers in that band, crossing the heterojunction. By gradually varying the composition of a semiconductor alloy a graded band gap can be produced as first suggested by Kroemer (1957). The effect of taking account of the *p* and *n* doping on

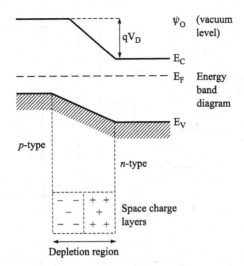

Figure 1.36 The conduction band minimum E_C and the valence band maximum E_V are continuous across the *p-n* junction. The energy band gap E_g is unchanged and the Fermi energy E_F is constant in equilibrium. This results in a built-in potential jump, V_D, across the junction.

either side of the heterojunction is to produce a built-in potential jump as in *p-n* junctions (Fig. 1.37b).

Injection laser action in a semiconductor was first achieved in heavily doped GaAs *p-n* (homo)junctions (Fig. 1.38a). Laser (Light Amplification by Stimulated Emission of Radiation) action in a semiconductor requires two device conditions. Firstly there must be an inverted carrier population, i.e. in this case, a density of hole-electron pairs in the active region that is far above that for thermal equilibrium. Secondly, the light must be reflected back and forth through the device. The first *p-n* junction and DH lasers had cleaved and therefore parallel mirror faces at either end. A standing wave was set up so the beam travelled repeatedly through the inverted population. The beam photons cause hole-electron pair stimulated emission of photons coherent with the beam. Above a threshold current density the gain per unit length in the active region exceeds all the losses. A laser beam thus builds up and some is emitted by (partial) transmission through the mirror face.

The first GaAs *p-n* junction lasers required such high threshold currents for laser emission (hundreds of amps per cm^2) that they failed in very short times (minutes for the first experimental GaAs devices, even when operated only with short, widely separated current pulses at liquid nitrogen temperature). Threshold current values were high due to electron and photon losses from the layer in which laser action occurs at the junction. H. Kroemer (1963) suggested, soon after, that carrier confinement in a low band gap material clad by wide-gap heterojunctions (back-to-back), would make population inversion and laser action possible at much lower current densities (as explained in the caption of Fig. 1.38b). Such devices are known

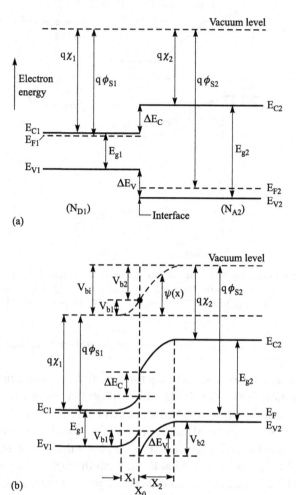

Figure 1.37 (a) The energy band edges of semiconductors, S_1 and S_2, not in contact, that differ in band gap are plotted relative to the common vacuum level (zero of potential). S_1 is donor doped so its Fermi level E_{F1} is near E_{c1} while S_2 is acceptor doped so E_{F2} is near E_{v2}. The materials of the heterojunction have different band gaps, $E_{g1} \neq E_{g2}$, so there are discontinuities, ΔE_V and ΔE_C in the band edges. These must be determined for each pair of materials. This is the situation before the materials are brought into contact. (b) On contact, transfer of charge occurs to create a 'built-in' potential jump, V_{bi}, across the heterojunction with the result shown. (After Sze 1985; *Semiconductor Devices: Physics and Technology.* Copyright 1985. Reprinted with permission of John Wiley & Sons, Inc.)

as double heterostructure (DH) lasers. DH lasers were first produced and shown to be capable of CW (continuous wave, i.e. continuously emitting) operation at room temperature by Alferov *et al.* (1971) and Hayashi *et al.* (1970). Compared to the simple *p-n* junction, the DH design reduces both carrier and photon losses due to

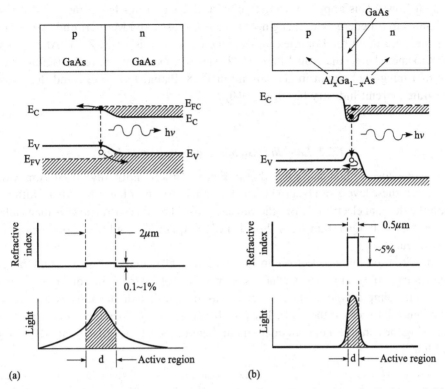

Figure 1.38 Diagrams comparing (a) a heavily doped p-n junction laser in GaAs, and (b) a double heterostructure (DH) laser. The dashed lines represent the Fermi level under large forward bias. (After Sze 1985; *Semiconductor Devices: Physics and Technology*. Copyright 1985. Reprinted with permission of John Wiley & Sons, Inc.) The cross hatched levels in the conduction band are filled with electrons and the white levels in the valence band contain holes. Under forward bias both are injected into the active region which, in (b) is a narrow layer of GaAs between wider gap $Al_xGa_{1-x}As$. The carriers form an inverted population resulting in light amplification by stimulated emission (laser action). In (b) the large conduction (valence) band edge step on the left (right) prevents injected electrons (holes) escaping to the left (right), respectively. The large change in refractive index between the GaAs and the alloy layers keeps the photons in the active region.

the abrupt changes at the cladding layer boundaries of the active layer. This increases the recombination efficiency and reduces the threshold current. Hence the first DH lasers could operate CW at room temperature, for over 100,000 hours (ten years), making the first generation of fibre optic telecommunications economic. Telecommunications laser operating lives, extrapolated from accelerated life testing, are now so long (centuries) that they will be replaced as obsolete long before they fail.

H. Kroemer made another early proposal for a heterostructure device to improve device performance. This was for HBTs (heterojunction bipolar transistors) that use a wider-energy-gap emitter layer to produce band offsets that limit unwanted current flows. This approach to designing device structures has come to be known as band gap (or energy band) engineering. See Capasso and Margaritondo (1987) for a systematic account. For their pioneering work in this field, Zh. I. Alferov and H. Kroemer shared the Nobel Prize in Physics in 2000 with J. S. Kilby (Kilby made the first integrated circuit in germanium in 1958. Planar technology and the silicon integrated circuit industry began in 1960).

1.7.2 Low-dimensional structures for devices

The active layer in DH lasers like that in Fig. 1.38b were normally 0.1–0.3 μm wide. However, these layers can be grown much thinner (10 nm or less, i.e. with a width less than the thermal electron de Broglie wavelength). The thin central layer is then called a quantum well. Quantum wells show beneficial quantum confined properties as we now discuss.

The situation of an electron in a quantum well can be approximated by an elementary problem in wave mechanics, namely that of a particle in a one-dimensional box. In the simplest model the potential walls are infinitely high so the electron wave functions fall to zero at the walls. The possible energy states then have wave functions satisfying the condition that an integral number of half wavelengths fit into the box of width w, i.e.

$$\ell\lambda/2 = w \tag{1.21}$$

where ℓ is an integer. In the case of the infinitely deep well it is easy to show (e.g. Jaros 1989, Appendix 2) that the energy of the electron states is given by:

$$E = \frac{\hbar^2}{2m}\left(\frac{\pi\ell}{w}\right)^2 \tag{1.22}$$

Since the walls of the potential well are, in reality, of finite height, the wave functions do not fall to zero at the walls but tail off exponentially into the potential barrier as shown in Fig. 1.39, so Equations (1.21) and (1.22) are only first approximations.

A valuable feature of quantum wells is that, as shown by Equation (1.22), the electron energy levels depend on w, the width of the well. By growing the well to a chosen width, the energies of the confined states and hence the magnitude of the radiative energy transitions can be designed so the light is emitted at the wavelength needed for a practical application.

In reality, wells are of finite depths and the mathematics is a bit more complicated. However, the principle that properties can be altered by growing quantum wells to particular widths remains valid. For a good introductory account of the wave mechanics of heterojunctions and quantum confined structures see Jaros (1989).

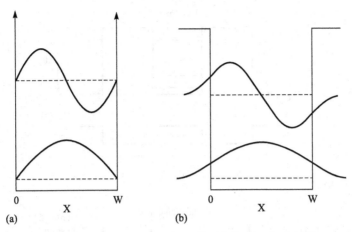

Figure 1.39   Schematic diagram of a ground state and first excited state energy levels and associated wave functions for (a) an infinitely deep square potential well, and (b) a square potential well of finite depth.

The energies of the discrete quantum states in the quantum well are determined by the band gaps of the two materials (and hence the depth of the well) as well as the width of the well. Thus the emitted wavelength can be varied over a certain range by altering the well width and moved to other ranges by using different pairs of materials.

Consider a double heterostructure laser (Fig. 1.38b) in which the central, active layer is a quantum well. The electrons and holes injected from the $Al_xGa_{1-x}As$ will be trapped in the quantum confined states in the GaAs well. These are close to one another and so form excitons even at higher temperatures. The epitaxial growth technology had by this time so advanced that the quantum well material could be grown essentially free of both unwanted impurities and extended defects. Virtually no non-radiative recombination mechanisms were then competing with stimulated radiative recombination. The emissive efficiency of such semiconductor lasers was therefore very high. These lasers were the first successful example of the incorporation of quantum confined structures in devices. In typical semiconductor fashion, lasers employing quantum wells rapidly became cheaper with time and were designed and produced for many applications. Their two chief fields of application are in transmitting the signals along telecommunications optical fibres and reading and writing data to CDs and DVDs. At the time of writing, moreover, semiconductor lasers are displacing other types (gaseous, liquid and non-semiconductor solid) from other applications one after another.

Modulation doping (Stormer *et al.* 1978) is important in this field. Consider a three-layer structure like that in Fig. 1.40. Suppose it is donor doped in the outer layers. The electrons will be trapped in the central well while the charged donor impurities are in the outer, barrier layers. Consequently the carrier mobility

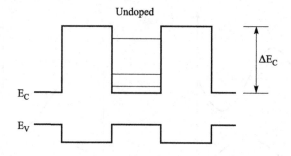

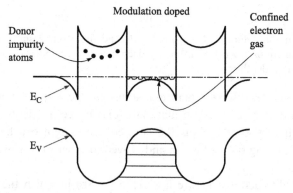

Figure 1.40   Band edge variation of undoped and n-doped GaAs–AlGaAs quantum wells.

in the plane of the central layer (normal to the plane of the figure) is not reduced by charged impurity scattering. The mobility increases, due to modulation doping, only by a few times at room temperature in GaAs, but at low temperatures it is increased by three orders of magnitude or more. HEMTs (high electron mobility transistors) are FETs (field effect transistors) that employ high carrier mobility, modulation-doped layer structures. They can be used in high frequency MMICs (monolithic microwave integrated circuits) of GaAs or InP (as can the HBTs mentioned above) for use in e.g. satellite TV receivers and mobile phones.

Besides layers thin enough to be quantum confined in one dimension (a quantum well) it is possible to produce strips of a thin layer that are narrow enough to provide quantum confinement in a second dimension (a quantum wire). It is also possible to make small volumes that are quantum confined in all three dimensions (a quantum dots). Quantum wires and dots also exhibit new properties that are predictable from simple quantum mechanical models. At the time of writing they are being developed and studied both for their intrinsic interest and electronic and optoelectronic device applications.

1.7.3 Multiquantum well materials

Quantum wells are, conceptually, the simplest type of quantum confined structure. However, the first proposal in this field, by Esaki and Tsu (1970), was for the growth of periodic, epitaxial layer 'superlattices'. The abrupt changes in composition would produce regular variations of energy gap and periodic square potential wells in both band edges (Figs. 1.41 and 1.42). These were like those of the simple quantum mechanical Kronig-Penney model. Esaki and Tsu pointed out that if the period of the superlattice were less than the de Broglie wavelength of the electrons, the material would be a new type of one-dimensional crystal. The periods of these man-made epitaxial structures are larger than the crystal lattice parameters of the materials of the layers, so the corresponding reciprocal space distances are smaller. Hence new Brillouin 'minizones' would be introduced.

The first such structures were grown by VPE of GaAs-GaP alloy layers of differing composition and periods down to 11 nm (Blakeslee 1971). They did not show the predicted properties. They were sold as resolution test specimens for scanning electron microscopes, however! Later, MBE (molecular beam epitaxy), a highly evolved method of evaporation in ultra-high vacuum, was developed for the purpose by A.Y. Cho (Fig. 1.43). It successfully grew superlattices in the GaAs-AlAs system.

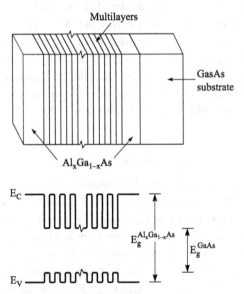

Figure 1.41 Diagram of a GaAs/Al$_x$Ga$_{1-x}$As superlattice and of the square-well, one-dimensional energy band to which such multilayer structures give rise. (After Dingle 1976. Reprinted from *Physics of Semiconductors*, Proc. 13th Internat. Conf. Rome, Optical properties of semiconductor superlattices, pages 965–74. Copyright 1976, with permission from Elsevier.)

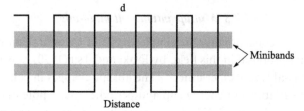

Figure 1.42 Potential profile of a superlattice structure (e.g. GaAs/AlGaAs) with the first two minibands marked.

This success was due to the achievement of atomically smooth, abrupt interfaces between the layers (Chang and Esaki 1979, Chang and Giessen 1985). To reproducibly grow layers of such small thicknesses, computerized operation of the shutters, controlling the fluxes of constituent elements and doping impurities reaching the growth surface, is essential (Fig. 1.43). MBE growth kits usually also include instrumentation for *in situ* monitoring. These may include mass spectrometers for identifying any residual gases, reflection high-energy electron diffraction (RHEED − marked as HEED in Fig. 1.43), cameras for surface structure studies and Auger analysers for surface impurity detection. The structures so produced exhibited the detailed behaviour predicted by the increasingly sophisticated theoretical treatments. Later it was shown that superlattices and other quantum confined structures could also be grown by MOCVD. The superlattice work was cited as part of that for which Esaki shared in the Nobel Prize in Physics for 1973. Later a number of device applications emerged, notably the quantum cascade laser to be discussed in Section 1.7.4 (Capasso *et al.* 2002). Consequently, the whole field of quantum confined structures rapidly expanded and advanced (Chang and Giessen 1985, Dingle 1987, Reid 1992, Jain 2000, Cho *et al.* 2001, Capasso *et al.* 2002).

1.7.4 *Quantum cascade lasers*

The photon energy emitted by DH and QW lasers is given by the energy band gap in the active layer or QW (plus the energies of the confined states in the QW case). This does not apply to quantum cascade lasers (QCLs). These devices emit photons with energy equal to the difference between two localized levels in quantum well (QW) active layers (Fig. 1.44), an idea suggested by Kazarinov and Suris in 1971. These energies tend to be small compared to band gap energies. The QCL design is thus good for the longer wavelengths (lower photon energies) of the infrared. The alternative for infrared lasers is to use semiconducting compounds with small band gaps (e.g. the lead chalcogenides, PbS, PbSe and PbTe). The technological limitations of such compounds, however, are severe. Infrared lead chalcogenide lasers have been available for 20 years but emit only milliwatt beams, have small

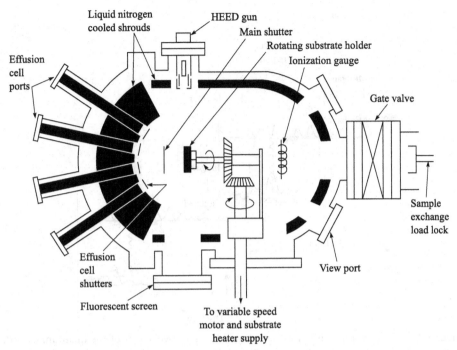

Figure 1.43 Schematic diagram of a computer controlled molecular beam epitaxy system, with analytical instruments used to produce quantum-confined structures including epitaxial superlattices. The effusion cells evaporate, as molecular beams, the constituent elements of the material to be grown and the desired dopants. [After Cho 1995, in *MRS Bulletin*, Vol. **20**, No. 4 (1995) p. 21, Figure 1. Reproduced by permission of the *MRS Bulletin*.]

tuning ranges and cannot operate at room temperature. They also tend to have short working lives due to thermal cycling.

QCL lasers employ doped superlattices to inject, via their minibands, electrons into the active quantum well regions. Successful QCLs were first demonstrated in 1994 by the research groups of Capasso and Cho (Faist *et al.* 1994). The discussion of the QCL here follows that of Capasso *et al.* (2002). It is included both because of its intrinsic importance and as an illustration of the sophistication of band gap engineering that device designs have now attained.

Fig. 1.45 illustrates the actual form of the QCL shown schematically in Fig. 1.44. The injector plus active region structure can be repeated anything from 1 to 100 times but usually there are from 20 to 35 repetitions for devices emitting in the wavelength range from 4 to 8 μm. The multiple stage design and the consequent multi-photon emission enable QCLs to emit beams with up to a watt of power. (The other contributing factor is that, since these devices are made of wider gap materials they can tolerate higher operating temperatures and thus employ larger drive

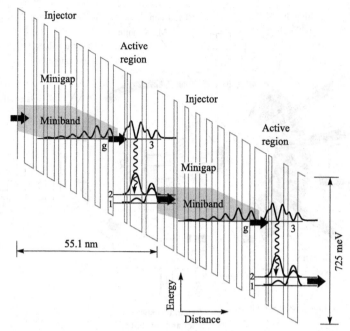

Figure 1.44   The conduction band edge plotted through two stages of the quantum cascade laser with the actual structure shown in Figure 1.45. The square of the modulus of the relevant electron wave functions are also shown. The QCL is operated under bias, producing the slope down from left to right. A field of 70 kV/cm is needed to inject the electrons through the superlattices and the narrow barriers, from the ground state of the minibands, g, into the upper localized states, 3, of the active quantum wells. Light is emitted in the transitions from level 3 to 2. A photon is produced in each of the active regions as an electron cascades down the diagram, giving many photons per electron in the drive current. This multiplication gives high power laser beams for relatively low currents. (After Capasso *et al.* 2002, *IEEE Journal of Quantum Electronics*, Vol. **38**, pp. 511–32; © 2002 IEEE.)

currents than can the competing lead chalcogenide infrared lasers.) The device of Fig. 1.45 (b) is a DFB (distributed feedback) laser. That is, it is one in which the Bragg grating selects the wavelength, λ, satisfying the reflection condition that $\lambda = 2\mu d$ where μ is the effective refractive index of the waveguide and d is the repeat distance of the grating. This permits only this one of all the possible modes of the laser to be reflected and amplified and suppresses all the others. The emission wavelength of the laser can then be tuned by varying the temperature which changes μ (so λ increases with T).

Since the energy levels for quantum wells can be obtained by solving Schrödinger's equation, the width and shape of the wells can be designed so the energy-level difference leads to the emission of the desired wavelength. Thus, the QCL can be designed to emit at any wavelength within a wide range of the infrared spectrum employing the same combination of semiconductors in the active region.

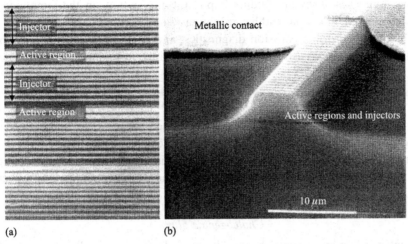

(a)   (b)

Figure 1.45   (a) Transmission electron micrograph through the epitaxial layer structure of the injector superlattices and the quantum wells of the active regions in the quantum cascade laser represented in Figure 1.44. (b) Scanning electron micrograph of the device structure of the QCL. The epitaxial layer structure of (a) lies horizontally at the depth marked 'Active regions and injectors'. The parallel lines across the top of the laser ridge constitute a Bragg grating that provides 'distributed feedback'. This selects one sharp wavelength only for laser emission along the ridge. At the top of the micrograph can be seen the metallic contact layer that has been removed over the lower portion of the top surface of the device. Current flows vertically through the device. (After Capasso *et al.* 2002, *IEEE Journal of Quantum Electronics*, Vol. **38**, pp. 511–32; © 2002 IEEE.)

In addition, the design can be arranged to maximize the matrix element for the optical transition and optimize the lifetimes of the quantum-well states to provide the population inversion required for laser action. (As discussed in Section 1.7.1, an inverted population and reflection of the light back-and-forth through the active region are required for laser action, i.e. continuous light amplification.) In the present case, the inverted population is achieved by making the electron relaxation time from state 3 to 2, τ_{32}, greater than the lifetime in state 2, τ_2. Level 3 will then be more populated than level 2. The energy separation of states 2 and 1 is made equal to that of an optical phonon. This ensures that τ_2 is small (0.3 picoseconds) whereas τ_{32} is large (2.6 ps in the case of the laser shown in Figs. 1.44 and 1.45). A number of other subtleties are incorporated into the design of this laser.

In contrast to *p-n* junction and DH lasers (Section 1.7.1) that involve both the electrons and holes, the operation of the QCL involves only one type of charge carrier (electrons). It is thus a unipolar device. The QCL is presently the only type of semiconductor laser functioning in the mid-infrared range and at room temperature. This spectral range is important since it is that where the rotation and vibration spectra of molecules occur. QCLs therefore are important for the analysis of traces of gaseous impurities. They are finding important applications in the remote sensing of environmental, industrial and hazardous gases and pollutants in applications such as pollution monitoring, emission diagnostics, and engine combustion, since many chemicals have characteristic absorption 'fingerprints' in the mid-infrared region.

1.7.5 Non-crystalline semiconductors

Although this book is concerned with extended defects in nearly perfect crystalline semiconductors, non-crystalline semiconductors need to be mentioned. This is both because of their practical importance and for certain lessons that can be learned from them, e.g. concerning passivation to eliminate the electrical and optical effects of defects.

Chalcogenide glasses

Covalent bonding in compounds and alloys of elements of the B subgroups of the Periodic Table with valences around 4, had been found consistently to lead to tetrahedrally coordinated adamantine crystal structures as we discussed above. Therefore the discovery that, in mixtures of several of these elements, glassy phases occurred over wide ranges of composition was surprising to many workers. With hindsight, it ought not to have been, since the only two elements that can be obtained as non-crystalline solids in bulk at room temperature, S and Se, are VIB 'chalcogens'. These are, of course, components of many crystalline semiconductors, e.g. II-VIs like ZnS and CdSe and lead chalcogenides like PbSe. For the early fundamental work on non-crystalline and liquid semiconductors see Ioffe (1960) and Ioffe and Regel (1960). Later there was extensive research on the composition regions in which alloys of chalcogenides of metals occur in the glassy state, on their properties (Kolomiets 1964), and on their possible device applications, which, however, came to nothing in the end. This was accompanied by basic work on the energy band theory of 'amorphous' semiconductors (Mott and Davis 1979). This was motivated by the fact that without crystalline periodicity, Bloch's Theorem does not apply and the earlier band theory could not be used for noncrystalline semiconductors. N. F. Mott and P. W. Anderson shared the Nobel Prize in Physics in 1977 for this work. In recent years interest in this field has centred on hydrogenated amorphous *a*-Si, the only such material to find commercial application.

Non-crystalline silicon for solar cells

Silicon (or germanium) evaporated onto non-heated substrates condenses as a non-crystalline or 'amorphous' assembly of atoms. The local environment of a Si atom is mainly one of tetrahedral bonds but there is no translational symmetry since the angles vary randomly about the tetrahedral value (109°28′) and there are many unpaired electrons (dangling bonds). Prior to 1974 it was believed that attempts at donor or acceptor doping would fail, due to the ease with which the loose structure might accommodate the dopant atoms, e.g. by providing five Si neighbours to a five-valent phosphorus atom (Mott 1969). Moreover, the large density of dangling bonds should provide a high density of states near the mid-gap position which could tend to pin the Fermi level, independently of doping.

However, it was shown that doping, both *n* and *p* type, was possible (Spear and Le Comber 1975) when the unpaired electrons were passivated by incorporating a sufficient amount of hydrogen into the structure. This made *p-n* junction devices possible and led to a great expansion in the field of hydrogenated amorphous silicon (*a*-Si:H).

a-Si:H can be prepared by methods ranging from sputtering Si in a hydrogen atmosphere to deposition from d.c. and r.f. glow discharges in silane (SiH_4) gas. Layer thicknesses from μms to mms can be produced.

The energy gap of *a*-Si:H is about 1.7 eV (depending somewhat on the amount of H present) whereas in crystalline Si it is about 1.1 eV. In crystalline Si the gap is empty apart from impurity, defect and surface states but the gap in *a*-Si:H has a continuous low-density distribution of states. The theory of amorphous semiconductors is based on the concepts of localization, band tailing and 'mobility edges'. The lack of translational symmetry means that the usual band structure theory is not applicable, as mentioned previously. Hence, in optical absorption, the indirect energy gap of silicon which requires phonon assistance near the absorption edge for electron excitations, is not a factor. Thus optical absorption is high and thin films, of order half a micron, can be used in solar cells whereas crystalline silicon needs thicknesses of 20 microns or more. Moreover the energy gap of *a*-Si:H of about 1.7 eV is better matched to the solar spectrum than that of crystalline Si. However, the transport properties are much poorer.

The disorder distribution, hydrogen content, bonding and defects in *a*-Si:H vary with the method and parameters of preparation. However, it can be shown generally that when variations in potential are present in a structure there is an associated range of electron wave functions and energies. The states furthest from the centre of this band are associated with the greatest, and therefore least likely, variations. Hence the density of states will 'tail' towards the band edges, but unlike crystalline materials, it will not cut off. The tails from two bands derived from separate atomic levels will overlap, so the 'band gap' is not empty of states.

The absence of long-range order is expected to result in some degree of localization of wave functions. The first demonstration of the existence of disorder-induced localization was by Anderson (1958) who analysed the problem of a three-dimensional array of square wells of random depths, spread over a range of energy V_0. Defining U_0 as the bandwidth that originates from the overlap of the ground state wave functions when the random potential is removed, Anderson showed that above a critical ratio of V_0 to U_0, an electron placed on one of the wells would not diffuse away. The critical ratio changes somewhat with the type of lattice and recent estimates suggest that its value may be around two. This condition for localization is not normally satisfied for most states in an amorphous semiconductor near the centre of the band where $V_o < U_o$, but near the band edges, where V_o is larger, it can be satisfied. Thus one has a picture of extended states within the band but localized states in the tails. There is thought to be a characteristic energy separating the two groups, known as the mobility edge due to the reduced mobility of electrons in localized states. These states also show a stronger electron phonon interaction than extended states, which may further modify their properties.

Despite uncertainties in theory and materials characterization a-Si:H solar cells have gone into use in watches and hand calculators and into limited use for generating terrestrial solar power. At the time of writing, terrestrial solar power is growing strongly and mono- and poly-crystalline and amorphous Si cells are competing with several types of photovoltaic cells of II-VI and other semiconducting compounds.

1.8 Materials development and materials competition

Today, demands for new properties for new device applications usually lead to the development of new quantum-confined epitaxial heterostructures. Originally, however, many new families of semiconducting compounds and their alloys (see Section 1.3 and Section 1.6) were explored and developed. These developments were motivated by the hope that these new materials would have superior properties for device manufacture.

Germanium was developed at the Bell Laboratories in the USA as the necessary preliminary to the search for the holy grail of a solid state amplifier. This was found in the form of the type A, point-contact transistor in 1947. Ge then remained the dominant semiconducting material for over a decade. The development of planar technology for silicon integrated circuits in 1960, which took advantage of the excellence of SiO_2 as an insulating layer, led to the displacement of Ge. Later GaP light-emitting diodes (LEDs) and GaAs LEDs and lasers and microwave integrated circuits reached production. There is also industrial interest in e.g. InP and alloy systems like $Ga_xIn_{1-x}As_yP_{1-y}$ for lasers and photodetectors for fibre optic telecommunications. Most recently GaN has emerged as an important material for GaN-based heterostructures for blue and ultraviolet LEDs and lasers and for high-brightness LEDs for general electric lighting use.

What determines which materials are developed and win the competition between alternative materials for particular applications? The rest of this book is concerned with materials defects of various kinds and the means for their diagnosis and cure. This section deals with the reasons for applying this information to additional materials and devices. Advances in science open up new possibilities and economic demand determines the resources made available for particular programmes.

The major epochs of semiconducting materials development and competition began with the perception of a need for some type of device and determination of the properties necessary for it to function. Investigation often suggested alternative materials and consideration had to be given to problems of purification, crystal growth and processing as well as probable costs, time scales and safety problems such as toxicity or explosion hazards. Individual judgements and experience and the magnitude of the resources available also influenced materials selection. These points are illustrated by actual developments.

1.8.1 The advent of germanium

By the end of the Second World War unreliable, delicate, power-consuming vacuum valves (tubes) limited advances in electronics. The first publicized computer, ENIAC (electronic numerical integrator and calculator) of the University of Pennsylvania (1946), had 18,000 valves consuming 150 kilowatts. There were two full-time engineers on valve testing and replacement but it was out of action more often than not. (A previous machine, built by the British Secret Services to break the codes used in messages produced by the German ENIGMA machines in the Second World War, was, of course, never publicized nor were details of its design or performance published.)

A programme aimed at producing a compact, robust, low-power-consuming amplifying device was therefore initiated at the Bell Telephone Laboratories. During the war radar 'detectors' of silicon had been developed. These crude devices were similar to the galena (mineralogical PbS) 'cats whisker' point contact rectifiers used for early radio reception. (K. F. Braun's discovery of point contact rectification in minerals like galena and other contributions to early radio development were rewarded by the award of a share of the Nobel Physics Prize for 1909. He shared this prize with G. Marconi.). Although variable, these Si detectors were then the best solid state devices available and this drew attention to the IVB elements. Germanium was selected, rather than silicon, because it had a lower melting point and was less strongly chemically bonded, in its oxide for example. It should, therefore, be easier to purify and grow, as good quality crystals, than silicon. (Ironically, these two properties proved to be decisive advantages of Si in its later competition with Ge.) This was necessary for the study of the basic physics of semiconductors. The success of this endeavour and the invention of the point contact transistor at the end of 1947 gave rise to the whole field of solid state electronics.

1.8.2 Technology feedback

A material like germanium, in a dominant position in production, is difficult to displace. Existing plant and equipment, designed for one material, is not suitable for new competitive materials. Moreover, people in the industry, and their customers, have acquired much knowledge and experience of the dominant material and devices and are reluctant to abandon this and move on. (Philips continued producing Ge devices long after other firms had abandoned them for Si.) Therefore, with few exceptions, the firms that led in the production of one type of material or device did not lead in the development of the next. Materials technology feedback is the second conservative factor. Practical industrial experience leads to cumulative small advances, sometimes by trial and error, in the technology of growth, processing, packaging (encapsulation) and testing of the material and devices. The established material therefore tends to increase its technological lead over competitors. Alternative materials can emerge despite this, due to superior properties, from research and development groups not committed to the old material. Thus GaN and SiC have recently become of intense practical interest as the result of impressive development work carried out, in spite of general disbelief in these materials, by groups at the laboratories of the Nichia Chemical Company in Japan and the Yoffe Research Institute in Russia, respectively.

1.8.3 The triumph of silicon

The very properties that originally militated against silicon in the selection of germanium by the Bell laboratory team came to be its chief advantages. The higher melting point, originally seen as making the crystal growth and zone refining technology more difficult, implies a wider forbidden energy gap. Hence silicon does not become intrinsic until higher temperatures and Si devices can work hotter than germanium. Silicon's strongly bonded oxide, originally seen as an obstacle to purification and surface cleanliness, is a decisive advantage for masking in photolithographic processing, for surface passivation and for use in metal oxide semiconductor (MOS) devices. The slow introduction of discrete silicon transistors and diodes in the late 1950s was followed by the development of planar technology by the Fairchild company in 1959. This led to the rise of the integrated circuit industry from 1960 and the many 'Fairchildren' firms (so called because people leaving Fairchild set them up) in 'Silicon Valley'. Huff (2002) published a full and readable historical account of the development of crystalline silicon technology for integrated circuit (IC) microelectronics, with full references to the original literature.

1.8.4 Integration

Component devices were integrated into a 'chip' of silicon originally, mainly to satisfy the demand for greater reliability in large systems. Repeated American failures in attempted satellite launches around the time of the first sputnik provided the strongest motivation, but later demand from the computer industry became the driving force for the integrated circuit (IC) industry. Integration eliminates many of the contacts that have always been a main source of failures in electronic systems. Now the impressive fall in cost per component or function is more important.

Historically the cost of integrated circuits, per circuit function, has fallen about 25% per year according to the International Technology Roadmap for Semiconductors. This is due to the uniquely rapid progress along the 'learning curve' of the IC industry. In all manufacturing industries, with increasing experience, efficiency rises so costs are reduced in constant (non-inflated) terms by 20 to 30% each time their total (cumulative) output doubles. The cost of integrated circuits has fallen by 28% for each doubling of the total number of units ever produced. The uniquely rapid growth of output in the microelectronics industry results in the rapid relative and even absolute fall in prices.

This fall in cost per component continues, due partly to the high consumption of existing electronic consumer products. Moreover, entirely new types of consumers goods are made possible by the advances in semiconductor technology. These new goods have included most recently, CD audio and DVD video players (plus other applications of these optical recording technologies), digital photography (camcorders and still cameras), mobile phones and digital radios. These industries in turn consume large volumes of ICs and other devices.

At the present, personal computers have microprocessors and memory chips containing of the order of hundreds of millions of transistors each. In addition, vast numbers are employed in mainframe computers and in major, rapidly developing systems like telecommunications, air traffic control and the Internet. Hence the numbers of transistors or functional units consumed continues to rise rapidly. According to reports (1998), in 1997 it was estimated that more than 100 transistors (including those in integrated circuits) were produced every day for every person on the planet.

In 1965, Gordon Moore, a co-founder of the Intel corporation, noticed that the number of components in the largest integrated circuits (ICs) had doubled every year since the first ICs appeared in 1959. This became known as Moore's Law. The rate of increase slowed to a doubling every 18 months in the late 1970s and has remained constant ever since (see Fig. 1.46). Various forms of similar plots have been published since. (However, the inevitable development of nanoscale devices may well lead to departure from that law and a levelling off in Fig. 1.46.)

Growth in consumption of chips leads to impressive totals for Si crystal production. According to Brice (1986) the world production of single crystal silicon in that year was 2000 tonnes making 1000 tonnes of slices (wafers). (Half the material is lost in machining: removing the tops and tails from the single crystal boules, turning them down into cylinders and sawing between the slices.) Taking into account the density of Si and the thickness of the slices (0.5 mm) the consumption of slices in 1986 amounted 1 km square of highly pure and perfect Si about the thickness of a razor blade.

Today the manufacture of photovoltaic (PV) cells is beginning to consume a tonnage of Si comparable to that used in integrated circuit production. According to Aulich and Schulze (2002) about 14,000 tonnes of EG (electronics grade) silicon was produced in 2001. Of this total, about 4000 tonnes went to PV production and the output of this industry is growing rapidly. About 90% of all terrestrial solar panels at present employ Si PV cells.

The rising scale of integration is made possible partly by the reduction in the line width (minimum size of areas on ICs). It has fallen from hundreds of microns in 1960 to $0.18\,\mu m$ in 1999 and $0.090\,\mu m$ (90 nm, which made possible a maximum of 180 million transistors per chip) in 2003. Hence the area of chips has not increased in proportion to the scale of integration (number of components on an IC). DRAM (dynamic random access memory) chip sizes, for example, increased by $1.4\times$ for each $4\times$ increase of bit capacity (Roadmap). Circuit component shrinkage can bring other benefits too, such as faster or lower power operation. How much further line widths can be reduced is uncertain at the time of writing. The ITRS (International Technology Roadmap for Semiconductors) intention is to aim for line widths of 65 nm in 2005, corresponding to 380 million transistors per chip, 45 nm (1.5 billion transistors/chip) in 2007 and 32 nm (3.01 billion transistors/chip) in 2009. There is a widespread agreement that some form of physical limit must be reached eventually. As devices become smaller, even single electrically active extended defect can produce unacceptably large percentage changes in device properties affecting the performance

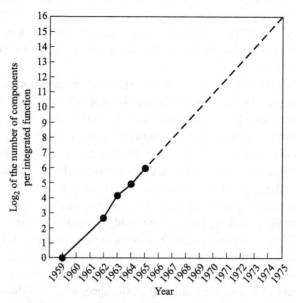

Figure 1.46   The original plot of Moore's Law (after Moore 1965). [Moore, G. E. (1965). Cramming more components onto integrated circuits. *Electronics*, Vol. **38**, No. 8, April 19.]

of the largest ICs. Hence the continued prevention of defect introduction during processing becomes ever more important as the line width shrinks and the scale of integration increases.

While the size of the largest individual IC chips rises slowly, the diameter of the wafers in which they are processed also increases. Industry used 100 mm (4 inch) diameter Si wafers in the early days; 200 mm (8 inch) slices were still in fairly widespread use at the time of writing while industry was moving on to 300 mm (12 inch) diameter wafers. The cost of processing is virtually independent of wafer diameter and line width so the material and processing costs per function and per chip fall with slice size.

The investment and technology feedback effects and the economics of increasing scales of integration operate powerfully in favour of silicon. Moreover, it has many inherent advantages such as the simplicity of elemental silicon compared to binary compounds or 3- or 4-component semiconductor alloys. It has a very advantageous oxide. Also, silicon is the commonest solid element in the earth's crust so raw material supply is no problem and prices will be stable. So it is unlikely that Si will be ousted from its dominance of microelectronics. (However, in the long run there are possibilities for the emergence of molecular electronic or purely photonic computers.) Nevertheless, materials development and competition continues and several compounds have reached industrial use. On the basis of discrete device applications, discussed below, GaAs became established and technology feedback led to the development of the capability to produce GaAs ICs. The higher electron mobility in GaAs and the occurrence of semi-insulating GaAs make possible superior circuits for high frequencies so they have a market niche as monolithic microwave integrated circuits (MMICs) for wireless applications like mobile phones. The scale of GaAs integration is rising too but lags well behind Si.

At present, there is increasing interest in heterostructures produced by growing epitaxial Si-Ge alloy layers on silicon. These make possible band gap engineering in Si based devices and ICs. Higher frequency operation has been made possible by these structures, increasing the competitive strength of Si versus GaAs or InP. The way in which Si workers so often find ways to fight back against challenges led some III-V materials workers to suggest, only half humorously, the formation of S^3 – the Society for the Suppression of Silicon.

There is also great interest in the possible use of LEDs, especially GaN-based, for general lighting applications. Even here Si is fighting back. It can be made to emit light either by employing Si quantum dots or by producing so-called porous Si by electrolytic etching treatments. While Si is unlikely ever to become an efficient and versatile light-emitter, it is not impossible that it may emit well enough for some optoelectronic niche applications.

In recent years methods for the epitaxial growth of diamond have emerged. This resulted in interest in semiconducting diamond, e.g. for (very) high temperature device applications. Thus although Si dominates and epitaxial semicondutor heterostructure research and development is extremely active, the exploration and development of new semiconducting materials continues.

1.8.5 The survival and application of the III-V compounds

III-V compound materials reached industrial production due to the demand for devices based on properties that the dominant material does not have. This occurred in the case of the III-V compounds proposed by Welker (1952) (while Goryunova and Obuchov (1951) had independently reported that InSb is a semiconductor). III-V devices were not maufactured in any numbers until the later 1970s. During the 1960s and early 1970s pessimists pointed to the difficulties of purification, control of non-stoichiometry (deviations from exactly equal numbers of atoms of the two elements in a binary compound, for instance) and doping in compounds. Many felt they were too great for practical devices to be achieved. Moreover, even if they were, the value of those manufactured would not justify the sums spent over decades on III-V materials research and development, according to those seeking reasons to cut research budgets.

III-V compound materials research survived this period due to interest in device phenomena neither Ge nor Si exhibit. Visible light has photon energies $h\nu > 1.8\,\text{eV}$ so electron-hole recombination can lead to visible emission only in materials with $E_g > 1.8\,\text{eV}$. Si and Ge have gaps narrower than this. (Visible light can, however, be emitted by Si, due e.g. to special quantum confined effects in porous Si and in Si quantum dots.) For many years GaP was the only wider gap semiconductor in which good *p-n* junctions could be made (Bergh and Dean (1976)) so the demand for light-emitting diodes and alpha-numeric displays kept the effort on GaP and $GaAs_yP_{1-y}$ going. Similarly, the direct gap of GaAs meant it could exhibit laser action. The interest in optical fibre communications motivated work in this field. Moreover the two types of conduction band minima in GaAs (Fig. 1.28) meant it could be used in transferred electron devices (Gunn oscillators), unlike Si or Ge. Demand for such microwave generators for mini-radars further strengthened GaAs research in the crucial period.

The rising level of III-V compound technology arising from this work and the technology feedback once industrial production got under way, made many other III-V materials and device structures practical including GaAs- and InP-based alloy double heterostructure lasers and photodetectors for fibre optic communications as well as GaAs and InP integrated microwave circuits.

1.8.6 Current developments

Much work was done on ternary and quaternary III-V alloys for energy band engineering. Applications included lasers and photodetectors for longer infrared wavelengths for fibre optic communications and lasers of shorter wavelengths for optical information storage. This comprises at the time of writing CD (compact disc) for audio and computer use and DVD (digital video disc, for recording films and D Versatile D for computer storage applications).

Only red, yellow and green LEDs can be made of the well-developed GaP and $GaAs_xP_{1-x}$ alloys. Hence there was intense interest following Nakamura's (1994) demonstration that good blue LEDs and, later, lasers could be made in GaN/GaInN structures. These devices were successful although, initially, the material contained very high densities of defects (see Section 5.1). For reviews of the early developments see Nakamura (1998) and other chapters in the same volume. Besides the possibility of extending LED and laser emission wavelengths to the blue and ultraviolet, GaN LEDs can be made so bright that LEDs are taking over niche markets from incandescent lights (e.g. in traffic lights). In the longer run much of the electric light market may be taken over. This is due to the low power consumption and long lives of LEDs with the consequent elimination of the need to periodically replace the lights.

There is interest in II-VI compounds, and active development of II-VI devices continues. Thin film solar cells based on various forms of II-VI and related material junctions have now reached the market for terrestrial photovoltaic power generation. For the state of the art in the II-VI field see the Proceedings of the II-VI Conferences published from time to time in the *Journal of Crystal Growth*.

References

Alferov, Zh. I., Andreev, V. M., Garbuzov, D. Z. *et al.* (1971). Investigation of the influence of the AlAs-GaAs heterostructure parameters on the laser threshold current and the realization of continuous emission at room temperature. *Soviet Physics Semiconductors*, **4**, 1573–75.

Anderson, P. W. (1958). Absence of diffusion in certain random lattices. *Physical Review*, **109**, 1492–505.

Aulich, H. and Schulze, F.-W. (2002). Silicon supply for solar PV. *Renewable Energy World*, **5**, 49–59.

Aven, M. and Prener, J. S. (1967). *Physics and Chemistry of II-VI Compounds*. Amsterdam: North-Holland.

Austin, I. G., Goodman, C. H. L. and Pengelly, A. E. S. (1956). New semiconductors with the chalcopyrite structure. *Journal of the Electrochemical Society*, **103**, 609–10.

Bachmann, K. J. (1995). *The Materials Science of Microelectronics*, Chapter 6. Weinheim: VCH.

Bergh, A. A. and Dean, P. J. (1976). *Light Emitting Diodes*, Oxford: Clarendon Press.

Blakemore, J. S. (1982). Semiconducting and other major properties of gallium arsenide. *Journal of Applied Physics*, **53**, R123–R181.

Blakemore, J. S. (1985). *Solid State Physics*, 2nd edn. Cambridge: Cambridge University Press.

Blakeslee, A. E. (1971). Vapor growth of a semiconductor superlattice. *Journal of the Electrochemical Society*, **118**, 1459–63.

Borchers, H. and Maier, R. G. (1963). Uber die ternare halbleitende kristallart $ZnSnAs_2$ und den aufbau des driestoffsystems Zinc – Zinn – Arsen. *Metall*, **17**, 775–80; Quasibinaare zustandsdiagramme der halbleitenden kristallart InAs mit $ZnSnAs_2$, $ZnGeAs_2$ und $CdGeAs_2$. *Metall*, **17**, 1006–10.

Brice, J. C. (1986). *Crystal Growth Processes*. New York: Wiley.

Buerger, M. J. (1978). *Elementary Crystallography. An Introduction to the Fundamental Geometrical Features of Crystals*. Cambridge, Mass.: MIT Press.

Capasso, F. and Margaritondo, G. (eds.) (1987). *Heterojunction Band Discontinuities: Physics and Device Applications*. Amsterdam: North-Holland.

Capasso, F., Paiella, R., Martini, R. *et al.* (2002). Quantum cascade lasers: ultrahigh-speed operation, optical wireless communication, narrow linewidth and far-infrared emission. *IEEE Journal of Quantum Electronics*, **38**, 511–32.

Chang, L. L. and Esaki, L. (1979). Semiconductor superlattices by MBE and their characterizaton. *Progress in Crystal Growth and Characterization*, **2**, 3–14.

Chang, L. L. and Giessen, B. C. (eds.) (1985). *Synthetic Modulated Structures*. New York: Academic Press.

Chelikowsky, J. R. and Cohen, M. L. (1976). Nonlocal pseudopotential calculations for the electronic structure of eleven diamond and zinc-blende semiconductors. *Physical Review*, **B14**, 556–82.

Cho, A. Y. (1995). Molecular beam epitaxy from research to manufacturing. *MRS Bulletin*, **20**(4), 21–8.

Cho, A. Y., Sivco, D. L., Ng, H. M. *et al.* (2001). Quantum devices, MBE technology for the 21st century. *Journal of Crystal Growth*, **227–228**, 1–7.

Dash, W. C. (1958). The growth of silicon crystals free from dislocations. In *Growth and Perfection of Crystals*, eds. R. H. Doremus, B. W. Roberts and D. Turnbull (New York: Wiley), pp. 361–85.

Davies, G. J. and Williams, R. H. (eds.) (1994). *Semiconductor Growth, Surfaces and Interfaces*. London: Chapman & Hall.

Dingle, R. (1976). Optical properties of semiconductor superlattices. In *Physics of Semiconductors* ; Proc. 13th Internat. Conf. Rome, ed. F.G. Fumi (Amsterdam: North-Holland), pp. 965–74.

Dingle, R. (ed.) (1987). Applications of multiquantum wells, selective doping and superlattices. In *Semiconductors and Semimetals*, Vol. **24**. New York: Academic Press.

Dismukes, J. P. and Ekstrom, L. (1965). Homogenous solidification of Ge-Si alloys. *Transactions of the Metallurgical Society of AIME*, **233**, 672–80.

Esaki, L. and Tsu, R. (1970). Superlattice and negative differential conductivity in semiconductors. *IBM Journal of Research and Development*, **14**, 61–5.

Faist, J., Capasso, F., Sivco, D. L. *et al.* (1994). Quantum cascade laser. *Science*, **264**, 553–6.

Finkelnburg, W. (1950). *Atomic Physics*. New York: McGraw-Hill.

Goodman, C. H. L. (1957). A new group of compounds with diamond-type (chalcopyrite) structure. *Nature*, **179**, 828–9.

Goodman, C. H. L. and Douglas, R. W. (1954). New semiconducting compounds of diamond type structure. *Physica*, **20**, 1107–9.

Goryunova, N. A. and Obuchov, A. P. (1951). *Zhur. Tekh. Fiz.*, **21**, 237.

Goryunova, N. A. (1965). *The Chemistry of Diamond Like Semiconductors.* Cambridge: The MIT Press.

Goryunova, N. A., Kesamanly, F. P. and Nasledov, D. N. (1968). Phenomena in solid solutions. In *Semiconductors and Semimetals*, Vol. **4**, *Physics of III-V Compounds*, eds. R. K. Willardson and A. C. Beer (New York: Academic Press), pp. 413–58.

Hayashi, I., Panish, M. B., Foy, P. W. and Sumski, S. (1970). Junction lasers which operate continuously at room temperature. *Applied Physics Letters*, **17**, 109–11.

Herman, M. A. and Sitter, H. (1989). *Molecular Beam Epitaxy: Fundamentals and Current Status.* New York: Springer-Verlag.

Holt, D. B. (1966). Misfit Dislocations in Semiconductors. *Journal of Physics and Chemistry of Solids*, **27**, 280–95.

Huff, H. R. (2002). An electronics division retrospective (1952–2002) and future opportunities in the twenty-first century. *Journal of the Electrochemical Society*, **149**, S35–S58.

Hume-Rothery, W. and Raynor, G. V. (1954). *The Structure of Metals and Alloys.* 3rd edn. London: Institute of Metals.

Hurle, D. T. J. and Rudolph, P. (2004). A brief history of defect formation, segregation, faceting, and twinning in melt-grown semiconductors. *Journal of Crystal Growth*, **264**, 550–64.

Ioffe, A. F. (1960). *Physics of Semiconductors.* London: Infosearch.

Ioffe, A. F. and Regel, A. R. (1960). Non-crystalline, amorphous and liquid electronic semiconductors. *Progress in Semiconductors*, **4**, 237–91.

Jain, S. C. (2000). *Compound Semiconductor Strained Layers and Devices.* Boston: Kluwer Academic Publishers.

Jaros, M. (1989). *Physics and Applications of Semiconductor Microstructures.* Oxford: Clarendon Press.

Kazarinov, R. F. and Suris, R. A. (1971). Amplification of electromagnetic waves in a semiconductor superlattice. *Soviet Physics Semiconductors*, **5**, 707–9.

Kelly, A. and Groves, G. W. (1970). *Crystallography and Crystal Defects.* London: Longman.

Kittel, C. (1996). *Introduction to Solid State Physics*, 7th edn. New York: Wiley.

Kleinman, W. and Phillips, J. C. (1960). Crystal potential and energy bands of semiconductors. III self-consistent calculations for silicon. *Physical Review*, **118**, 1153–67.

Kolomiets, B. T. (1964). Vitreous semiconductors I. *Physica Status Solidi*, **7**, 359–72; and Vitreous semiconductors II. *Physica Status Solidi*, **7**, 713–31.

Kroemer, H. (1957). Quasi-electric and quasi-magnetic fields in nonuniform semiconductors. *RCA Review*, **18**, 332–42.

Kroemer, H. (1963). A proposed class of heterojunction injection laser. *Proceedings of the IEEE*, **51**, 1782–3.

Madelung, O. (1964). *Physics of III-V Compounds.* New York: Wiley.

Mahajan, S. (2004). The role of materials science in microelectronics: past, present and future. *Progress in Materials Science*, **49**, 487–509.

Mahajan, S. and Sree Harsha, K. S. (1999). *Principles of Growth and Processing of Semiconductors.* New York: McGraw-Hill.

Matthews, J. M. (ed.) (1975). *Epitaxial Growth*, Vols A and B. New York: Academic Press.

Miller, A., MacKinnon, A. and Weaire, D. (1981). Beyond the binaries — the chalcopyrite and related semiconducting compounds. *Solis State Physics*, **36**, 119–75.

Moore, G. E. (1965). Cramming more components onto integrated circuits. *Electronics*, Vol. **38**, No. 8, April 19.

Mooser, E. and Pearson, W. B. (1960). The chemical bond in semiconductors. *Progress in Semiconductors*, **5**, 103–39.

Mott, N. F. (1969). Conduction in non-crystalline materials III. Localized states in a pseudogap and near extremities of conduction and valence bands. *Philosophical Magazine*, **19**, 835–52.

Mott, N. F. and Davis, E. A. (1979). *Electronic Processes in Non-crystalline Materials*. 2nd edn. Oxford: Clarendon Press.

Mullin, J. B. (2004). Progress in the melt growth of III−V compounds. *Journal of Crystal Growth*, **264**, 578–92.

Nakamura, S. (1994). Growth of $In_xGa_{(1-x)}N$ compound semiconductors and high power InGaN/AlGaN double heterostructure violet-light-emitting diodes. *Microelectronics Journal*, **25**, 651–9.

Nakamura, S. (1998). III-V nitride-based short-wavelength LEDs and LDs. In *Group III Nitride Semiconductor Compounds*, ed. B. Gil (Oxford: Clarendon Press), pp. 391–416.

Pamplin, B. R. (ed.) (1980). *Crystal Growth*. 2nd edn. Oxford: Pergamon.

Parker, E. H. C. (ed.) (1985). *The Technology and Physics of Molecular Beam Epitaxy*. New York: Plenum Press.

Parthe, E. (1966). *Crystal Chemistry of Tetrahedral Structures*. New York: Gordon & Breach.

Pashley, D. W. (1991). The epitaxy of metals. In *Processing of Metals and Alloys*, ed. R. W. Cahn, Materials Science and Technology: A Comprehensive Treatment, Vol. **15** (Weinheim: VCH), pp. 290–328.

Pfann, W. G. (1952), Principles of zone melting. *Journal of Metals*, **4**, 747–53.

Phillips, F. C. (1971). *An Introduction to Crystallography*. 4th edn. London: Longman.

Phillips, J. C. (1973). *Bonds and Bands in Semiconductors*. New York: Academic Press.

Rees, H. D. and Gray, K. W. (1976). Indium phosphide: A semiconductor for microwave devices. *IEEE Journal of Solid State and Electron Devices*, **1**, 1–8.

Reid, M. (ed.) (1992). Nanostructural systems. In *Semiconductors and Semimetals*, Vol. **35**. New York: Academic Press.

Shay, J. L. and Wernick, J. H. (1975). *Ternary Chalcopyrite Semiconductors: Growth, Electronic Properties and Applications*. Oxford: Pergamon.

Shockley, W. (1950). *Electrons and Holes in Semiconductors*. Princeton: van Nostrand.

Spear, W. E. and Le Comber, L. G. (1975). Substitutional doping of amorphous silicon. *Solid State Communications*, **17**, 1193–6.

Stormer, H. L., Dingle, R., Gossard, A. C., Wiegmann, W. and Logan, R. A. (1978). Electronic properties of modulation-doped $GaAs-Al_xGa_{(1-x)}As$ superlattices. In *Physics of Semiconductors 1978*. Proceed. 14th Internat. Conf. Phys. Semicond. Edinburgh. Conf. Series No. 43 (Bristol: Inst. Phys., 1979), pp. 557–60.

Stradling, R. A. and Klipstein, P. C. (eds.) (1990). *Growth and Characterization of Semiconductors*. New York: Adam Hilger.

Stringfellow, G. B. (1989). *Organometallic Vapor-Phase Epitaxy: Theory and Practice*. Boston: Academic Press.

Sze, S. M. (1981). *Physics of Semiconductor Devices*. New York: Wiley.

Sze, S. M. (1985). *Semiconductor Devices. Physics and Technology*. New York: Wiley.

Verma, A. R. and Krishna, P. (1966). *Polymorphism and Polytypism in Crystals*. New York: Wiley.

Vurgaftman, I., Meyer, J. R. and Ram-Mohan, L. R. (2001). Band parameters for III−V compound semiconductors and their alloys. *Journal of Applied Physics*, **89**, 5815–75.

Wang, C. C. and Alexander, B. H. (1955). Hardness of germanium-silicon alloys at room temperature. *Acta Metallurgica*, **3**, 515–16.

Welker, H. (1952). Uber neue halbleitende verbindungen. *Zeitschrift fur Naturforschung A: A Journal of Physical Sciences*, **7**, 744–9.

Welker, H. and Weiss, H. (1956). Group III − group V compounds. In *Solid State Physics*, eds. F. Seitz and D. Turnbull, Vol. **3**, pp. 1–78.

Wilkes, P. (1973). *Solid State Theory in Metallurgy*. Cambridge: Cambridge University Press.

Woolley, J. C. (1962). Solid solution of III-V compounds. In *Compound Semiconductors, Vol. 1. Preparation of III-V Compounds*, eds. R. K. Willardson and H. L. Goering (New York: Reinhold), pp. 3–20.

Ziman, J. M. (1972). *Principles of the Theory of Solids*. 2nd edn. (Cambridge: Cambridge University Press).

Further Reading

Semiconductors

Mayer, J. W. and Lau, S. S. (1990). *Electronic Materials Science: For Integrated Circuits in Si and GaAs*. New York: Macmillan Publishing.

Pierret, R. F. and Neudeck, G. W. (eds.) (1989). *Modular Series on Solid State Devices*. Reading, Mass.: Addison-Wesley.

Seeger, K. (1999). *Semiconductor Physics: An Introduction*. New York: Springer-Verlag.

Streetman, B. G. (1995). *Solid State Electronic Devices*. Englewood Cliffs, N.J.: Prentice-Hall.

Sze, S. M. (1981). *Physics of Semiconductor Devices*. New York: Wiley.

Wilson, J. and Hawkes, J. F. B. (1998). *Optoelectronics: An Introduction*. Englewood Cliffs, N.J: Prentice-Hall.

Yu, P. Y. and Cardona, M. (1996). *Fundamentals of Semiconductors: Physics and Materials Properties*. New York: Springer.

Semiconductor growth

Brice, J. C. (1986). *Crystal Growth Processes*. New York: Wiley.

Davies, G. J. and Williams, R. H. (eds.) (1994). *Semiconductor Growth, Surfaces and Interfaces*. London: Chapman & Hall.

Herman, M. A. and Sitter, H. (1989). *Molecular Beam Epitaxy: Fundamentals and Current Status*. New York: Springer-Verlag.

Lewis, B. and Anderson, J. C. (1978). *Nucleation and Growth of Thin Films*. New York: Academic Press.

Pamplin, B. R. (1975). *Crystal Growth*, International Series of Monographs in *The Science of the Solid State*, Volume **6**. New York: Pergamon Press.

Parker, E. H. C. (ed.) (1985). *The Technology and Physics of Molecular Beam Epitaxy*. New York: Plenum Press.

Stradling, R. A. and Klipstein, P. C. (eds.) (1990). *Growth and Characterization of Semiconductors*. New York: Adam Hilger.

Stringfellow, G. B. (1989). *Organometallic Vapor-Phase Epitaxy: Theory and Practice*. Boston: Academic Press.

2

An introduction to extended defects

Defects in the broadest sense are all forms of deviation from perfection in the crystal structures defined in the previous chapter.

Ideal crystal structures. To appreciate the full diverse range of types of defect we start with the approach used in crystallography. This begins with the concept of a lattice which is an array of points in space with the defining property that the surroundings of every point are the same as those of every other point. That is, from any point the operation of the translation group (of vectors):

$$t_i = n_{i1}a_1 + n_{i2}a_2 + n_{i3}a_3 \qquad (2.1)$$

carries us to the other points of the lattice. Here a_1, a_2, and a_3 are the fundamental lattice vectors (along the crystal axes) and the n's are integers. All possible crystal structures result from placing identical motifs or basis units (atoms, groups of atoms or molecules) at each point of one of the 14 possible lattices. The atoms are ideally located at rest at their correct sites and in their minimum energy states. Moreover, since the set of translation vectors extends out without limit, the ideal, perfect crystal structure is infinite.

Defects. All forms of deviation from such unexcited, perfect, infinite crystal structures are defects. Defects can be grouped into sets with so much in common, theoretically and experimentally that their study constitutes independent research fields. *Transient defects* are elementary quanta of excitation of the atoms or the crystal as a whole such as phonons, polarons, magnons, etc. which move through the crystal and can be created and decay. Real crystals are finite and therefore have free surfaces which thus are unavoidable defects. *Free surfaces*, their structure, contamination, passivation, etc., is another field with its own theory, experimental techniques and technology. *Point defects* are atomic sized and include impurity atoms, atoms on the wrong crystal sites ('antisite defects'), vacant sites, interstitial atoms, etc. Because impurity doping determines device properties and because the interactions of the doping atoms with other point defects can alter their effects, this is an enormous field with its own literature. Most books on semiconductors that mention defects at all refer primarily or exclusively to point defects.

Since transient and point defects and surfaces are already well served by specialist literature including numerous books, they will not be dealt with here.

2.1 Basic definitions

We now define those defects with which we are concerned and briefly introduce the basic ideas about dislocations.

2.1.1 Extended (structural) defects in semiconductors

Extended or structural defects are permanent deviations from structural perfection that is from the correct placement of atoms in real crystals.

Extended defects are usefully classified by the number of dimensions in which they are extended, that is, of greater than atomic size, into volume, surface and line defects. Only point defects are of low enough formation energies to occur in thermodynamic equilibrium, i.e. to be formed by thermal activation. At any temperature, therefore, there are equilibrium concentrations of 'native' point defects like vacancies (unoccupied crystal structure sites) and interstitials (atoms in positions between structure sites). Native point defects are so described to distinguish them from those involving 'foreign', i.e. impurity atoms.

All the larger defects, those with which we are concerned, occur only as the result of accidents of growth and processing, i.e. the 'history' of the specimen. They are described as grown-in or process-induced defects, respectively. A major aim of semiconductor technology is to reduce their density below the thresholds above which they affect materials properties or device performance parameters. It is an outstanding achievement of semiconductor technology that macroscopic crystals of Ge, Si and GaAs can be grown with zero dislocation density. This has been attained for no other materials.

'Defects' is a pejorative term. Many non-uniformities are deliberately introduced such as differently doped regions (patterns of p- and n-type conductivity in integrated circuits) and epitaxial structures like QWs (quantum wells), QWrs (quantum wires) and QDs (quantum dots). These are essential for many optoelectronic devices and are not thought of as defects although they are deviations from crystal perfection. Similarly, impurity atoms in a crystal are regarded as point defects if they are undesired but as donors, acceptors, or radiative recombination centres, etc. if they were deliberately added to control transport or luminescent properties.

One class of defects is often overlooked. These we have called point defect *maldistributions* (see Chapter 6). They are macroscopically varying, unintended and undesirable distributions of impurities and other point defects in semiconductor crystals. These defects combine the character of point and volume defects.

2.1.2 Yield and reliability

All defects are undesirable but some are more disastrous than others. Defects that cause devices to malfunction and be rejected are sometimes referred to as 'fatal'. They are a major factor in the low yields encountered in the early part of the 'learning curve' of production of a new device. Other defects may not cause immediate rejection but initiate some failure mechanism to shorten device-operating life. This reliability problem can be more economically serious than the yield problem as whole systems may fail as a consequence and maintenance and repair are expensive.

2.1.3 Defect characterization

Defects are always undesirable, if only as an indication that the production technology is not fully understood and controlled. Hence it is necessary to have techniques to locate, identify and determine the origin of the defects present. This process is described as (micro)structural characterization. The numerous forms of microscopy (optical, scanning laser, transmission electron, scanning electron, scanning tunnelling, etc.) are used. Defects in semiconductors are a problem because of their electronic and optoelectronic effects. Thus, it is necessary to be able to determine the electronic properties of defects. This is because, by virtue of its type or precise location, the defect may have no effect on device operation. Even if the defects are harmful, it is often possible to passivate them (i.e. render them electrically inactive). A number of techniques are also available for the electronic characterization of defect microstructure such as SEM-EBIC (scanning electron microscope electron-beam-induced current), SEM-CL (cathodoluminescence) and LBIC (laser- or light-beam-induced current). These techniques will be treated in the next chapter.

2.2 Types of extended defect in semiconductors

The defining characteristics of the extended defects are as follows.

2.2.1 Volume defects

Any volume differing in structure, composition, orientation or state variables (e.g. magnetic or polarization alignment) from the rest of the crystal (the matrix) is a volume defect.

Precipitates differ in composition from the matrix.

Second Phase Grains differ in crystal structure and may also differ in composition from the matrix.

Grains in polycrystalline material are volumes differing only in orientation from their surroundings. 'Included grains' in otherwise single crystal material are isolated grains. *Twins* are regions differing in orientation, in particular crystallographically simple ways, from the matrix. Twins are grains with low energy interfaces so they are readily formed in growth and are often the last to be eliminated as crystal growth techniques are improved.

Domains are regions differing from their neighbours in e.g. electron spin alignment (ferromagnetic domains) or site-occupation-function 'phase' (antiphase or inversion domains), etc.

2.2.2 Surface defects

Surface or area defects are free surfaces or interfaces between distinguishable volumes. Interfaces that are deliberately created like contacts, *p-n* and heterojunctions are not considered to be defects (unless something goes wrong!).

The (free) surface is present in all specimens. A decisive advantage of Si over other semiconductors is that the technology exists to passivate the surface (reduce the surface recombination velocity to a negligible value) by oxidizing it in well-defined ways.

The *interfaces* surrounding all types of volume defect are surface defects, e.g. twin and domain boundaries.

Stacking faults are planes across which the stacking sequence of the crystal structure is altered.

2.2.3 Line defects (dislocations)

Line defects are particularly important for three reasons:

(1) The formation energy of defects decreases as the number of dimensions in which they are extended is reduced. The formation energies of volume and area defects are so large that they are (relatively!) easy to avoid or eliminate. They are only found in semiconductors, therefore, in the early stages of the technology of new materials or devices or in devices that must be made with low cost techniques such as polycrystalline Si solar cells for terrestrial power generation. Hence the largest, most important defects that are likely to be generally encountered are line defects.
(2) They are all fundamentally related. All one-dimensional (line) defects are dislocations and their conceptual unity makes a systematic treatment possible as will be shown below. The elastic energy and crystallography of dislocations are understood but their atomic-core-dependent, electronic and optical properties are still under study, especially in semiconductors.
(3) Dislocations interact strongly with, or play a role in generating, all the other types of defect. Hence they form a logical starting point for the study of extended defects.

2.3 Dislocations, plastic deformation and slip systems

Dislocations were proposed to account for the unexpectedly low critical resolved shear stresses above which single crystals of metals undergo permanent, plastic deformation. It was seen that the plastic deformation of crystalline materials occurs by the slip (or glide) of specific crystallographic planes of atoms over their neighbours in particular crystallographic directions (Fig. 2.1). Diamond- and sphalerite-structure semiconductors, for example, slip on {111} planes in ⟨110⟩ directions. The relations of these planes and directions are conveniently represented by means of Thompson's (1953) tetrahedron (Fig. 2.2).

To shear all the atoms in a plane simultaneously over those in the adjacent plane all the connecting bonds must be broken and reformed. This would require stresses much greater than the observed critical resolved shear stresses (the minimum stresses required to cause plastic slip at the rates imposed in testing machines). The hypothesis was that a small slipped area nucleates and spreads over the plane. The boundary of the slipped area is a dislocation (Fig. 2.3). Only bonds along the line of the dislocation must be broken to move it, so the shear stress for slip is greatly reduced. In metals there are no directed bonds and the inherent, viscous (Peierls-Nabarro) stress needed to move a dislocation is low. In semiconductors, however, covalent bonds must be broken, the Peierls-Nabarro stress is high and deformation requires relatively high temperatures and low strain rates. Corresponding to the broken bonds in dislocation cores, there can be partially filled levels deep in the forbidden energy gap, which give rise to unique electrical and optoelectronic effects.

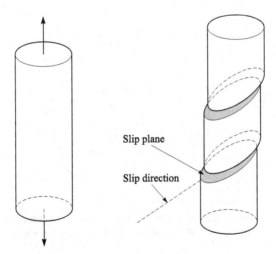

Figure 2.1   Slip in a single crystal undergoing plastic deformation. The slip plane and slip direction are shown.

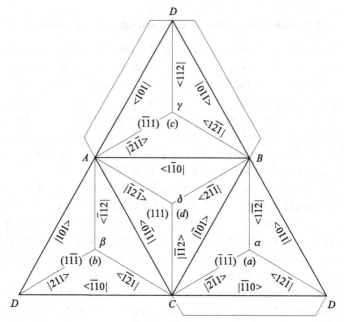

Figure 2.2 When cut out and folded this diagram forms Thompson's (1953) tetrahedron. The faces represent the four {111} slip planes of materials with crystal structures based on the f.c.c. lattice such as the semiconducting elements with the diamond cubic structure and the compounds with the sphalerite structure. The edges of the tetrahedron represent the ⟨110⟩ slip directions. The bisectors of the angles at the apexes of the triangular faces are the ⟨112⟩ directions of the Burgers vectors of partial dislocations in materials with these structures. (After Thompson 1953, *Proceedings of the Physical Society of London*, **B66**, pp. 481–92.)

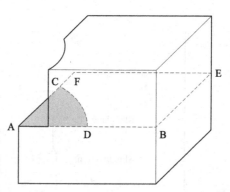

Figure 2.3 Slip has occurred over the cross-hatched area and is spreading over the slip plane ABEF by the outward movement of the line, CD, bounding the slipped area. This line is a (slip) dislocation.

A useful empirical rule was found early in the study of the subject. It is that covalently bonded adamantine semiconductors will deform plastically (under indentations) at macroscopic rates only at temperatures above about one-half their melting point T_m in degrees Kelvin (Churchman *et al.* 1956). This 'brittle-ductile' transition temperature, $T_c \sim 0.5\,T_m$, depends strongly on the imposed strain rate, the doping and the nature of the applied stresses. Plastic deformation and the introduction of large numbers of dislocations therefore only occurs readily under stress, in general, at high temperatures, but such conditions can occur during crystal growth or device processing. Churchman *et al.* (1956) also reported evidence suggesting that deformation twinning was involved in indentation. At room temperature slices of most semiconductors are brittle and Si can be cleaved into chips in the manner of glass 'cutting'. During crystal growth, large thermal gradients and differential thermal contraction can produce slip. Such thermal strains must be avoided in order to grow dislocation-free crystals (Dash 1958, Section 2.3.9). Large, slip-inducing thermal gradients must also be avoided during device processing steps such as oxidation, diffusion and post-implantation annealing.

In-diffusion of dopants that are very different in atomic size from Si can also cause slip when there is a high surface concentration of the dopant and a steep concentration gradient into the Si. The resulting steep change of lattice parameter produces stresses that can exceed the plastic limit at the elevated diffusion temperatures, e.g. in the emitter regions of bipolar transistors. This and the additional stress concentrations at the edges of windows in the SiO_2 mask often result in many 'emitter edge' dislocations.

Deformation can also occur under conditions of high hydrostatic pressure and localized stressing such as occur in mechanical polishing and indentation even at room temperature. Thus thermocompression bonding of Au wire leads to chips can introduce dislocations as can mishandling of slices. The results of this are often referred to as 'mechanical damage'.

2.3.1 The Burgers vector

Dislocations are characterized by a pair of vectors: **l**, along the dislocation line and **b**, the Burgers vector defined by the Burgers circuit procedure.

The alignment and length of the Burgers vector are fixed but its direction (sense) is given by the doubly arbitrary FS/RH (finish to start/right-handed) sign convention, via the following steps (Fig. 2.4).

(i) Select an arbitrary positive direction along the dislocation line: **l**. (Fig. 2.4a)
(ii) Execute a closed atom-to-atom path, in locally perfect crystalline material, around the dislocation line in the clockwise sense when viewed in the **l** direction, i.e. in the right-handed (RH) screw sense. This is the Burgers circuit in the real crystal (Fig. 2.4b).

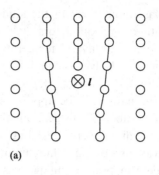

(a)

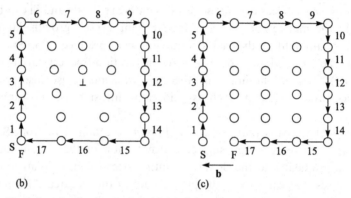

(b)

(c)

Figure 2.4   To define the Burgers vector of the dislocation in (a) draw a closed atom-to-atom path round the dislocation line as in (b) and a corresponding path in a perfect reference crystal as in (c). The closure failure FS is **b**.

(iii) Execute the corresponding series of atom-to-atom steps to form a Burgers circuit in a perfect reference crystal (Fig. 2.4c). (The Burgers circuit in the real crystal (Fig. 2.4b) has to be in good material, i.e. drawn sufficiently far from the dislocation line so that it is clear which interatomic step in the reference crystal corresponds to each one around the dislocation.)

This Burgers circuit in the reference crystal will not close. The closure failure, taken in the FS sense – from the finish atom to the start atom – is the Burgers vector.

Conventions other than FS/RH were used in the past, so care is necessary in reading the early literature and some elementary textbooks. The sense of **l** is arbitrary for any one dislocation but not when comparing two or more dislocations. For example, parallel dislocations should have the same sense for **l** but if a dislocation curves round from one direction to another the sense of **l** must be continuous so that **l** is oppositely directed on opposite sides of a dislocation loop.

2.3.2 Unit dislocations

The dislocations with the shortest lattice translation vector **b** as Burgers vector have the minimum elastic energy as will be shown below. These 'unit' dislocations therefore predominate and this determines the slip direction which must lie along **b**.

2.3.3 Edge and screw dislocations

The angle between **l** and **b** determines the form of the core structure and elastic field of dislocations. Edge dislocations are those having **b** perpendicular to **l** while screw dislocations have **b** and **l** parallel or antiparallel. In covalently bonded crystals, like the semiconducting adamantine materials, dislocations tend to run in low index crystallographic directions in the slip plane (the Peierls-Nabarro potential troughs; see Section 2.5.2 and Fig. 2.24 below). Thus in diamond and sphalerite structure materials, 30° and 60° dislocations (these are the angles between **b** and **l**) are important.

Edge dislocations can be produced by shearing a portion of a slip plane in a direction perpendicular to the (dislocation) boundary line (Fig. 2.5). The dislocation is the edge of a partial plane of atoms, hence the name. If the incomplete plane, known as an 'extra half-plane', lies above the slip plane, i.e. the plane containing **l** and **b** (as in Fig. 2.5), the dislocation is said to be a positive edge. If for such a dislocation, **l** is taken into the plane of the figure, **b** points from right to the left. If the dislocation is a negative edge, the extra half-plane extends down from the slip plane and the Burgers vector points from left to right for **l** into the plane of the figure.

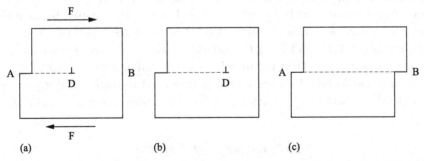

(a)   (b)   (c)

Figure 2.5   (a) The top half of the crystal sheared as shown at A, relative to the half below the slip plane. The shear propagated only to the boundary at D. This is an edge dislocation line running normal to the plane of the figure and to the direction of the shear. (b) The dislocation has moved further. (c) The shear has spread across the whole slip plane. The dislocation has escaped through the surface to produce a slip step.

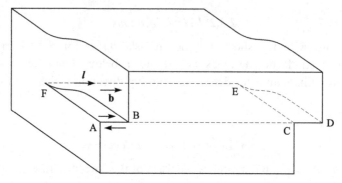

Figure 2.6 In the case of a screw dislocation, the crystallographic shear, AB, is parallel to the dislocation line, FE, bounding the slipped area, AFEC, on the slip plane (cf. Figure 2.3).

Screw dislocations can be regarded as produced by slip over a portion of a plane such as ACEF in which the shear, AB, is parallel to the dislocation line FE (Fig. 2.6). Burgers circuits round such dislocations are helical, i.e. screw dislocations turn the previously parallel crystal planes into a spiral ramp, hence the name of this type of dislocation. If the circuit round the dislocation advances into the crystal when l is in that direction, the dislocation is a right-hand screw and l and **b** are parallel. If the circuit round the dislocation advances out of the crystal when l is into it, the dislocation is a left-hand screw and l and **b** are anti-parallel.

2.3.4 The shear produced by the movement of dislocations

A moving dislocation produces a certain shear displacement of the material on one side of the slip plane relative to that on the other, but if the dislocation moves back the original situation must be restored. Hence the shear must reverse when the direction of motion of the dislocation reverses so the shear produced depends on **v**, the direction of movement (velocity) as well as l and **b**. The relation between these quantities is given by Bilby's (1950) right-hand rule. Let the first finger of the right hand point in the direction l along the dislocation and the second finger indicate the direction of motion (**v**). Then the thumb indicates that side of the slip plane (that containing l and **b**) on which the crystal is translated through the Burgers vector relative to the remainder of the crystal, which is regarded as remaining at rest.

2.3.5 The energy of dislocations

One branch of the study of dislocations that is essentially complete is the elasticity theory. Much of this was worked out as an intellectual game concerning discontinuities in elastic continua, before anyone thought of dislocations as occurring in crystals, or found that they exist and are important. The theory was initiated by a school of Italian applied mathematicians. Dislocations behaving according to elasticity theory are therefore often referred to as Volterra dislocations, in honour of the founder of the subject.

Dislocations in crystals are treated using elasticity theory although, near the dislocation line, atom displacements are large so Hooke's Law does not apply. It is assumed that at distances greater than some cut-off radius r_0 from the dislocation, the material behaves elastically. Inside the cylinder of radius r_0, called the core, other methods must be used.

The elastic strain energy can be most easily found for screw dislocations because they have cylindrically symmetrical shear strain fields. Cylindrical coordinates r, θ and z are used and a cylindrical shell is considered (Fig. 2.7). The shell has been cut along ABCD, sheared through b and the cut rebonded, producing a uniform shear strain. By considering the shell recut and rolled out flat, this shear strain is seen to be:

$$\gamma = b/2\pi r \tag{2.2}$$

By Hooke's Law the corresponding elastic stress can be written as:

$$\tau = Gb/2\pi r \tag{2.3}$$

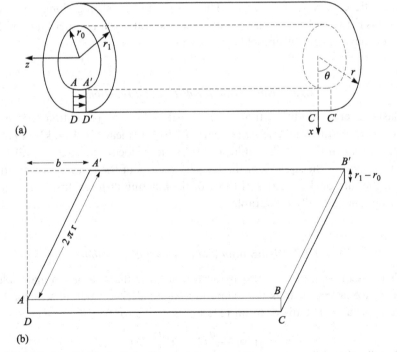

Figure 2.7 (a) A cylindrical shell of material surrounding a screw dislocation line. (b) The shell, cut along the slip plane, ABCD, and rolled out flat.

where G is the shear modulus. The elastic energy of the shell is the work done in displacing the faces of a crystal of unit length along z through a distance b and is half the stress times the final strain:

$$dE = \frac{1}{2}\frac{Gb}{2\pi r}b \tag{2.4}$$

To obtain the elastic energy of the dislocation we integrate from r_0 to an appropriate outer limit r_1:

$$E = \int_{r_0}^{r_1} \frac{Gb^2\,dr}{2\pi r} \tag{2.5}$$

It is easily shown that the elastic energy of edge dislocations is given by the same expression multiplied by about 3/2 and the energies of intermediate types of dislocations have intermediate values of the multiplying constant. Thus the total energy of any type of dislocation can be written in the form:

$$E_{\text{total}} = E_{\text{core}} + \frac{CGb^2}{4\pi}\ln\left(\frac{r_1}{r_0}\right) \tag{2.6}$$

where $C \approx 1$. The elastic fields of dislocations extend to macroscopic distances (r_1) so the elastic energy term is large. This also leads to strong, long-range elastic interactions between dislocations and between dislocations and other types of defects which also have elastic strain fields.

2.3.6 Frank's rule for dislocation reactions

The elastic energy of dislocations is proportional to b^2 (Equation 2.5) and so, approximately, is the total energy (Equation 2.6). This leads to Frank's (1949) rule. This states that any crystallographically possible dislocation reaction will tend to occur if the sum of the squares of the Burgers vectors of the resultant dislocations is less than the sum of the squares of those of the reacting dislocations. Such reactions are (elastic) energetically favourable.

2.3.7 Dissociation and extended dislocations

Frank's rule can be applied to the dissociation of unit dislocations in materials with an f.c.c. space lattice such as the diamond and sphalerite structure semiconductors (and f.c.c. metals). The dissociation reaction is of the form:

$$\frac{a}{2}[110] \rightarrow \frac{a}{6}[211] + \frac{a}{6}[12\bar{1}] \tag{2.7}$$

and is energetically favourable since the energy which is proportional to b^2 is

$$\frac{a^2}{4}[1^2 + 1^2 + 0] > \frac{a^2}{36}[2^2 + 1^2 + 1^2] + \frac{a^2}{36}[1^2 + 2^2 + 1^2]$$

$$\text{i.e. } \frac{a^2}{2} > \frac{a^2}{6} + \frac{a^2}{6} \tag{2.8}$$

Expression (2.7) represents the dissociation of a unit dislocation with the Burgers vector on the left into two Shockley 'partial' dislocations with the Burgers vectors on the right (see Thompson's tetrahedron, Fig. 2.2). These are the vectors $\mathbf{b}_1$, $\mathbf{b}_2$ and $\mathbf{b}_3$ respectively in Fig. 2.8. The unit dislocation Burgers vector $\mathbf{b}_1$ is such that the passage of the dislocation results in a shear (Bilby's right-hand rule) that carries the atoms from one site to another of the same type, e.g. A to A, so the stacking sequence across the slip plane is unaltered. The partial dislocation Burgers vectors are such that the atoms are sheared into a different type of site, e.g. A to B or vice versa and back again. Thus the passage of one partial dislocation alters the stacking sequence and that of the second restores it, so between the two there will be a strip of stacking fault. For example, in Fig. 2.8b the planes above the slipped plane have all been carried into the next position in the cycle of A, B and C stacking positions by the passage of the 'lead' partial dislocation with Burgers vector $\mathbf{b}_2$. The 'trailing' partial dislocation with Burgers vector $\mathbf{b}_3$, on passing a given point, shears the atom planes all back into their correct stacking positions so restoring the unfaulted diamond or sphalerite structure.

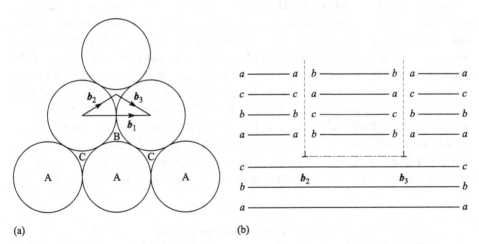

(a) (b)

Figure 2.8 The dissociation of unit dislocations in f.c.c. lattice based crystals. (a) The Burgers vector, $\mathbf{b}_1$ of a unit dislocation and $\mathbf{b}_2$ and $\mathbf{b}_3$ of partial dislocations into which the unit dislocation can split in accordance with equation (2.7). (b) The area between the two Shockley partial dislocations (viewed end-on and indicated by the inverted T symbols) is a stacking fault.

The two partial dislocations elastically repel each other. The stacking fault has a certain amount of energy per unit area. It therefore exerts a surface tension, pulling the partial dislocations together, to minimize the total stacking fault energy. The equilibrium width of the extended dislocation is the separation of the partials at which the elastic repulsion, which falls as the separation increases, just balances the constant surface tension of the fault. The stacking fault energy in Si, Ge and GaP is relatively high so the equilibrium spacing is small, about 5 nm.

Under certain circumstances one partial dislocation can be made to glide away from the other. This widens the stacking fault and this process plays the vital role in phase changes, deformation twinning and the production of polytypes in ZnS as will be discussed in Section 4.1.5 and Section 4.5.1.

Extrinsic stacking faults and Frank partial dislocations

The stacking faults formed by the dissociation of unit dislocations are not the only possible type. In addition stacking faults can be formed by inserting or removing an area of one double atom {111} plane. If part of a {111} plane is removed the fault is described as intrinsic. If part of a {111} plane is inserted an extrinsic fault results. The boundaries of these types of faults are Frank partial dislocations. These processes actually occur. Condensation of an excess of vacancies onto a {111} plane as the crystal cools from its growth temperature actually removes an area of the plane and creates an intrinsic fault. Conversely, condensation of an excess of interstitial atoms can insert an area of a new {111} (double atom) plane between the two neighbouring planes, to create an extrinsic fault.

The Burgers vectors of the Shockley partial dislocations, which bound the stacking faults in extended unit dislocations as discussed above, lie in the slip plane, as did that of the original unit dislocation. The Burgers vectors of Frank partials, which bound intrinsic and extrinsic stacking faults, are perpendicular to the plane of the stacking fault. The magnitude is equal to the change of spacing produced by one close packed double atom layer, i.e. Frank partials have Burgers vectors of the type $\frac{1}{3}\langle 111 \rangle$. By convention, negative Frank partial dislocations are those that bound intrinsic stacking faults while positive Frank partials bound extrinsic faults. Frank partials are in edge orientation; however, since their Burgers vector is not contained in a slip plane but is normal to one, they cannot glide. They can move only by diffusion, increasing or decreasing the area of the intrinsic or extrinsic fault.

2.3.8 The crystallography of slip

Since the elastic energy of dislocations is proportional to b^2 (Equation (2.5)), the predominant minimum-energy 'unit' dislocations are those with $\mathbf{b}$ = the shortest lattice translation vector of the structure which is $a/2\langle 110 \rangle$ in the diamond and sphalerite structures (and the f.c.c. structure of many metals). Since slip occurs by dislocation motion, this is the slip direction. In elementary accounts, with reference mainly to metals, it is argued that the slip plane can be expected to be the one that is most densely packed and so the most widely spaced, as this is easiest to shear. In the case of the diamond and sphalerite structures, however, there are actually two types of {111} slip planes: those between e.g. an *a* plane of atoms and the α plane above it, called type I planes, and those between the *a* plane and the γ plane of atoms below it, called type II planes, in the ... *cγ aα bβ cγ* ... stacking sequence (Fig. 2.9). Dislocations that would shear type I planes are now known as the shuffle set following the terminology of Hirth and Lothe (1968). The shuffle set were the type considered by Shockley (1953) and Hornstra (1958). The latter showed that, if unit dislocations of the type I planes in the diamond structure dissociate, the stacking fault must lie in a type II plane. That dislocations were of the shuffle type and produced slip on type I planes was accepted throughout the literature prior to the work of Hirth and Lothe (1968) who suggested that dislocations that shear type II planes could move more readily, without having to 'shuffle', so they termed them the glide set dislocations.

It was then found by weak beam transmission electron microscopy that dislocations in semiconductors do dissociate and move in extended form. Therefore, it was argued, slip must take place over those planes on which the stacking faults occur, i.e. the type II planes. However, Hornstra (1958) first showed that the originally assumed 'shuffle' set cores could dissociate by forming stacking faults on an adjacent type II plane. Thus the occurrence of stacking faults in dislocation cores does not exclude the possibility that their cores are essentially of shuffle set, dissociated form. Subsequently, experimental evidence and theoretical treatments have led to a consensus that the glide and shuffle set dislocation cores are not mutually exclusive and that which type predominates depends on the material and the plastic deformation conditions. This is discussed in detail in Section 4.1.2.

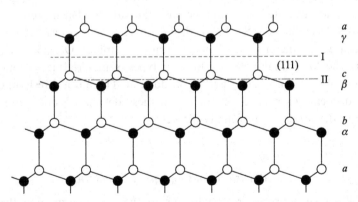

Figure 2.9   Type I and type II {111} slip planes in the diamond and sphalerite crystal structures. 'Glide' and 'shuffle' set dislocations have extra half planes that end on type II and type I planes, respectively.

The geometry of slip in the f.c.c., diamond and sphalerite structures can be visualized by means of Thompson's (1953) tetrahedron (Fig. 2.2). The faces are the {111} slip planes, the edges are the ⟨110⟩ slip directions and the bisectors of the corners of the triangles are the directions of the partial dislocation Burgers vectors for dissociation on that plane. Dislocation lines, l, in semiconductors tend to run along the minimum energy directions in the slip plane, i.e. the Peierls-Nabarro potential troughs (Fig. 2.24).

2.3.9 Growth of macroscopic dislocation-free crystals

Dash (1958) developed the only method yet capable of growing macroscopic dislocation-free crystals. This is a development of the Czochralski (Cz) method (Fig. 1.1). In Cz growth a seed crystal is dipped in a melt of the material in a vacuum or protective atmosphere contained in a furnace. The seed is drawn up while rotating so molten material freezes on the seed. If the conditions are right (a flat liquid-solid interface and slow, vibration-free withdrawal) a single crystal can be obtained.

To grow dislocation-free material, Dash had to eliminate the dislocations in the Cz seeds and introduce no new ones. No dislocation-free material then existed so seed crystals containing high densities of dislocations had to be used. Dislocations cannot simply end in the interior of a crystal, as it would create impossible local strains. The elimination of the dislocations in the seed was achieved by initially raising the temperature so the diameter of the crystal was reduced. This is referred to as 'necking down'. The dislocations in the seed then 'grew out', i.e. terminated at the surface of the solid where the cross section decreased. To avoid introducing fresh dislocations radial thermal gradients were kept low by careful heater design and controlled operation of the crystal 'puller'. This avoided differential thermal contraction and the resulting 'quenching' strains, which would have produced plastic deformation and generated large numbers of dislocations. Dash grew large boules of dislocation-free Ge and Si. Later dislocation-free GaAs was also produced. Such material is commercially available at a price premium. Often a low density of dislocations (far lower than available in other types of material) is acceptable and it is such material that is normally used in production. There is a threshold density of about 10^4 dislocations per cm^2 below which these defects do not cause excessive difficulties in the production of many devices.

2.3.10 Dislocations and grain boundaries

Frank (1949) gave a procedure, analogous to that for determining the Burgers circuit, which determines the Burgers vector content of a section of any grain boundary. So, formally, all grain boundaries can be regarded as resolvable into dislocations. However, in many cases the required dislocation density is so high that the dislocation cores overlap and this approach is not useful.

Small angle grain boundaries (those with relative misorientations across the interface $< \sim 10°$), however, can be resolved into walls of discrete dislocation lines. The simplest case is the symmetric tilt boundary shown in Fig. 2.10. This is a boundary between grains related by a tilt about an axis in the grain boundary that is a possible dislocation line direction. It can be seen (Shockley and Read 1949, 1950) that the misorientation across this boundary is:

$$\theta = b/D \qquad (2.9)$$

where θ is the tilt angle and b is the magnitude of the Burgers vector of the dislocations with a spacing D. Vogel *et al.* (1953) demonstrated that this relation was obeyed by small-angle boundaries in germanium (Fig. 3.1). They measured D between dislocation sites marked by etch pits (see Chapter 3), determined θ by

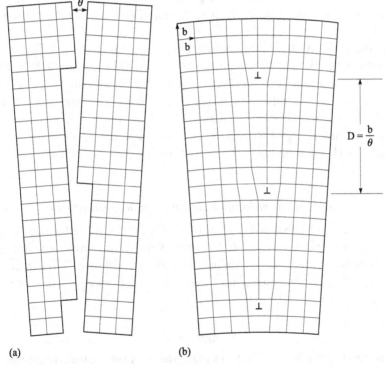

(a)   (b)

Figure 2.10 Schematic diagram of a small-angle symmetric tilt grain boundary in a simple cubic structure. (a) The grains are related by a tilt, θ, about an axis in the boundary that is a possible dislocation line direction. (b) The boundary consists of a vertical wall of parallel unit edge dislocations, running parallel to the viewer's line of sight. [After Shockley and Read 1949. Reprinted with permission from *Physical Review*, **75**, p. 692, 1949 (http://publish.aps.org/linkfaq.html). Copyright 1949 by the American Physical Society.]

x-rays and, knowing that $b = a/2\langle 110 \rangle$, found that their experimental data fitted Equation (2.9). This famous result was the first quantitative confirmation of any prediction of dislocation theory. For $\theta > {\sim}10°$ to $15°$, D becomes comparable with b, i.e. with interatomic distances, and the dislocation cores are no longer separate.

Shockley and Read (1950) used the elasticity theory of (Volterra) dislocations to obtain for the energy per unit area of grain boundary, E:

$$E(\theta) = \tau\theta(A - \ln\theta) \tag{2.10}$$

for a misorientation angle θ. The elastic coefficient τ was given explicitly in terms of the known elastic parameters of the crystal material, while the quantity A, which is related to the unknown dislocation core energy, was chosen to give the best fit with experimental results. This gave good results for orientation differences up to about $15°$ (Read 1953).

Hornstra (1959, 1960) showed that the atomic core structures of large angle grain boundaries in the diamond structure could be analysed by building ball-and-wire models. The occurrence of several of the grain boundary core structures predicted by Hornstra has been confirmed by high resolution TEM. This will be discussed in Chapter 4.

Grain boundaries (GBs) can be usefully analysed in terms of coincidence lattices. If both lattices be continued through the GB interface, a fraction $1/\Sigma$ of the sites will be found to coincide. Friedel (1926) showed that Σ can only be an odd integer (for a recent account of this theorem see Holt (1984)). The interface will contain a network of coincidence sites, each surrounded by an area of coherence (good matching) and low energy bonding. Separating the coherent areas are interfacial dislocations. Such grain boundaries are classified by the value of the index Σ.

2.3.11 Misfit dislocations in epitaxial heterostructures

We saw in Section 1.1.1 that epitaxy was originally only of limited academic interest (for a review see Pashley 1956). It became an important semiconductor growth technology, however, both for Si planar devices and III-V devices. Now there is great interest in epitaxial heterostructures that implement energy band engineering (see Sections 1.7.1 to 1.7.4).

Epitaxy

To understand the two basic conditions that, when not fully satisfied, can lead to the introduction of defects during epitaxial growth, we return to basics. Royer (1928) coined the term epitaxy from the Greek roots *epi* (meaning on) and *taxis* (ordered) for the phenomenon of oriented overgrowth. He reviewed the literature of such overgrowths in minerals and in his own studies of the growth of alkali halides on one another from aqueous solution. From this he deduced rules for epitaxy. One has stood the test of time. It states that two epitaxial materials always have, at the interface, parallel lattice planes with meshes that are 'identical or quasi-identical' (see also Pashley 1956).

Epitaxy is oriented overgrowth. That is, one material grows on another in one orientation so that if the substrate is monocrystalline so is the epilayer deposited. Cubic materials, like two diamond structures or a sphalerite- and a diamond-structure semiconductor, usually grow epitaxially in parallel orientation. For further details see the reviews by Pashley (1956, 1991), Davies and Williams (1994) and Mahajan and Sree Harsha (1999).

More complicated orientation relations occur for cubic semiconductors grown on complex, lower-symmetry substrate materials, e.g. in the growth of silicon on insulator (SOI) substrates. Fig. 2.11 shows a mesh-matching diagram from one of the earliest papers on epitaxial Si on sapphire (SOS). SOS occurs in five different orientations depending on the surface orientation of the sapphire substrate. The most useful, (001) Si on ($\bar{1}$012) sapphire with [100] Si parallel to [1$\bar{2}$10] sapphire is shown in Fig. 2.11.

If mesh matching is to be possible, there must be (1) symmetry matching so the meshes are the same shape, and (2) lattice parameter matching so the meshes are the same size. Matching meshes ensures that there will be a dense set of coincidence sites in the interface, i.e. sites in which atoms of the two materials are closely placed and can form strong, minimum-energy bonds. Such interfaces are energetically favourable which is why epitaxy occurs. Research in the 1950s and 1960s showed that virtually any material could be grown epitaxially on any other once the best growth technique and conditions were found. Thus it is not necessary for epitaxy, as Royer (1928) thought, that the two materials be of similar bonding, or crystal structures or lattice parameters. The more closely related the two materials are in these respects, however, the more easily epitaxial deposits of high crystalline perfection can be obtained.

Any interfacial symmetry mismatch makes possible the occurrence of domains of different crystal variants. This is exemplified in semiconductors by the occurrence of antiphase or polarity-reversal domains in the epitaxial growth of GaAs on (001) oriented Ge or Si substrates (see Section 4.5). Any lattice parameter mismatch makes possible the occurrence of interfacial misfit dislocations. These defects need not occur, however. Methods have been found to avoid introducing misfit dislocations and antiphase domains during epitaxy as will be discussed in Sections 4.5 and 4.6.

Pseudomorphism, the 15% rule and the Frank and van der Merwe model

The percentage mismatch between the lattice parameter of the epitaxial film, a_f, and that of the substrate, a_s, (see Fig. 2.12a) is:

$$m = \frac{100(a_f - a_s)}{a_s} \qquad (2.11)$$

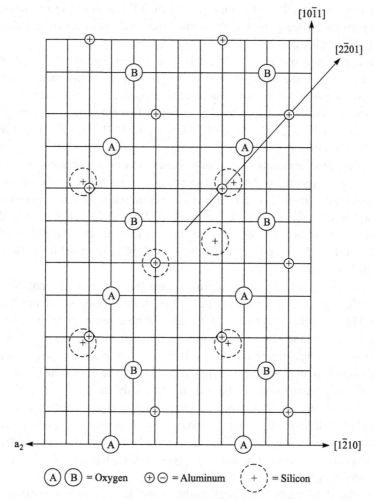

Figure 2.11   Mesh-matching at the interface of epitaxial (001) Si on the ($\bar{1}$012) surface of sapphire (i.e. SOS – silicon on sapphire) with [100] Si ∥ [1$\bar{2}$10] sapphire. The unit mesh of (001) Si is the square array of the (large dashed circle) Si atom symbols. They match the square array of four small +ve circle symbols for Al atoms. The lattice mismatch along [1$\bar{2}$10] is about 12% and along [10$\bar{1}$1] it is about 4% (after Nolder and Cadoff 1965). A and B indicate two types of lattice site and the + and − sites are at different heights (for additional information see Nolder and Cadoff 1965).

This is usually quoted as a percentage but is not standardized. It may be calculated relative to a_f, i.e. with a_f in the denominator rather than a_s, or relative to the average of the two. This quantity is positive if $a_f > a_s$ and the larger parameter film material is compressively strained to match the substrate. If $a_f < a_s$, the misfit is negative and the smaller lattice parameter film is strained in tension.

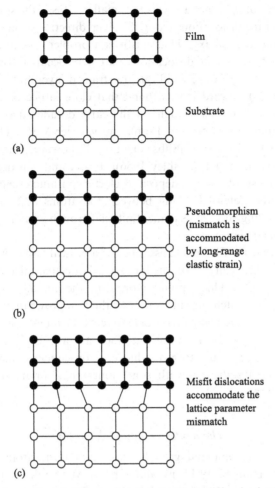

Figure 2.12 (a) An epitaxial film and substrate that differ in lattice parameter can accommodate this mismatch either by (b) long-range elastic strain in the film that brings it into registry with the substrate (pseudomorphism) or (c) by the introduction of misfit dislocations.

The simplest geometry is the parallel orientation of cubic system substrate and epifilm (as in Fig. 2.12a), which is the usual case for semiconducting materials. This involves a sort of Vernier scale matching at the interface as shown schematically in Fig. 2.12b. Then there will be an integer P such that:

$$Pb = (P+1)a = p$$

$$\text{so} \quad p = ba/(b-a) \tag{2.12}$$

where b is the larger of a_f and a_s and a is the smaller. This is the spacing of dangling bonds and hence of the dislocations along the given direction in simple ball-and-wire models for fully relaxed epitaxy. Thus misfit is a problem because of the possible electronic effects of any misfit dislocations that may occur. Growth of misfitting materials can, however, produce good quality strained layer epitaxy (Chapter 4).

From Royer (1928) onward it was found that close lattice parameter matching was beneficial. In fact, it was thought in the early decades that epitaxy was not possible if the mismatch was >15%. Frank and van der Merwe (1949a, 1949b) set out to account for the 15% rule. [Subsequent work, however, showed that epitaxy could occur for any mismatch (Pashley 1956)]. Frank and van der Merwe (1949a, 1949b) modelled the substrate and deposit as elastic continua except at the epitaxial interface. A periodic potential acting between the atoms adjoining the interface represented the bonding across this plane. Epitaxial growth was simulated by varying the thickness of the epilayer, h.

The result was that for film thicknesses below a critical value, h_c, it was energetically favourable to elastically strain the film so its lattice parameter matched that of the substrate (Fig. 2.12b). This is pseudomorphism [the film adopts a false (pseudo) form (morphology) to match the substrate]. As the film grows thicker than a critical value, h_c, it becomes energetically favourable to relax the long-range strain in the film by forming misfit dislocations in the interface as in Fig. 2.12c. This happens gradually as the film continues to grow thicker. Consequently, many films end up in a partially relaxed state, i.e. with some long-range strain and partial misfit dislocation relaxation.

The modes of epitaxial growth

Many TEM studies of epitaxial growth showed that the atomic monolayer by monolayer growth assumed by Frank and van der Merwe did indeed occur sometimes. However, other modes of epitaxial growth also occurred. Bauer and Poppa (1972) classified them as shown in Fig. 2.13. There are two extremes. One is the two dimensional, layer-by-layer growth of Frank and van der Merwe (Fig. 2.13a). The other is the Volmer-Weber mode (Fig. 2.13b) in which three-dimensional crystallites nucleate on the initial substrate surface and grow to form the complete epilayer. In the Stranski-Krastanow mode (Fig. 2.13c) initial growth is layer-by layer but later three-dimensional nucleation takes over.

This was rationalized by Bauer and van der Merwe (1986), as follows. Let the surface energies of the substrate and epitaxial film be γ_s and γ_f, respectively, and that of the interface be γ_{in} (see Fig. 2.14). Then if

$$\Delta\gamma = \gamma_f + \gamma_s - \gamma_{in} \leq 0 \tag{2.13}$$

the epilaxial film will 'wet' the substrate, i.e. it is energetically favourable for monolayer (Frank and van der Merwe) growth to occur (Fig. 2.13a), so long as this

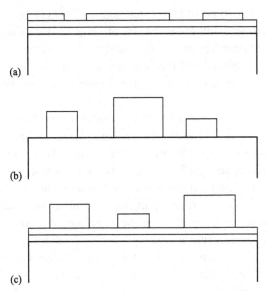

Figure 2.13 The three modes of epitaxial growth: (a) the Frank and van der Merwe monolayer (2D) growth mode, (b) the Volmer-Weber (3D) growth mode, and (c) the Stranski-Krastanow mode. (After Pashley 1991).

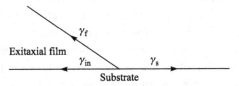

Figure 2.14 Schematic diagram of the edge of an epitaxial nucleus on a substrate. The energies, γ, of the film, γ_f, surface, γ_s, and interface, γ_{in}, act as surface tensions as shown.

relation holds. However, monolayer growth involves elastic strain of the layers (Fig. 2.12), which is included in the surface energy values. If this contribution is large and increases with thickness, eventually Equation (2.13) will cease to hold and three-dimensional nucleation will begin, giving Stranski-Krastanow growth (Fig. 2.13c). If relation (2.13) does not hold, initially growth will be by Volmer-Weber nucleation (Fig. 2.13b) from the start. For a more detailed discussion see Pashley (1991).

Ball-and-wire modelling of misfit dislocations

The first heterojunction to be studied, GaAs grown on Ge, had a lattice mismatch near zero (Anderson 1960, 1962). Interface states were neglected in Anderson's basic account of the construction of energy band diagrams for heterojunctions.

Oldham and Milnes (1964) pointed out that e.g. Si/Ge (111) heterojunctions would, ideally, be expected to contain an array of edge dislocations. They therefore suggested that, for mismatches greater than about 1%, a sufficiently large density of states could arise, due to misfit dislocations at the interface, to cause band bending and strongly affect device characteristics (for further detail see the book by Sharma and Purohit 1974).

The dislocations first suggested by Frank and van der Merwe and represented in Fig. 2.12c, are called perfect misfit dislocations. These are pure edge dislocations with their Burgers vector in the film/substrate interface. This type has the maximum effectiveness in relieving the misfit strain. Following Hornstra (1958), they can be investigated by constructing ball-and-wire models as shown in Fig. 2.15. Suppose that, due to the difference in lattice parameters, a and b, the repeat distances in the two semiconductors in direction $[hkl]$ in the interface are c and d. Then there will always be an integer p such that $p = nd = (n + 1)c$, so there is a Vernier condition. That is, at intervals p there are coincidence sites, i.e. positions at which the atoms in the two materials coincide vertically in the model of Fig. 2.15c. Midway between these sites the atoms of the two materials are so out of register that a dangling bond occurs in the material with the smaller lattice parameter. This 'Vernier matching' occurs in all directions in the interface plane. The result for a {111} interface is shown in a three-dimensional model in Fig. 2.16.

The existence of misfit dislocations in epitaxial heterojunctions and other epitaxial interfaces was established by many transmission electron microscope studies (for reviews see Matthews 1975, 1979). However, the dislocations did not immediately appear at thickness $h = h_c$. This is partly due to the limited energetic driving force for misfit dislocation formation when h exceeds h_c by only a small amount. It is also partly due to the difficulty of generating the required dislocations and moving them into position. Later treatments deduce a lower limit $h_1 \approx h_c$ at which the first misfit dislocations begin to appear and a greater thickness h_2 by which they reach the number required to accommodate the misfit completely. Films of thicknesses $h_c < h < h_2$ are generally only partially relaxed. This will be discussed in more detail in Section 4 6.2.

In addition, TEM studies found that a number of different mechanisms operated in different cases (Matthews 1975, 1979). Most of them did not produce the perfect misfit dislocations of Figs. 2.12 and 2.15. This will be discussed in Section 4 6.2.

2.3.12 Dislocations and point defects

Dislocations can emit or absorb vacancies and interstitial atoms (native point defects). They are then said to act as sources or sinks, respectively, for these defects. Edge dislocations incorporate point defects at the edge of the extra half-plane. This lowers the energy of the dislocation plus vacancy or interstitial. It creates jogs as

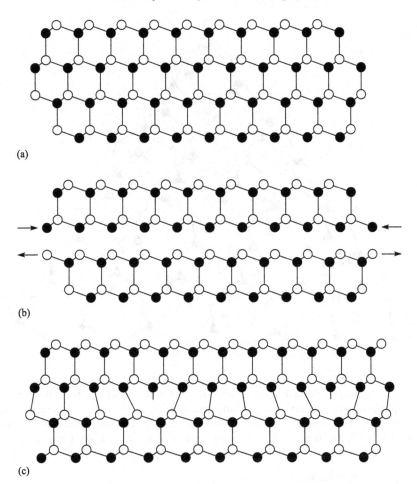

(a)

(b)

(c)

Figure 2.15 Construction of perfect misfit dislocations at a (111) epitaxial heterojunction. (a) A defect-free single crystal projected on the (01$\bar{1}$) plane (b) is separated over the (111) interface plane and the lattice parameter is changed on either side to the values for the substrate, c, and epilayer, d, where the repeat distance is $p = 5c = (5 + 1)d$. (c) When the two crystals are bonded together as in Figure 2.12(c) there is a Vernier of period p. The dangling bonds occurring at a spacing p mark the positions of successive misfit dislocations. At the coincidence sites, midway between the dislocations, bonding is unstrained and so of minimum energy. (After Holt 1966.)

shown in Fig. 2.17 so the dislocation becomes ragged. The emission or absorption of point defects produces 'climb' of the dislocation line up or down out of the slip plane.

Screw dislocations, it should be remembered, turn the whole crystal into a spiral ramp instead of a set of parallel atomic planes. Hence the result of their absorption or emission of vacancies is that the dislocation line climbs out onto the spiral ramp into which the crystal was turned by the screw and are, therefore, turned

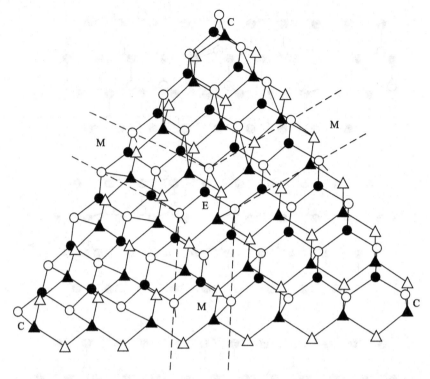

Figure 2.16 Three-dimensional ball-and-wire model of a unit mesh of a (111) heterojunction between sphalerite-structure crystals with lattice parameters $d = 6c/7$. The three rows of atoms with dangling bonds running in the directions indicated by broken lines constitute the cores of the dislocations: ○ atoms in the upper (film) crystal; △ atoms in lower (substrate) crystal. There is a coincidence site, C, at each corner of the model. (After Holt 1966.)

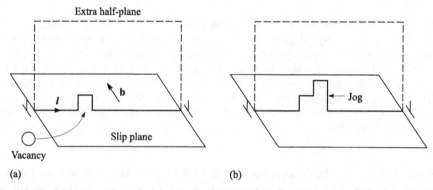

Figure 2.17 (a) Absorption of one vacancy and (b) then more causes the extra half-plane of an edge dislocation to be reduced in area. This is described as the dislocation climbing up out of the slip plane. Jogs, i.e. short lengths of dislocation line running up out of the slip plane, are formed. The reverse process, the absorption of atoms, i.e. the emission of vacancies, increases the area of the half-plane causing the dislocation to climb down from the slip plane.

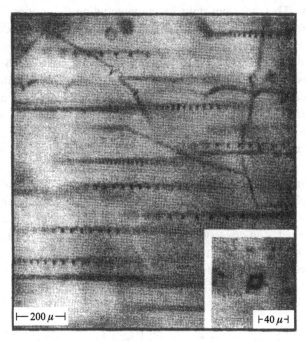

Figure 2.18 Helical dislocations in Au diffused Si. The inset is a view along the axis of a typical helix to demonstrate the polygonal structure. This is a transmission infrared light micrograph. (After Dash 1960. Reprinted with permission from *Journal of Applied Physics*, **31**, pp. 2275–83. Copyright 1960, American Institute of Physics.)

into helical dislocations. Screw dislocations in material cooled from a high temperature contain supersaturated concentrations of vacancies or interstitials that have to disappear to return the system to equilibrium. Screw dislocations are also turned into helices in this way in certain cases of diffusion, as shown in Fig. 2.18.

Dislocations thus make it possible for crystals to rapidly attain equilibrium defect densities on heating or cooling. It was found beneficial to have a few dislocations present since it allows the high densities of native point defects at elevated temperatures to disappear on cooling, only making the dislocations ragged. In zero density material, however, the excess native point defects either react with impurities to form complexes that can alter their electronic effects, or cluster to form numerous small areas of stacking fault with bounding partial dislocations, often with more serious effects than a few dislocations.

Jogs and kinks and point defect generation

It is important to distinguish jogs from kinks. Kinks are short lengths of dislocation in the slip plane. They differ from most of the dislocation line in that they do not lie in the Peierls potential minima, which define the directions in which the dislocations run. They are important for understanding the motion of dislocations especially in semiconductors, as we shall see in Section 2.5.4 and Section 4.2.

Jogs (Fig. 2.17) are short lengths of dislocation line running up or down out of the slip plane (defined by l the dislocation line direction and **b** the Burgers vector). Jogs in screw dislocations (Fig. 2.19a) are important since they generate point defects when the dislocation glides during plastic deformation. The jog is at right angles to the screw line direction, l, and therefore is normal to **b** since this is parallel to l. Hence the jog is a short length of edge dislocation. That is, it is the edge of a strip of extra half-plane. Movement of the screw dislocation necessitates lengthening or shortening the extra half-plane (shown dotted). This requires energy so jogs are low mobility dragging points along screw dislocations. Screw dislocations are consequently less mobile than edges. Dragging a short (one or a few atoms in length) jog along with a screw dislocation leaves a string of one form of point defect or the other. This is called 'non-conservative' dislocation motion. The vacancies, divacancies (vacancies on the adjacent sites of a basis unit) etc. are left on the slipped planes, whence they then tend to diffuse away.

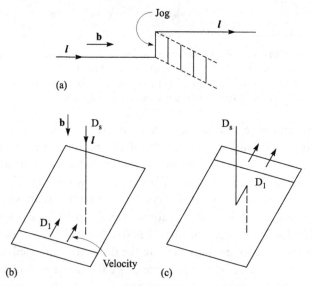

(a)

(b)   (c)

Figure 2.19   (a) A jog in a screw dislocation (a right-hand screw dislocation with l and **b** parallel is shown here) is necessarily of edge type (since l in the jog is normal to **b**). The extra half-plane is of width equal to the length of the jog as indicated by the cross-hatched area running from the plane of the diagram away from the observer. Therefore, if the screw dislocation glides from the plane of the paper, away from the viewer, the extra half-plane must get shorter, i.e. atoms must be emitted or vacancies must be absorbed. (b and c) When, during plastic deformation, a dislocation D_1 gliding over the plane shown, cuts through a threading screw dislocation, D_s, the shear across the slip plane produces a jog. In this way intersecting dislocations generate jogs and these in turn generate point defects (see the caption of a) or dipoles.

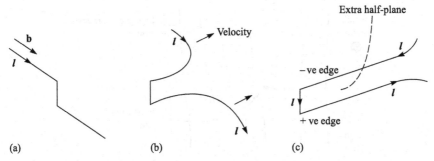

Figure 2.20 (a) A long jog in a screw dislocation (b) acts as a dragging point. The moving screw dislocation therefore leaves two growing lengths of line stretching back to the jog. (c) These are edge dislocations of opposite sign and constitute a dislocation dipole, with the extra half-plane lying between them in this case.

Jogs are produced during plastic deformation when moving dislocations on other slip planes cut through screw dislocations (Fig. 2.19b and c).

If the jog is of macroscopic length, however, it may not move (Fig. 2.20b). Instead, in such a case, two long parallel dislocations run back from the later position of the screw, one to the top of the jog, the other to the bottom (Fig. 2.20c). This configuration is called a *dislocation dipole*. The opposite signs of the two dislocations of the dipole may be such that the region between them is an extra part-plane of atoms (as in Fig. 2.20c) or, in the reverse case, an area of missing atoms. In the latter case the planes of atoms on either side of the dipole relax into the area between the two parallel edge dislocations.

The point defects and dipoles, left on the slip plane after screw dislocations have swept over it, are collectively referred to as 'debris'. It is now known that they make major contributions to the mechanical and electrical effects of plastic deformation (see e.g. Section 4.7 and Section 5.9.1). Together with the interaction of dislocations with impurity atoms debris is one of the reasons that the intrinsic electronic properties of dislocations are still somewhat uncertain. It is also one of the complications that foiled most attempts to put dislocations to use in semiconductor devices.

Impurities and the Cottrell atmosphere

Dislocations and point defects interact elastically since they all strain the lattice locally. Similarly, impurity atoms, whether larger or smaller than the atoms they replace, can lower the energy of the system by segregating to dislocations with an edge component as shown in Fig. 2.21. Atoms that are larger than the host atom move into the region of tensile strain where the lattice is expanded. Those that are smaller move into the region of compressive strain. This elastic attraction results in the concentration or segregation of impurities around dislocation lines in Cottrell atmospheres (Cottrell and Bilby 1949).

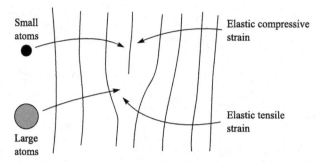

Figure 2.21   There is a region around the edge of the extra half-plane of edge dislocations which is in compression. This is a region of low energy sites for impurity atoms of smaller diameter than the host atoms. Similarly the region of tensile strain below the edge of the extra half-plane attracts large impurity atoms.

The related Suzuki effect is the preferential solution of an impurity in the crystal structure of the stacking fault layer due to the fact that there the atomic surroundings are those of the other structure (wurtzite/sphalerite in sphalerite/wurtzite respectively). This lowers the energy of the stacking fault. It, therefore, exerts a lower surface tension pull to oppose the mutual elastic repulsion between the bounding partial dislocations. Consequently the width of dissociated dislocations increases when the Suzuki effect occurs.

Both the Cottrell and Suzuki mechanisms lower the energy of the dislocation affected. Extra energy is, therefore, required to move the dislocation away from this low energy position. These effects are said to pin the dislocation in place and this requires a higher shear stress to cause plastic slip. Once free of pinning the dislocation can move at a lower stress. Such drops in the required flow stress are called yield points.

Impurity segregation can lead to the nucleation of precipitates, i.e. larger aggregates of impurities or of second phase material. Precipitates are often found at the nodes of dislocation networks, along dislocation lines or dotted about the stacking faults in heavily doped or contaminated semiconductors. Deliberate precipitation to make dislocations visible is called decoration. Decoration formed the basis for the observation of defects in Si by transmission infrared light microscopy (see Section 3.3.1 and the famous example of such a micrograph in Fig. 2.29).

The segregation of impurities to dislocations and other defects in semiconductors often greatly changes the electrical and luminescence properties of dislocations and grain boundaries. The result can be either stronger effects (see Section 5.5.8 and Section 5.6.4) or the elimination of such effects. The latter case is referred to as passivation of the defect. Any metallic precipitates that decorate extended defects can cause localized 'microplasma' breakdown in reverse biased *p-n* junctions and consequent excess electrical noise and soft reverse junction characteristics.

2.4 Electrical effects of dislocations in semiconductors

Dislocations break the symmetry of the crystal structure. As this determines the form of the electronic energy band structure of the material (Section 1.5) we must expect the electronic energy states to be altered locally. In fact, two types of effect appear to be important. Firstly, as Shockley first pointed out on the basis of model of Fig. 2.22, the configurations of the atoms in the cores of dislocations with edge components in the diamond structure are such that it appears likely that some of them will have uncompleted, broken or dangling bonds, i.e. there is no atom in position to complete the bond so it ends in space. These bonds can be expected to have energies between those of the bonding orbitals that correspond to valence band states and the anti-bonding orbitals that correspond to conduction band states. Thus, dangling bonds should give rise to states near the middle of the band gap, i.e. deep levels acting as recombination centres. Secondly, the elastic strain fields of the dislocations mean that the distance between atoms will be altered so (Section 1.2.1 and Fig. 1.5) the separation of some electronic states will be different. That is, electronic states may be pulled down out of the conduction band a short way into the band gap. Such shallow levels tend to act as traps. Thus both deep and shallow gap states may arise from dislocations. In addition, dislocations attract misfitting atoms, i.e. atoms of sizes different from those of the host material. The presence of higher concentrations of

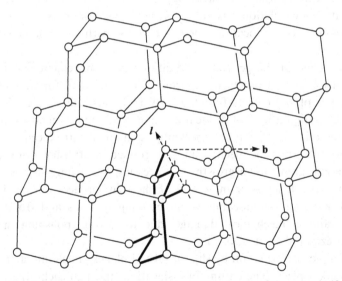

Figure 2.22   Bonds dangling from the atoms along the edge, **l**, of the extra half-plane (heavy lines) in the core of what is now described as an undissociated 'shuffle set' 60° dislocation (see Section 4.1; after Shockley 1953). **l** is the dislocation line direction and **b** is the Burgers vector. [Reprinted with permission from *Physical Review*, **91**, p. 228, 1953 (http://publish.aps.org/linkfaq.html). Copyright 1953 by the American Physical Society.]

doping atoms in and near the dislocation cores must be expected to alter the defect states of the dislocation. In fact there is strong evidence that in silicon dislocations are only electrically active when 'decorated' by traces of heavy metal contaminants (see Section 5.5.8 and Section 5.6.4).

Dislocation lines are often found to be charged negatively, indicating the presence of deep trap states presumably associated with broken bonds. Such states can also act as acceptors (or donors) and so affect the carrier density. Broken bonds are chemically reactive as well. These effects are discussed in detail in Chapter 5.

2.5 Plasticity of semiconductors

When a semiconductor is subjected to a steadily rising stress, say in tension, it first deforms elastically. That is it at first deforms in accordance with Hooke's Law. If the stress is removed it reverts to its initial length. If the stess continues to rise, and the temperature is sufficiently high, the semiconductor will begin to deform permanently. This is plastic deformation. The plasticity (plastic behaviour) is of considerable research interest because only in semiconductors it is possible to compute stress-strain and creep curves, i.e. to predict deformation behaviour from basic, if semi-empirical, dislocation velocity-stress and generation rate relations. This is the *microdynamical theory of plasticity*. The study of dislocation movement (velocity-stress relations) is called *dislocation dynamics* and this is another advanced branch of dislocation theory that is perhaps best understood in semiconductors.

Some knowledge of the generation, multiplication, movement and interactions of dislocations is also of practical importance. It is essential for the identification and elimination of the defects found in new materials and devices. Specific and characteristic configurations of dislocations and other defects tend to appear each time a new material or growth or processing technique is introduced.

Plastic deformation is the mechanism of introduction of dislocations, etc. due to excessive thermal gradients in growth or processing (quenching strains). Deformation can also arise from poor mounting of slices in furnaces during diffusion or oxidation, from poor thermocompression bonding of leads and from heating of material containing 'mechanical damage' due to sawing, polishing, mishandling (tweezer squeezes and scratches), etc.

During complex deformation by slip on several planes, much interaction between dislocations takes place. The cutting of dislocations through each other, where slip planes intersect, generates jogs in screw dislocations (Figs. 2.19 and 2.20). Their subsequent movement generates many point defects and dipoles by the mechanisms considered above. This 'debris', left behind dislocations, is now known to make a significant contribution to the electrical effects of plastic deformation.

2.5.1 Atomic mechanisms of dislocation motion in semiconductors

Dislocation core structures in covalently bonded semiconductors are more important and better defined than in other types of materials. The cores are narrow, tend to be crystallographically aligned rather than curved and have major electronic and plastic effects.

The first model (Fig. 2.23a) of a dislocation core in diamond-structure semiconductors was put forward by Shockley (1953). He suggested that dislocations with an edge component such as this 60° dislocation would contain a row of atoms along the core each with an unsatisfied 'dangling' bond. This model was the first to be arrived at by constructing and considering real ball-and-wire models. This approach has been very useful for semiconductors. It is based on the fact that the stiff and brittle nature of the wires in the physical models is not an unreasonable representation of the directional sp^3 covalent bonds in the cores of the real dislocations. This approach was used later to suggest the form of more complex core structures in both dislocations and grain boundaries by Hornstra and others. This simple Shockley model will suffice for the present. We will discuss more sophisticated models and the evidence for them in Chapters 4 and 5.

2.5.2 The Peierls stress and the double-kink mechanism

Peierls (1940) modelled a crystal containing a dislocation as two elastic continua meeting at the slip plane across which a periodic potential applied. The width of the core, *w*, i.e. the region in which the displacements of the atoms are too large for linear elastic theory to apply, depended on the periodic potential representing the

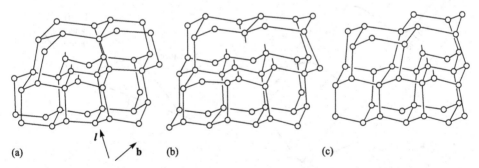

(a)   (b)   (c)

Figure 2.23 (a) The model of a 60° (angle between **l** and **b**) dislocation in germanium or silicon put forward by Shockley (1953). Note the row of unconnected or 'dangling' bonds in the core. (b) and (c) as the dislocation moves the core configuration changes and the numbers of broken and deformed bonds and so the core energy will vary periodically and strongly. This statement remains true, although this model is now known not to be realistic. [Reprinted with permission from *Physical Review*, **91**, p. 228, 1953 (http://publish.aps.org/linkfaq.html). Copyright 1953 by the American Physical Society.]

interatomic forces acting. This model led to an expression for the shear stress necessary to overcome the intrinsic lattice resistance of an otherwise perfect crystal to the movement of a dislocation. This is the Peierls (or Peierls-Nabarro) stress, τ_P:

$$\tau_P = \frac{2\mu}{(1-v)} e^{-2\pi w/b} \tag{2.14}$$

where μ is the elastic shear modulus, v is Poisson's ratio, b is the Burgers vector and w is the width of the dislocation core. This exponential dependence on the width w emphasized the importance for dislocation mobility of the dislocation core, i.e. of the interatomic forces involved, as will be discussed more fully in Section 4.2.1. Due to the periodic bonding force acting across the slip plane, the energy of the dislocation core, i.e. the Peierls potential, $U_P(x)$, varies periodically as the dislocation moves. [Clear accounts of the Peierls stress will be found in Cottrell (1953), pp. 62–4, and Hull and Bacon (1984), pp. 215–20.]

In semiconductors, the width, w, of dislocations is small, compared to f.c.c. metals, due to the strongly directional character of covalent bonds. Hence, according to the Peierls model, i.e. Equation (2.14), the Peierls stress is relatively large. The necessity to break and reform bonds when moving a dislocation (Fig. 2.23) means that the Peierls potential, $U_P(x)$, varies much more strongly with distance than in crystallographically similar materials like the f.c.c. metals. That is, the Peierls troughs (Fig. 2.24) are relatively deep in semiconductors. Dislocations will therefore tend to lie in the Peierls troughs, which run in low index directions.

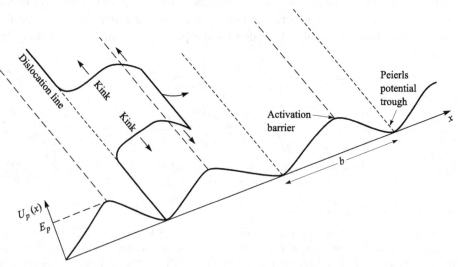

Figure 2.24 The 'Peierls potential' of a dislocation, $U_P(x)$, varies with a repeat distance, $\approx b$, in the direction of motion. Dislocations in covalently bonded materials tend to lie in the deep Peierls potential troughs. Dislocations move through the thermal activation of double-kink loops into the next trough.

The Peierls stress is the maximum slope of the Peierls potential:

$$\tau_P = 1/b \, (dU_P/dx)_{max} \tag{2.15}$$

where b is the length of the Burgers vector. This is the stress that would be required to move a straight unit dislocation forward over the peaks in U_P. It was found that τ_P, calculated by early models for undissociated dislocations, was greater than the observed intrinsic movement stress for dislocations, so dislocations do not move in this way (Haasen and Alexander 1968). Instead they are believed to move by the nucleation and growth of double-kink loops like that shown in the dislocation running along the left-most minimum in Fig. 2.24. The kinks run relatively easily, i.e. at a lower stress, along the dislocation, carrying it into the next trough. Dislocations in semiconductors are dissociated so the process in Fig. 2.24 occurs in each of the partial dislocations. The double-kink mechanism is believed to determine the dynamics of dislocations, i.e. the dislocation velocity-stress relation, as will be outlined in Chapter 4.

2.5.3 The force on a dislocation

Dislocations are lines along which the atoms in crystals are incorrectly located. They are not material objects so the concept of the force on a dislocation requires careful definition.

We saw above that elasticity theory led to an expression for the energy of a dislocation per unit length:

$$E_{total} = E_{core} + \frac{CGb^2}{4\pi} \ln\left(\frac{r_1}{r_0}\right) \tag{2.16}$$

where C is a parameter that varies from 1 for screw dislocations to $1/(1-v) \approx 3/2$ for edges (v is Poisson's ratio). This energy is large since the elastic strain field extends an effective distance r_1 equal to half the distance to the next dislocation line which in semiconductors is generally at least of the order of μms. Dislocations therefore tend to shorten, i.e. they have a line tension given by the increase in energy per unit increase in length (see Equation 2.16). In non-covalently bonded materials, the elastic component of the energy is dominant so the energy is almost isotropic. In covalently bonded materials the core energy is significant and depends on the numbers and types of strained, wrong and broken bonds per unit length. Hence it is strongly orientation dependent and dislocations in semiconductors tend to align with the Peierls potential troughs. Dislocations in the diamond structure elements tend to lie along ⟨110⟩ directions in the {111} glide plane in specimens deformed slowly at relatively low temperatures (Fig. 2.25).

An external load does work in deforming a crystal and this occurs by moving dislocations. Hence work is done in moving those dislocations. This leads to a

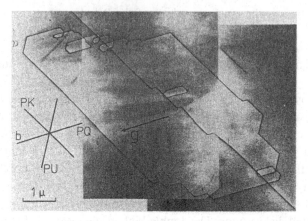

Figure 2.25   Dislocations in the primary glide plane of Si after deformation for a few seconds at the relatively low temperature of 520° C and a stress of 20 kg mm^{-2}. All the dislocations are aligned along the three $\langle 110 \rangle$ directions in the $\{111\}$ glide plane marked PK, PQ and PU. (Transmission electron micrograph from Wessel and Alexander 1977. Reprinted with permission from *Philosophical Magazine*, **35** (1977), pp. 1523–36; Taylor & Francis Ltd., the Journal's web site: http://www.tandf.co.uk/journals.)

definition of the force acting as follows. Consider a segment of a dislocation line of length ds that moves a distance dx under the influence of a shear stress τ, due to the external load. If the dislocation is swept across the whole area, A, of the slip plane, an offset of magnitude b would be produced between the portions of the crystal on either side. In sweeping the element of area, therefore, a shear strain of $(ds dx/A)b$ is produced. Since the applied force producing the stress is $A\tau$, the work done when the increment of slip occurs is given by $dW = A\tau(ds dx/A)b = \tau(ds dx)b$.

The force per unit length acting on a dislocation can thus be defined to be the work done when unit length of the dislocation moves unit distance, i.e.

$$F = \frac{dW}{ds dx} = \tau b \tag{2.17}$$

In addition to the line tension and any external force, dislocations are subject to forces arising from interactions with other defects and surfaces. The elastic interactions are fully covered in the standard textbooks on dislocations (Hirth and Lothe 1968, Hull and Bacon 1984) so they need only be mentioned here.

The elastic interactions between dislocations and other defects can be represented by adding a force due to these 'internal stresses' to that due to the external load. That is:

$$F = (\tau_{\text{ext}} + \tau_{\text{int}})b \tag{2.18}$$

A widely used semi-empirical expression for the effect of the dislocations that accumulate in the crystal during deformation is:

$$\tau_{int} = C\rho^{1/2} \tag{2.19}$$

where C is a constant and ρ is the steadily increasing density of dislocations present. This expression represents the temperature-independent contribution to the work hardening, i.e. the increase in flow stress with increasing deformation (Seeger 1957, Nabarro *et al.* 1964). The elastic attraction of dislocations for impurity atoms leads to the formation of Cottrell atmospheres (excess concentrations of impurities) around dislocation lines. These lower the energy of the system and so 'pin' the dislocations in position, i.e. impede subsequent movement.

2.5.4 Semiconductor dislocation dynamics

The form of the dislocation velocity-stress relation varies characteristically from materials of one type of bonding to another (Fig. 2.26). These relationships can be roughly approximated by power laws of the form:

$$v = B_1\tau^m \tag{2.20}$$

where τ is the stress, v the velocity and B is a constant for a given temperature and doping concentration. The covalently bonded semiconductors have the lowest power dependence of dislocation velocity on stress of any class of materials (m is close to 1).

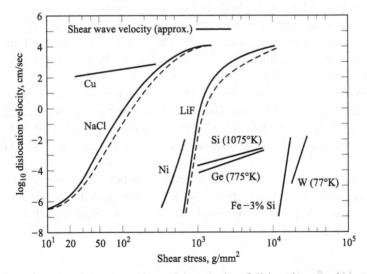

Figure 2.26 The form of the dependence of the velocity of dislocations on the applied stress for materials with different types of bonding. (After Gilman 1969, *Micromechanics of Flow in Solids*, © 1969 McGraw-Hill, reproduced with permission of The McGraw-Hill Companies.) The dislocation velocity increases more slowly with stress for semiconductors than for the other types of material and the velocity is strongly temperature-dependent.

Dislocation velocity is measured as a function of stress by the repeated pitting technique. This involves etching (immersing the crystal in a special solvent) to mark the position of the dislocation by a surface etch pit (to be discussed in Chapter 3). Then stressing to a value τ for a time t at a temperature T and re-etching marks the new position of the dislocation a distance x further on (Chaudhuri *et al.* 1962, Kabler 1963, Schafer 1967). The etchant produces pits at the points at which dislocation lines emerge because the rate of dissolution is faster down the high-energy core. Re-etching after moving the dislocations produces new pointed-bottom pits at the new dislocation sites. The pits at the former sites of the dislocations are widened but not deepened by re-etching so they can readily be recognized.

Chaudhuri *et al.* (1962) and Schafer (1967) found in this way that the dislocation velocity in Ge could be expressed as

$$v = B_0\tau^m \exp(-E_d/kT) = B_1\tau^m \tag{2.21}$$

where B_0 and m are constants, B_1 is constant for constant temperature and E_d is the activation energy for dislocation movement.

The outcome of dislocation velocity-stress measurements on Ge, Si, InSb and GaAs by many workers was that:

$$v \approx B_1\tau^{1.5} \tag{2.22}$$

Moreover, applying the microdynamical theory of plasticity to the data on macroscopic deformation of Ge confirmed that m = 1.5 and suggested the same approximate power law for Si, InSb and GaAs (see Section 4.4).

Values of the activation energy E_d in Equation (2.21) are: 2 eV for screw and about 2.5 eV for 60° dislocations in Si. They are about 1.4 eV for both 60° and screw dislocations in Ge, about 0.7 eV for In 60° and 0.8 eV for Sb 60° dislocations in InSb, and about 1.1 eV for As 60° and 1.3 eV for Ga 60° dislocations in GaAs. (The non-centrosymmetry of the sphalerite and wurtzite structures of many semiconducting compounds and alloys makes possible the occurrence of two forms of any dislocation that are of opposite polarity or chemical type like the In- and Sb- and Ga- and As-60° dislocations mentioned here. This will be discussed in Section 4.1.3.)

The low and strongly temperature-dependent mobility characteristic of dislocations in semiconductors (Fig. 2.26) is due to the directional character of covalent bonding and is accounted for by double-kink loop theories as will be discussed further in Chapter 4.

2.5.5 *Modification of dislocation behaviour by the electronic environment*

As we saw in Section 2.4 and shall see in more detail in Chapter 5, the dislocation core is associated with localized defect states (associated with broken bonds or the elastic strain) so the dislocation line will generally be charged negatively. The broken bonds are chemically reactive as well. Thus we can expect that there will be electrostatic interactions between the charged dislocation lines and ionized point defects and short-range chemical bonding effects may occur. These chemical reactions modify or reduce (neutralize or passivate) the electrical and luminescence effects of dislocations.

The long-range electrostatic interaction strengthens the attraction of the dislocation to its Cottrell impurity atmosphere and strengthens the pinning effect opposing dislocation motion. The charge states of the dislocation and point defects can be altered by light of appropriate photon energies. These effects are believed to be responsible for the photoplastic and electroplastic effects in the more ionic II-VI compounds and similar effects in ionic crystals (Ossipiyan *et al.* 1986, see Section 4.3). For example, applying an electric field across e.g. a ZnS crystal under compressive deformation exerts a force on the charged dislocations and can reversibly alter the stress required to keep the crystal deforming at the same rate. This is known as the electroplastic effect (Fig. 2.27). The effect is a large one. The flow stress was increased and decreased by about 15% by the field in Fig. 2.27.

It is also found that doping can increase or decrease the flow stress (hardness) of III-V semiconductors and Si. In metallic materials only alloy hardening occurs. It is thought that these effects result from changes in the core bond structure due to changes in the Fermi level which increase or decrease the mobility of the dislocation (Hirsch 1985). The effects of surface-active agents on the surface charges and band bending are believed to underlie the ill-understood chemomechanical effects on the

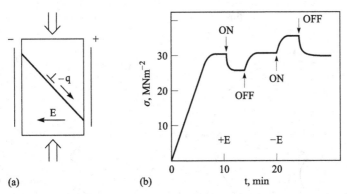

Figure 2.27 The electroplastic effect in ZnS. (a) The application of an electric field E to a II-VI crystal under constant strain rate plastic deformation in compression exerts a force on the mobile, negatively charged dislocations. (b) The resultant effect on the stress-strain curve of a 4×3×1 mm ZnS crystal deformed at a rate of 10 μm/min with an applied potential difference of 2 kV. Oppositely directed fields decrease and increase the required flow stress reversibly. (After Ossipiyan *et al.* 1986. Reprinted with permission from *Advances in Physics*, **35**, pp. 115–88; Taylor & Francis Ltd., the Journal's web site: http://www.tandf.co.uk/journals.)

deformation of semiconductors (i.e. effects on the mechanical properties of immersion in surface active liquids, see Section 4.2.3).

2.5.6 The Frank-Read source and the z-mill

There must be dislocation multiplication mechanisms. Firstly, the observation that slip steps on the surfaces of deformed crystals are optically visible (so they must be of the order of the wavelength of light, i.e. $10^{-1}\,\mu\text{m}$ in height) means that 10^3 or more dislocations move over individual atomic planes (since each produces an offset of the slip plane of the order of atomic sizes, i.e. $10^{-1}\,\text{nm}$). Secondly, during deformation, the dislocation density in the material rises from near zero to $10^6\,\text{cm}^{-2}$ or higher.

To understand the operation of the basic mechanism for the multiplication of the grown-in dislocations, we must first consider the forces on a curved dislocation in equilibrium as shown in Fig. 2.28. Dislocations have a certain elastic energy per unit length given by Equation (2.6) with $C=1$ for a screw and the same expression

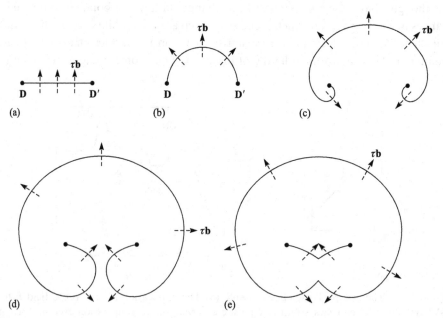

Figure 2.28 Diagram of successive positions of the dislocation in the operation of a Frank-Read source: (a) the source is a segment of dislocation pinned at both ends. (b) It is bowed out under the critical value of the stress to reach the semicircular form. (c and d) Increasing the stress beyond the critical value continues the bowing until the lines meet and (e) annihilate to reform the pinned segment and form a complete loop which continues to expand under the stress.

with $1 < C < {\sim}1.5$ for other types of dislocation, i.e. for other values of the angle between **l** and **b**. Hence we can write:

$$E_{el} = \alpha Gb^2 \qquad (2.23)$$

where $\alpha = (C/4\pi)\,\ln(r_1/r_0)$. (Frank and Read were considering dislocation sources in metals where only the elastic energy is important. Of course, in semiconductors the core energy is also significant. This results in the tendency for dislocations to align with low energy crystallographic directions in the slip plane. Consequently, in semiconductors the loops thrown off by a Frank-Read source are not circular as drawn in Fig. 2.28 but tend to be straight and crystallographic as shown in Fig. 2.29). Thus, it costs energy to lengthen dislocations so they tend to straighten. That is, there is an effective line tension, T, with units of energy per unit length, given by this expression, so:

$$T = \alpha Gb^2 \qquad (2.24)$$

Hence a segment of curved dislocation (Fig. 2.28) experiences a force acting toward the centre of curvature, tending to straighten it. It is composed of the x components of the tensions shown and this must be balanced by an applied stress τ_0. The two forces are given by $\tau_0 bdl$ to the right and $2T\sin(d/2)$ (by geometry from Fig. 2.28) $\approx Td\theta$ for small θ, to the left.

In equilibrium:

$$Td\theta = \tau_0 bdl \qquad (2.25)$$

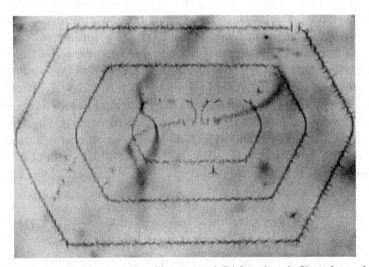

Figure 2.29 A Frank-Read source in a silicon crystal. Dislocations in Si can be made visible in a transmission infrared microscope image by precipitating Cu on them. This image shows loops of dislocation thrown out by a Frank-Read source in silicon (Dash 1957). The loops are strongly aligned in the $\langle 110 \rangle$ Peierls potential troughs in the slip plane (compare Figs. 2.24 and 2.25).

Using the fact that $d\theta = dl/R$, substituting for T from Equation (2.24) and rearranging, the stress required to bend a dislocation to a radius R is:

$$\tau_0 = T/bR = \alpha Gb/R \qquad (2.26)$$

The large numbers of dislocations involved in producing macroscopic plastic deformation can be emitted by Frank-Read sources (Frank and Read 1950) (Fig. 2.28). These are lengths, L, of dislocations pinned at both ends. They can be segments of grown-in three-dimensional networks of dislocations. Both ends are then dislocation nodes of the network. Suppose the resolved shear stress on the slip plane of the segment rises steadily as in a mechanical test. It can be seen from Equation (2.26) that the radius of curvature R to which the dislocation bends will decrease. When the stress reaches the value corresponding to $R = L/2$, the dislocation will have been bent into a semicircular arc (Fig. 2.28b). This occurs for a stress

$$\tau_{max} \approx Gb/L \quad (\text{since } \alpha \approx 1/2) \qquad (2.27)$$

At this or any higher stress the dislocation will continue to expand as shown in Fig. 2.28, throwing out dislocation loops and reconstituting the original segment. (The two dislocations approaching each other in Fig. 2.28d annihilate, so an independent dislocation loop is generated for each revolution of the Frank-Read source.) This can explain the common observation of slip lines with step heights corresponding to 10^3 or more dislocations having swept across a single slip plane. The clearest evidence of the occurrence of Frank-Read sources was found in Si by Dash (Fig. 2.29).

The z-mill is a mechanism for creating new Frank-Read sources on planes adjacent to that on which slip first takes place (Fig. 2.30). A fast-moving dislocation on the first slip plane cross-slips onto an inclined slip plane and again onto a plane parallel to the first. The segment of the dislocation lying in the new, parallel plane acts as a Frank-Read source to produce slip on the new plane. This can account for the observed spread of single slip planes into slip bands (sets of closely spaced, parallel slip lines). The minimum spacing of slip lines in bands calculated for this mechanism is stress dependent and is found to agree with the observed values in deformed crystals.

Although these multiplication mechanisms are known to operate in semiconductors, they do not play a determining role in the onset of deformation as they do in metals. This is because of the low stress dependence of dislocation velocity in these covalently bonded materials.

2.5.7 Process-induced defects

In addition to defects introduced during crystal growth like grown-in dislocations others may be process induced. During the early years some of these were introduced by rough handling, i.e. by too great pressure applied at too high a temperature in thermocompression bonding of wire leads to chips (e.g. Brantley and Harrison 1973)

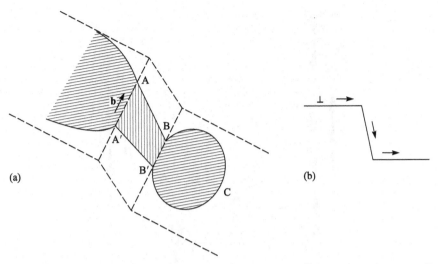

Figure 2.30   The z-mill creates new Frank-Read sources on adjacent planes by cross slip as shown in (a) oblique and (b) edge-on views.

or by too tight application of tweezers in handling slices or bare devices. In practice, this can be resolved by replacing manual handling with automation. Mechanical polishing of wafers involved dragging abrasive particles across the surface which introduced rows of dislocation loops and microcracks, as shown in Fig. 2.31. Now, however, to achieve smooth and planar surfaces, wafers are commonly chemomechanically polished. This process employs both mechanical abrasion and polish-etching to preferentially remove damage as it occurs (for more on chemomechanical effects see Section 4.2.3). A promising alternative is tribochemical polishing which removes material by friction-activated chemical dissolution (Hah and Fischer 1998, Muratov and Fischer 2000) employing a hard polishing disk without abrasives.

Large numbers of dislocations can be introduced during the in-diffusion of dopant impurities. This is likely to occur in cases where the impurity atoms differ considerably in size from those of the semiconductor and the desired dopant concentration is large. In such a case, at the diffusion front the lattice parameter can change steeply. This produces a large stress, which can exceed the elastic limit and result in plastic deformation. Such diffusion-induced misfit dislocations often occurred at the edges of emitter regions (see Fig. 2.32). Such emitter-edge dislocations are also avoided in modern processing. Later forms of defect introduced in processing are discussed in Section 4.8.2.

2.5.8 The brittle-ductile transition

At room temperature the diamond and sphalerite-structure semiconductors are so brittle that e.g. silicon slices can be separated into chips by scribing and cleaving

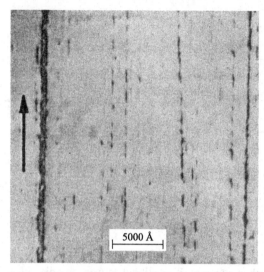

Figure 2.31 Transmission electron micrograph of rows of dislocation half loops introduced into silicon by abrasive 0.25 μm diamond particles dragged over the surface in the [110] direction arrowed. [After Stickler and Booker 1963. Reprinted with permission from *Philosophical Magazine*, **8** (1963), pp. 859–76; Taylor & Francis Ltd., the Journal's web site: http://www.tandf.co.uk/journals.]

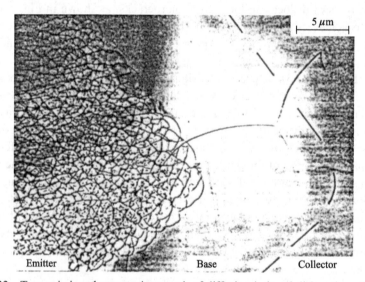

Figure 2.32 Transmission electron micrograph of diffusion-induced dislocations throughout the emitter region of an *npn* transistor. One dislocation runs from the emitter to the collector forming an emitter-collector pipe. (After Heydenreich 1985. Reprinted from *Crystal Growth of Electronic Materials*, Electron microscopical characterization of electronic materials, pages 325–41. Copyright 1985, with permission from Elsevier.)

processes much like the 'cutting' of panes of window glass. Early studies of the indentation hardness of these materials showed that they began to be ductile (i.e. deform plastically instead of undergoing brittle fracture under rapid stressing) at temperatures above $\sim 0.5\,T_m$ in degrees Kelvin (Churchman *et al.* 1956, see Section 2.3). With increasing temperature they become increasingly plastic until, above about 2/3 of their melting points, they are as ductile as the f.c.c. metals (Haasen and Alexander 1968). The gradual transition is due to the large activation energies for dislocation motion and the consequent slow increase in capability for fast dislocation motion and macroscopic deformation at practical rates under stress. Churchman *et al.* also reported evidence suggesting that deformation twinning was involved in plastic indentation.

Some plastic deformation can occur under indentations even at temperatures at which these materials are, macroscopically, totally brittle (see e.g. Hill and Rowcliffe 1974). This is because the compressive pressure below the indentation suppresses cracking. The mechanisms of plastic flow under these conditions involve twinning and perhaps phase transformations especially at low temperatures (Pirouz 1987, Pirouz and Ning 1995 and Pirouz *et al.* 1999). These plus 'dislocation cracking' (formation of microcracks which rejoin but with offsets of the walls, constituting dislocations – see Allen 1959) are most probably also involved in the 'mechanical damage' produced by mishandling, sawing and mechanical polishing (Stickler and Booker 1963, see Fig. 2.31), and by improper thermocompression bonding (Brantley and Harrison 1973), etc. Annealing of mechanically damaged surfaces allows changes in the defect configurations, under the influence of the internal stresses, since the dislocations become more mobile at the high annealing temperatures. Deformation at higher temperatures occurs at lower stresses for the same strain rate.

Practical solutions to the problems of damage are well established. Mechanical polishing must be followed by chemical polishing with a reagent that dissolves the surface layers away without spoiling the surface smoothness and flatness, until all the damage is removed. The damage extends to a depth that is a few times the diameter of the abrasive particles used in the final mechanical polishing process. Si is therefore often 'chemomechanically' polished, i.e. with abrasive in a chemical polish to prevent damage building up. ('Syton' is a well-known brand of chemomechanical polish for silicon.)

Quenching strains are avoided in crystal growing and slice processing, by keeping the rates of temperature change below the deformation threshold level.

References

Allen, J. W. (1959). On a new mode of deformation in indium antimonide. *Philosophical Magazine*, **4**, 1046–54.

Anderson, R. L. (1960). Germanium-gallium arsenide heterojuctions. *IBM Journal of Research and Development*, **4**, 283–7.

Anderson, R. L. (1962). Experiments on Ge-GaAs heterojunctions. *Solid-State Electronics*, **5**, 341–51.

Bauer, E. and Poppa, H. (1972). Recent advances in epitaxy. *Thin Solid Films*, **12**, 167–85.

Bauer, E. and van der Merwe, J. H. (1986). Structure and growth of crystalline superlattices: from monolayer to superlattice. *Physical Review*, **B33**, 3657–71.

Bilby, B. A. (1950). Static models of dislocations. *Journal of the Institute of Metals*, **76**, 613–27.

Brantley, W. A. and Harrison, D. A. (1973). Localized plastic deformation of GaP and GaAs generated by thermocompression bonding. *Journal of the Electrochemical Society*, **120**, 1281–4.

Chaudhuri, A. R., Patel, J. R. and Rubin, L. G. (1962). Velocities and densities of dislocations in germanium and other semiconductor crystals. *Journal of Applied Physics*, **33**, 2736.

Churchman, A. T., Geach, G. A. and Winton, J. (1956). Deformation twinning in materials of the A4 (diamond) structure. *Proceedings of the Royal Society*, **A238**, 194–203.

Cottrell, A. H. (1953). *Dislocations and Plastic Flow in Crystals*. London: Oxford University Press.

Cottrell, A. H. and Bilby, B. A. (1949). Dislocation theory of yielding and strain ageing. *Proceedings of the Physical Society*, **A62**, 49–62.

Dash, W. C. (1957). The observation of dislocations in silicon. In *Dislocations and Mechanical Properties of Crystals*, eds. J. C. Fisher, W. G. Johnston, R. Thomson and T. Vreeland, Jr (New York: Wiley), pp. 57–68.

Dash, W. C. (1958). The growth of silicon crystals free from dislocations. In *Growth and Perfection of Crystals*, eds. R. H. Doremus, B. W. Roberts and D. Turnbull (New York: Wiley), pp. 361–85.

Dash, W. C. (1960). Gold-induced climb of dislocations in silicon. *Journal of Applied Physics*, **31**, 2275–83.

Davies, G. J. and Williams, R. H. (eds.) (1994). *Semiconductor Growth, Surfaces and Interfaces*. London: Chapman & Hall.

Frank, F. C. (1949). Answer by Frank in discussion of a paper by N. F. Mott that introduced what became known as Frank's rule. *Physica*, **15**, 131–3.

Frank, F. C. and van der Merwe, J. H. (1949a). One-dimensional dislocations. I Static theory. *Proceedings of the Royal Society*, **A198**, 205–16.

Frank, F. C. and van der Merwe, J. H. (1949b). One-dimensional dislocations. II Misfitting monolayers and oriented overgrowth. *Proceedings of the Royal Society*, **A198**, 216–25.

Frank, F. C. and Read, W. T. (1950). Multiplication processes for slow moving dislocations. *Physical Review*, **79**, 722–3.

Friedel, G. (1926). *Lecons de Cristallographie* (Paris: Berger Levrault), pp. 250–2.

Gilman, J. J. (1969). *Micromechanics of Flow in Solids*. New York: McGraw-Hill.

Haasen, P. and Alexander, H. (1968). Dislocations and plastic flow in the diamond structure. *Solis State Physics*, **22**, 27–158.

Hah, S. R. and Fischer, T. E. (1998). Tribochemical polishing of silicon nitride. *Journal of the Electrochemical Society*, **145**, 1708–14.

Heydenreich, J. (1985). Electron microscopical characterization of electronic materials. In *Crystal Growth of Electronic Materials*, ed. E. Kaldis (Amsterdam: North-Holland), pp. 325–41.

Hill, M. J. and Rowcliffe, D. J. (1974). Deformation of silicon at low temperatures. *Journal of Materials Science*, **9**, 1569–76.

Hirsch, P. B. (1985). Dislocations in semiconductors. *Materials Science and Technology*, **1**, 666–77.

Hirth, J. P. and Lothe, J. (1968). *Theory of Dislocations*. New York: McGraw-Hill.

Holt, D. B. (1966). Misfit dislocations in semiconductors. *Journal of Physics and Chemistry of Solids*, **27**, 1053–67.

Holt, D. B. (1984). Polarity reversal and symmetry in semiconducting compounds with the sphalerite and wurtzite structures. *Journal of Materials Science*, **19**, 439–46.

Hornstra, J. (1958). Dislocations in the diamond lattice. *Phys. Chem. Solids*, **5**, 129–41.

Hornstra, J. (1959). Models of grain boundaries in the diamond lattice I. Tilt about ⟨110⟩. *Physica*, **25**, 409–22.

Hornstra, J. (1960). Models of grain boundaries in the diamond lattice. II. Tilt about ⟨001⟩ and theory. *Physica*, **26**, 198–208.

Hull, D. and Bacon, D. J. (1984). *Introduction to Dislocations*. 3rd edn. Oxford: Pergamon.

Kabler, M. N. (1963). Dislocation mobility in germanium. *Physical Review*, **131**, 54.

Mahajan, S. and Sree Harsha, K. S. (1999). *Principles of Growth and Processing of Semiconductors*. New York: McGraw-Hill.

Matthews, J. W. (1975). Coherent interfaces and misfit dislocations. In *Epitaxial Growth Part B*, ed. J. W. Matthews (New York: Academic Press), pp. 559–609.

Matthews, J. W. (1979). Misfit dislocations. In *Dislocations in Solids*, Vol. **2**, ed. F. R. N. Nabarro (Amsterdam: North-Holland), pp. 461–545.

Muratov, V. A. and Fischer, T. E. (2000). Tribochemical polishing. *Annual Review of Materials Science*, **30**, 27–51.

Nabarro, F. R. N., Holt, D. B. and Basinski, Z. S. (1964). Plasticity of pure single crystals. *Advances in Physics*, **13**, 193.

Nolder, R. and Cadoff, I. (1965). Heteroepitaxial silicon–aluminium oxide interface. Part II – orientation relations of single-crystal silicon on alpha aluminium oxide. *Transactions of the Metallurgical Society of AIME*, **233**, 549–56.

Oldham, W. G. and Milnes, A. G. (1964). Interface states in semiconductor heterojunctions. *Solis State Electronics*, **7**, 153–65.

Ossipiyan, Yu. A., Petrenko, V. F., Zaretskii, A. V. and Whitworth, R. W. (1986). Properties of II-VI semiconductors associated with moving dislocations. *Advances in Physics*, **35**, 115–88.

Pashley, D. W. (1956). The study of epitaxy in thin surface films. *Advances in Physics*, **5**, 173–240 (plus plates 1 to 10).

Pashley, D. W. (1991). The epitaxy of metals. In *Processing of Metals and Alloys*, ed. R. W. Cahn. Materials Science and Technology: A Comprehensive Treatment, Vol. **15** (Weinheim: VCH), pp. 289–328.

Peierls, R. E. (1940). The size of a dislocation. *Proceedings of the Physical Society of London*, **52**, 34–7.

Pirouz, P. (1987). Deformation mode in silicon, slip or twinning? *Scripta Metallurgica*, **21**, 1463–8.

Pirouz, P. and Ning, X. J. (1995). Partial dislocations in semiconductors: structure, properties and their role in strain relaxation. In *Microscopy of Semiconducting Materials* 1995. Conf. Series No. 146 (Bristol: Inst. Phys.), pp. 69–77.

Pirouz, P., Samant, A. V., Hong, M. H., Moulin, A. and Kubin, L. P. (1999). On deformation and fracture of semiconductors. In *Microscopy of Semiconducting Materials* 1999. Conf. Series No. 164 (Bristol: Inst. Phys.), pp. 61–6.

Read, W. T. (1953). *Dislocations in Crystals*. New York: McGraw-Hill.

Royer, L. (1928). Recherches experimentales sur l'epitaxie ou orientation mutuele de cristaux d'especes differentes. *Bulletin of the French Society of Mineralogy*, **51**, 7–159.

Schafer, S. (1967). Messung von versetzungsgeschwindigkeiten in germanium. *Physica Status Solidi*, **19**, 297.

Seeger, A. (1957). The origin of the small angle scattering of x-rays in plastically deformed metals. *Acta Metallurgica*, **5**, 24–8.

Sharma, B. L. and Purohit, R. K. (1974). *Semiconductor Heterojunctions*. Oxford: Pergamon Press.

Shockley, W. (1953). Dislocations and edge states in the diamond crystal structure. *Physical Review*, **91**, 228.

Shockley, W. and Read, W. T. (1949). Quantitative predictions from dislocation models of crystal grain boundaries. *Physical Review*, **75**, 692.

Shockley, W. and Read, W. T. (1950). Dislocation models of crystal grain boundaries. *Physical Review*, **78**, 275–89.

Stickler, R. and Booker, G. R. (1963). Surface damage on abraded silicon specimens. *Philosophical Magazine*, **8**, 859–76.

Thompson, N. (1953). Dislocation nodes in face-centred cubic lattices. *Proceedings of the Physical Society of London*, **B66**, 481–92.

Vogel, F. L., Pfann, W. G., Corey, H. E. and Thomas, E. E. (1953). Observations of dislocations in lineage boundaries in germanium. *Physical Review*, **90**, 489–90.

Wessel, K. and Alexander, H. (1977). On the mobility of partial dislocations. *Philosophical Magazine*, **35**, 1523–36.

Symbols

Latin symbol	Meaning of the symbol
a_1, a_2, and a_3	are the fundamental lattice vectors (along the crystal axes)
b	the Burgers vector of a dislocation
b_1, b_2, and b_3	the Burgers vectors of a unit dislocation and of the partial dislocations into which the unit dislocation can split
C	is the constant in Equation (2.6): $E_{\text{total}} = E_{\text{core}} + \frac{CGb^2}{4\pi} \ln\left\{\frac{r_1}{r_0}\right\}$ The second term expresses the elastic strain energy of a dislocation. $C \approx 1$ and varies with the type of dislocation edge to screw.
D	the spacing of parallel dislocations in a small-angle grain boundary
E	the elastic energy of a dislocation
E_{total}	the total (core + elastic) energy of a dislocation
E_{core}	the inelastic core energy of a dislocation

Latin symbol	**Meaning of the symbol**
E_d	the activation energy for the motion of a dislocation in a semiconductor
F	the force acting on a dislocation in a crystal under stress
G	the shear modulus
h_c	the critical thickness for an epitaxial film at which it becomes energetically favourable to introduce misfit dislocations
l	arbitrary positive direction along a dislocation line
n_{il}	integers (for $l = 1$ to 3)
r_0	the cut-off radius from the dislocation, inside which the material does not behave elastically (obey Hooke's Law)
r_1	outer limit to which the elastic strain field of a dislocation extends
R	radius to which a dislocation is bent under stress
t_i	translation group (of translation vectors) of a lattice
T	line tension of a dislocation (= elastic energy per unit length)
$U_P(x)$	the Peierls potential is the varying energy of a dislocation as it moves through the crystal
v	the velocity of a dislocation

Greek symbol	**Meaning of the symbol**
γ	elastic shear strain
θ	the angle of misorientation across a small-angle grain boundary
v	Poisson's ratio
ρ	the density of dislocations (no./unit area)
τ	elastic shear stress
τ_P	the Peierls(-Nabarro) stress is the intrinsic stress required to move a dislocation through perfect crystalline material.
τ_{ext}	the stress on a dislocation due to the external forces (loading) of the crystal
τ_{int}	the stress on a dislocation due to other defects and to quenched-in internal stresses

3

Characterization of extended defects in semiconductors

3.1 Introduction

This chapter outlines the principles, advantages and limitations of the methods in use for the characterization of extended defects and should enable the reader to appreciate the experimental results presented. For additional accounts see Brundle *et al.* 1992, Yacobi *et al.* 1994, Schroder 1998, and Runyan and Shaffner 1998.

Characterization methods can be classified as (i) either surface or bulk techniques, as (ii) either destructive or non-destructive methods, and as (iii) either requiring the application of contacts or not. We have excluded the free surface from consideration on the grounds that, like point defects, its study constitutes a large specialized field already covered by many publications. We therefore also omit surface microscopy and analysis techniques here. Generally, non-destructive techniques are preferred as are contactless ones. However, in practice neither is a very important factor as generally one or a few specimens can be sacrificed for destructive examination and contacts can usually be applied.

Characterization techniques are essential for failure analysis and quality control of semiconductor materials and devices. Often failure modes are associated with manufacturing process-induced defects or with defect-dependent device degradation in service.

Electrical measurements for the analysis of a wide range of semiconductor transport properties such as, for example, *resistivity (conductivity)*, *Hall effect* and *capacitance-voltage* measurements are made on whole bulk specimens and devices. The net influence on these properties of all the defects present then appears in the results. One form of capacitance-voltage measurements, however, has been developed for use in the SEM as a microcharacterization technique, that is, one that determines a physical property with a high spatial resolution so the physics of selected defects can be investigated. This is scanning deep level transient spectroscopy (SDLTS). It enables the traps associated with the particular resolved defects to be analysed. Similarly the EBIC (electron beam induced current) and CL (cathodoluminescence) modes of the SEM enable the dislocation properties responsible for the contrast (visibility) of the defects to be determined. Such microcharacterization techniques combine the advantages of microscopy with those of physical property measurements and are particularly important for semiconductors.

3.2 Microscopy techniques

In semiconductor technology, it is essential to control the undesirable defect densities present in the material. The main objectives in this case are (i) to detect defects and measure their densities, (ii) to identify them, and (iii) to establish their origin so they may be eliminated. For this the various types of microscopy are crucial. It is important to have at least an elementary understanding of the physical principles on which any form of microscopy operates and the properties of defects to which it is sensitive, in order to critically appreciate the evidence that it provides. The accounts of the techniques given here are intended to give such an introduction to the essentials.

3.3 Light microscopy

The original technique for producing enlarged images of objects is that of light microscopy (LM). (The term 'optical microscopy' cannot be recommended since electron microscopes employ electron optics and ion microscopes employ ion optics.) Standard textbooks show that LM spatial resolution is limited by diffraction. Due to this, the image of a bright object point is a diffraction pattern consisting of a bright central spot (Airey's disk) surrounded by diffraction rings. Rayleigh suggested the standard definition. This is that two points in the object are just resolved when, in the image, the central spot of one falls on the first dark ring of the other diffraction pattern. This definition leads to the expression that the limit of resolution (smallest resolvable separation of object points) is given by:

$$\delta = \frac{\lambda}{2NA} \tag{3.1}$$

where λ is the wavelength of the light and the numerical aperture, $NA = \mu \sin \theta$. Here μ is the refractive index and θ is the semi-angular aperture of the objective lens of the microscope. The values of NA for the best objective lenses are such that the limit of resolution for LMs, using visible light (λ from 350 to 750 nm, say 550 nm, the peak eye-response value) is about 0.2 μm. No detail finer than this can be laterally resolved by a light microscope. Actually the value given by Equation (3.1) is the minimum diffraction limited value. In addition, any lens aberrations will increase this value, i.e. will degrade the performance of the microscope. The level of development of lens systems for light is such that lens aberrations are not significant. (Recently a number of techniques have been developed that can reduce this limit somewhat.) LMs have the advantages of economic prices and ease of use.

Moreover they have a variety of possible enhancements such as oblique illumination and Nomarski interference contrast that make features of interest more readily visible. For more detail, see e.g. Hayes (1984) and Chapter 1 of Goodhew and Humphreys (1988).

3.3.1 Etching, decoration and stress birefringence

Three means by which defects are made visible in light microscopy are etching, decoration and stress birefringence. *Etching* is the slow, controlled dissolution of a crystal in, generally, an acid solution, so pits are produced where dislocations and grain boundaries reach the surface. To develop a reliable dislocation etchant is a lengthy process but recipes for hundreds of etching solutions for a wide range of semiconductors have been published [Johnston 1962, Faust 1962, Holmes 1962, Amelinckx 1964 (pp. 15–50), Bogenschutz 1967 and Runyan and Shaffner 1998 Chapter 2]. Some of the best etches for the more important semiconductors are listed in Table 3.1.

Historically, etching was the first technique developed for rendering dislocations visible in the earliest single crystal Ge. Here the dislocation densities were, for the first time, low enough so the defects could be resolved by LMs. The earliest quantitative confirmation of any prediction of dislocation theory came through observations on low angle grain boundaries in Ge. By measuring the angle of misorientation across the boundary, θ, and the etch pit spacing (Fig. 3.1), D, and knowing the magnitude of the Burgers vector, **b**, it was shown that Equation (2.9) was obeyed, thus both confirming the model of Shockley and Read (Section 2.3.10) and the one-to-one correspondence of etch pits and dislocations.

Decoration is the deposition of many small precipitates along dislocations, stacking faults and related defects during suitable heat treatments of heavily metal-doped semiconductors. Metallic precipitates will usually be opaque at wavelengths to which a semiconducting host material is transparent. Then not only can precipitates be seen in dark contrast due to absorption but also the defects that they decorate.

This technique was most effectively used in silicon by Dash (1957, 1958) (Fig. 2.29). Silicon is transparent in the near infrared so transmission infrared (IR) microscopy was carried out using an ordinary light microscope plus an image converter for observations and infrared film for recording results. Accidental decoration of dislocations is frequently observed in as-grown crystals of transparent semiconducting compounds, e.g. ZnTe (Lynch *et al.* 1963) and CdS (Drebeen 1965).

Stress Birefringence in semiconductors depends on the fact that most have cubic crystal structures. The dielectric properties of cubic materials are isotropic as are all their second-rank 'matter' tensor properties (Kelly *et al.* 2000). This means that unstressed cubic system crystals placed between crossed polarizers and analysers show complete extinction, i.e. they transmit no light. Any internal stresses render

Table 3.1. *Some defect etches*

Etchant	Material	Composition, etc.	Reference
CP-4[1]	Ge	5 HNO_3: 3 HF: 3 CH_3COOH: 0.2 Br_2	Camp, P.R. (1955) *J. Electrochem. Soc.*, **102**, 586–93
Sirtl etch[2]	Si	Solution of CrO_3 in water mixed with HF	Sirtl, E. and Adler, A. (1961). *Zeitschrift fur metallkunde*, **52**, 529–31.
Secco etch[3]	Si	Solution of $K_2Cr_2O_7$ in water mixed with HF	Secco d'Aragona, F. (1972). *J. Electrochem. Soc.*, **119**, 948–51
Wright etch[4]	Si	60 ml conc. HF (49%), 30 ml conc. HNO_3 (69%), 30 ml 5M $CrCO_3$ (1 g $CrCO_3$/2 ml H_2O), 2 g $Cu(NO_3)_2 \cdot 3\ H_2O$ (reagent grade), 60 ml conc. (glacial) acetic acid and 60 ml deionized water	Wright Jenkins, M. (1977) *J. Electrochem. Soc.*, **124**, 757–62
AB etch[5]	GaAs	A solution of 2 ml water, 8 mg $AgNO_3$, 1 g CrO_3 and 1 ml HF	Abrahams, M. S. and Buiocchi, C. J. (1965). *J. Appl. Phys.*, **36** 2855–63

[1] This was first reported in a US Patent by R. D. Heidenreich in 1952. It was the first and is one of the best dislocation etchants.

[2] The Sirtl etch is first CrO_3 based etchant. It works well only on {111} surfaces.

[3] This etchant works for all surface orientations and reveals dislocations, etc.

[4] In mixing the solution, the best results are obtained by first dissolving the $Cu(NO_3)_2$ in the given amount of water, otherwise the order of mixing is not critical. This etch works for both {001} and {111} surfaces *on p*- and *n*-type Si. It reveals dislocations, stacking faults, swirl defects and striations. The etch rate is about 1 µm/min at room temperature.

[5] The components are added in the order listed. If the order is changed, difficulty is experienced in dissolving all the $AgNO_3$. During etching the solution is continuously stirred. This etch reveals dislocations whether normal, inclined or parallel to the surface. (It is the only etch that does the latter in any material.) It works for the {111}Ga, $\{\overline{111}\}$As, {001} and {110} surfaces. It also produces pits at the precipitates along decorated dislocations. It removes about 2.5 µm/min at room temperature.

them optically active so they transmit some light in the region of the elastic strain field. Thus the elastic stress fields around dislocations and other defects are made visible in LM. Stress birefringence is sometimes referred to as the elasto-optic effect, piezobirefringence or photoelasticity.

Stress birefringence was introduced in half a dozen classical studies of Si in the late 1950s and early 1960s (for a review see Amelinckx 1964 pp. 105–8). Later theoretical work (Fathers and Tanner 1973a and b, Tanner and Fathers 1974, Jenkins and Hren 1976, Booyens and Basson 1980a,b) made it possible to calculate defect stress

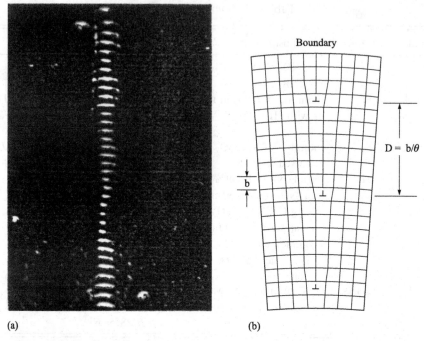

(a)   (b)

Figure 3.1   (a) Reflection light microscope image of etch pits delineating the dislocations in a low angle grain boundary in Ge (after Vogel *et al.* 1953) and (b) the Shockley-Read model for such a boundary (cf. Figure 2.10). [Reprinted with permission from *Physical Review*, **90**, pp. 489–90, 1953 (http://publish.aps.org/linkfaq.html). Copyright 1953 by the American Physical Society.]

birefringence contrast for anisotropic as well as isotropic materials and for any orientation of the dislocation line. It was found that dislocations could readily be observed by stress birefringence in GaP (Matthews *et al.* 1977, Hilgarth 1978, Iqbal 1980) and InP (Fig. 3.2) (Elliott and Regnault 1981, Elliott *et al.* 1982, Stirland *et al.* 1983). In addition, the availability of defect contrast theory, including computer-generated contrast plots (Jenkins and Hren 1976) makes possible the complete identification of dislocations (Burgers vector, line direction and slip plane) and the quantitative analysis of internal stress fields due, e.g. to quenching strains in liquid-encapsulated Czochralski (LEC) GaP crystals (Kotake *et al.* 1980) or to misfit at heterojunctions (Ahearn *et al.* 1976, Reinhart and Logan 1973, Booyens and Basson 1980a).

Given the availability of reliable contrast theory, the advantages of stress-birefringence observations using light microscopes are that the method is applicable to large crystals with a low defect density, is non-destructive and it is cheaper and quicker than the alternative, x-ray topography (Section 3.9). This is true even for infrared transparent crystals like InP using modern infrared image converters and film or TV cameras and monitors. The drawback of the technique is that it can only

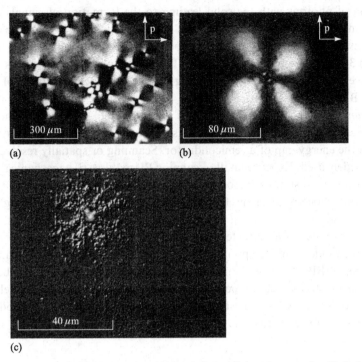

(a)

(b)

(c)

Figure 3.2   (a) Stress Birefringence transmission infrared light micrograph of InP showing the stress fields surrounding several inclusions, and (b) around an inclusion surrounded by many dislocations. The directions p are those of the polarizer and analyser. This type of defect was found in profusion in LEC InP. (c) The form of the dislocation etch pit pattern around these defects as shown here led to them being called grappes (French for a bunch of grapes). These defects will be discussed in Section 4.8.1. [After Elliott and Regnault 1981; in *Microscopy of Semiconducting Materials* 1981. Conf. Series No. 60, eds. A. G. Cullis and D. C. Joy (Bristol: Institute of Physics), pp. 365–70.]

be applied to materials that are transparent to visible, near infrared or ultraviolet radiation. (Fortunately, this includes all the materials of interest for optoelectronics, optical communications and integrated optics.) Moreover, stress birefringence is not effective for materials in which the elasto-optic coefficients, the refractive index and the Burgers vector are small (Matthews *et al.* 1977).

These LM techniques are now somewhat outdated and neglected but they remain fast, inexpensive and relatively simple.

3.4 Scanning laser beam microscopy techniques

The advantage of scanning is that a signal is generated in a form suitable for data and image processing. Beams other than electrons can be scanned over specimens to generate the signals and the resultant techniques are collectively referred to as scanned image microscopy (Ash 1980). Scanning laser techniques are based on the interaction of photons with the solid leading to the formation of signals especially photoluminescence (PL) and laser beam induced current (LBIC) analogous to cathodoluminescence (CL) and electron beam induced current (EBIC) (discussed in Section 3.7.2), respectively, in the SEM. The ability to analyse specimens in air and the absence of charging in non-conducting samples are the advantages of scanning PL and LBIC over the SEM techniques,

Photoluminescence (PL) spectroscopy is a widely used, sensitive, contactless and non-destructive technique for the analysis of dopant and impurity energy levels present in the energy gap of a semiconductor. Scanning or spatially resolved PL has been less widely used. Experimentally, scanning PL requires (i) a mechanical scanning specimen stage, (ii) a stationary, focused laser beam (often via a LM) with appropriate emission, (iii) a spectrometer (preferably a double monochromator for high spectral resolution), and (iv) as detector a photomultiplier or a solid-state detector with suitable wavelength response characteristics. Photomultiplier tubes offer good sensitivity in the visible range, while Ge photodiodes are used in the near infrared range.

LBIC, like EBIC (discussed in Section 3.7.2), reveals the electrical barriers in devices (*p-n* junctions, heterojunctions and Schottky barriers) and any electrically active defects that may be present. It can have a performance just as good as SEM EBIC (De Vittorio *et al.* 1997, 1999).

3.4.1 Quenched infrared beam induced current (Q-IRBIC)

One important advantage of LBIC is that it can readily be combined with flood illumination of another wavelength. Cavallini and her group have shown infrared beam induced current (IRBIC) and especially quenched IRBIC (QIRBIC) to be very promising techniques for the study of dislocations in Si (Castaldini *et al.* 1987, Castaldini and Cavallini 1989, and Cavallini and Castaldini 1991). In this technique, the sample is scanned from above by an IR beam while it is illuminated all over (from below) by infrared light of a wavelength that can be varied. The infrared flood illumination is of an energy corresponding to transitions in the forbidden energy gap of the sample (Si). As the infrared flood wavelength is varied, absorption occurs when the photon energy equals that for a transition between a dislocation energy level and either the conduction or valence bands or between two dislocation levels. The effect on the electrical activity (LBIC contrast) of any defect may be observed. This provides direct information on the energy levels and bands associated with dislocations in semiconductors.

3.5 Electron microscopy

The three forms of electron microscope are, in the order in which they were developed to commercial success, the transmission electron microscope (TEM), the scanning electron microscope (SEM) and the scanning transmission electron microscope (STEM). They have become the most powerful tools available for observing and determining the micro- and nano-structures of materials and in particular of semiconducting materials and devices.

3.5.1 Types of instrument and their advantages

Transmission electron microscopes (TEMs) were developed first and are closely analogous to transmission light microscopes. Both use a series of lenses to produce real images of the object. Here 'image' is used in the technical meaning of the term in geometrical optics. The limit of resolution of TEMs therefore, like that of light microscopes, is given by Equation (3.1), $\delta = \lambda/2NA$. The wavelengths, λ, of electrons accelerated through readily available potential drops are much less than those of visible light. The visible spectrum runs from about 350 to 750 nm. For 100 kV electrons, $\lambda = 3.7$ pm. Thus a reduction in the limit of resolution of a factor of 10^5 might be expected from Equation (3.1) on going from light to 100 kV electron 'illumination'. This provided the motive for the development of the TEM in the 1930s, for which Ernst Ruska belatedly received the Nobel Prize for Physics in 1986 (he shared it with Gerd Binnig and Heinrich Rohrer who developed the scanning tunnelling microscope, see Section 3.10.1). The whole of this increase in resolution has not been attained, however, because the aberrations, particularly the chromatic and spherical aberrations, of electromagnetic lenses still severely limit performance. Small angular apertures, and so small NA values, have to be used. These are 10^{-3} or 10^{-2} radians for electromagnetic lenses as against about a radian for light lenses of the highest resolving powers. However, by about 1980, technical advances reduced the limit of resolution of the best available TEMs to about 2 Å. This made atomic resolution just possible, that is, it can be attained but only with difficulty, by the best instruments and most skilled microscopists, examining the best specimens under optimum conditions.

Electron beams of variable energy can readily be finely focused and scanned by the use of electron guns, electromagnetic lenses and scan coils. This makes possible a new type of instrument, the scanning electron microscope (SEM) that does not form real optical images but (enlarged) scanned video pictures. It offers much new information about materials because electron irradiation of a solid excites several types of signal (Fig. 3.3) providing the basis for the modes of the SEM with resolutions from about 2 nm to about 1 μm depending on the signal/mode.

These signals provide valuable, often complementary, information on the material. The backscattered electrons are emitted with energies close to those of the primary electrons, i.e. with little or no energy loss. The secondary electrons are those emitted with low energies (< 50 eV). The energy dissipation by the primary electrons also results (i) in the emission of characteristic x-ray photons, and (ii) in the generation of

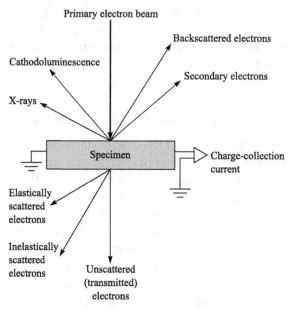

Figure 3.3 Major types of signals produced as a result of the interaction of the primary electron beam with a semiconductor.

electron-hole pairs that may lead to the generation of cathodoluminescence (beam induced light emission) or to charge-collection signals in semiconductor devices. In thin specimens, electrons may be transmitted undeviated or scattered elastically (with no energy loss) or inelastically (with energy loss). The inelastically scattered electrons can be employed for quantitative analysis in *Electron Energy Loss Spectroscopy* (EELS). *Transmitted electrons* (for samples of the order of 100 nm thick) are used in transmission electron microscopes (TEMs) and scanning transmission electron microscopes (STEMs).

Both SEMs and TEMs can be configured as multi-mode instruments to provide complementary information on the structure, composition and physical properties of solid-state materials and semiconductor devices. By using a field emission electron gun, adequate beam currents can be focused into much smaller spots than in conventional SEMs and TEMs (with thermionic emission guns). Hence FEGSTEMs (field emission gun scanning transmission electron microscopes) can also give atomic resolution but with simple, intuitively interpretable contrast, as we shall see below.

3.6 Transmission electron microscopy

Transmission electron microscopy (TEM) employs electrons transmitted through thin samples, $\sim 1\,\mu m$ to 100 nm or less in thickness, depending on the TEM beam energy, the material and the problem under study. The transmitted beams of electrons are focused to an image by the objective, intermediate and projector electromagnetic electron-optical lenses, shown schematically in Fig. 3.4, in a way completely analogous to the (transmission) light microscope. Two condenser lenses at the top of the electron optical column control the electron beam 'illumination' so it is as coherent and parallel as possible. The electron beam energy is typically between 100 keV and 1 or more MeV. The specimen is situated inside the pole piece of the objective lens where space is limited but stages are available to provide tilting, rotation, heating, cooling, etc. In the transmitted electron images the crystal structure and defects in solid-state materials can be brought into contrast, i.e. made visible.

The objective, intermediate, and projector lenses can focus either the magnified image or the diffraction pattern of the specimen on the fluorescent screen or digital camera. This is the second great advantage of the TEM. (High spatial resolution was the first.) This is an advantage because the diffraction pattern makes it possible to set up the diffraction conditions required to produce one of several well-defined types of contrast and allows the interpretation of the detailed contrast in the image.

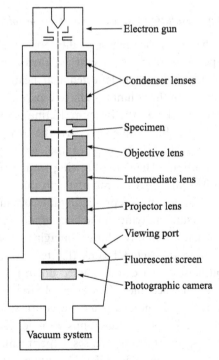

Figure 3.4 The basic components of a transmission electron microscope are mounted in an evacuated electron optical column.

Inserting the *selected area diffraction* aperture in the image plane of the objective lens gives the diffraction pattern of the selected area (SADP) from which the crystalline structure and exact orientation can be determined. Diffraction patterns can also be obtained with a high spatial resolution by means of the *convergent-beam diffraction* technique. For detailed accounts of TEM and electron diffraction see Hirsch *et al.* (1977) (the classic monograph), Loretto and Smallman (1975), Joy *et al.* (1986), Goodhew and Humphreys (1988) and Spence (1988).

There are three contrast mechanisms in TEM images. In order of decreasing conceptual simplicity these are mass–thickness contrast, lattice imaging or atomic contrast, often referred to as high-resolution TEM (HRTEM), and diffraction contrast.

Mass–thickness contrast arises simply from the fact that thicker or denser regions absorb more electrons than do thinner or less dense regions. The brightness (number of electrons/second) of image points varies accordingly. This contrast is not generally of interest. In fact one aim in the thinning specimens for TEM examination is to obtain a sufficiently uniform or slowly and smoothly varying thickness so mass-thickness contrast does not obscure, e.g. diffraction contrast.

3.6.1 High resolution transmission electron microscopy

In early lattice imaging the use of the highest-quality objective lenses allowed the use of sufficiently wide apertures to pass several diffracted beams as well as the undeviated beam. Suppose that the undeflected 000 beam and the $h_1k_1l_1$ Bragg reflected beam are passed down the column of the microscope. These beams interfere to produce a set of fringes parallel to the Bragg reflecting planes $h_1k_1l_1$ in the image. If the magnification of the final image is M, the fringes have the spacing $M \times d(h_1k_1l_1)$ where $d(h_1k_1l_1)$ is the spacing of the $h_1k_1l_1$ planes in the crystal. This is 'direct lattice resolution'. The fringes arising from different pairs of beams, e.g. 000 plus $h_1k_1l_1$ and 000 plus $h_2k_2l_2$ combine to produce contrast in the form of regular arrays of dark and bright spots representing the columns of atoms parallel to the electron beam in the crystal and lying in the lines of intersection of the $h_1k_1l_1$ and $h_2k_2l_2$ planes. This is termed 'atomic resolution' imaging or simply high-resolution TEM (HRTEM). The more diffracted beams (limited by the aberrations of the electromagnetic electron lenses) that can be passed down the column to contribute to the image, the sharper this becomes. However, obtaining reliably interpretable atomic resolution in this way is not easy as the contrast in the image changes radically with the smallest changes of focus. Thus, atomic resolution TEM requires unusually thin specimens, special objective lenses, experimental skill and theoretical assistance in interpretation. Nevertheless HRTEM makes it possible to image the atomic core structures of defects as well as quantum confined materials and device structures (Spence 1988). Atomic resolution of a more easily interpreted kind (called z contrast) is available in STEM as we shall see below.

3.6.2 Diffraction contrast transmission electron microscopy

This, the most useful method for the study of defects, employs a form of contrast with no analogue in light microscopy. In this technique, a small objective aperture is placed in the back focal plane of the objective lens, i.e. in the plane of the diffraction pattern. It is positioned so only one beam is passed. The TEM is thus operated below the minimum condition (two beams to interfere) for resolution of detail in the image according to the Abbe theory (Hayes 1984) leading to Equation (3.1). For diffraction contrast imaging, the specimen must be carefully oriented so only two beams are bright (strongly excited) in the electron diffraction pattern, i.e. two beam conditions are obtained. This is assumed in the interpretive diffraction-contrast theory. One beam only, either the undeviated 000 or the *hkl* diffracted beam, is passed through the objective aperture to be focused to form the image. The undeviated beam forms a bright field image, i.e. most points are bright with only localized regions dark. The diffracted beam gives a dark field image with only the localized, *hkl* Bragg reflecting regions bright. These 'images' are maps or displays of the varying intensity of the particular beam, 000 in the 'bright-field' pictures or *hkl* in the 'dark-field' pictures, over the exit surface of the thin specimen. This variation (contrast) arises from changes in the diffraction conditions so the technique is said to produce diffraction contrast.

The variations of intensity in a single-phase, single crystal specimen are due to small changes in the local orientation of the *hkl* reflecting planes due to the long-range elastic strain fields of defects. Strong *hkl* diffraction occurs where these lattice planes are strained into the exact Bragg reflecting orientation, e.g. near a dislocation. The diffracted electrons are excluded from the bright-field image so such dislocations are observed as dark lines on a bright background. The dark-field image is obtained by tilting so the beam of *hkl* diffracted electrons travels down the optic axis. Those dislocations that influence the *hkl* planes appear as bright lines in the *hkl* dark-field image.

A detailed theory was early developed for interpreting diffraction contrast micrographs (Hirsch *et al.* 1977). This theory and the great resolving power of the TEM made the diffraction contrast method the most powerful of all for the identification of structural defects. The method compares the detailed contrast of the defect in a number of diffraction contrast micrographs recorded under precisely defined conditions with that calculated for particular types of defect under these conditions. Procedures for identifying most types of crystalline defects in this way have been developed and tested (Head *et al.* 1972 and Goodhew and Humphreys 1988).

TEMs have three disadvantages for semiconductor studies. (1) The instruments are expensive and the highly developed experimental techniques require considerable time and experience to exploit. (2) The method is destructive. Moreover the specimens are thin and the magnifications great. Hence the volume of material examined in a typical micrograph is only of the order of 10^{-11} to 10^{-12} cm^3 and may well not be typical. Dislocation densities are below about 10^4 to 10^5 cm^{-2} for all good semiconducting material and devices. So there are no defects to be seen in a field of view at the magnifications ($\sim 10^4 \times$) used in most diffraction contrast microscopy. (3) TEMs are insensitive to many large (device-scale) inhomogeneities such as doping variations but most sensitive to defects such as dislocations. This selective sensitivity can attract more attention to certain defects than their real importance in semiconductors justifies.

3.6.3 *The weak-beam diffraction contrast technique*

A further limitation of the two-beam diffraction contrast technique arises from the long range of the strain fields of defects and the strong excitation of the two beams. The result is that the widths of the contrast regions of defects are large (50 to 100 times larger than the limits of resolution of modern TEMs). Hence, much detail near the core of dislocations is obscured by excessively strong contrast. This limitation is removed by the 'weak-beam' method.

In this technique the specimen is tilted so that it is set well away from the exact Bragg reflecting condition for the selected reflection. The *hkl* beam is then weakly excited, usually by arranging that some other order of reflection from the (*hkl*) planes, *mh mk ml*, where m is a small integer, 3 say, is exactly, i.e. strongly, excited. The weak *hkl* beam is then focused to produce the micrograph. Instead of two-beam conditions, multiple beam conditions apply and have to be taken into account in the contrast theory. The emergence of detail in the region near the core of the defect can be dramatic. An early application of the technique revealed the unexpected form of the dissociation of dislocations (Section 2.3.7) in Si (Fig. 3.5). The unit dislocations were seen to be dissociated but repeatedly constricted back to the unit dislocation form at short intervals along their lengths.

The weak-beam diffraction-contrast TEM technique has experimental difficulties, but these are less severe than those of direct lattice resolution of atomic detail. For more detail see Loretto and Smallman (1975).

3.6.4 *The convergent-beam electron diffraction technique*

Convergent-beam electron diffraction (CBED) has been employed to analyse the microcrystallography of materials for several decades (for reviews, see e.g. Spence and Zuo 1992, Steeds 2002, Vijayalakshmi *et al.* 2003). When a parallel beam of

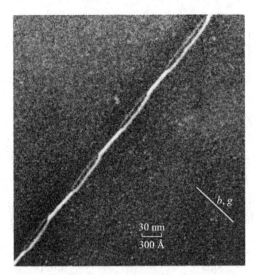

Figure 3.5   Weak-beam diffraction contrast TEM micrograph revealing the core structure of a dislocation in Si. It can be seen that it is repeatedly dissociated into two well-resolved partial dislocations and then constricted back into a unit dislocation at short intervals along its length. (After Ray and Cockayne 1971.)

electrons illuminates a thin crystal and the microscope focuses the diffraction pattern on the viewing screen a selected-area diffraction pattern (SADP) of spots is seen. The CBED technique employs a highly convergent beam to produce from crystals transmission electron diffraction patterns in the form of arrays of disks instead of sharp spots (see Fig. 3.6). The contrast within the disks (e.g. networks of lines, intensity variations and their symmetry) is examined for useful information about the local crystallography.

Thus, using a convergent beam for CBED, information can be obtained on the crystal point and space groups, lattice parameters, and crystal orientation, as well as local lattice strains in nanometer-scale volumes (e.g. Steeds and Vincent 1983, Spence and Zuo 1992, Vijayalakshmi *et al.* 2003, and references therein). This high spatial resolution (on the order of the electron beam spot size) is obtained because the diffracting area is that under the finely focused electron-beam. In contrast, the resolution of standard selected area diffraction (SAD) is given by the large area irradiated by the parallel beam. This is determined by the aberrations of the objective lens and the selected-area aperture. CBED patterns include diffracted beams from higher order Laue zones (HOLZ). Having non-zero components along the incident beam direction, HOLZ diffraction vectors can be used to identify components of lattice displacement fields, and this can be used to determine the Burgers vectors of dislocations (Carpenter and Spence 1982).

Large-angle convergent-beam electron diffraction (LACBED) (Tanaka *et al.* 1980) is widely employed to analyse defects in materials (Morniroli 2002).

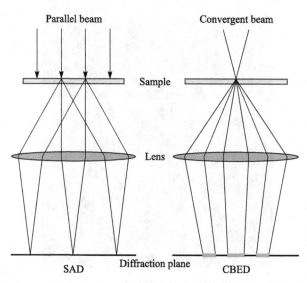

Figure 3.6　Two types of electron diffraction: (a) selected area diffraction (SAD) and (b) convergent beam electron diffraction (CBED).

For LACBED the converging beam focus is moved out of the specimen plane. Thus, the shadow image of the illuminated area is superimposed on the diffraction pattern, effectively providing concurrent information on both the direct and reciprocal lattices. Defects typically generate a displacement field and modify locally the orientation of the lattice planes. This disturbs the image/diffraction information and can be used to analyse the defects. The LACBED method can determine the Burgers vector **b** of a dislocation (Cherns and Preston 1986) since it allows the observation of numerous HOLZ lines. For a straight perfect dislocation, **b** is given by an empirical rule (Cherns and Preston 1986) since the higher-order reflections split near a perfect dislocation into n fringes where $\mathbf{g}\cdot\mathbf{b}=n$ and **g** is the reciprocal lattice vector of the HOLZ line. Thus, the Burgers vector can be derived using observations in three linearly independent **g**-vectors (Cherns and Morniroli 1994).

Numerous reports have appeared of applications of CBED and LACBED to analyse dislocations, nanopipes, grain boundaries, stacking faults, and antiphase boundaries (e.g. Carpenter and Spence 1982, Zhu and Cowley 1983, Tanaka *et al.* 1991, Cherns and Morniroli 1994, Morniroli 1997, Cherns *et al.* 1997, Morniroli 2002, Morniroli *et al.* 2003).

3.7 Scanning electron microscopy

Scanning electron microscopes (SEMs) provide a set of microscope and micro-characterization techniques. Because of the range of physical properties of the specimen on which information can be obtained, these techniques are particularly valuable for the study of semiconducting and optoelectronic materials and devices.

In the SEM a finely focused electron beam, produced by the electron-optical column, is swept by the scan coils in a raster fashion over the specimen surface (Fig. 3.7).

3.7.1 Signal formation and resolution in the SEM

The images are formed using as signal one of the types of energy produced by the dissipation of the electron beam energy in the material (Fig. 3.3). Any such signal can be detected and turned into an electrical signal and amplified (Fig. 3.7). It is displayed as video signal on a CRT (cathode ray tube) scanned in synchronism with the scanning of the beam over the specimen to produce the micrograph. This results in a one-to-one correspondence between picture points on the display CRT screen (the image) and points scanned on the specimen. The magnification is $M = L/l$ where L is the side of the square area scanned on the display CRT screen and l is that of the area scanned on the specimen. M can be continuously varied from about $10\times$ to $300\,000\times$, by varying the currents in the SEM scanning coils.

The first type of imaging to be studied was that in which the signal was provided by detecting the numerous low energy secondary electrons. This produces

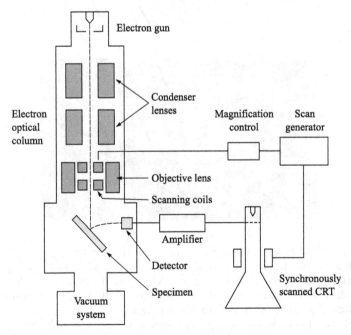

Figure 3.7   The basic components of a scanning electron microscope.

secondary electron images (SEI). The initial motivation for developing the SEM was the early discovery that in addition to its high spatial resolution in SEI (2 nm or less), this mode of operation of the SEM also has a much larger depth of field than the light microscope, giving uniquely descriptive micrographs (Fig. 3.8) of surface topography. (The depth of field is the height over which the specimen surface in the magnified image is in focus.)

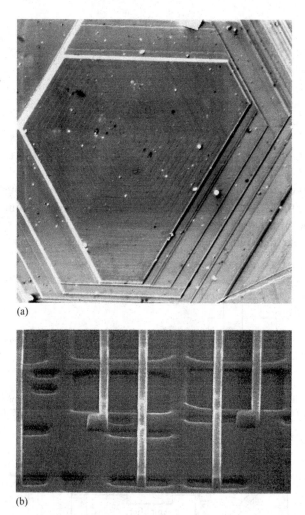

(a)

(b)

Figure 3.8 (a) An SEM secondary electron image (SEI) of a vapour grown platelet crystal of ZnS with a stepped hexagonal pyramid morphology. A hexagonal spiral growth step, originating at a screw dislocation at the centre, is visible on the top surface. The field of view is 110 μm wide. (After Brada *et al.* 1999.) (b) SEI showing 0.24 μm gates on silicon. At the time this was the 'state-of-the-art' (minimum technologically attainable) line width. [(After Brinkman 1997.) In *Microscopy of Semiconducting Materials* 1997 (Institute of Physics Conference Series), 157, 1–12.]

Instead of being video displayed to form a micrograph the signal data can be processed to obtain quantitative values of the properties controlling the strength of the signal generated at the particular point (microcharacterization). As the SEM thus provides a set of relevant modes of operation it is one of the most indispensable tools in semiconductor characterization. An additional advantage is that the SEM can accommodate and examine macroscopic specimens (e.g. whole semiconductor wafers and devices) with little or no special sample preparation, unlike TEMs or FEGSTEMs.

Both the elastic and inelastic electron scattering mechanisms are important in signal formation. Elastic (no energy loss) scattering of the incident electrons by the nuclei of the atoms in the sample gives rise to high-energy backscattered electrons (Fig. 3.3). In the backscattered mode (using these electrons as signal), therefore, there is atomic number contrast. Inelastic interactions result in the emission of secondary and Auger electrons and of characteristic x-rays, and in the generation of electron–hole pairs and thermal effects. The electron–hole pairs can recombine to produce cathodoluminescence (CL) or be separated to produce electron beam induced current (EBIC) signals. Electron channelling affects the rate of electron energy loss, so channelling contrast can be seen in several types of images (signals).

As a result of elastic and inelastic scattering in the material, the original trajectories of the incident electrons are randomized. The effective range, R_e, of electron penetration depends on the electron–beam energy E_b as $R_e = (k/\rho) E_b^\alpha$, where k and α depend on the atomic number of the material and on E_b (empirical values can be found in Everhart and Hoff 1971) and ρ is the density of the material. From this one can estimate the *generation* (or *excitation*) *volume* in the material which determines the resolution of the 'bulk' modes: the CL, EBIC and x-ray modes. This last is also known as electron probe microanalysis (EPMA). The *generation factor* (i.e. the number of electron–hole pairs generated per incident beam electron) is given by $G = E_b(1-\gamma)/E_i$, where E_i is the ionization energy (i.e. the energy required for the formation of an electron–hole pair) and γ represents the fractional electron beam energy loss due to the backscattered electrons. The ionization energy E_i is related to the energy gap of the material, and if unknown, can be taken to be roughly $E_i \approx 3E_g$. For example, for Si, $E_i = 3.63$ eV; thus, one incident 30–keV electron can generate in Si about 8000 electron–hole pairs in the excitation volume that is several microns in diameter.

A useful method for the treatment of the interaction of the beam with the specimen is that of Monte Carlo electron trajectory simulation (for details see Joy 1995). Since the mechanisms of signal formation are different for the SEM modes, their spatial resolution, quantifiability and sensitivity vary substantially. For further details see Goldstein *et al.* (1981), Holt and Joy (1989), Yacobi and Holt (1990), Yacobi *et al.* (1994).

3.7.2 *The modes of scanning electron microscopy*

The emissive mode of the SEM was the first to be developed and is the most widely used. In bulk samples, secondary electrons originate from material within about 100 Å of the specimen surface. Little beam spreading occurs in this distance so the signal is derived from the area of the incident beam spot, which therefore determines the resolution. This is 2 nm or less, depending on the instrument available. The *secondary electron image* (SEI) mode shows the surface topography and is widely used for observing small three dimensional objects and the detail on solid surfaces, e.g. on ICs (Fig. 3.8).

The SEI mode can also be used to detect any electric and magnetic fields present at the surface of the material. Since the emission of the low-energy secondary electrons depends on the electrical potential at the electron-beam impact point, it is possible to image and measure variations in the electrostatic potential. This *voltage contrast* is employed in the characterization of semiconductor devices. Voltage contrast with high-frequency stroboscopy is especially useful in the analysis of integrated circuits. Using such stroboscopic voltage contrast in the SEM, or special purpose 'e-beam testers', an integrated circuit device can be monitored operating at its normal frequency. This allows one to locate any faulty regions in the working device.

Under appropriate operating conditions, *backscattered electrons* provide crystallographic information due to electron channelling, which is based on the fact that the backscattered electron yield varies as the angle of incidence of the scanned electron beam passes through the Bragg angle to crystal lattice planes. By this means, SEM pictures can be obtained consisting of series of bands and fine lines forming *electron channelling patterns* (ECPs) which depend on the crystal structure and orientation of the sample.

The x-ray mode or electron probe microanalysis (EPMA) was also developed early and is widely used but not for semiconductors. *Characteristic x-rays* are emitted due to electronic transitions between inner-core levels. They can be used to identify the particular chemical element present and its concentration, i.e. the composition. The sensitivity of the x-ray mode is too poor, however, to detect doping impurities or small variations in the composition of semiconductor alloys.

The *cathodoluminescence* (CL) and *charge-collection* (CC) modes provide microcharacterization of the luminescent and electronic properties of semiconductors and semiconductor devices. CL offers a contactless and non-destructive characterization tool, whereas, in the CC mode, electrical contacts to the semiconductor device allow monitoring of the electrical signal, the so-called electron beam induced current (EBIC).

The *charge-collection* (CC, or EBIC), and *cathodoluminescence* (CL) modes of the SEM are especially valuable for the study of the electronic properties of defects, which often have a profound effect on both (i) the electronic and optoelectronic properties of semiconductors, and (ii) the performance of semiconductor devices.

In EBIC, electron–hole pairs generated in any depletion region in a specimen, or within minority carrier diffusion range of it, are separated by the built-in electric field and the resultant charge-collection current is used as signal. EBIC is routinely employed in the evaluation of *p-n* junction and Schottky-barrier characteristics and in the analysis of defects such as dislocations (for various charge-collection geometries, see Fig. 3.9).

EBIC contrast is due to variations in charge-collection efficiency and this arises from recombination at electrically active defects. In such measurements, regions with higher carrier recombination efficiency will appear darker than regions with lower carrier recombination rates. This provides a means for the determination of the concentration and distribution of electrically active defects and detection of subsurface defects and damage. EBIC can also be used for measuring the minority carrier diffusion length and lifetime, the surface recombination velocity, and the width and depth of depletion zones. These capabilities make EBIC very valuable for the failure analysis of semiconductor devices.

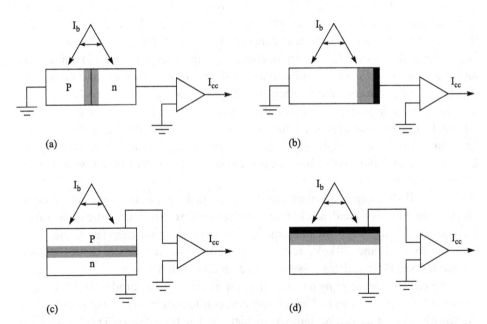

Figure 3.9 Illustration of charge-collection geometries: (a) and (b) illustrate perpendicular *p-n* junction and Schottky barrier geometries, respectively; (c) and (d) show planar *p-n* junction and Schottky barrier geometries, respectively.

REBIC (remote EBIC) is a variant in which no charge collecting barrier such as a *p-n* junction is present and contacts are applied a macroscopic distance (mm or cm) apart. It has been used very successfully for the microcharacterization of very high resistivity electronic materials such as semi-insulating II-VI compounds and electroceramics (for a review see Holt 2000). Charge collection by local fields around dislocations and grain boundaries gives contrast in REBIC.

Cathodoluminescence (CL) is the emission of photons in the ultraviolet, visible and near-infrared ranges due to the radiative recombination of electron−hole pairs generated by the incident electron beam. CL micrographs or spectra can be produced. In CL microscopy, the luminescence of the specimen is displayed as video signal on the CRT. Detecting one or all CL wavelengths gives monochromatic or panchromatic images, respectively. In CL spectroscopy, the electron beam is stationary or scans a small area, so luminescence spectra from selected areas are obtained.

CL photons are generated by electronic transitions between the conduction and the valence bands, and/or levels due to impurities and defects located in the energy gap. The rate of CL emission is proportional to the radiative recombination efficiency, $\eta = \tau/\tau_{rr} = [1 + (\tau_{rr}/\tau_{nr})]^{-1}$, where τ is the minority carrier lifetime and τ_{rr} and τ_{nr} are the radiative and non-radiative recombination times, respectively. Therefore, in the observed CL intensity one cannot separate radiative and non-radiative processes quantitatively. In general, η depends on the temperature, the particular dopants and their concentrations, and the presence of defects. Some applications of CL microcharacterization of defects include (i) uniformity analysis of semiconductors, i.e. obtaining the distributions of defects and impurities, (ii) degradation studies of optoelectronic devices, and (iii) depth-resolved studies of defects in ion-implanted samples and of interface states in heterojunctions. In the CL mode, dislocations appear as dots if viewed end on (for threading dislocations) or as lines (for misfit dislocations). In the former case, examples of the so-called CL *dark dot* and *dot and halo* dislocation contrast in GaAs (doped with Te) are presented in Fig. 3.10.

CL (and EBIC) depth profiling can be performed by varying the electron-beam energy. The range of incident electron penetration (excitation depth) can be varied from about 10 nm to several micrometers for electron-beam energies in the range between about 1 and 30 keV, respectively. (For more details on the EBIC and CL modes, see Holt and Joy 1989 and Yacobi and Holt 1990).

In the *electron acoustic mode*, the chopped electron beam produces intermittent localized heating, and thermal expansion and contraction result in the propagation of acoustic waves that can be detected by piezoelectric transducers. This technique is useful in detecting subsurface defects in solid-state materials. (For more details see, e.g. Holt and Joy, 1989).

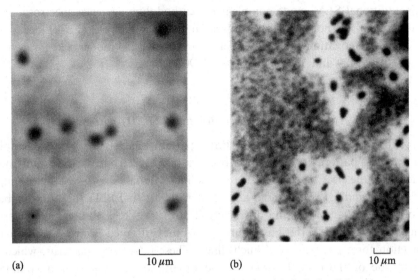

(a)   10 μm   (b)   10 μm

Figure 3.10   CL micrographs of Te-doped GaAs showing (a) *dark dot* dislocation contrast in GaAs doped with a Te concentration of 10^{17} cm^{-3}, and (b) *dot and halo* contrast in GaAs doped with a Te concentration of 10^{18} cm^{-3}. The contrast is due to variations in the doping concentration around dislocations.

3.7.3 Scanning deep level transient spectroscopy

Deep level transient spectroscopy (DLTS) uses capacitance-voltage (C-V) measurements on device structures, e.g. *p-n* junctions, Schottky barriers, and metal-oxide-semiconductor (MOS) and metal-insulator-semiconductor capacitors. Voltage biasing pulses are used to fill the deep levels in the depletion region. Thermal emission of captured carriers then restores thermal equilibrium. In DLTS repetitive filling and emptying of the traps takes place. For levels located in the space-charge region of, for example, a *p-n* junction or Schottky barrier, the relaxation will result in a measurable current or capacitance transient, with the rate of decay depending on the energy of the deep level and also on temperature. Thus, by monitoring the time constant of the transients as a function of the excitation pulse repetition rate, or the rate window, at different temperatures, the energy level and concentration of deep levels can be derived. Capacitance transient measurement allows one to differentiate between electron and hole traps from the sign of the signal, which is independent of the rate window. However, in the case of current transients, the sign of the signal depends on the rate window and is the same for electrons and holes. Scanning deep level transient spectroscopy in the SEM (SDLTS) combines the advantages of DLTS with the resolving power of the SEM. The price of this resolution is the decrease in the strength of the transient signal obtained from the small beam-excited volume.

This additional SEM mode complements cathodoluminescence spectroscopy for the assessment of non-radiative centres in semiconductors. (For more details on microcharacterization of semiconductors using the SEM modes, see, for example, Holt and Joy, 1989 and Yacobi and Holt 1990.)

3.7.4 Electron backscattering diffraction (EBSD) pattern analysis

Electron backscattering diffraction (EBSD) pattern analysis in the SEM employs the diffraction patterns produced by backscattered electrons as the result of the interaction of the electron beam with a crystalline material (see, for example, Dingley *et al.* 1993, Randle 1993, Randle and Engler 2000). In EBSD the electron beam falls on a tilted sample so the diffracted electrons strike a fluorescent phosphor screen to one side. The variations of intensity of the diffracted electrons as a function of direction form patterns of Kikuchi lines. These diffraction patterns, which are characteristic of the crystal structure and orientation of a μm sized volume of the sample, are then analysed by computer software to derive crystallographic information.

With stepper-motor driven specimen stages and appropriate software, this can rapidly give orientation maps of polycrystalline material. This reveals any crystallographic texture (preferred crystal orientations) and gives grain sizes and grain boundary misorientations. This capability led to the widespread adoption of EBSD. EBSD can also distinguish different materials and together with x-ray microanalysis in the SEM, identify different phases.

3.8 Field emission gun scanning transmission electron microscopy

In *scanning transmission electron microscopy* (STEM), a field emission gun (FEG) is used rather than a thermionically emitting source as in conventional TEMs and SEMs. The FEG beam can be focused to give adequate current in a much finer spot. The price for the resultant improved resolution is that the gun and often the whole microscope must be operated in ultra high vacuum. (SEMs also are available incorporating FEGs and giving improved SEI resolution.) In FEGSTEM the fine electron beam spot is scanned in a square raster over the surface of the thin sample. A transmitted signal can be detected, as in the SEM, such as transmitted electrons, x-rays, cathodoluminescence, and EBIC. The signal is amplified and fed to the grid of a synchronously scanned display CRT so it modulates the brightness of the CRT.

EBIC measurements in STEM are often referred to as STEBIC and can provide a powerful means for investigating the electrical activity of dislocations in semiconductors (e.g. Brown and Humphreys 1996).

3.8.1 Z contrast in FEGSTEM

By the use of annular detectors placed in the transmission electron diffraction pattern it is possible to image only the inelastically scattered electron signal. In a highly magnified image formed from these electrons a pattern of bright spots appears. These are the columns of the inelastically scattering atoms running up through the specimen parallel to the incident beam. Brighter and less-bright spots represent columns of higher and lower atomic number. This atomic number (so-called 'Z') contrast interpretation is direct and intuitive (see Fig. 3.11) (Pennycook and Jesson 1991, Browning *et al.* 2001).

In addition, for electron energies greater than about 100 keV, energy losses of the transmitted electrons are characteristic of the elements present in the material; this is the basis for *electron energy-loss spectroscopy* (EELS) that can provide chemical compositional and structural information. These modes together with CL and EBIC make STEM a truly analytical tool for the examination of structure and physical properties; and together with high-resolution imaging of defects, this allows one to correlate defect atomic structure and properties. Some of the limitations of STEM include (i) the low signals levels compared to bulk SEM analysis, and (ii) the fact that, since the volumes analyzed are relatively small, the results may be highly atypical, unless these volumes are carefully selected or the

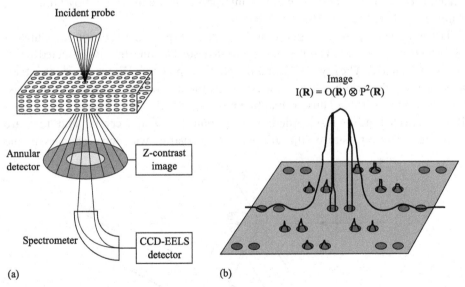

Figure 3.11   (a) Schematic diagram of a STEM with an annular electron detector as well as one for EELS. (b) The Z-contrast in the image can be interpreted as a simple convolution of the experimental probe and the object function. [(After Browning *et al.* 2001.) In *Microscopy of Semiconducting Materials* 2001 (Institute of Physics Conference Series), 169, 1–12.]

observations are repeated in different regions (this is equally a problem for TEM), and (iii) the sample preparation (especially cross-sectional sample) that is destructive and time consuming (as it is for TEM).

3.9 X-ray topography

X-rays can only be focused with difficulty and x-ray microscopy is not yet commercially available although development continues (Burge *et al.* 1987, Michette and Potts 1994). What is in use is x-ray topography (XRT). X-ray topographs are recorded at the same size as the specimen and photographically enlarged to give up to 50× magnification. The limit of resolution is of the order of a μm. The attractions of the x-ray topographic techniques are that they (i) are non-destructive, (ii) can image the whole area and thickness of the slices and whole devices characteristic of solid state electronics, and (iii) have available a sophisticated interpretive theory for defect contrast. This is similar to that of diffraction contrast in transmission electron microscopy (Section 3.6.2).

3.9.1 Techniques and applications of x-ray topography

There are a number of different x-ray topography (XRT) techniques. For detailed accounts of XRT methods and contrast interpretation see the reviews by Lang 1978, Authier 1978, Tanner and Bowen 1980.

The most widely used technique is Lang topography, the principle of which is illustrated in Fig. 3.12. The sample is oriented so that one Bragg diffracted beam is strongly excited. The crystal, C, and the photographic film, F, are moved together, as shown by the two-headed arrow, through the stationary x-ray beam and past the stationary slit S. Thus a line-by-line map of the intensity in the chosen Bragg reflected beam is recorded photographically as the crystal and film are translated. The intensity of the diffracted beam varies with local changes in the

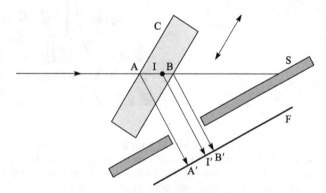

Figure 3.12   Lang's projection x-ray topographic technique.

orientation of the reflecting planes due to elastic strain. Defects therefore appear in contrast due to their elastic fields, just as in the analogous diffraction contrast in TEM.

X-ray topographs can be recorded covering large processed semiconductor slices containing integrated circuits. If this is done repeatedly during processing, it can be seen when and where defects are introduced. The original Lang technique, however, cannot do this, as the internal stresses in such large thin crystals and/or the stresses introduced by mounting the crystals in the topographic camera suffice to change the orientation of the crystal so only in a few narrow areas will it be in the Bragg reflecting condition and be recorded. For such cases Schwuttke developed a scanning oscillation topography (SOT) technique. The crystal is scanned by translation and both the film and the crystal are also oscillated simultaneously about the normal to the plane containing the incident and reflected beams, i.e. about the normal to the plane of the diagram of Fig. 3.12. The angle of oscillation is made large enough to cover the whole reflecting range of the crystal. In this way the entire slice can be recorded on one exposure. Schwuttke (1970) applied this technique systematically in a valuable study of process-induced defects in silicon slices.

The low resolution of this technique means that only low defect densities in highly perfect crystals can be distinguished. The smaller devices in currently processed slices could not be resolved by the SOT technique, either. This is the opposite of TEM which, due to the small thinned areas examinable at high magnifications and resolutions, can see nothing unless high densities of the defect are present at that particular (thinned) depth in the material. The main practical disadvantage of XRT is the long times taken to record the images (hours to days). This can be reduced by using x-ray sensitive video cameras to produce images rapidly with somewhat reduced resolution. Alternatively, the very intense x-radiation from synchrotrons can be used to produce topographs with exposure times of seconds (see the review by Tanner 1977).

3.10 Scanning probe microscopy

Scanning tunnelling microscopy (STM) uses electron tunnelling from the material to a fine mechanically scanned probe. It is capable of imaging surfaces on the atomic scale and won the 1986 Nobel Prize for Physics for its inventors, Gerd Binnig and Heinrich Rohrer. The development of several other forms of scanning probe microscopy (or microscopy without lenses) followed rapidly.

3.10.1 Scanning tunnelling microscopy

In the scanning tunnelling microscope (STM) the electron tunnelling current between an ultrasharp tip and the sample is measured. The tip is made of conductive material and can be moved in three dimensions by piezoelectric translators. The tip is positioned about 10 Å above the sample surface, so that at an operating potential difference of the order of millivolts, a tunnelling current of about 1 nA is detected. The tunnelling current depends exponentially on the distance between the tip and the sample surface. Two operating modes can be used: the constant current and constant height methods. In the former mode, by using a feedback circuit to change the tip height z (separation from the surface) by applying the voltage to the z-controlling piezoelectric element, the tunnelling current, I_t, is kept constant throughout the scan. As the tip is scanned across the specimen surface in the x and y directions, the voltage that controls the tip height is recorded. This can be displayed as an image that reveals the surface topography with atomic resolution. In the constant height mode, the tip travels in a horizontal plane above the sample, and the tunnelling current varies depending on the topography and local surface electronic properties of the sample. The constant current mode is more suitable for irregular surfaces, but the measurement requires a longer time. The constant height mode is faster, but requires smooth surfaces. The basic components employed in the STM are shown in Fig. 3.13.

The tunnelling current depends on s, the gap spacing between the tip and the sample surface, and also on the applied voltage V and the effective barrier height ϕ, and can be expressed as

$$I = C(V/s)\exp(-As\phi^{1/2}) \tag{3.2}$$

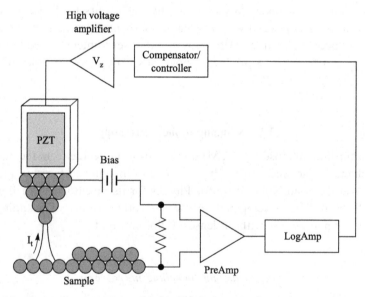

Figure 3.13 Basic components and principle of operation of a scanning tunnelling microscope (STM). (For more details, see, for example Grigg and Russell 1994.)

where C is a constant, $\approx 1.025\,(\text{eV})^{-1/2}\,\text{Å}^{-1}$, and ϕ is an effective work function $[\phi = (\phi_1 + \phi_2)/2$, where ϕ_1 and ϕ_2 are the work functions of the tip and sample]. For ϕ of a few eV, the tunnelling current changes by about an order of magnitude for each Angstrom change in s. The surface electronic structure can be analysed using scanning tunnelling spectroscopy (STS), which essentially involves determining peaks in dI/dV (obtained from the tunnelling current—voltage spectra), which can be related to the surface electronic density of states. Thus, since the tunnelling current depends on the local electronic structure, spectroscopic information on an atomic scale can be derived. Also, by studying the dependence of current on distance (barrier width), ϕ can be derived. Thus, in addition to the atomic—scale analysis of surface structure, any changes in ϕ as a function of lateral position on the sample surface can be determined. These can be related, for example, to a chemical bond of adsorbed species to provide information on local surface contamination.

3.10.2 *Atomic force microscopy and other SPM techniques*

Fine scanning probes are used in several other forms of microscopy. For STM, the surface of the material has to be electrically conductive. *Atomic force microscopy* (AFM) can be employed in the analysis of both conductors and insulators. In AFM a tip, such as a diamond crystal fragment attached to a flexible cantilever, is deflected due to the interaction force between the tip and the sample surface. The atomic interaction force, experienced by the tip, can be derived from the deflection of the cantilever that can be measured employing either electron tunnelling or optical detection. In one form of optical detection a diode laser beam is reflected off the cantilever to a position-sensitive photodetector. The deflection, and thus the interaction force, is controlled by a feedback system. This allows one to record the topography of the sample surface by keeping the force constant in a manner analogous to the constant current-mode in STM.

The interatomic force deflecting the cantilever is commonly due to the Van der Waals interaction. This is repulsive in *contact mode* (i.e. with the tip-to-sample separation less than a few Angstroms) and attractive in *non-contact mode* (for tip-to-sample separation in the range between about 10 and 100 Angstroms).

The basic principles of scanning probe microscopy (SPM) are also employed in different ways in electrostatic force microscopy (EFM), magnetic force microscopy (MFM), ballistic-electron-emission microscopy (BEEM), scanning Kelvin probe microscopy (SKPM), scanning tunnelling luminescence (STL, also known as photoemission STM, i.e. PE-STM, see Section 3.10.3), and scanning capacitance microscopy (SCM). Several of these techniques have provided useful information on the distribution and density of dislocations and the nature of the dislocation charges (see below).

Other SPM techniques include scanning ion-conductance microscopy, scanning thermal microscopy, scanning probe luminescence microscopy and near-field scanning optical microscopy, which provide information on a wide range of materials properties with the nanometer-scale spatial resolution. These techniques are likely to be increasingly used due to the continuing development of quantum well, quantum wire and quantum dot devices. (For more details, see Wiesendanger 1994, Wiesendanger and Güntherodt 1995, and Gustafsson *et al.* 1998.)

For a review of the family of STM based microscopies with emphasis on their applications to the study of materials problems see Gimzewski (1993).

3.10.3 Scanning tunnelling luminescence (STL) microscopy

Electrons from an STM tip, tunnelling into the conduction band of a semiconductor, can give rise to radiation characteristic of the material and build up images of the varying luminescent intensity. Equipment for the study of low-dimensional semiconductor structures by this STL (scanning tunnelling luminescence) technique is shown in Fig. 3.14. The signal here is a form of EL (electroluminescence, i.e. light generated by electrical current flow in a solid) rather than CL as no high-energy,

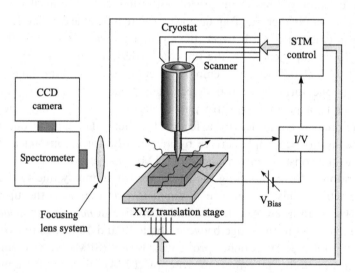

Figure 3.14   Schematic diagram of an STL microscope. The STM is immersed in a cryogenic liquid in a cryostat. A lens, near an optical window in the cryostat, focuses an image of the sample-tip region on the entrance slit of a monochrometer fitted with a CCD camera. Standard STM electronics is used and the electrical feed-throughs provide voltages to the piezo-scanner, the coarse XYZ movements of the sample stage and a current to the tip. (After Gustafsson *et al.* 1998. Reprinted with permission from *Journal of Applied Physics,* **84,** pp. 1715–75. Copyright 1998, American Institute of Physics.)

external beam of electrons is used to excite the emission. It is referred to as STL by some working in the field (Montelius *et al.* 1992, Gustafsson *et al.* 1998) and as PE-STM (photon emission-scanning tunnelling microscopy) by others (Dumas *et al.* 1994a and b).

The resolution might be expected to be limited by the diffusion length of the electrons, which are injected with energies of a few eV and rapidly thermalized. However, due to the rapid fall in the injected electron density as diffusion occurs, the resolution of STL will be commensurate with the size of the tip injecting area and Berndt and Gimzewski (1992) reported resolving a bright spot in CdS about 2.5 nm across. In particular cases, resolution may be determined by other phenomena such as quantum confinement (in porous Si – Dumas *et al.* 1994a and b) as will be discussed below.

Information can be extracted from STL in four ways (Gimzewski *et al.* 1988) as follows:

(i) *isochromat spectroscopy*, in which the energy of the photons detected is kept constant while the tip to sample voltage V_p is varied. This can be done for any chosen isochromat energy (wavelength).

(ii) *fluorescence spectroscopy*, in which the energy of the electrons, i.e. V_p, is kept constant while the wavelength (energy) spectrum of the emitted photons is recorded.

(iii) *luminescence spectroscopy*, which is a special case of fluorescence spectroscopy in which, in semiconductor samples, the injected electrons thermalize and fall to the bottom of the conduction band or a recombination level before dropping to a lower level or to the valence band to emit a photon.

(iv) *spatial mapping* (i.e. monochromatic or panchromatic microscopy), in which a significant feature of the chosen type of spectrum (i to iii) or all of it, respectively, is selected and recorded to form a high resolution image of the intensity of the light as a function of position (x, y, z), displayed in various ways.

Berndt and Gimzewski (1992) carried out a study of cleavage faces of CdS that showed that combining STL spectral analysis with luminescence imaging had the same advantages as in the case of SEM CL.

Abraham *et al.* (1990) used STL to image the interfaces of quantum wells in a surface emitting laser structure of alternating GaAs and AlAs layers with a nanometer resolution. Using a photomultiplier tube with a GaAs photocathode they obtained signal levels of a few counts per second per nA of tunnelling current. A drawback of the technique was, therefore, that image collection times were 10 to 20 min.

Alvarado *et al.* (1991) used STL for cross-sectional examination of MBE grown, GaAs/$Al_{0.38}Ga_{0.62}As$ MQW (multiquantum well) material as shown in Fig. 3.15. Each of the four sets of quantum wells (labelled A to C with the lowest set unlabelled) in this specimen consists of pairs of QWs of individual widths 2, 5, 10, 20, 50 and 100 nm. A, B and C mark the 100 nm QWs in three of the sets with doping

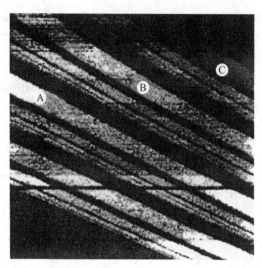

Figure 3.15   Panchromatic STL image of a cross-section through four sets of different GaAs/Al$_{0.38}$Ga$_{0.62}$As QWs with different doping density at 300 K. The area of the image is $1.73 \times 1.73\,\mu m^2$. (After Alvarado *et al.* 1991. Reprinted with permission from *Journal of Vacuum Science and Technology*, **B9**, pp. 409–13. Copyright 1991, American Institute of Physics.)

densities of $p = 10^{19}$, 3×10^{18} and $10^{18}\,cm^{-3}$, respectively. It can be seen that the luminescent intensity decreases with the doping density. It was possible to image the luminescence from wells down to 10 nm width but not below. For a review of other studies of quantum-confined structures by STL and related techniques see Gustafsson *et al.* (1998).

The results of studies of extended defects using SPM techniques are discussed in Section 5.7. For a review of basic concepts in STL see Gimzewski (1995).

Electron injection from a scanning mechanical probe can also be used to excite electrical responses, which leads to an analogue of EBIC with a better resolution than that of EBIC in the SEM (Kazmerski 1991, Koschinski *et al.* 1993, 1994a,b).

3.10.4 *Near-field scanning optical microscopy*

The principles of scanning probe microscopy have also been applied to the development of new methods for the analysis of the optical properties of materials. Near-field scanning optical microscopy (NSOM also sometimes known as SNOM) allows achieving a spatial resolution exceeding the far-field diffraction limit (see Section 3.3) by employing the near-field collimation of light (for a review, see for example Hsu 2001 and references therein). This method employs a subwavelength-size aperture (for light emission or collection), which is mechanically scanned (in a raster pattern) above the sample surface in the near-field of an object and the light transmitted through the sample is collected with a microscope objective in the far field (see Fig. 3.16). The common light source is based on a subwavelength-size aperture at the end of a metal-coated tapered optical fibre (light from a laser is emitted through this aperture).

Illumination, collection, and reflection modes of operation can be employed (see Fig. 3.17). Resolutions (determined by the aperture size) of less than 50 nm can be obtained. Examples of applications include defect analysis, optical spectroscopy (luminescence, Raman), near-field optical beam induced current (NOBIC), and localized photoconductivity imaging.

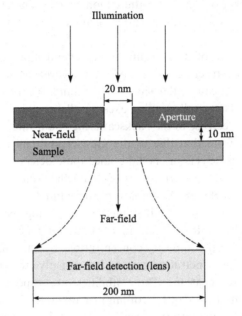

Figure 3.16   Schematic diagram of a near-field scanning optical microscope (NSOM).

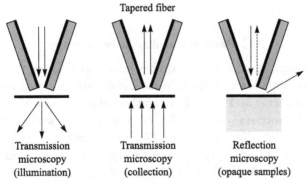

Figure 3.17   Schematic diagram of the illumination, collection, and reflection modes of NSOM.

3.11 Rutherford backscattering

Rutherford backscattering Spectrometry (RBS) is a depth profiling technique which yields quantitative information on elemental composition or impurity concentration without the need for standards. RBS employs high-energy ion bombardment of the material (typically with He ions with energies on the order of 1 MeV) and the measurement of the energy of the backscattered ions. In RBS, some of the incident high-energy ions are backscattered, transferring some energy to the stationary atom of a target in accordance with the laws of conservation of energy and momentum. Consequently the energy of the backscattered ion is determined by the masses of the incident ion and target atom, and the scattering angle so the mass of the target atom can be obtained.

An important feature of this technique is channelling, which can provide information about the structure of the material, as well as whether the impurity atoms occupy lattice or interstitial sites. Channelling occurs when the sample is oriented with one of its crystallographic axes parallel to the incident beam, and it results in a significant decrease in the backscattering count rate. The deviation of the channelling spectrum from that due to the ideal crystal can be employed for the investigation of various defects in crystals, since these defects predictably influence the channelling spectra. (The deterioration of channelling due to defects in crystals is referred to as dechannelling). Examples of dechannelling by defects such as dislocations, twins, and interstitials affecting the channelling spectrum are presented in Fig. 3.18 (for details, see Revesz and Li 1994 and references therein).

RBS can depth profile the impurity concentration (by varying the incident beam energy and hence the ion penetration range). It is sensitive to possible damage, and to specific lattice site location. These abilities are especially useful for the (i) examination (quantitative depth profiling) of dopants and defects in thin-film structures (including superlattices) and ion-implanted semiconductors, and (ii) investigations related to the near-surface processing of semiconductor devices.

3.12 Positron annihilation spectroscopy

Positron annihilation spectroscopy (PAS) has been shown during the past decade to provide useful information on defects in semiconductors (e.g. Krause-Rehberg and Leipner 1999, Leipner *et al.* 1999). The interaction of positrons with solids in general involves several processes, such as backscattering, channelling, thermalization, diffusion and trapping (for details see Krause-Rehberg and Leipner 1999, and references therein). As an energetic positron approaches the surface of a solid, depending on its initial energy and the material's properties, it may undergo backscattering or it may penetrate the material. The entering positron, travelling certain distances (this is referred to as channelling) and rapidly losing its energy,

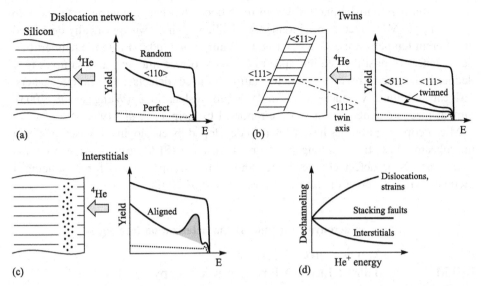

Figure 3.18   The effect of various types of defect on the channelling spectra: (a) dislocations, (b) twins, (c) interstitials; (d) illustration of the energy dependence of dechannelling due to various types of defect (After Revesz and Li 1994.)

thermalizes within a few picoseconds. Continuing its motion at thermal velocities, the positron diffuses through the lattice. During diffusion, positrons may be trapped by lattice defects (e.g. vacancies and dislocations), and then annihilate with electrons. In most cases, the annihilation results in the emission of γ-rays.

The annihilation rate at a given site (or the reciprocal of the lifetime) is proportional to the local electron density seen by the positron. Thus, from annihilation parameters, particular defects (e.g. vacancies, dislocations) can be investigated. For a defect-free crystal, the lifetime (of a few 100 ps) is a specific material characteristic. However, in material containing (negatively charged) defects that can trap positrons, that lifetime becomes longer due to the lower local electron density in these defects. Thus, the positron lifetime measurement facilitates the identification and the determination of the concentration of these defects. The positron source that is typically employed in such measurements is the ^{22}Na radioactive isotope sandwiched between the reference and the sample. The positron lifetime is measured as the time difference between the generation of the positron, detected by the appearance of the γ-quantum in the source, and the annihilation γ-quanta. Two additional techniques, which are related to the momentum conservation in the annihilation process, are the measurement of the angular spread of the collinear γ-quanta and of the Doppler shift of the annihilation energy (for details, see Krause-Rehberg and Leipner 1999).

Applications of positron annihilation spectroscopy in the investigations of semiconductors, including plastically deformed materials, were described by

Krause-Rehberg and Leipner (1999) in their book that lists numerous references to early studies. More recent measurements on silicon, which was plastically deformed at different temperatures, were reported by Wang *et al.* (2003). From the temperature dependences of positron lifetime spectra for as-deformed and annealed samples they deduced the presence of such deformation-induced defects as vacancy clusters, dislocation-bound vacancies and undisturbed dislocations (Wang *et al.* 2003). Similar measurements on GaAs were reported by Leipner *et al.* (1999).

Microscopic capabilities have been also developed by employing a pulsed positron microbeam; thus, the scanning positron microscope (SPM) can provide both the local analysis of defect characteristics and a microscopic means for producing a lifetime image of defect distributions (see e.g. Kögel 2002).

Common acronyms for (micro) characterization techniques

AFM:	Atomic Force Microscopy
BEEM:	Ballistic Electron Emission Microscopy
CL-SEM:	Cathodoluminescence Scanning Electron Microscopy
DLTS:	Deep Level Transient Spectroscopy
EBIC:	Electron Beam-Induced Current
EFM:	Electrostatic Force Microscopy
FEGSTEM:	Field Emission Gun Scanning Transmission Electron Microscopy (the FEG prefix is passing out of use since high-vacuum TEMs with scanning attachments are no longer referred to as STEMS)
QIRBIC:	Quenched Infrared Beam Induced Current
LBIC:	Laser Beam Induced Current
LM:	Light Microscopy
MFM:	Magnetic Force Microscopy
NSOM:	Near-field Scanning Optical Microscopy
PAS:	Positron Annihilation Spectroscopy
RBS:	Rutherford Backscattering Spectrometry
SDLTS:	Scanning Deep Level Transient Spectroscopy
SKPM:	Scanning Kelvin Probe Microscopy
SEM:	Scanning Electron Microscopy
SPM:	Scanning Probe Microscopy
STEM:	Scanning Transmission Electron Microscopy (ultra high vacuum instruments)
STL:	Scanning Tunnelling Luminescence (also known as PE-STM, i.e. photoemission STM)
STM:	Scanning Tunnelling Microscopy

TEM: Transmission Electron Microscopy
XRT: X-ray Topography

References

Abraham, D. L., Veider, A., Schönenberger, Ch., Meier, H. P., Arent, D. J. and Alvarado, S. F. (1990). Nanometer resolution in luminescence microscopy of III-V heterostructures. *Applied Physics Letters*, **56**, 1564–6.

Abrahams, M. S. and Buiocchi, C. J. (1965). Etching of dislocations on the low-index faces of GaAs. *Journal of Applied Physics*. **36**, 2855–63.

Ahearn, J. S., Ball, C. A. B. and Laird, C. (1976). Stress-induced birefringence of mismatching III-V heterojunctions. *Physica Status Solidi*, **A38**, 315–20.

Alvarado, S. F., Renaud, P. H., Abraham, D. L. *et al.* (1991). Luminescence in scanning tunnelling microscopy on III-V structures. *Journal of Vacuum Science and Technology*, **B9**, 409–13.

Amelinckx, S. (1964). *The Direct Observation of Dislocations*. New York: Academic Press.

Ash, E. A. (ed.) (1980). *Scanned Image Microscopy*. London: Academic Press.

Authier, A. (1978). Contrast of image in x-ray topography. In *Diffraction and Imaging Techniques in Materials Science*, vol. II Imaging and Diffraction Techniques, ed. S. Amelinckx, R. Gevers and J. Van Landuyt. (Amsterdam: North-Holland), pp. 481–520.

Berndt, R. and Gimzewski, J. K. (1992). Injection luminescence from CdS $(11\bar{2}0)$ studied with scanning tunnelling microscopy. *Physical Review*, **B45**, 14095–9.

Bogenschutz, A. F., Krusemark, W., Locherer, K. H. and Mussinger, W. (1967). Activation energies in chemical etching of semiconductors in HNO_3-HF-CH_3COOH. *Journal of the Electrochemical Society*, **114**, 970–3.

Booyens, H. and Basson, J. H. (1980a). The application of elastobirefringence to the study of strain fields and dislocations in III-V compounds. *Journal of Applied Physics*, **51**, 4368–74.

Booyens, H. and Basson, J. H. (1980b). The analysis of dislocations in strained III-V semiconductor crystals using elastobirefringence. *Journal of Applied Physics*, **51**, 4375–8.

Brada, Y., Holt, D. B. and Napchan, E. (1999). Growth spirals and morphology in vapour deposited Zns. *Institute of Physics Conference Series* (164), 697–702.

Brinkman, W. F. (1997). The materials basis behind the telecommunications revolution. In *Microscopy of Semiconducting Materials* 1997, Institute of Physics Conference Series (157), pp. 1–12.

Brown, P. D. and Humphreys, C. J. (1996). Scanning transmission electron beam induced conductivity investigation of a $Si/Si_{1-x}Ge_x/Si$ heterostructure. *Journal of Applied Physics*, **80**, 2527–9.

Browning, N. D., Arslan, I., Ito, Y., James, E. M., Klie, R. F., Moeck, P., Topuria, T. and Xin, Y. (2001). Application of atomic scale STEM techniques to the study of interfaces and defects in materials. *Journal of Electron Microscopy*, **50**, 205–18.

Brundle, C. R., Evans, C. A. Jr and Wilson, S. (eds.) (1992). *Encyclopedia of Materials Characterization*. Boston: Butterworth-Heinemann.

Burge, R. E., Michette, A. G. and Duke, P. J. (1987). X-ray microscopy and x-ray imaging. *Scanning Microscopy*, **1**, 891–900.

Camp, P. R. (1955). A study of the etching rate of single-crystal germanium. *Journal of the Electrochemical Society*, **102**, 586–93.

Carpenter, R. W. and Spence, J. C. H. (1982). Three-dimensional strain-field information in convergent-beam electron diffraction patterns. *Acta Crystallographica*, **A38**, 55–61.

Castaldini, A. and Cavallini, A. (1989). Imaging of extended defects by quenched infra-red beam induced currents (Q-IRBIC). In *Point and Extended Defects in Semicondoctors*, eds. G. Benedek, A. Cavallini and W. Schröter (New York: Plenum Press), pp. 257–68.

Castaldini, A., Cavallini, A. and Gondi, P. (1987). IRBIC semiconductor defect pictures. *Bulletin of the Academy of Sciences of the USSR Division of Physical Science*, **51**, 77–80.

Cavallini, A. and Castaldini, A. (1991). Developments of IRBIC and QIRBIC in defect studies: A review. *Journal de Physique*, **C6**, 89–99.

Cherns, D. and Preston, A. R. (1986). Convergent beam diffraction studies of crystal defects. In *Proceedings of the 11th International Congress on Electron Microscopy, Kyoto*, eds. T. Imura, S. Maruse and T. Suzuki (Tokyo: The Japanese Society of Electron Microscopy), pp. 721–2.

Cherns, D. and Morniroli, J. P. (1994). Analysis of partial and stair-rod dislocations by large-angle convergent-beam electron-diffraction. *Ultramicroscopy*, **53**, 167–80.

Cherns, D., Young, W. T. and Ponce, F. A. (1997). Characterisation of dislocations, nanopipes and inversion domains in GaN by transmission electron microscopy. *Materials Science and Engineering*, **B50**, 76–81.

Dash, W. C. (1957). The observation of dislocations in silicon. In *Dislocations and Mechanical Properties of Crystals*, eds. J. C. Fisher, W. G. Johnston, R. Thomson and T. Vreeland (New York: Wiley), pp. 57–67.

Dash, W. C. (1958). The growth of silicon crystals free from dislocations. In *Growth and Perfection of Crystals*, eds. R. H. Doremus, B. W. Roberts and D. Turnbull (New York: Wiley), pp. 361–82.

De Vittorio, M., Cingolani, R., Mazzer, M., Napchan, E. and Holt, D. B. (1997). Sub-micron characterization tool for fast investigation of defects and morphology of semiconductor devices. *Physics of Low-Dimensional Structures*, **12**, 63–8.

De Vittorio, M., Cingolani, R., Mazzer, M. and Holt, D. B. (1999). Sub-micron photocurrent mapping of heterostructures by miro-probe optical beam induced current. *Review of Scientific Instruments*, **70**, 3429–31.

Dingley, D. J., Randle, V. and Baba-Kishi, K. Z. (1993). *Atlas of Backscattering Patterns*. Bristol: Adam Hilger.

Drebeen, A. (1965). Microstructures in CdS:Au single crystals. *Journal of the Electrochemical Society*, **112**, 493–6.

Dumas, P., Gu, M., Syrykh, C. *et al.* (1994a). Photon spectroscopy, mapping and topography of 85% porous silicon. *Journal of Vacuum Science and Technology*, **B12**, 2064–6.

Dumas, P., Gu, M., Syrykh, C. *et al.* (1994b). Nanostructuring of porous silicon using scanning tunnelling microscopy. *Journal of Vacuum Science and Technology*, **B12**, 2067–9.

Elliott, C. R. and Regnault, J. C. (1981). Birefringence studies of defects in III-V semiconductors. In *Microscopy of Semiconducting Materials* 1981. Conf. Series No. 60, eds. A. G. Cullis and D. C. Joy (Bristol: Institute of Physics), pp. 365–70.

Elliott, C. R., Regnault, J. C. and Wakefield, B. (1982). Applications of polarised infrared microscopy in the evaluation of InP and related compounds. In *Proc. Int. Symposium on GaAs and Related Compounds*, Albuquerque. Conf. Series No. 65 (Bristol: Institute of Physics), pp. 553–60.

Everhart, T. E. and Hoff, P. H. (1971). Determination of kilovolt electron energy dissipation vs. penetration distance in solid materials. *Journal of Applied Physics*, **42**, 5837–46.

Fathers, D. J. and Tanner, B. K. (1973a). Optical contrast of inclined boundaries in birefringent magnetic materials. *Philosophical Magazine*, **27**, 17–34.

Fathers, D. J. and Tanner, B. K. (1973b). Line defects in barium titanate observed by polarized light microscopy. *Philosophical Magazine*, **28**, 749–70.

Faust, J. W. (1962). Etching of the III-V intermetallic compounds. In *Compound Semiconductors 1, Preparation of III-V Compounds*, eds. R. K. Willardson and H. L. Goering. New York: Reinhold.

Gimzewski, J. K. (1993). Scanning tunnelling microscopy. *Journal de Physique*, **C3**, 41–8.

Gimzewski, J. K (1995). Photon emission from STM: concepts. In *Photons and Local Probes: Proc. NATO Advanced Research Workshop* 1995, pp. 189–208.

Gimzewski, J. K., Riehl, B., Coombs, J. H. and Schlittler, R. R. (1988). Photon emission with the scanning tunnelling microscope. *Zeitschrift für Physik*, **B72**, 497–501.

Goldstein, J. I., Newbury, D. E., Echlin, P. *et al.* (1981). *Scanning Electron Microscopy and X-Ray Microanalysis*. New York: Plenum Press.

Goodhew, P. J. and Humphreys, F. J. (1988). *Electron Microscopy and Analysis*. London: Taylor & Francis.

Grigg, D. A. and Russell, P. E. (1994). Scanning probe microscopy. In *Microanalysis of Solids*, eds. B. G. Yacobi, D. B. Holt and L. L. Kazmerski (New York: Plenum Press), pp. 389–447.

Gustafsson, A., Pistol, M. E., Montelius, L. and Samuelson, L. (1998). Local probe techniques for luminescence studies of low-dimensional semiconductor structures. *Journal of Applied Physics*, **84**, 1715–75.

Hayes, R. (1984). *Optical Microscopy of Materials*. Glasgow: Internat. Textbook Co.

Head, A. K., Humble, P., Clareborough, L. M., Morton, A. J. and Forward, C. T. (1972). *Computed Electron Micrographs and Defect Identification*. Amsterdam: North-Holland.

Hilgarth, J. (1978). Direct observation of dislocations in GaP crystals. *Journal of Materials Science*, **13**, 2697–702.

Hirsch, P. B., Howie, A., Nicholson, R. B., Pashley, D. W. and Whelan, M. J. (1977). *Electron Microscopy of Thin Crystals* (2nd edn). Huntington, NY: Krieger Publishing.

Holmes, P. J. (1962). Practical applications of chemical etching. In *The Electrochemistry of Semiconductors*, ed. P. J. Holmes. London: Academic.

Holt, D. B. (2000). The remote electron beam induced current analysis of grain boundaries in semiconducting and semi-insulating materials. *Scanning*, **21**, 28–51.

Holt, D. B. and Joy, D. C. (eds.) (1989). *SEM Microcharacterization of Semiconductors*. London: Academic Press.

Hsu, J. W. P. (2001). Near-field scanning optical microscopy studies of electronic and photonic materials and devices. *Materials Science and Engineering*, **33**, 1–50.

Iqbal, M. (1980). Birefringence observations of strain and plastic deformation in GaP. *Journal of Materials Science*, **15**, 781–4.

Jenkins, D. A. and Hren, J. J. (1976). Quantitative piezobirefringence studies of dislocations in transparent crystals. *Philosophical Magazine*, **33**, 173–80.

Jenkins, M. W. (1977). New preferential etch for defects in silicon-crystals. *Journal of the Electrochemical Society*, **124**, 757–62.

Johnston, W. G. (1962). Dislocation etch pits in non-metallic crystals, with bibliography. *Progress in Ceramic Science*, **2**, 1–76.

Joy, D. C. (1995). *Monte Carlo Modelling for Electron Microscopy and Microanalysis*. Oxford: Oxford University Press.

Joy, D. C., Romig, A. D. and Goldstein, J. I. (1986). *Principles of Analytical Electron Microscopy*. New York: Plenum.

Kazmerski, L. L. (1991). Specific atom imaging, nanoprocessing and electrical nanoanalysis with scanning tunnelling microscopy. *Journal of Vacuum Science and Technology*, **B9**, 1549–56.

Kelly, A., Groves, G. W. and Kidd, P. (2000). *Crystallography and Crystal Defects*. Chichester: Wiley.

Kögel, G. (2002). Microscopes/microprobes. *Applied Surface Science*, **194**, 200–9.

Koschinski, P., Dworak, V., Kaufmann, K. and Balk, L. J. (1993). Prospects of an application of a scanning tunnelling microscope to electron beam induced current (EBIC) investigations. In *Defect Recognition Image Proc. in Semicond. and Devices Conf., Santander*. IoP Conf. Series No. 135 (Bristol: Institute of Physics), pp. 65–8.

Koschinski, P., Kaufmann, K. and Balk, L. J. (1994a). EBIC-investigations of grain boundaries in diamond films with SEM and STM. In *Proceedings of ICEM 13, Paris*, pp. 1121–2.

Koschinski, P., Dworak, V. and Balk, L. J. (1994b). High resolution electron beam induced current measurements in a scanning tunnelling microscope on GaAs-MESFET. *Scanning Microscopy*, **8**, 175–80.

Kotake, H., Hirahara, K. and Watanabe, M. (1980). Quantitative photoelastic measurement of residual stress in LEC grown GaP crystals. *Journal of Crystal Growth*, **50**, 743–51.

Krause-Rehberg, R. and Leipner, H. S. (1999). *Positron Annihilation in Semiconductors*, Vol. **127** of Springer Series 'Solid-State Sciences'. Berlin: Springer-Verlag. (For early studies of semiconductors by numerous authors, see references therein.)

Lang, A. (1978). Techniques and interpretation in x-ray topography. In *Diffraction and Imaging Techniques in Materials Science Vol. II Imaging and Diffraction Techniques*, eds. S. Amelinckx, R. Gevers and J. Van Landuyt (Amsterdam: North-Holland), pp. 407–79.

Leipner, H. S., Hubner, C. G., Staab, T. E. M., Haugk, M., Krause-Rehberg, R. (1999). Positron annihilation at dislocations and related point defects in semiconductors. *Physica Status Solidi*, **A171**, 377–82.

Loretto, M. H. and Smallman, R. E. (1975). *Defect Analysis in Electron Microscopy*. London: Chapman & Hall.

Lynch, R. T., Thomas, D. G. and Dietz, R. E. (1963). Growth and decoration of ZnTe crystals. *Journal of Applied Physics*, **34**, 706–7.

Matthews, J. W., Plaskett, T. S. and Blum, S. E. (1977). Optical birefringence images of dislocations in large gallium phosphide crystals. *Journal of Crystal Growth*, **42**, 621–4.

Michette, A. G. and Potts, A. W. (1994). X-ray microscopy. In *Microanalysis of Solids*, eds. B. G. Yacobi, D. B. Holt and L. L. Kazmerski (New York: Plenum Press), pp. 233–46.

Montelius, L., Pistol, M.-E. and Samuelson, L. (1992). Low-temperature luminescence due to minority carrier injection from the scanning tunnelling microscope tip. *Ultramicroscopy*, **42–44**, 210–14.

Morniroli, J. P. (1997). Accurate measurement of grain boundary misorientation by large angle convergent beam electron diffraction. *Interface Science*, **4**, 273–83.

Morniroli, J.-P. (2002). *Large-angle convergent beam electron diffraction (LACBED). Applications to crystal defects.* (Paris: Societe Francaise des Microscopies, English translation.)

Morniroli, J. P., No, M. L., Rodriguez, P. P. *et al.* (2003). CBED and LACBED: characterization of antiphase boundaries. *Ultramicroscopy*, **98**, 9–26.

Pennycook, S. J. and Jesson, D. E. (1991). High-resolution Z-contrast imaging of crystals. *Ultramicroscopy*, **37**, 14–38.

Randle, V. (1993). *The Measurement of Grain Boundary Geometry*. Bristol: Adam-Hilger.

Randle, V. and Engler, O. (2000). *Introduction to Texture Analysis: Macrotexture, Microtexture and Orientation Mapping*. Amsterdam: Gordon & Breach.

Ray, I. L. F. and Cockayne, D. J. H. (1971). The dissociation of dislocations in silicon. *Proceedings of the Royal Society*, **A325**, 543–54.

Reinhart, F. K. and Logan, R. A. (1973). Interface stress of $Al_xGa_{1-x}As$ - GaAs layer structures. *Journal of Applied Physics*, **44**, 3171–5.

Revesz, P. and Li, J. (1994). Applications of megaelectron-volt ion beams in materials analysis. In *Microanalysis of Solids*, eds. B. G. Yacobi, D. B. Holt and L. L. Kazmerski (New York: Plenum Press), pp. 179–215.

Runyan, W. R. and Shaffner, T. J. (1998). *Semiconductor Measurements and Instrumentation*. New York: McGraw-Hill.

Schroder, D. K. (1998). *Semiconductor Material and Device Characterization* (2nd edn). New York: Wiley.

Schwuttke, G. H. (1970). Silicon material problems in semiconductor device technology. *Microelectronics and Reliability*, **9**, 397–412.

Secco d'Aragona, F. (1972). Dislocation etch for (100) planes in silicon. *Journal of the Electrochemical Society*, **119**, 948–51.

Sirtl, E. and Adler, A. (1961). Chromsaure-flusssaure als spezifisches system zur atzgrubenentwicklung auf silizium. *Zeitschrift fur metallkunde*, **52**, 529–31.

Spence, J. C. H. (1988). *Experimental High Resolution Electron Microscopy*. Oxford: Oxford University Press.

Spence, J. C. H. and Zuo, J. M. (1992). *Electron Microdiffraction*. New York: Plenum.

Steeds, J. W. (2002). Convergent beam electron diffraction. *Advances in Imaging and Electron Physics*, **123**, 71−103.

Steeds, J. W. and Vincent, R. (1983). Use of high-symmetry zone axes in electron-diffraction in determining crystal point and space-groups. *Journal of Applied Crystallography*, **16**, 317−24.

Stirland, D. J., Hart, D. G., Clark, S., Regnault, J. C. and Elliott, C. R. (1983). Characterization of defects in InP substrates. *Journal of Crystal Growth*, **61**, 645−57.

Tanaka, M., Saito, R., Ueno, K., Harada, Y. (1980). Large-angle convergent-beam electron-diffraction. *Journal of Electron Microscopy*, **29**, 408−12.

Tanaka, M., Terauchi, M. and Kaneyama, T. (1991). Identification of lattice-defects by convergent-beam electron-diffraction. *Journal of Electron Microscopy*, **40**, 211−20.

Tanner, B. K. (1977). Crystal assessment by x-ray topography using synchrotron radiation. *Progress in Crystal Growth and Characterization*, **1**, 23−56.

Tanner, B. K. and Bowen, D. K. (eds.) (1980). *Characterization of Crystal Growth Defects by X-Ray Methods*. NATO Adv. Study Inst. Series B: Physics Vol. **63** (New York: Plenum Press).

Tanner, B. K. and Fathers, D. J. (1974). Contrast of defects under polarized light. *Philosophical Magazine*, **29**, 1081−94.

Vijayalakshmi, M., Saroja, S. and Mythili, R. (2003). Convergent beam electron diffraction − a novel technique for materials characterisation at sub-microscopic levels. *SADHANA-Academy Proceedings in Engineering Sciences*, **28**, 763−82.

Vogel, F. L., Pfann, W. G., Corey, H. E. and Thomas, E. E. (1953). Observations of dislocations in lineage boundaries in germanium. *Physical Review*, **90**, 489−90.

Wang, Z., Leipner, H. S., Krause-Rehberg, R., Bodarenko, V. and Gu, H. (2003). Defects properties in plastically deformed silicon studied by positron lifetime measurements. *Microelectronic Engineering*, **66**, 358−66.

Wiesendanger, R. (1994). *Scanning Probe Microscopy and Spectroscopy*. Cambridge: Cambridge University Press.

Wiesendanger, R. and Güntherodt, H.-J. (eds.) (1995). *Scanning Tunneling Microscopy II*. New York: Springer-Verlag.

Yacobi, B. G. and Holt, D. B. (1990). *Cathodoluminescence Microscopy of Inorganic Solids*. New York: Plenum.

Yacobi, B. G., Holt, D. B. and Kazmerski, L. L. (eds.) (1994). *Microanalysis of Solids*. New York: Plenum Press.

Zhu, J. and Cowley, J. M. (1983). Micro-diffraction from stacking faults and twin boundaries in FCC crystals. *Journal of Applied Crystallography*, **16**, 171−5.

4

Core structures and mechanical effects of extended defects specific to semiconductors

4.1 Atomic core structure of dislocations

Extended defects disturb the crystal structure over many atomic distances in one or more dimensions. Those that continue to be of practical importance are interfaces and dislocations.

Extended defect cores in semiconductors are of relatively high energy due to the directional character of tetrahedral covalent bonds. If such a bond is bent from its tetrahedral direction by the displacement of the neighbouring atom, the bond energy rises rapidly. Hence sp^3 bonds resist bending as if they were elastically stiff. If the neighbour atom is too far off the tetrahedral direction or at too great a distance, the bond energy would be too high and a broken or 'dangling' bond occurs. Hence, extended semiconductor defects can be modelled using plastic spheres with tetrahedrally drilled holes or protrusions and wire or plastic straws to connect them. Since such connections are also stiff and brittle, ball-and-wire (or caltrop-and-spoke) models can give insight, through their ease or difficulty of construction, into the likelihood of occurrence of particular atomic core arrangements in dislocations and grain boundaries. Such modelling was introduced by Hornstra (1958, 1959, 1960).

The high energetic cost of broken and strained bonds leads to a tendency for dislocations and grain boundaries to be crystallographically aligned to minimize the number of such bonds in the core. This contrasts with the curved or arbitrarily directed defects seen in many metals. Thus, dislocations in covalently bonded semiconductors are constrained to lie in deep crystallographic Peierls troughs (see Section 2.5.2 and Fig. 2.24). The bonding and consequent defect core structures largely determine the mechanism of movement of dislocations and so the plastic properties of semiconducting materials as will be discussed below.

In semiconductors, dislocation dynamics depends on the atomic core structure and this can be affected by interactions with charge carriers. Consequently influences like doping and illumination alter dislocation mobilities and so give rise to, e.g. electroplastic and photoplastic effects (see Section 4.2 and Section 4.3). The discovery of simple semi-empirical relationships for dislocation numbers and

163

velocities made it possible to develop a successful microdynamical theory of the plasticity of semiconductors. These topics, unique to semiconducting materials, are discussed in this chapter.

Defects also alter electrical conductivities and other optical and electronic properties (these phenomena are discussed in Chapter 5).

As just discussed, dislocation core structures in covalently bonded semiconductors are more important than in other types of materials. Firstly, dislocation cores make a relatively large contribution to the total energy of the dislocation, i.e.

$$E_{\text{total}} = E_{\text{core}} + \frac{CGb^2}{4\pi}\ln\left(\frac{r_1}{r_0}\right) \tag{4.1}$$

which appeared previously as Equation (2.6) in Section 2.3.5. E_{core} is not negligible, compared to the second term, the elastic energy, as usually assumed in elementary accounts of dislocations. Secondly, the Peierls potential troughs in covalent semiconductors are relatively deep (see Fig. 2.24) so dislocations in semiconductors tend to be crystallographically aligned, as in Figs. 4.1 and 4.2, to minimize this energy. Thirdly, dislocation cores in semiconductors are narrow, compared to those in the f.c.c. metals although these materials have the same slip system. The Peierls model showed the intrinsic stress required to move a dislocation through a perfect lattice, the Peierls (or Peierls-Nabarro) stress, to be given by:

$$\tau_P = \frac{2\mu}{(1-\nu)}e^{-2\pi w/b} \tag{4.2}$$

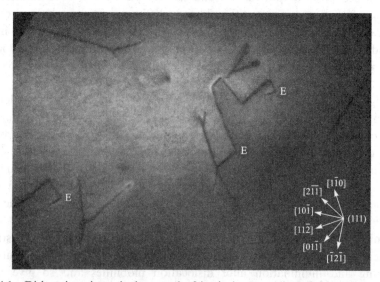

Figure 4.1 Dislocations in a single crystal of intrinsic germanium deformed to a total of 0.3% strain, in creep, at 470° C under a stress of 1.7 Kg mm^{-2}. At the elbows, marked E, line tensions are unbalanced. The magnification is 27,000×. (After Holt and Dangor 1963.)

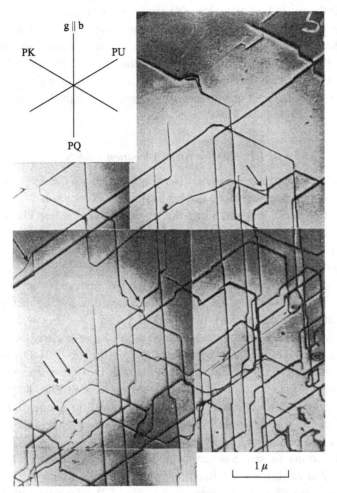

Figure 4.2 Dislocations in the primary slip plane of Si deformed at a relatively high stress (40 Kg mm^{-2}) and low temperature (520° C) for a few seconds. The dislocations are aligned with the ⟨110⟩ directions PK, PU and PQ. (After Wessel and Alexander 1977. Reprinted with permission from *Philosophical Magazine*, **35** (1977), pp. 1523–36; Taylor & Francis Ltd., the Journal's web site: http://www.tandf.co.uk/journals.)

which appeared previously as Equation (2.14) in Section 2.5.2, where μ is the elastic shear modulus, ν is Poisson's ratio, b is the Burgers vector and w is the width of the dislocation core. That is, τ_P increases exponentially as the width of the dislocation core falls, so the dislocation dynamics of tetrahedrally bonded semiconductors and f.c.c. metals are extremely different (see Fig. 2.26 and Section 4.2.1).

Ideas about dislocation atomic-core structures in semiconductors began with the Shockley (1953) 60°-dislocation model (Fig. 2.23). 60° dislocations have Burgers vectors, **b**, and line directions, **l**, in ⟨110⟩ directions at 60° to each other. This model assumed that the sp^3 hybrid orbitals are so strongly directional that it is energetically preferable to have core bonds broken or 'dangling' (not bonded to a second atom) rather than have many bonds outside the core strained from the tetrahedral direction. It appeared that the atoms in the row along the edge of the extra half-plane in dislocations with an edge component are likely to have dangling bonds, since such atoms have 3, not 4, nearest neighbours as required for tetrahedral bonding.

4.1.1 The Hornstra dislocation models

In materials with relatively deep Peierls troughs, such as the covalently bonded, tetrahedrally coordinated semiconducting materials, dislocation lines, seen on an atomic resolution scale, will run along the troughs, except where kinks carry them from one trough to another over the energy barriers (Figs. 2.24 and 4.3). Seen more macroscopically, e.g. in TEM images, clearly, curved segments of dislocation must contain high densities of kinks, while straight lengths have low kink densities.

Early TEM observations of dislocations in Ge, deformed at relatively low temperatures in creep (see Fig. 4.1) and in Si deformed at low temperatures by the two temperature method (Wessel and Alexander 1977), see Fig. 4.2, showed that they were straight and crystallographically aligned particularly with the ⟨110⟩ directions, i.e. the edges of Thompson's tetrahedron (see Fig. 2.2). Dislocations also sometimes run along ⟨112⟩ directions. It can be seen that the dislocations can turn sharply around corners, i.e. exhibit elbows as shown in both Figs. 4.1 and 4.2. This shows the strength of the alignment since, at such points, the line tensions of the two dislocations meeting at the point, due to the total energy per unit length of the dislocations (see Equation 4.1) are, otherwise, not balanced.

Since dislocations in semiconductors align with the ⟨110⟩ or ⟨112⟩ directions only, the number of distinct combinations of line direction with the ⟨110⟩ Burgers vector is small (see Thompson's tetrahedron Fig. 2.2). Hornstra (1958) produced

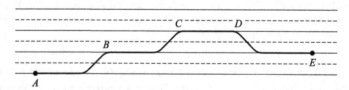

Figure 4.3 Diagram of a dislocation (heavy line) in a material with relatively deep Peierls troughs (light horizontal lines) between potential maxima (dashed lines). The dislocation line runs along the Peierls troughs except where kinks, at *B*, *C* and *D*, carry it from one trough to the next. At lower magnifications, dislocations will appear curved if they contain many kinks per unit length but straight if they contain few. (After Hirsch 1981. Reprinted from *Defects in Semiconductors*, Electronic and mechanical properties of dislocations in semiconductors, pages 257–71. Copyright 1981, with permission from Elsevier.)

ball-and-wire models of the core structures of all these dislocations in the diamond structure, including the important 60° dislocation of Fig. 2.22. Such models, made in real ball-and-wire (or plastic caltrop and spoke) structures, are useful in exploring the likely core structures of dislocations. Difficulties in connecting the wires or spokes indicate the relative severity of bond strain and hence the energetic likelihood of alternative core configurations. This method was then also applied to many types of grain boundaries in diamond-structure materials by Hornstra (1959, 1960).

These dislocation models, like that of Shockley, were based on the implicit assumption that slip takes place between the widely spaced type I {111} planes rather than the closely spaced type II {111} planes, since only 1/3 as many bonds have to be broken or strained in type I slip as in type II (see Fig. 4.4).

Observation shows that just three basic types of unit dislocation with $\mathbf{l} = \langle 110 \rangle$ predominate, identified by the angle between **b** and **l** as screws (0°), edges (90°) and 60° dislocations. Hornstra's model of the edge dislocation is shown in Fig. 4.5 and his models of two possible core forms for the screw dislocation in Fig. 4.6. The core structure of the 60° dislocation, considered by Shockley and Hornstra, is shown in Fig. 2.22 and in Fig. 4.9a below.

Hornstra's work led to two important new ideas. Firstly, it showed that there were often ways to reconnect broken or strained bonds to reduce the core energy, e.g. from the energy of the screw dislocation core configuration of Fig. 4.6a to that of Fig. 4.6b, which contains double bonds 1–2, 3–4 and 5–6 but less strained bonds. Such bond rearrangements are believed to be the major reason that the observed density of defect states, N_d, is orders of magnitude less than one per atomic distance along a 60° dislocation line, as in the original Shockley model of Fig. 2.22.

Secondly, Hornstra showed that dissociation, i.e. the splitting of a unit dislocation into a pair of partial dislocations separated by a strip of stacking fault, was possible

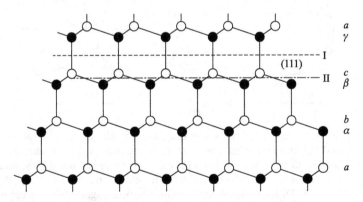

Figure 4.4   The (111) stacking sequence $a\,\alpha, b\,\beta, c\,\gamma, a\ldots$ in the sphalerite structure, or if both black and white sites are occupied by atoms of the same chemical element, in the diamond structure. The two types of possible slip plane, I and II, are shown.

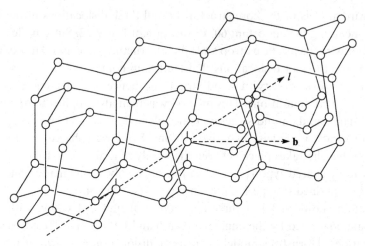

Figure 4.5 Hornstra's model of the (shuffle set) edge dislocation with the glide plane {111} in materials with the diamond cubic structure. The dislocation line direction is *l* and **b** is its Burgers vector. (After Hornstra 1958. Reprinted from *Journal of Physics and Chemistry of Solids*, **5**, Dislocations in the diamond lattice, pages 129–41. Copyright 1958, with permission from Elsevier.)

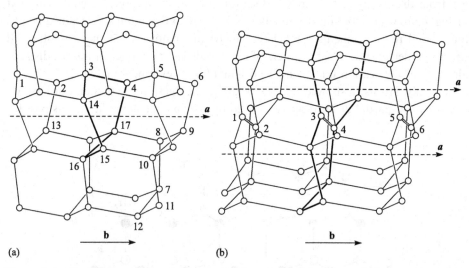

Figure 4.6 Hornstra's models of the screw dislocation in the diamond structure in (a) its simplest form and (b) an alternative form with double bonds, 1–2, 3–4, etc. and less bond strain. Corresponding atoms in the two cases have the same number. *a* is the line direction and **b** is the Burgers vector. (After Hornstra 1958. Reprinted from *Journal of Physics and Chemistry of Solids*, **5**, Dislocations in the diamond lattice, pages 129–41. Copyright 1958, with permission from Elsevier.)

in diamond-structure materials. In crystals with the f.c.c. translation lattice, like semiconductors with the diamond or sphalerite structure, a unit dislocation with a $\langle 110 \rangle$ Burgers vector can split into two partial dislocations with $\langle 112 \rangle$ Burgers vectors as follows:

$$\frac{a}{2}[011] \rightarrow \frac{a}{6}[121] + \frac{a}{6}[\bar{1}12] \tag{4.3}$$

The square of the original Burgers vector is greater than the sum of the squares of the Burgers vectors of the partial dislocations. Hence by Frank's rule this dissociation reaction is favourable since it lowers the dislocation elastic strain energy.

Hornstra showed how stacking faults could form to allow an analogue of dissociation in the diamond or sphalerite structure, as shown for a 60° dislocation in Fig. 4.7. This involves altering the configuration and rebonding the atoms in two stacking sequence planes above and below a type II plane, adjacent to the dislocation. After such rotating of atom pairs and rebonding as shown by the dotted lines in Fig. 4.7a or alternatively in Fig. 4.7b the resultant stacking fault layer of material contains a type II {111} plane (these layers are contained between the pairs of dashed lines in Figs. 4.7c and 4.7d.

Ball-and-wire modelling showed that stacking faults of reasonable energy could only occur between the closely spaced pairs of {111} planes of atoms, i.e. on type II planes. Weak beam TEM observations later showed that most dislocations in silicon (Ray and Cockayne 1971), germanium (Häusserman and Shaumburg 1973) and III-V compounds (Gottschalk *et al.* 1978) are in fact split into two Shockley partial dislocations in accordance with Equation (4.3), with a strip of stacking fault between. In most glide dislocations these stacking faults were found to be intrinsic, although some exceptions were found (Cullis 1973, Gomez *et al.* 1975 and Wessel and Alexander 1977).

Alexander (1979) pointed out that the defects shown in Figs. 4.7c and 4.7d are not well described as dissociated 60° dislocations. They are better described as 'associations' of the 60° dislocation D with strips of stacking fault bounded by partial dislocations, an idea due to Haasen and Seeger (1958). The driving force for the formation of the stacking fault ribbon is the relaxation of the elastic stress field near the core of dislocation D, to lower the energy. Hence, the sum of the Burgers vectors of D and the partial on the side of D should equal that of the Shockley partial at the same side in the case of the splitting of the shuffle set dislocation represented in Equation (4.3). That is, we can represent the association of Fig. 4.7 as:

$$\frac{a}{6}[1\bar{1}\bar{2}] - -stacking - fault - -\frac{a}{6}[\bar{1}12]$$

$$+D : \frac{a}{2}[011] \tag{4.4}$$

$$Sum : \frac{a}{6}[121]$$

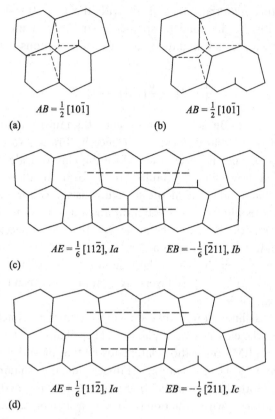

Figure 4.7 A 60° dislocation in the diamond cubic lattice (compare to Fig. 2.22) drawn projected onto the plane perpendicular to the line direction, **l**. In (a) and (b) the solid lines show the initial configuration of the atoms and the bonds. The short vertical dangling bond marks the dislocation core position each case. The dotted lines in (a) and (b) indicate two possible ways in which the atoms could rearrange and re-bond, to allow dissociation into extended dislocations containing stacking faults lying parallel to the slip plane. The stacking faults occur either in the plane containing the dangling bonds, as shown in (a) and (c), or in the plane above it, as in (b) and (d). (c) and (d) show the extended dislocation cores obtained when atom rearrangement and re-bonding has occurred for a width of three successive atom pair ribbons in each case. The faults lie between the dashed horizontal lines, i.e. in closely spaced type II {111} planes in both cases. Below each dislocation the Burgers vector is given together with Hornstra's notation: *I* (intrinsic) *a*, *b*, and *c*. *Ib* and *Ic* both correspond to a single partial in the f.c.c. structure. (After Hornstra 1958. Reprinted from *Journal of Physics and Chemistry of Solids*, **5**, Dislocations in the diamond lattice, pages 129–41. Copyright 1958, with permission from Elsevier.)

4.1.2 The glide and shuffle sets of dislocations

Hirth and Lothe (1968) suggested that slip takes place between the more closely spaced type II planes (see Figs. 4.4 and 4.8), rather than between the widely spaced type I planes, as previously assumed. They, therefore, drew attention to an alternative set of dislocation models. The dislocations resulting, e.g. from the removal of the 1-5-6-4 slab of material in Fig. 4.8 they called the 'glide set' as they believed them to be the more mobile. Dislocations resulting, e.g. from the removal of the 1-2-3-4 slab, the original Shockley-Hornstra type, Hirth and Lothe named the 'shuffle set' as they believed their glide requires a characteristic, complicated, cooperative shuffling movement of atoms. Since 1968 a great deal of work, both experimental and theoretical, has been done in the attempt to determine which of the possible sets of dislocation atomic cores, 'glide' or 'shuffle' actually occur in particular semiconductors in any given experimental conditions. The present consensus is that it is neither always the one nor always the other but sometimes one and sometimes the other, as will be discussed in the remainder of this section.

Removing the slabs of different depths outlined by the dashed lines in Fig. 4.8, running into the plane of the diagram in the appropriate direction, and closing up, leads to the formation of the unit (undissociated) shuffle and glide set 60° dislocations whose ball-and-wire models are represented in Fig. 4.9.

After 'dissociation' in the manner shown in Fig. 4.7, the so-called shuffle set 60° dislocation consists of what Alexander (1979) called an 'association' of the

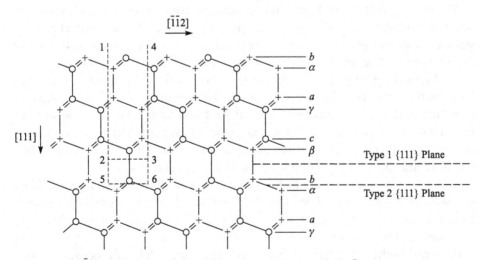

Figure 4.8 ($1\bar{1}0$) projection of the diamond structure. The symbols $\bigcirc$ and $+$ represent atoms in and below the plane of the diagram respectively. Vertical slabs of material cut to different depths, either 1-2-3-4 or 1-5-6-4 can be removed before closing up the crystal to form dislocations with some edge character. (After Hirth and Lothe 1968.)

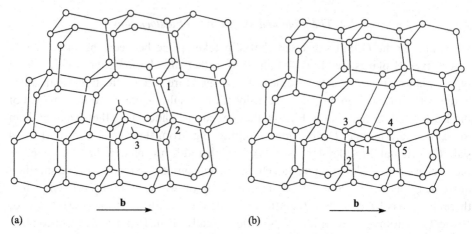

Figure 4.9 The two forms of 60° dislocation: (a) the shuffle set form (after Hornstra 1958. Reprinted from *Journal of Physics and Chemistry of Solids*, **5**, Dislocations in the diamond lattice, pages 129–41. Copyright 1958, with permission from Elsevier) and (b) the glide set form (after Hirth and Lothe 1968).

shuffle set dislocation with two shuffle set partial dislocations and a ribbon of stacking fault on an adjacent glide set (type II) {111} plane.

Alexander (1979) published the illustrations of three-dimensional ball-and-wire models of dissociated glide set 60° and screw dislocations shown in Figs. 4.10 and 4.11.

Alexander (1979) pointed out that, comparing the case of the dissociated 60° shuffle set dislocation, i.e. the association of the 60° shuffle set dislocation D with the stacking fault and partials as represented in Fig. 4.7 and Equation (4.4) and the dissociated 60° glide set dislocation is in the nature of the partial dislocations.

The hypothesis that glide set partial dislocations are the more mobile was a major factor in the early belief that the weak beam TEM evidence, i.e. that dislocations in Si etc. move while dissociated, means that these dislocations are of the glide set type. In fact, later experimental and theoretical studies of the mechanism(s) of dislocation movement and their effect(s) on the plastic deformation behaviour of semiconductors have helped to clarify the glide versus shuffle set question.

At first, many workers were inclined to accept that dislocations in semiconductors were probably of the glide rather than the shuffle set, as previously assumed. However, reasons were soon put forward for doubting that they were simply always or entirely of the glide set.

Hirth and Lothe (1968) had originally pointed out that glide set dislocations could be transformed into the shuffle set by absorption of point defects. Absorbing (single) vacancies will move the dislocation line up from the depth corresponding to the glide form to that for the shuffle set. We may term the step up in such cases as a demi-jog.

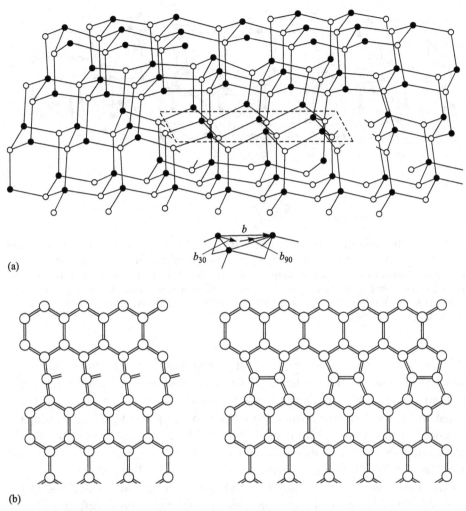

(a)

(b)

Figure 4.10   (a) Ball-and-wire model of a dissociated 60° glide-set dislocation in the diamond or sphalerite structure. The lower diagram gives the Burgers vectors of the 30° and 90° partial dislocations at the left and right boundaries of the stacking fault, respectively. Possible rearrangements or reconnections of the bonds in the cores of the partials have not been represented here. (After Alexander 1979.) In the core of the 90° partial, at the right, the atoms of the two rows with dangling bonds probably form reconnected or 'substituting' bonds as suggested by the results of EPR studies. In the core of the 30° partial, the atoms with dangling bonds, in the row below the left-hand dashed boundary of the stacking faulted area in (a), and at the left in (b), might well also interact in pairs to form reconnected bonds as shown at the right in (b). (After Schröter and Cerva 2002.)

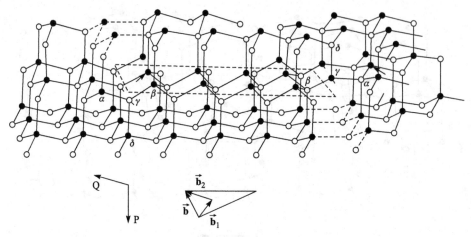

Figure 4.11   Ball-and-wire model of a dissociated glide-set screw dislocation in the diamond structure. The Burgers vectors of the 30° partial dislocations, $\mathbf{b_1}$ and $\mathbf{b_2}$, and of the unit dislocation $\mathbf{b}$ are given in the lower diagram. The left-hand 30° partial dislocation is clearly of the same type as in Fig. 4.10a and probably the dangling bonds reconnect as in Fig. 4.10b. The atoms with dangling bonds in the cores of both the left and right-hand partials are labelled β. (After Alexander 1979.)

This is to distinguish it from the result of adding a divacancy to move the line up from one glide set position to the next. The step in this case is a jog. Conversely, the absorption of (single) interstitial atoms (or, equivalently, the emission of vacancies) would move the line down from the shuffle to the glide depth, i.e. through a demi-jog. In this way two forms of shuffle set perfect 60° dislocation, the shuffle interstitial (S_I) and the shuffle vacancy (S_V) form can, in principle, be produced from the unit 60° glide set dislocation shown in Fig. 4.12 (Blanc 1975).

Blanc (1975) gave a preliminary thermodynamic account of the coupling of the glide and shuffle forms of dislocation lines through native point defects and gave rough estimates of the energies involved. He concluded that point defect-dislocation equilibria were likely to be important for the core structures of dislocations and hence for the semiconductor properties affected by them.

This idea was further considered by Louchet and Thibault-Desseaux (1987, 1989) who also cautioned against allowing the 'glide' versus 'shuffle' terminology to prejudice one unduly in favour of the glide set. They put forward ball-and-wire models of dissociated G, S_I and S_V 60° dislocations, with and without bond reconstruction and discussed these in the light of their HREM studies of dissociated 60° dislocations in Si. They concluded that contrast simulations of the central atomic row in the image of 30° partial dislocations in Si suggested that the density of S_I and S_V sites along the core lies between 1/2 and 3/2 of that for G sites. The 90° partial dislocation images, however, were consistent with mainly G cores, with some S_V sites

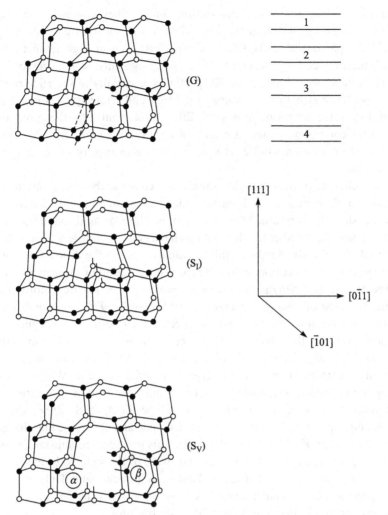

Figure 4.12 Models of the three conceivable configurations of a perfect 60° dislocation in the diamond or sphalerite structure with Burgers vector $a/2[01\bar{1}]$ (without bond reconstructions). G is the glide set form with the extra half plane indicated by the dashed lines. (S_I) and (S_V) are the shuffle set interstitial and vacancy forms obtained by adding a row of interstitial atoms or of vacancies, respectively, to the G line. S_I is the form of dislocation core originally considered by Shockley and Hornstra. (After Blanc 1975. Reprinted with permission from *Philosophical Magazine*, **32** (1975), pp. 1023–32; Taylor & Francis Ltd., the Journal's web site: http://www.tandf.co.uk/journals.)

but few S_I ones. In sphalerite structure semiconducting compounds less HREM work had been done but the results were consistent with a similar interpretation.

The approach of dislocation-point defect energetics was subsequently adopted by, e.g. Justo *et al.* (2000), who used it to calculate the likelihood of shuffle set core configurations along 30° partial dislocations in Si. They concluded that the free energy of vacancy formation in the core of such a dislocation is reduced by more than 1 eV compared with bulk sites. Nevertheless, even at high temperatures, the predicted thermal concentration of the shuffle segments, consisting of rows of vacancies in the core, is low, so the 30° partial is essentially in the glide position.

In the sphalerite structure, of course, the black and white spheres in models like those of Fig. 4.12, represent atoms of different elements. In the more ionically bonded II-VI compounds, these are ions of opposite charge. Thus, moving up or down from the glide to a shuffle set core involves a change of ionic charge on the dislocation line.

This line charge, it was thought, should affect a number of properties of the dislocations and, hence, several phenomena accompanying slip. Study of these phenomena should, therefore, help to determine the sign and magnitude of these charges and indicate whether the dislocations have glide or shuffle set cores. In purely ionic crystals like the alkali halides, this net ionic charge in the dislocation core is the sole determinant of dislocation charge. Then, determination of the sign and direction of charge flow accompanying dislocation movement in plastic deformation can determine the type of dislocation involved (Whitworth 1975). Speake *et al.* (1978) made such observations on ZnS and interpreted their results as evidence that the dislocations were of the glide set. However, although the II-VI compounds are partly ionic, they are electronic semiconductors so the dislocation charge changes as electrons are trapped or emitted. Therefore, Petrenko and Whitworth (1980), following the approach of Speake *et al.*, determined the sign of dislocation current flow in monocrystals of *p*-type ZnTe and *n*-type ZnO, ZnS, ZnSe, CdS, CdSe and CdTe and applied a model for the combined ionic and electronic line charge in the II-VIs, developed by Kirichenko *et al.* (1978). This provided circumstantial evidence that the charges carried through these crystals were of the signs to be expected if the cores were glide rather than shuffle set. That is, as Louchet and Thibault-Desseaux (1987) suggested, the concentration of G sites along both the 30° and 90° partial dislocations appears to be larger than that of shuffle sites.

All the experimental evidence on the chemical and physical differences between polar-opposite edge dislocations (α and β type) in sphalerite-structure materials, prior to Hirth and Lothe (1968), was plausibly and self-consistently interpreted in terms of the shuffle set model. They must be reconsidered for their bearing on a possible glide set interpretation (see Section 5.1.9).

It is now recognized that there are many types of kinks, jogs and demi-jogs possible in dislocation cores and that they have many possible atomic bond configurations and energies; so much so that one worker in this field wrote of its 'bottomless complexity' (Bulatov 2001).

In the remainder of Section 4.1.2 we discuss the experimental and theoretical work bearing directly on the question of whether dislocations in semiconductors are of the glide or shuffle set type. This can be investigated experimentally, by

TEM (see below) or theoretically (see below). Due to the strongly directional covalent bonding in semiconductors, dislocation core structures determine the defect mobility. Hence dislocation dynamics and plastic deformation studies provide indirect evidence on the core form (Section 4.2 and Section 4.4).

Dislocation core structures also play roles in determining the electrical and luminescent properties of dislocations (Chapter 5).

Weak beam TEM evidence of dissociation

The earliest TEM observations of dislocations in semiconductors, before the weak-beam diffraction contrast technique was developed, showed the need to resolve and identify the core structures of dislocations and their detailed modes of movement. It was found that dislocations in Ge, introduced by creep at relatively slow rates, were strongly aligned with low index crystallographic directions ($\langle 110 \rangle$ and $\langle 112 \rangle$, Holt and Dangor 1963) and, similarly, dislocations in Si, deformed before examination, take the form of hexagonal loops with the segments aligned in $\langle 110 \rangle$ directions (Wessel and Alexander 1977). Moreover, those observed in a heated straining stage in a high-voltage TEM, moved slowly and smoothly in this form (Sato and Sumino 1977). This strongly suggested that dislocations in covalently bonded semiconductors align in relatively deep Peierls potential troughs and move by the double kink mechanism (Section 2.5.2). The weak-beam technique was developed precisely in order to resolve these dislocation core structures (Section 3.6.3).

Weak-beam TEM studies show that dislocations are dissociated and move as extended (dissociated) dislocations in these materials but direct tests of the glide or shuffle set character of the core structure have not yet been devised. Atomic resolution (HRTEM) analyses of the core structures of dissociated dislocations are as yet inconclusive. We briefly discuss this TEM evidence here and in the following section.

Ordinary 'strong beam' transmission electron microscopy results in such strong diffraction contrast at dislocations that any near-core structure is obscured (see Section 3.6.3). It was only with the development of the 'weak beam' TEM technique that the structure of the narrowly dissociated dislocations in semiconductors could be resolved (Ray and Cockayne 1970, 1971, 1973, Cockayne and Hons 1979). This narrowness is due to the high stacking fault energy, which resists the separation of the partials, due to their elastic repulsion. Ray and Cockayne (1971) made systematic observations on unit dislocations with $a/2\langle 110 \rangle$ Burgers vectors and found that they were dissociated into Shockley partial dislocations with a separation of 7.5 ± 0.6 nm for those in edge orientation and 4.1 ± 0.6 nm for those in screw orientation. They calculated the intrinsic stacking fault energy from these dissociation widths using anisotropic elasticity theory and obtained 51 ± 5 mJ m^{-2}. They also found that dislocations in Si specimens that had been annealed at high temperature were constricted, back to the unit form, at intervals along their lines as can be seen in Fig. 4.13.

Meingast and Alexander (1973) examined (100) specimens of Ge using the weak beam technique. They observed groups of dissociated dislocations and found that

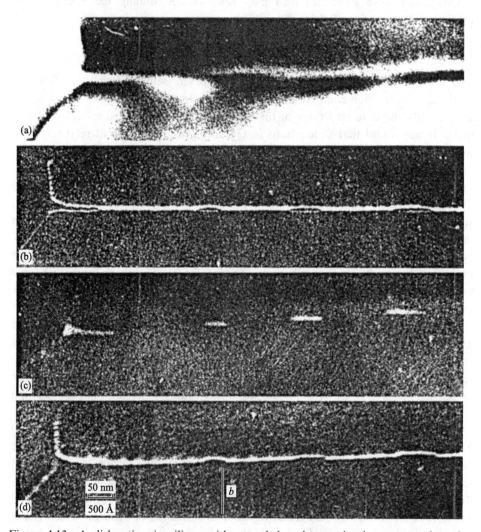

Figure 4.13 A dislocation in silicon with extended and constricted segments along its length. (a) A strong-beam $2\bar{2}0$ dark-field micrograph in which the local changes cannot be seen. (b) A weak-beam $2\bar{2}0$ dark-field micrograph showing both partials in the dissociated segments. (c) A weak-beam $11\bar{1}$ dark-field micrograph for which g·b = 0 for the total Burgers vector, so the constricted unit dislocation segments are out of contrast (invisible). The stacking faults in the extended segments appear in bright contrast. (d) A weak-beam $0\bar{2}2$ dark-field micrograph in which the partial dislocation with Burgers vector $a/6[2\bar{1}\bar{1}]$ is out of contrast. (After Ray and Cockayne 1971.)

they could move without re-associating (constricting back to unit dislocations). Häussermann and Schaumburg (1973) found that all the dislocations in Ge deformed at a low temperature (520° C) and stress (2 Kp mm^{-1}) i.e. in creep, were dissociated over at least a part of their lengths. They found a partial dislocation separation of 54 ± 11 nm for edge dislocations.

It was also observed by weak-beam TEM that dislocations in Si and Ge are dissociated both at room and elevated temperatures before and after moving (Gomez and Hirsch 1977) so there was little doubt that they actually moved in dissociated form. Some workers regard this fact as the chief evidence for believing that dislocations in semiconductors have glide set core structures.

Gomez et al. (1975) used the weak beam TEM technique to measure the separations of the partial dislocations in dislocations in near-screw orientations. They found that these separations fell into two distinct groups, one wide and one narrow. The narrow separations were in line with those expected from the measurements previously reported, which were all made on dislocations with line to Burgers vector angles of 30° or more. These narrow-dissociation near-screw dislocations were found to contain intrinsic stacking faults. The more widely dissociated category of near screw dislocations contained extrinsic stacking faults. Their measurements are shown together with those of Ray and Cockayne in Fig. 4.14. It can be seen that one group of low angle separations like the large-angle dislocation measurements correspond to low values of the stacking fault energies (the lower, solid curves) while the other falls between the much higher (about twice as large) values of the two upper, broken curves. They proposed a model to explain why both forms of dissociation should occur in near-screw orientation dislocations but not in near-edge orientation dislocations.

The conclusion that both intrinsic and extrinsic faults occur in diamond structure semiconductors and that the two fault energies are not very different (by a factor of about 2 in Fig. 4.14) is in agreement with some earlier observations of Ray and Cockayne (1971). They applied the weak-beam TEM technique to threefold dislocation nodes in Si and found that all the nodes were extended. In a hexagonal network, if all the threefold nodes are dissociated, alternate ones must contain intrinsic and extrinsic stacking faults showing that intrinsic and extrinsic stacking fault energies in Si are comparable. This finding has not yet been taken into account in terms of dislocation core structure models.

Gomez et al. (1975) plotted their values for the dissociation separations in Si against the angle between l and b along with those of Ray and Cockayne. Whereas Ray and Cockayne values all fell between the theoretical curves corresponding to fault energies of 46 and 56 mJ m^{-2} those of Gomez et al. all fell outside these limits. Thus the fault energy of Si is not at present understood.

Another problem of interpretation is presented by the work of Gai and Howie (1974) on GaP shown in Fig. 4.15. It can be seen that the variation in fault width (partial dislocation separation) with angle between l and b, i.e. dislocation type, is

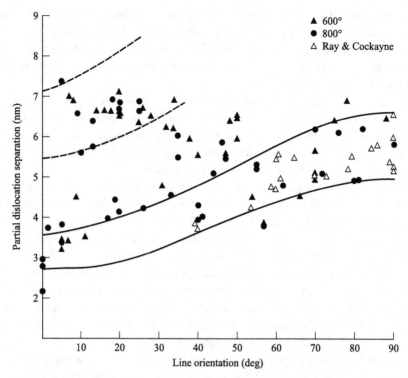

Figure 4.14 The measured separations of dislocations in Ge plotted against the angle between the dislocation line direction, **l**, and the Burgers vector, **b**. Measurements were made on specimens deformed at two temperatures as indicated by the circular and triangular points and the earlier measurements of Ray and Cockayne (1970, 1971, 1973) on high-angle (near-edge) dislocations are also shown. The theoretical curves are based on anisotropic elasticity theory (for the repulsion between the partials) and assumed stacking fault energies of 52 and 68 mJ m^{-2} for the lower, solid curves and 23 and 28 mJ m^{-2} for the upper, broken curves. (After Gomez *et al.* 1975. Reprinted with permission from *Philosophical Magazine*, **31** (1975), pp. 105–13; Taylor & Francis Ltd., the Journal's web site: http://www.tandf.co.uk/journals.)

much less than predicted by theory as given by the solid curves. Later, it was shown that dislocations in III-V compounds with Burgers vector $1/2\langle 1\bar{1}0\rangle$ on {111} planes are generally dissociated (Gomez and Hirsch 1978, Gottschalk *et al.* 1978).

Another important result was reported by Alexander (1976) and Wessel and Alexander (1977). They found that in Si deformed under creep conditions of relatively high stress and low temperature, practically all the dislocations had lines running accurately in ⟨110⟩ directions as shown in Fig. 4.2, so that only screw and 60° dislocations occurred. An early paper using the original strong-beam technique similarly reported that in Ge deformed in creep the dislocations were generally accurately aligned along ⟨110⟩ and ⟨112⟩ directions so they were again of the small set

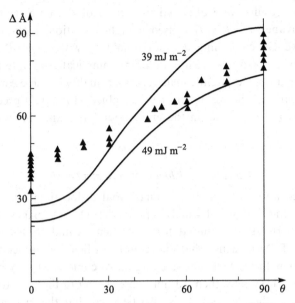

Figure 4.15 Separations of partials measured by the weak-beam TEM technique versus the line direction – Burgers vector angle, θ, for GaP. The curves are from anisotropic elasticity theory for stacking fault energies of 39 and 49 mJ m^{-2}. (After Gai and Howie 1974. Reprinted with permission from *Philosophical Magazine*, **30** (1974), pp. 939–43; Taylor & Francis Ltd., the Journal's web site: http://www.tandf.co.uk/journals.)

of types for which the line directions were the edges and corner bisectors of the Thompson tetrahedron (Fig. 2.2).

This type of observation must be significant for the mobility and the mechanisms of movement of dislocations in covalently bonded crystals; if nothing else than for the fact that the number of cases that must be modelled is small. Moreover, both the nature of the core, that is of the partials and of the stacking fault in extended dislocations and the constrictions at intervals along their lengths (Fig. 4.13), appear to be important for understanding dislocation dynamics (Packeiser and Haasen 1977).

HRTEM evidence on dislocation cores

High resolution transmission electron microscopy (HRTEM), i.e. atomic resolution TEM, is possible but not easy and the interpretation of the images is not straight-forward (see Section 3.6.1). Hence no one has been able to produce completely convincing evidence of the glide or shuffle set character of the cores of dislocations yet – but the capabilities of the technique are approaching the level required to resolve the relevant details of the dislocation core structure (Kolar *et al.* 1996 and Cai *et al.* 2004).

The best early results were obtained in studies of the core structures of grain boundaries. Krivanek *et al.* (1977) used a high-resolution, high-voltage (500 kV) TEM to observe Ge containing a grain boundary with a ⟨110⟩ tilt axis and a misorientation of 38°56.6′ but with a structure complicated by intersections with a number of twin planes. Despite the relatively poor quality of these early micrographs and the complications, the results closely resembled Hornstra's predictions. Much later HRTEM work was done on the grain boundaries and this will be discussed in Section 4.5.3.

Calculated dislocation core structures

As was discussed above, it was first thought that dislocations should be confined to the shuffle set, i.e. the set of widely separated {111} slip planes since the lattice resistance or Peierls stress required for movement would be lower in this case. The weak beam TEM evidence that dislocations in Si were dissociated, even when moving, together with the fact that stacking faults could exist only on the glide set (closely spaced {111}) planes then led opinion to swing to the view that dislocations were exclusively of the glide set. It is also believed that the broken bonds of the earliest Shockley and Hornstra models are rearranged to minimize the energy and the numbers of trap states (see the calculations of Marklund 1979, the discussion of Hirsch 1985 and the reviews by Duesbery and Richardson 1991, Schröter and Cerva 2002 and Cai *et al.* 2004). It was also found that impurities and dislocation cores could interact to alter, passivate or create dislocation states (to be discussed in Section 5.5.8 and see, e.g., the references and calculations of Heggie *et al.* 1989 and Jones *et al.* 1993). It is, therefore, now thought dangling bonds only occur at occasional 'special sites' along dislocation lines (Hirsch 1981, 1985). The detail in ESR signals led to the conclusion that major effects were also due to what are probably point defects such as vacancies or divacancies left as debris behind the moving dislocations and this conclusion is supported by DLTS (deep level transient spectroscopy) results as will be discussed in Section 5.9.1.

Later theoretical work therefore concentrated almost exclusively on glide partial dislocations. However, arguments were put forward to show that dislocations could exist in both the glide and shuffle sets. By adding or removing vacancies, partial dislocation cores could move from positions and structures of the one set to the other (Louchet and Thibault-Desseaux 1987) which Justo *et al.* (2000) describe as the shuffle-glide coexistence idea. HREM evidence indicates that significant concentrations of intrinsic point defects do occur in the core of partial dislocations (Olsen and Spence 1981, Bourret *et al.* 1983).

Justo *et al.* (2000), therefore, calculated the energetics of glide and shuffle partial dislocations in Si for a series of cases of vacancy concentrations from zero for the purely glide core to 100% for the entirely shuffle structure. Of the most important partials, the 30° and 90° types, they decided to concentrate on the former for three reasons. Firstly it has a well-established core structure (Csányi *et al.* 1998) whereas the 90° partial does not (e.g. Bigger *et al.* 1992, Bulatov *et al.* 1997, and Csányi *et al.* 1998). Secondly, both experiment and theory suggest that the mobility of the extended dislocations is limited by that of the slower 30° partials. Thirdly, evidence indicates that the electrically active centres found by ESR are related to vacancies in the cores of 30° partials. Their conclusion was that although the equilibrium concentration of vacancies in a 30° glide partial core is considerably higher than in the bulk, it is still low (about one in 10^4 sites at the melting point). Thus they concluded that 30° partial dislocations in Si belong to the glide set.

In addition to the deep dislocation bands of states believed to arise from dangling bonds and/or impurity decoration, shallow states are believed to arise from the deformation-potential effect of the strain fields of dislocations (e.g. Claesson 1979) or from the rearranged, strained bonds formed to eliminate dangling bonds (e.g. Jones 1979).

Dislocation motion

Sato and Sumino (1977) used a high voltage TEM (HVTEM) fitted with a heated tensile straining stage to directly observe dislocation motion in silicon undergoing plastic deformation. The advantage of HVTEM is that thicker films can be observed in transmission than if lower voltage instruments are used. They found that the dislocations moved sufficiently slowly, controllably and uniformly to make it possible to take a series of still photographs for examination. The moving dislocations were found to be straight or formed parts of hexagonal loops, aligned in ⟨110⟩ directions, suggesting that the Peierls mechanism operates during slip. No hold-ups at point obstacles were observed although both float- and Czochralsi-grown Si samples were examined after annealing at 1050° C for 24 hours to promote the formation of clusters of oxygen atoms. However, when stationary dislocations were set in motion, a segment first bowed out and then dragged the remainder out of its initial position suggesting some form of locking. But whether this locking was due to impurity atmospheres or to the Peierls mechanism Sato and Sumino (1977) could not tell.

Sato and Sumino (1977) also observed pile-ups, cross slip, generation from dislocation sources, formation of jogs and dipoles by intersecting dislocations and other forms and effects of dislocation interactions.

For authoritative reviews of the work on dislocation dynamics and the plastic deformation of semiconductors see Bullough and Tewari (1979), Weertman and Weertman (1980), Shoeck (1980), Duesbery (1989) and Cai *et al.* (2004).

George (1997) reviewed the mechanical behaviour of semiconductors, including the core structure of dislocations and glide processes, as well as dislocation mobilities. He emphasized the importance of computer simulation of atomistic processes for the determination of such microscopic quantities as core defect energies, which cannot be measured directly (George 1997).

Dislocation motion in semiconducting crystals may be enhanced by various electronic perturbations (see Maeda *et al.* 2000 and Section 5.3.3) such as radiation-enhanced dislocation glide (REDG), which is dislocation glide in a semiconductor enhanced by carrier injection (carriers can be injected due to laser light excitation, electron-beam irradiation, or current injection). This process can play an important role in the degradation of semiconductor devices.

Recent progress in elucidating some of the fundamental issues, related to dislocation core structure and energetics and their influence on dislocation motion in silicon, was outlined in a series of articles (see George and Yip 2001) that are briefly described below.

Bulatov (2001) pointed out the 'bottomless complexity of core structure and kink mechanisms of dislocation motion in silicon'. There are many types of kinks, jogs and demi-jogs in dislocation cores and they have many possible atomic bond configurations and energies. He also emphasized the importance of atomistic simulations for elucidating the relevant core mechanisms of dislocation motion and predicting the intrinsic dislocation mobility (Bulatov 2001).

Iunin and Nikitenko (2001) analysed the dependence of dislocation motion modes on stress and temperature. They proposed that the discrepancies between the theory (simple kink model) and experimental observations could be resolved (in part) by taking into account the effect of point defects on the formation and motion of kinks (Iunin and Nikitenko 2001).

Jones and Blumenau (2001) suggested an alternative theory to the model of dislocation glide by the nucleation and propagation of kinks. They proposed that obstacles, such as Si interstitial clusters bound to the line, control the dislocation mobility.

Yonenaga (2001) investigated the dynamic behaviour of dislocations in heavily doped Si. Suppression of the dislocation generation from a surface scratch was observed for Si doped with B, P, and As in excess of $10^{19}\,\mathrm{cm^{-3}}$. This was related to dislocation locking as the result of impurity segregation. It was also found that the velocity of dislocation increases with increasing dopant concentration (Yonenaga 2001).

Spence and Koch (2001) discussed a new electron diffraction technique in a STEM for deriving direct experimental evidence of dislocation core structures in silicon. They proposed to use this method for investigations related to the temperature dependence and activation energy of atomic processes at dislocation cores (Spence and Koch 2001).

4.1.3 Polarity in the sphalerite and wurtzite structures

The array of sites in the sphalerite structure is the same as that in the diamond cubic structure. However, the sites of the two f.c.c. lattices are occupied by atoms of elements A and B in AB sphalerite-structure semiconducting compounds. Hence such compounds, unlike diamond-structure semiconducting elements, are non-centrosymmetric and exhibit polarity. That is, neither opposite directions like [111] and $[\bar{1}\bar{1}\bar{1}]$ nor opposite planes such as (111) and $(\bar{1}\bar{1}\bar{1})$ are equivalent. Polarity attracted attention at the beginning of the study of III-V compounds, when great differences were found between the properties of opposite faces of {111} oriented slices. [The standard orientation of slices in the early days of the semiconductor industry was {111}. Later it switched to {100} for convenience of integrated circuit manufacture.]

Coster *et al.* (1930) first showed that bulk polarity could be determined in sphalerite-structure ZnS by x-ray diffraction. The method was treated in the monograph by James (1948, Chapter 4 and Appendix III). Interest was revived by the discovery of the strikingly different chemical etching and crystal growth behaviour of opposite faces of {111} slices of the III-V compounds (see Table 4.1). White and Roth (1959) and Warekois and Metzger (1959) independently applied the method of Coster *et al.* (1930) to determine the polarity of III-V compounds and many others used it on both III-V and II-VI compounds. References to this work will be found in the later accounts of polarity determinations by Burr and Woods (1971) for ZnSe, and Schmidt *et al.* (1973) for $Ga_{1-x}Al_xAs$ and $Ga_{1-x}Al_xSb$ alloy crystals. These crystallographic (bulk) polarity determinations do not determine the surface polarity.

Bulk polarity controls the signs of polar properties. For example, the piezoelectric effect produces surface charges when a sphalerite structure semiconductor is elastically strained. Thus the signs of the charges on opposite faces of {111} slices depend on the polarity as well as on the sign of the strain (e.g. tensile or compressive).

Polar differences between the opposite faces of {111} slices

Consider, to be specific, the sphalerite structure in Fig. 4.16a. It was assumed that the crystal always separates or terminates at a type I plane, so the top and bottom faces of a (111) slice of e.g. GaAs would be different. We take the A atoms to occupy the 0,0,0 f.c.c. lattice sites and the B atoms to occupy those displaced by 1/4, 1/4, 1/4 from them. Then the sense from *a* to α along the vertical bonds will be the [111] direction as shown. The top of the slice, to which [111] is the outward drawn normal, is the (111)A surface in which all the sites are occupied by A atoms with a single dangling bond. Conversely, the direction α to *a* is $[\bar{1}\bar{1}\bar{1}]$, the normal to the bottom $(\bar{1}\bar{1}\bar{1})$B surface which is solely occupied by B atoms with single dangling bonds. The properties of the (111)A and $(\bar{1}\bar{1}\bar{1})$B faces of III-V compounds were in fact found to be very different as shown in Table 4.1.

It was assumed the faces of (111) slices terminate on type I planes only, as shown in Fig. 4.16, because then the surface atoms are held in place by three bonds to the interior of the slice with only one bond dangling. Hence, type I termination should be

Table 4.1. *Physical, chemical and materials property differences between the (111)A and ($\bar{1}\bar{1}\bar{1}$)B faces of III-V compounds*

Property	Difference	References
Oxidation	The ($\bar{1}\bar{1}\bar{1}$)B faces anodically oxidize faster than do the (111)A	Dewald (1957), Lavine *et al.* (1958), Venables and Broudy (1959, 1960), Miller *et al.* (1961)
Dissolution by oxidizing reagents (etchants)	(1) Attack is faster on ($\bar{1}\bar{1}\bar{1}$)B faces than on (111)A faces (2) so ($\bar{1}\bar{1}\bar{1}$)B faces polish while dislocation etch pits are generally only produced on (111)A faces	(1) Gatos and Lavine (1960a,b) (2) Allen (1957), White and Roth (1959), Warekois and Metzger (1959)
Electrode potentials	(111)A faces are the more noble (less chemically reactive)	Gatos and Lavine (1960a)
Crystal growth	(1) Czochralski-grown GaAs and InSb crystals are of better quality if the seeds are dipped into the melt (111)A face down (2) (111) A faces grow faster than ($\bar{1}\bar{1}\bar{1}$)B faces	(1) Moody *et al.* (1960), Gatos *et al.* (1960), Mueller and Jacobson (1961) Hulme and Mullin (1962) (2) Ellis (1959), Steinmann and Zimmerli (1963), Ewing and Greene (1964), Ku (1963)
Mechanical damage resistance	(1) The depth and severity of mechanical polishing damage is greater under ($\bar{1}\bar{1}\bar{1}$)B than under (111)A faces (2) Hence InSb {111} slices polished down to < 10 µm thickness spontaneously bend so the ($\bar{1}\bar{1}\bar{1}$)Sb face is convex.	(1) Warekois *et al.* (1960), Gatos *et al.* (1961), Nicolaeva *et al.* (1975) (2) Hanneman *et al.* (1962)

energetically favoured over type II, for which surface atoms would have three dangling bonds but only one bond to the crystal. It was found that the (111)A and ($\bar{1}\bar{1}\bar{1}$)B surfaces differ greatly in (i) growth properties when used either dipped into the melt for Czochralski bulk growth or as the substrate surface for epitaxial film growth, in (ii) etching behaviour and in (iii) mechanical abrasion resistance, etc. (Table 4.1).

Thus, one side of {111} slices is an A atom face (i.e. the atoms are of the lower valence e.g. III-valent element) and the other is a B (e.g. V-valent) atom face. The polar face notation, (111) for A and ($\bar{1}\bar{1}\bar{1}$) for B, is that of Gatos and Lavine (1960a) and is equivalent to placing an A atom at the f.c.c. lattice sites (e.g. at 0,0,0) in the sphalerite structure and the B atoms at sites like 1/4, 1/4, 1/4. However, Dewald (1957), in the first important paper in this field, used the opposite convention, as did

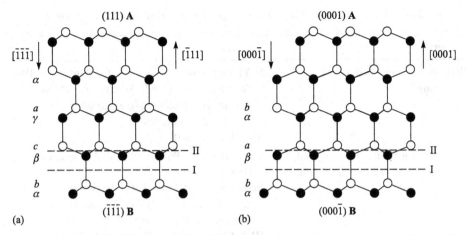

Figure 4.16   Schematic diagrams illustrating polarity in the (a) sphalerite and (b) wurtzite structures of semiconducting AB (e.g. III-V and II-VI) compounds. The open circles represent atoms of the lower valency (A) element and the black circles represent the higher valency (B) atoms. The stacking sequence positions of the planes of atoms (see Section 1.4.3 and Section 1.4.4) are indicated to the left of each drawing (from the bottom upwards: α, b β, c γ, a α, ... for sphalerite, and α, b β, a α, b β ... for wurtzite). The dashed lines marked I and II indicate the positions of the two types of planes at which the structures could terminate.

a number of other early workers. To avoid confusion, it is important to use the unambiguous notation (111)A and ($\bar{1}\bar{1}\bar{1}$)B. The oxidation and chemical attack rates of the polar faces are very different (see the first two rows of Table 4.1). As demonstrated directly by electrode potential measurements, the (111)A faces are the more noble (less chemically reactive) than the ($\bar{1}\bar{1}\bar{1}$)B faces (see row three in Table 4.1). Consequently, in sphalerite structure III-V compounds, dislocation etch pits are produced by etching reagents only on the (111)A faces (Allen 1957, White and Roth 1959, and Warekois and Metzger 1959). This can be understood as, due to the slow attack rate on A faces, the slightly faster attack down dislocation cores is sufficient to nucleate pitting. The attack on B faces, however, is so fast that minor differences like that due to dislocations are smoothed out, resulting in chemical polishing. Early bulk polarity determinations were often correlated with such etching studies. Etching, directed at determining which side chemically polished, therefore, became the usual, quick and easy way to identify the surface polarity of (111) III-V slices. Type I termination of the surfaces of III-V {111} slices, interpreted in such ways, was found to be consistent with all the findings on (111)A versus ($\bar{1}\bar{1}\bar{1}$)B properties (see Tables 4.1 and 4.2).

Many other properties of polar surfaces in III-V and II-VI compounds were also studied and some of the more valuable work of this kind is summarized in Table 4.2.

A number of reviews of the etching of semiconductors including reports of reagents for use in identifying polar surfaces and in producing pits to mark the sites

Table 4.2. *Studies of Polar Properties of III-V and II-VI Compounds*

Property	References
Etching including Dislocation Etching	**II-VI Compounds:** Woods (1960), Zare *et al.* (1961), Warekois *et al.* (1962), Wolff *et al.* (1964), Sturner and Bleil (1964), Sagar *et al.* (1968), Gezci and Woods (1972) **SiC:** Brack (1965)
Rates of growth, morphology and perfection of crystals grown on polar faces	**GaAs:** Booker (1962), Sheftal and Magumedov (1967) **InSb:** Faust and John (1962), Seidensticker and Hamilton (1963), Wolfe *et al.* (1965), Mueller and Jacobson (1961) **III-V Compounds:** Faust and John (1964)
Growth and structure of films epitaxially deposited on polar surfaces	**II-VI Compounds:** Weinstein and Wolff (1967), Morizumi and Takahashi (1970), Holt and Wilcox (1972), Holt (1974)
Thermal decomposition of A and B faces in vacuum	**InSb:** Haneman (1960), Haneman (1962b), Churchill and Watt (1969) **GaAs:** Millea and Kyser (1965) **III-V Compounds:** Haneman *et al.* (1964)
Polarity effects in alloy junction formation	**InSb p-n junctions:** Minamoto (1962), Barber and Heasell (1965a,b), Henneke (1965) **Heterojunctions:** Hinkley *et al.* (1964)
Hardness	**Wurtzite Structure Compounds Including BeO:** Cline and Kahn (1963) **Sphalerite Structure Compounds:** Usually the (111)A surface is harder than the $(\bar{1}\bar{1}\bar{1})$B surface Shimizu and Sumino (1970), Gatos and Lavine (1965), Maeda *et al.* (1977). The A face is harder than the B in n-type GaAs but in p-type GaAs the As face is the harder Hirsch *et al.* (1985).
Work function and sticking coefficient for oxygen on Si and C faces of SiC	Dillon (1960)
Photoelectric emission and the adsorption of oxygen and H_2O on polar faces of ZnTe	Dillon (1962)
Sign and magnitude of piezoelectric surface charges in III-V and II-VI compounds	**III-V:** Arlt and Quadriflieg (1968) **II-VI:** Berlincourt *et al.* (1963)

of dislocations threading up to the surface were listed in Section 3.3.1. A number of these etch polar dislocations differently so the two types can be distinguished. Reagents are available to do this reliably for virtually all III-V and II-VI compounds. Etching is also fast and simple so it has become the standard method for both surface and dislocation polarity determinations.

Polarity in the wurtzite structure is analogous to that in the sphalerite structure except that there is only one polar plane and direction. These are the basal plane (0001) and its polar opposite, (000$\bar{1}$), and their outward drawn normals, ⟨0001⟩ and its polar opposite ⟨000$\bar{1}$⟩ shown in Fig. 4.16b. [All the ⟨111⟩ directions and {111} planes in the sphalerite structure are polar. Thus, for example, the (1$\bar{1}$1) and ($\bar{1}$1$\bar{1}$) planes are of opposite polarity, as are their outward drawn normals, ⟨1$\bar{1}$1⟩ and ⟨$\bar{1}$1$\bar{1}$⟩.] (0001) slices of wurtzite structure compounds were assumed to terminate on type I planes, for the same energetic (bonding) reason as in the case of sphalerite structure semiconductors. This gives the structures of the opposite faces of (0001) oriented slices of wurtzite structure III-V and II-VI compounds shown in Fig. 4.16b. The sign convention for the polar directions and polar surfaces of such slices are defined similarly to those for the sphalerite structure as discussed above and as shown in Fig. 4.16b.

Direct evidence for type I termination, i.e. the models of Fig. 4.16, was provided by Brongersma and Mul (1973). They used noble gas ion reflection mass spectroscopy (NIRMS) also known as ion scattering spectroscopy (ISS). This mass analyses the outermost atomic layer of a crystal surface by measuring the noble-gas-ion energy loss on back reflection. They studied polar faces of sphalerite structure ZnS and wurtzite structure CdS in ultra high vacuum. Their results clearly showed the (111) and (0001) faces, as defined by x-ray diffraction and etching, to be the A faces (Zn and Cd respectively) and the ($\bar{1}$$\bar{1}$$\bar{1}$) and (000$\bar{1}$) faces to be the B faces (S in both cases).

Polar asymmetries in {100} faces

There is evidence that the atomic distribution along the [011] direction differs from that along [01$\bar{1}$]. [These are not polar opposite directions, which are at 180° to one another but directions at 90° to each other, lying in the (100) face.]

Olsen *et al.* (1974) found that, as shown in Fig. 4.17, the pits produced by Sirtl etching the two opposite sides of {001} slices of GaAs were elongated in ⟨110⟩ directions at right angles to each other. (The Sirtl etch is a CrO_3–HF–H_2O solution, see Sirtl and Adler 1961.) The morphology of these pits makes it possible to differentiate the ⟨011⟩ and ⟨01$\bar{1}$⟩ type directions in {100} faces.

The crystallography of the opposite polarity pits was explicitly identified by Faust and Sagar (1960) in Fig. 4.18, but they indexed the surface of the crystal as (001) not as (100) as Olsen *et al.* (1974) did. The directions of different polarity in the surface were therefore indexed, for example, on the bottom face as [$\bar{1}$10] and [$\bar{1}$10].

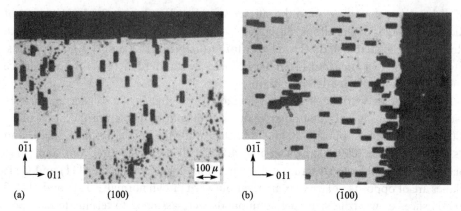

Figure 4.17   Sirtl etch pits on opposite sides of a {100} slice of GaAs. (After Olsen *et al.* 1974. Reprinted with permission from *Journal of the Electrochemical Society*, **121**, 1650, 1974. Copyright 1974, The Electrochemical Society.)

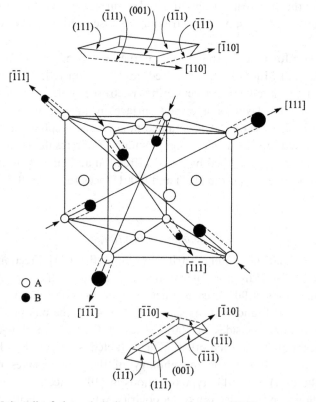

Figure 4.18   Unit cell of the sphalerite structure of many III-V compounds and sketches of the crystallographic orientations of pits produced on the top (0001) and bottom (000$\bar{1}$) faces of an {0001} slice. (After Faust and Sagar 1960. Reprinted with permission from *Journal of Applied Physics*, **31**, pp. 331–3. Copyright 1960, American Institute of Physics.)

When {100} slices were used as substrates for epitaxial growth of ternary III-V alloys, differences in lattice parameter and differences in thermal contraction on cooling after growth led to asymmetric elastic, plastic or fracture phenomena. Nagai (1972) examined slices of GaAs on which $Ga_xIn_{1-x}As$ or $GaAs_xP_{1-x}$ had been epitaxially deposited and found that they were anisotropically bent in the $\langle 110 \rangle$ and $\langle \bar{1}10 \rangle$ directions. When growth occurred on both sides of the substrate a 'saddle like' curvature was observed, i.e. the slices were convex about one $\langle 110 \rangle$ direction but concave about the other. Kressel (1975) suggested that this observation can be understood in terms of the fact that the misfit dislocations running in the two directions in the interface are different in atomic core structure. There is evidence that such α and β dislocations have different mobilities, i.e. that the critical stresses required to move them are different. Thus, although due to lattice misfit and thermal contraction, strain is applied equally along both $\langle 110 \rangle$ directions in these epitaxial specimens, plastic strain relief occurs more easily in one direction. Consequently, the other direction exhibits greater bending due to residual elastic strain. On the bottom of the substrate the symmetry is rotated by 90° and the axis of greater elastic bending is perpendicular to the one on the topside. This bending applied along perpendicular axes but from opposite sides results in the observed saddle-like bending of the specimen (Kressel 1975).

Olsen *et al.* (1974) also observed asymmetric, unidirectional cracking, like that in Fig. 4.19, in lattice mismatched III-V compounds grown epitaxially on GaAs substrates by chemical vapour deposition. When growth occurred on both sides of a {100} substrate, the cracks in the top face ran in one $\langle 110 \rangle$ direction, but on the

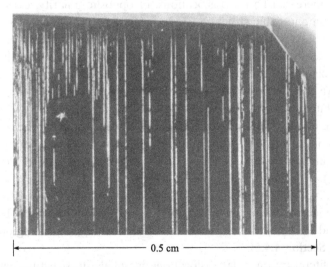

├──────────────── 0.5 cm ────────────────┤

Figure 4.19 Unidirectional cracks in $In_xGa_{1-x}P$ grown on a (100) GaAs substrate. The $In_xGa_{1-x}P$ is under tension. (After Olsen *et al.* 1974. Reprinted with permission from *Journal of the Electrochemical Society*, **121**, 1650, 1974. Copyright 1974, The Electrochemical Society.)

bottom in the orthogonal direction. Asymmetric cracking was readily observed in layers under tension like $In_xGa_{1-x}P/GaAs$ with $x < 0.49$ but, of course, rarely in those under compression, i.e. in layers with $x > 0.49$. Olsen *et al.* explained this asymmetry by a crack initiation mechanism due to Cottrell. In this, pairs of dislocations on intersecting slip planes combine to form sessile (immobile) dislocations in the epitaxial interface. The missing half-planes of the edge-type sessile misfit dislocations are incipient microcracks. This mechanism was first extended to the sphalerite structure by Abrahams and Ekstrom (1960) to explain cleavage on {110} planes. The slip dislocations involved in nucleating cracks in the orthogonal [110] and [1$\bar{1}$0] directions have to be A type in the one case and B type in the other. The difference in mobility of the A and B dislocations leads to operation of the mechanism and consequent cracking in the one direction but not in the other.

One III-V compound can be grown on another in (001) orientation, under such conditions of growth temperature, lattice mismatch and epilayer thickness, that cracking does not occur. However, the lattice mismatch and thickness can then be such that misfit dislocations (MDs) are introduced to relieve the long-range elastic stress (see Section 2.3.11). The difference in numbers of the two types of misfit dislocations of different polar character in (100) epitaxy will be discussed in Section 4.6.4.

Dislocation polarity in semiconducting compounds, polar bending and indentation rosettes

The sphalerite structure is non-centrosymmetric. Hence, as Haasen (1957) first pointed out, there can be two dislocations, of opposite polarity, corresponding to one in the diamond cubic structure. He argued that a (111) slice of InSb could be bent in two opposite polar senses to introduce majorities of edge dislocations of opposite sign as shown in Fig. 4.20(a and b) and Fig. 4.21. (See also the account of such plastic bending of diamond-structure materials in Section 5.2.4 and Fig. 5.9) When positive edge dislocations, for example, were introduced as shown (assumed to be of the 'shuffle set'), their extra half-planes would end in a row of Sb atoms with dangling bonds. These he termed β-dislocations. Bending in the opposite sense would result in In-edged α-dislocations. Such polar bending became a standard means for studying polar dislocations (e.g. Bell and Willoughby 1966, 1970, see Section 5.2.7)

Booyens *et al.* (1978) bent GaAs single crystals about a ⟨110⟩ axis, not a ⟨112⟩ axis as recommended producing single slip in Fig. 4.21. They, therefore, observed work hardening and studied polar differences in this hardening, depending on the sign of bending (see Section 4.4.5).

The early literature on polar dislocations in sphalerite structure materials used Haasen's (α versus β) shuffle set terminology. Once the possibility of glide or shuffle set core forms had been introduced (Hirth and Lothe 1968), it became clear, as

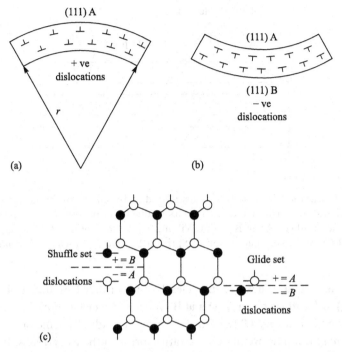

(a)   (b)

(c)

Figure 4.20   Polar bending, illustrating the relation between the crystallographic polarity, the chemical type of edge dislocations and their glide or shuffle set form. (a) and (b) Plastic bending of a {111} slice of a sphalerite-structure material in the two opposite senses relative to the specimen polarity introduces excesses of positive or negative edge dislocations. (c) The dangling bonds in the cores of the positive edge dislocations will be from B atoms if the core is of the shuffle set but from A atoms if the core is glide set. In negative edges the bonds will dangle from A atoms in shuffle set cores but from B atoms in glide-set cores. In the recommended, unambiguous terminology then, in (a) the positive dislocations are termed **B(s)** type if of the shuffle set but **A(g)** type if they are of the glide set. In (b) the negative edges will either be **A(s)** (shuffle set) or **B(g)** (glide-set cores).

shown in Fig. 4.20, that the chemical (polar) character of the dislocations, whether the dislocation-core dangling bonds be from A (lower valence) or B (higher valence) atoms, depended on whether shuffle or glide forms were assumed. Therefore, it was proposed at a meeting in 1979 that the α- and β-dislocation nomenclature be dropped for an unambiguous alternative. This was published by the Organizing Committee, including Haasen (*Journal de Physique* (1979) Colloque C6 Int. Symposium on Dislocations in Tetrahedrally Coordinated Semiconductors). They proposed that: '**To avoid confusion in nomenclature on dislocations in polar AB compounds the participants in the symposium recommended the use of the terms "A- or B-dislocation" for a dislocation with A or B atoms in the most distorted core positions . . . in the so-called shuffle set (s)**' or '**the glide set (g) of {111} planes . . . It is**

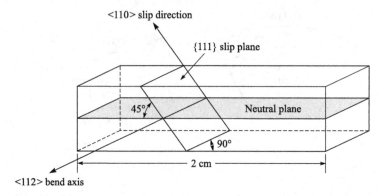

Figure 4.21   Orientation of crystals for plastic bending by single slip on the slip plane and in the slip direction shown. This avoids complications due to dislocations on different slip planes intersecting one another. [After Bell and Willoughby 1966. *Journal of Materials Science*, **1**, 1966, pp. 219–28. With kind permission of Springer Science and Business Media.]

hoped that future publications will state clearly whether they base their discussion on A (g) and B (g) dislocations or on A (s) and B (s) = α and β-dislocations.' (See the legend to the right and left of Fig. 4.20c.) This terminology gradually came into use but care must be exercised in interpreting the literature prior to the early 1980s. In this book we have used the original authors' interpretations of polar observations whether in terms of shuffle or glide forms.

An alternative method for producing separate regions in a semiconducting compound containing majorities of polar dislocations of the two types was later developed. This uses the wings or rays of dislocation etch rosettes arising from microhardness indentations. Microhardness is determined by pressing, under a constant load, a diamond indenter of any one of several shapes into the surface of the material. For example, the well-known Vickers hardness is measured using a pyramidal diamond indenter. This forces material to flow outward from the original point of contact of the indenter. The material is under compression, which is unfavourable for brittle fracture so even fairly brittle materials can be so indented. Dislocation etching reveals the pattern of dislocations that moved outward on the slip planes to accommodate the necessary plastic flow.

The patterns of pits are simple and resemble the petals of flowers and so are called rosettes. They reflect the symmetry of the material. That is, they are twofold rotationally symmetric on (100) surfaces as shown in Fig. 4.22. They are threefold rotationally symmetric on (111) faces as shown in Fig. 4.23. The analysis of indentation rosettes is too complicated to discuss in detail here (for the details in the case of {111} surfaces see Hirsch *et al.* 1985 and for {110} Schreiber *et al.* 1999). The point of interest here is that etch rosettes in the case of diamond cubic structure elements are symmetric (Fig. 4.22), i.e. the left and right vertical rows of etch pits are equal in length

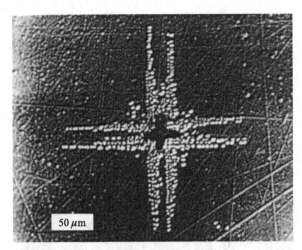

Figure 4.22   Typical dislocation etch rosette pattern around an indentation on the {100} surface of intrinsic Ge indented at 360° C with a 50 g load. Both the vertical arms and both the horizontal arms are of equal lengths. [After Warren *et al.* 1984. Reprinted with permission from *Philosophical Magazine*, A50 (1984), pp. L23 to L28; Taylor & Francis Ltd., the Journal's web site: http://www.tandf.co.uk/journals.]

as are the upper and lower horizontal ones. In contrast those in sphalerite structure compounds are asymmetric, as pointed out by Warren *et al.* (1984), exhibiting long and short arms (Fig. 4.23). This difference is essentially due to the difference in mobility between A- and B-edged dislocations (Hirsch *et al.* 1985). The long and short rows thus mark the positions of rows of dislocations of the two opposite polarities.

Etch pit rosettes around microhardness indentations thus reveal spatially separated regions containing dislocations of opposite polarity. Their differences in physical properties can then be studied by microscopical methods like SEM EBIC and CL (e.g. Schreiber *et al.* 1999 and see Section 5.6.3). Other advantages of the indentation method are as follows: (i) It is simple, requiring only a small, smooth area of single crystal and simple inexpensive equipment; (ii) It works at relatively low temperatures; (iii) It can be used to study the polar anisotropy of microhardness (Warren *et al.* 1987), i.e. of dislocation movement (Warren *et al.* 1984).

Polarity influences defects other than dislocations. To each Hornstra (1959, 1960) type of grain boundary there are two of opposite polarity in the sphalerite structure (Holt 1964, see Section 4.5.3). Antiphase boundaries (APBs) (also known as polarity reversal domain boundaries) are interfaces across which only the polarity (crystal structure site occupation) is reversed in sphalerite structure compounds (Holt 1969, see Section 4.5.3). These can obviously only occur in the sphalerite and wurtzite structures and have no corresponding defect type in the diamond cubic structure. An example of a polarity reversal domain in a wurtzite structure material is shown in Fig. 4.24.

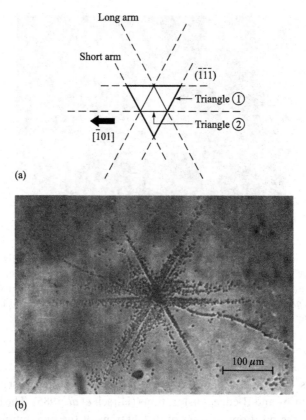

(a)

(b)

Figure 4.23   (a) Diagram giving the orientation of the figure on the semiconductor surface and identifying the main features, especially the occurrence of long and short arms or rays in the rosette. The sides of triangle 1 are parallel to the long arms, those of triangle 2 are parallel to the short arms. (b) Etch pit rosette in the ($\bar{1}\bar{1}\bar{1}$) Te face of p-type CdTe indented with a load of 2×10^{-3} N for 30 sec. There are, in this case, short and long arms of distinctly unequal lengths (see especially the arms extending downwards from the indentation). [After Braun and Helberg 1986. Reprinted with permission from *Philosophical Magazine*, **A53** (1986), pp. 277–84; Taylor & Francis Ltd., the Journal's web site: http://www.tandf.co.uk/journals.]

Osipiyan and Smirnova (1968, 1971) extended ball-and-wire crystal model analysis to shuffle set dislocations in the hexagonal wurtzite structure and found dislocations on the basal slip plane to be polar but not those on prismatic slip planes (see Section 4.1.5).

4.1.4 Dislocations in the sphalerite structure

Following Hornstra (1958), Holt (1962) produced ball-and-wire models of the core structures of the shuffle set dislocations having all the ⟨110⟩ and ⟨112⟩ line directions

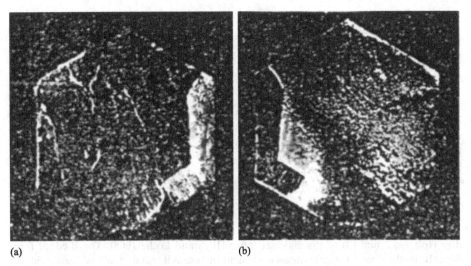

(a) (b)

Figure 4.24 Optical reflection micrographs of two matching cleavage surfaces showing polarity reversal domains in a wurtzite-structure of ZnO revealed by the different etching of the (0001) Zn and (000$\bar{1}$) O regions of the surface in 6% HCl. In (a) the greater part exhibits the etching behaviour of the Zn surface, the smaller that of the O surface. (b) Shows the other matching cleavage surface and exhibits the reverse behaviour. It was confirmed that the direction of the polar c-axis perpendicular to these surfaces was reversed between the two regions by observations of the sign of the piezoelectric voltage. (After Heiland and Kunstmann 1969. Reprinted from *Surface Science*, **13**, Polar surfaces of ZnO crystals, pages 72–84. Copyright 1969, with permission from Elsevier.)

in the sphalerite structure with the [110] Burgers vector. This showed that to each such dislocation in the diamond structure (except the screw) analysed by Hornstra, there were two types of dislocation of opposite polarity in the sphalerite structure, like the α- and β-edge (A(s) and B(s)) dislocations of Haasen (1957). The shuffle set unit 60° dislocation of Hornstra and Shockley takes the polar form in the sphalerite structure shown in Fig. 4.25a, if it is of shuffle set form, otherwise it is of the glide set form in Fig. 4.25b.

Dislocations of opposite polarity are expected to differ in their electronic properties (see e.g. the piezoelectric effects in Section 5.1.7). Interpreting electrical effects of deformation, which introduces dislocations of both polarities in sphalerite-structure compound semiconductors, therefore, must be more complex than in the case of Si. Moreover, the semiconducting compound material available is of poorer quality than the best silicon. It will have some degree of non-stoichiometry (see Section 6.3.2) and contain residual unwanted impurities. For these reasons much less work has been done on the effects of polar dislocations than on those of dislocations in elemental semiconductors like Si.

It was early suggested by Holt (1960) and independently by Haneman (1962a) that bonds dangling from the A and B atoms in the cores of dislocation

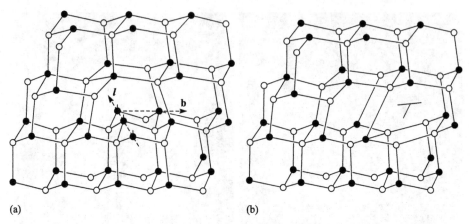

(a)   (b)

Figure 4.25   Ball-and-wire models of the unit 60° dislocation core (a) B(s) of the shuffle set (after Holt 1962) and (b) of the glide set (after Hirth and Lothe 1968). The B-edged form, where B is the (black) higher-valence element in an AB compound, is shown in (a). Interchanging the colours (elemental occupation of the two types of sites) turns this into the A-edged form. Similarly, inverting the polarity produces the other chemical form of model (b).

in AB semiconducting compounds (e.g. III-V or II-VI materials) would be different. Thus bonds dangling from e.g. Ga atoms in the cores dislocations, whether of A(s) or A(g) and bonds dangling from As atoms in the cores of B(s) or B(g) dislocations should contain different numbers of electrons. Holt and Haneman suggested that A-dangling bonds would contain 3/4 electron while B dangling bonds would contain 5/4. This corresponds to the 3 and 5 valence electrons on III and V element atoms, respectively, being equally shared among the 4 tetrahedral sp^3 bond of each atom. That is, this would correspond to neutral bonding (no net transfer of charge, i.e. no ionic contribution to the purely covalent bonding).

The bonds from III-valent atoms can be expected to contain 3/4 electron in the neutral dislocation while the bonds from the V-valent atoms will contain 5/4 electrons in the neutral dislocation. Note that while the +ve dislocations are B(s) (V-valent atom dangling bonds in Fig. 4.20) and the −ve dislocations are A(s) (III-valent atom dangling bonds), this reverses if the cores are glide rather than shuffle set as originally assumed. That is, if the dislocations are glide set the +ve dislocations will be α type and the −ve dislocations have β character. The V-valent dangling bonds will clearly be expected to be more donor in character and the III-valent more acceptor in nature so the sign or magnitude of the electrical effects of bending will depend on the polarity of the bending. The effects of polar bending were as expected on a shuffle set interpretation. This and a number of related observations are the strongest evidence that the cores of dislocations are in fact shuffle set in structure. Alternatively: (The relative size of the effects of polar bending of the two signs is one way to test for the predominance of glide or shuffle set core structures. The other is from the observation that dislocations move in extended form. Atomic model considerations lead theorists to deduce that the dislocation cores are of the shuffle set.)

The amplitude of the Peierls energy variation (Fig. 2.24) for moving dislocations in the (111) slip plane is greatest when they move in $\langle 112 \rangle$ directions, i.e. normal to the $\langle 110 \rangle$ directions. This is why the dislocations that have moved at relatively low temperatures are found to be aligned in the $\langle 110 \rangle$ directions (Fig. 2.25). The energy variation depends on the breaking and reforming of chemical bonds in the dislocation core during movement. Little is known about this.

By comparison with silicon, the study of the effects of impurity decoration on the electrical effects of dislocations in polar semiconducting compounds has hardly begun. Thus while there is some evidence that dislocations of polar character have different properties it is likely that as in the case of Si, these are impurity-decoration dominated. The reactions of impurity atoms with A- and B-edged dislocations should also be different.

4.1.5 Slip systems in the wurtzite structure

The most closely packed crystallographic planes, which are the most widely spaced, are normally the slip planes in any crystal structure. (However, see the shuffle versus glide set controversy for the sphalerite structure, Section 4.1.2.) The most widely spaced planes in the wurtzite structure, according to Osipiyan and Smirnova (1968) are the (0001) planes, then the type I prismatic planes which are those of the form $\{10\bar{1}0\}$, then the type II prismatic planes of the form $\{1\bar{2}\bar{1}0\}$. Since the elastic energy of a dislocation is proportional to b^2 (see Section 2.3.5), the shortest lattice translation vectors are the Burgers vectors of the lowest energy dislocations. These are, therefore, the slip directions. According to Osipiyan and Smirnova, in the wurtzite structure, the shortest translation vectors are $\mathbf{b_1} = \mathbf{a} = 1/3\langle\bar{1}2\bar{1}0\rangle$ and $\mathbf{b_2} = \mathbf{c} = [0001]$. In fact the slip systems found in CdS by Weinstein *et al.* (1965) include both these slip directions and one other, namely $\langle 10\bar{1}0 \rangle$.

The slip systems in the wurtzite structure are more complex than those in the diamond- and sphalerite-structures which are $\{111\}$ with $\langle 110 \rangle$. This is because the predominant system, i.e. the energetically most preferred slip system in the wurtzite structure, (0001) with $\langle\bar{1}2\bar{1}0\rangle$, has only one orientation of slip plane. In contrast, the diamond- and sphalerite-structures have four $\{111\}$ planes oriented as the faces of a tetrahedron (Thompson's tetrahedron Section 2.3 and Fig. 2.2). This provides a sufficient number of slip systems to accommodate any possible deformation. The single predominant (0001) basal plane in the wurtzite structure, however, does not. Therefore, other slip systems frequently operate during deformation. Thus the slip systems found in the wurtzite structure (Weinstein *et al.* 1965) are (0001) with $\langle\bar{1}2\bar{1}0\rangle$ (predominant), (0001) with $\langle 10\bar{1}0 \rangle$, $\{10\bar{1}0\}$ with $\langle\bar{1}2\bar{1}0\rangle$, $\{10\bar{1}0\}$ with [0001] and $\{1\bar{2}\bar{1}0\}$ with $\langle 10\bar{1}0 \rangle$ which are illustrated in Fig. 4.26. Despite the complexity of the possible

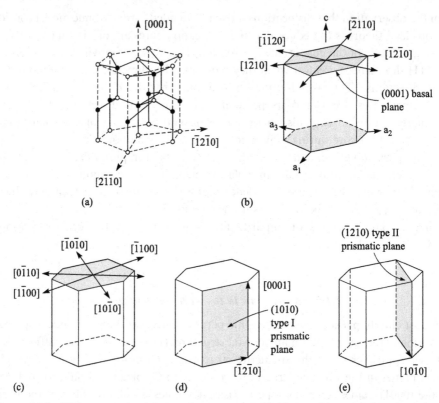

Figure 4.26 Slip systems in the wurtzite structure. (a) The positions of the atoms in the hexagonal structure unit cell. (b) The predominant basal plane slip system. (c) The secondary basal plane slip system. (d) The type I prismatic slip system. (e) The type II prismatic slip system.

types of dislocations in the wurtzite structure there has been a revival of interest in defects in this structure since wurtzite-structure GaN rose to prominence as a material for blue and ultraviolet high brightness LEDs and lasers.

The types of dislocations that arise by combining the possible crystallographic Burgers vectors and line directions in materials with the wurtzite structure are not only more varied than in the case of the diamond and sphalerite structures, they are also more complex than in the simpler but related close packed hexagonal (c.p.h.) structure of a number of metals. Much work was done on the defect crystallography and plastic deformation of these metals (Partridge 1967) but less on those in the wurtzite structure semiconducting compounds. It is therefore useful to consider first the results of the analysis of dislocation geometry in the c.p.h. metals, as it provides useful background for the crystallography of the less studied dislocations in wurtzite structures.

A bi-tetrahedron was introduced by Berghezan *et al.* (1961) as the analogue of the Thompson tetrahedron for the case of the hexagonal close packed metals.

This is shown in Fig. 4.27a with the modified notation of Nabarro *et al.* (1964). Its relation to the unit cells of the wurtzite structure is shown in Fig. 4.27b. As the vector $\Phi\Theta$ is of the length c, and the vectors AB etc. are of length a, the geometry of the bi-tetrahedron varies with the value of the c/a ratio from one h.c.p. metal to another. Fortunately because of the tetrahedral covalent bonding in the semi-conducting compounds with the wurtzite structure, the c/a ratio is always very close to its ideal value of $\mu = 3/8\, c = 0.375\, c$ in these materials. The parameter μ gives the positions of B relative to A atoms.

Thus, in the upper primitive hexagonal cell of the wurtzite structure in Fig. 4.27b there are four atoms at positions with coordinates (in terms of the lattice parameters a_1, a_2 and c as follows:

A atoms occur at positions: $0,0,0$ (*a* site) and $2/3, 1/3, 1/2$ (*b* site)

B atoms occur at positions: $0,0,1/2-\mu$ (α site) and $2/3, 1/3, 1-\mu$ (β site)

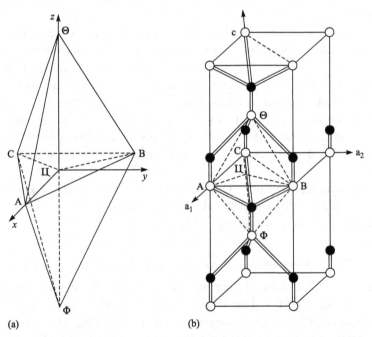

(a)   (b)

Figure 4.27 (a) The bi-tetrahedron of Berghezan, Fourdeaux and Amelinckx (BFA) (1961). A, B and C are atomic sites in the basal plane and Θ and Φ are like atomic sites in the planes above and below. Joins of letters from the same alphabet are Burgers vectors of unit dislocations; joins of letters from different alphabets are Burgers vectors of partial dislocations. (The centre point is labelled with the Cyrillic letter ц.) (b) Two primitive hexagonal cells of the wurtzite structure showing the bonding of the atoms in relation to the BFA bi-tetrahedron. (After Berghezan *et al.* 1961. Reprinted from *Acta Metallurgica*, **9**, Transmission electron microscopy studies of dislocations and stacking faults in a hexagonal metal-zinc, pages 464–90. Copyright 1961, with permission from Elsevier.)

where a, b, α and β are the positions of the atoms in the h.c.p. stacking sequence: $a\ \alpha$, $b\ \beta$, $a\ \alpha$ .... The 0,0,0 site is apex C and 2/3,1/3,1/2 site is apex Θ of the bi-tetrahedron. Note that Osipiyan and Smirnova draw the atom positions a and α upside down relative to the way that they were represented in Blank *et al.* (1964), i.e. here the α sites are above the a sites. Moreover, it is possible to represent the wurtzite structure by means of either the a, b stacking sequence used here or the equivalent a, c stacking sequence involving 0,0,0 and 1/3,2/3,1/2 positions. Therefore, the coordinates of the atom positions in the unit cell can be stated in several different but equivalent forms. For example, the form given in Blank *et al.* (1964) is different from that quoted here.

It can be seen in Fig. 4.27b that each corner of the bi-tetrahedron is at a like-atom site. Therefore, in the bi-tetrahedron, vectors that connect points denoted by letters of the same alphabet are the Burgers vectors of perfect unit dislocations and this is equally true for the c.p.h. and wurtzite structures. Vectors that join points denoted by letters of different alphabets are the Burgers vectors of partial dislocations in both structures.

In the case of the wurtzite structure there are added complications. Firstly, there are two slip systems, namely (c) and (e) in Fig. 4.26, which have $\langle 10\bar{1}0 \rangle$ slip directions. Slip on these systems must occur through the movement of unit dislocations with $\langle 10\bar{1}0 \rangle$ Burgers vectors, not partial Burgers vectors in $\langle 10\bar{1}0 \rangle$ directions like ${}_{II}A$, ${}_{II}B$ and ${}_{II}C$ and their reverses, which are the only $\langle 10\bar{1}0 \rangle$ vectors represented in the bi-tetrahedron. Osipiyan and Smirnova (1968), therefore, introduced a triangular prism as shown in Fig. 4.28. The vectors represented by the darker lines in this figure are the Burgers vectors for perfect unit dislocations in wurtzite structure materials.

Later, Osipiyan and Smirnova (1971) put forward a more complicated analogue of the bi-tetrahedron which specifies the Burgers vectors of the partial dislocations which we will discuss below (see Fig. 4.33).

Ball-and-wire models of dislocations in the wurtzite structure

The core structures of some dislocations in materials with the wurtzite structure were studied, using the crystal model approach, by Abrahams and Dreeben (1965), Weinstein *et al.* (1965) and by Blistanov and Geras'kin (1970). Their results were in general agreement with those of the more complete and systematic later analyses of Osipiyan and Smirnova (1968, 1971). We shall only report the main points that emerged from the latter work.

Osipiyan and Smirnova (1968) began by considering the dislocations with vectors of the types $\mathbf{a} = 1/3a\langle 1\bar{2}\bar{1}0 \rangle$, like as ΦA in Fig. 4.28, and $\mathbf{c} = c\,(0001)$, like Aa, where a is the basal plane lattice parameter and c is the height of the hexagonal unit cell. They implicitly assumed that the dislocations belonged to the shuffle set. For example, it can be seen in their model of the 60° dislocation (Fig. 4.29), that the

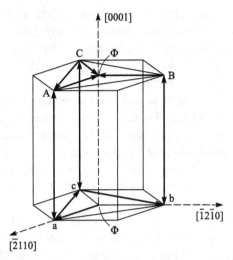

Figure 4.28 The triangular prism of unit dislocation Burgers vectors in wurtzite-structure materials introduced by Osipiyan and Smirnova (1968). The darker lines represent possible Burgers vectors. The lighter lines indicate the hexagonal cell of the structure. Vectors joining different Latin letters of the same case are of the form $\langle \bar{1}2\bar{1}0 \rangle$. Vectors joining a Greek and a Latin letter of the same case are of the form $\langle 10\bar{1}0 \rangle$. Any joins of upper and lower case forms of the same letter are vectors of the form $\langle 0001 \rangle$. (After Osipiyan and Smirnova 1968.)

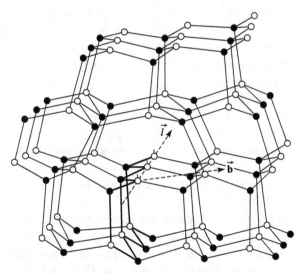

Figure 4.29 The A(s) basal plane 60° dislocation in the wurtzite structure. (Our convention takes white (alabaster) sites to be occupied by the A atoms of the AB compound. The B(s) 60° dislocation model is identical except that the colour of all the sites is reversed.) The line direction and Burgers vector are $\mathbf{l} = [2\bar{1}\bar{1}0]$ and $\mathbf{b} = 1/3[\bar{1}2\bar{1}0]$, respectively. (After Osipiyan and Smirnova 1968.)

dislocation occurs between widely spaced planes. (Hirth and Lothe first suggested the possibility of glide set dislocations, between closely spaced basal planes, only in 1968.) Osipiyan and Smirnova pointed out that any arbitrary dislocation line direction in the wurzite structure could be resolved into a combination of these two types of steps in $\langle \bar{1}2\bar{1}0 \rangle$ and $\langle 0001 \rangle$ directions.

Thus all the combinations of these two types of directions with Burgers vectors of these forms had to be considered. These combinations yield a screw dislocation with the Burgers vector **a**, a screw dislocation with the Burgers vector **c**, two types of edge dislocation and a 60° dislocation with various slip planes. Taking into account line directions made up of $\langle \bar{1}2\bar{1}0 \rangle$ and $\langle 0001 \rangle$ steps to produce dislocations aligned in $\langle \bar{1}010 \rangle$ and $\langle \bar{1}2\bar{1}3 \rangle$ directions gives an additional eight types for a total of 11 types of dislocations with Burgers vectors **a** or **c**.

In the wurtzite structure the most widely spaced plane is (0001). However, there are both widely and closely spaced pairs of atomic (0001) planes just as in the case of the sphalerite-structure {111} planes. (The glide versus shuffle set controversy for the sphalerite structure, discussed in Section 4.1.2, subsequently taught us to take seriously the possibility that some or all the dislocations actually occur and move between the more closely spaced (0001) pairs of planes lying between the widely spaced (0001) planes.) Osipiyan and Smirnova and others who modelled dislocation cores in the wurtzite structure considered only the shuffle set possibilities for dislocations with the (0001) slip plane. Models of glide set (0001) dislocations have not yet appeared.

Many of the features of shuffle set models of dislocations in the sphalerite structure appear also in shuffle set dislocation models. For example, the screw dislocation contains no broken bonds as can be seen in Fig. 4.30. Dislocations with the basal plane (0001) and some edge character are polar, like the 60°-dislocation in Fig. 4.29. Bond rearrangements to lower the energy by connecting dangling bonds are possible in the wurtzite structure just as in the diamond cubic and sphalerite structures. As in sphalerite, this is only energetically favourable if the dangling bonds are one A- and one B-type and so can form a 'right' A-B bond. Rearranging to form A-A or B-B 'wrong' bonds is unlikely, so polar dislocations would retain dangling bonds and remain electrically active. For discussions of possible bond rearrangements in wurtzite structure dislocations see Osipiyan and Smirnova (1968) and Blistanov and Geras'kin (1970).

Dislocations with prismatic slip planes and some edge character are like the type I prismatic plane edge dislocation shown in Fig. 4.31. That is, they are non-polar, containing equal numbers of A and B atoms in alternate sites. Rearrangement can form energetically favourable 'right', A to B bonds. Hence bond rearrangements to eliminate the dangling bonds are likely and should make these dislocations electrically inactive.

Similarly the type II prismatic plane pure edge dislocation, shown in Fig. 4.32, is non-polar. Bond rearrangement again should eliminate the dangling bonds and make

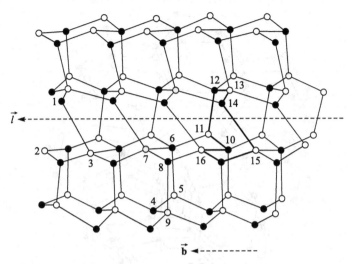

Figure 4.30 The screw dislocation with the [$\bar{1}2\bar{1}0$] Burgers vector. (After Osipiyan and Smirnova 1968.)

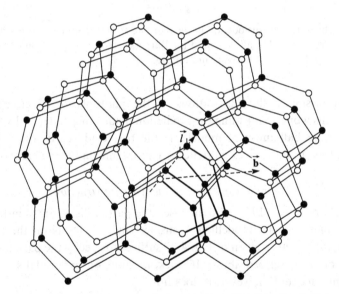

Figure 4.31 The pure edge dislocation in the type I prismatic glide plane, of the form {$10\bar{1}0$}, in the wurtzite structure. The line, *l*, runs in the ⟨0001⟩ direction and the Burgers vector, **b** in the ⟨$\bar{1}2\bar{1}0$⟩ direction. (After Osipiyan and Smirnova 1968.)

these dislocations electrically inactive. This is true for all type I and II prismatic plane dislocations with an edge component. Carlsson (1971) noted this general rule from the work of Osipiyan and Smirnova (1968) and of Blistanov and Geras'kin (1970) and he demonstrated the electrical activity of basal slip dislocations and the electrical

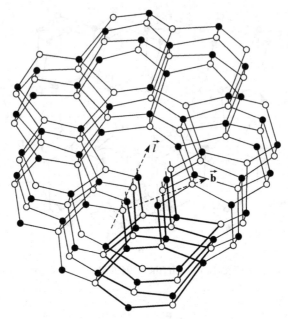

Figure 4.32 The pure edge dislocation in the wurtzite structure in the type II prismatic glide plane, of the form $\{1\bar{2}\bar{1}0\}$. The alignment of the Burgers vector, **b**, is $\langle\bar{1}100\rangle$ and that of the line direction, *l*, is $\langle0001\rangle$. (After Osipiyan and Smirnova 1968.)

inactivity of prismatic slip dislocations in the photoplastic effect, discussed in Section 4.3. This work, confirming the prediction made independently by two groups of workers using ball-and-wire models, is the best evidence of the value of this approach at least in the case of wurtzite structure semiconducting compounds.

Partial dislocations in the wurtzite structure

Osipiyan and Smirnova (1971) also analysed the partial dislocations in the wurtzite structure by ball-and-wire modelling. Consider stacking faults on the basal plane (0001). Then the bi-tetrahedron of Fig. 4.27 suffices to specify the partial dislocation Burgers vectors as described above. If stacking faults on pyramidal slip planes are also taken into account, however, it does not.

To specify all the partial dislocation Burgers vectors, Osipiyan and Smirnova (1971) introduced the additional points F, N and P shown in Fig. 4.33. These points are the centres of triangles like ABC and adjacent to it but of a different symmetry (apex down instead of apex up) as shown in Fig. 4.34c. That is F, N and P are positions analogous to σ, the centre of the bi-tetrahedron. A, B and C are atom sites and represent one of the stacking sequence positions, '*b*', say. σ is then '*a*' stacking position while F, N and P are '*c*' stacking positions. The wurtzite structure is produced by *a* α, *b* β, *a* α, *b* β... stacking, so translating atoms in this plane or

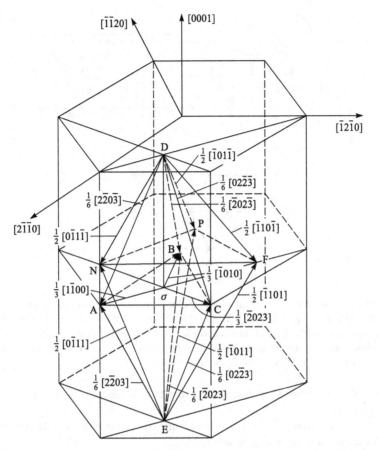

Figure 4.33 Representation of the partial dislocation Burgers vectors in the wurtzite structure. Note that the centre and the top and bottom apices of the bi-tetrahedron are lettered differently than in Figure 4.28. (After Osipiyan and Smirnova 1971. Reprinted from *Journal of Physics and Chemistry of Solids*, **32**, Partial dislocations in the wurtzite lattice, pages 1521–30. Copyright 1971, with permission from Elsevier.)

into it but to *a*- or *c*-type positions instead of *b*-type positions introduces faults. Vectors from A, B or C and ending at σ (or equivalently at F, N or P) are, therefore, the Burgers vectors of partial dislocations in the basal (stacking) plane.

There are a total of three types of partial dislocations in the wurtzite structure. These can be specified in terms of their Burgers vectors types as follows.

- The six vectors from σ to A, B and C and their reverses are those of basal plane Shockley partial dislocations,
- The six vectors from D (or equivalently from E) to A, B or C and their reverses are the Burgers vectors of Frank partial dislocations and correspond to closing up the crystal after removing a basal double-atom (*b* β) plane or to opening up the crystal to allow an extra

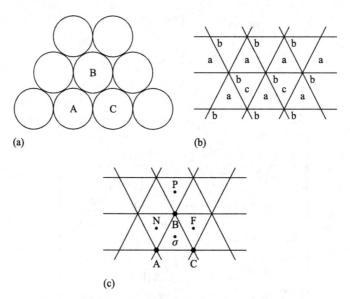

Figure 4.34 Atomic sites in a basal plane of the wurtzite structure. (a) Representation in terms of the close packing of spheres. A, B and C are the points so labelled in Figs. 4.34b and c. (b) Types of sites for stacking close-packed planes of atoms. (c) The sites that are labelled in the basal plane half-way up the unit cell in Fig. 4.33. It can be seen that A, B and C are 'b' sites (compare 4.34b), σ is an 'a' site and F, N and P are 'c' sites.

  basal double-atom plane to be inserted in order to produce intrinsic or extrinsic stacking faults, respectively.
- The twelve vectors from D or E to F, N and P and their reverses are the Burgers vectors of Shockley dislocations in type II prismatic planes.

The vectors from D or E to σ and their reverses are also possible Frank-partial-dislocation Burgers vectors. These vectors result, however, in placing atoms in wrong sites, whereas vectors such as DA, as compared with Dσ, do not. Since vectors like Dσ result in worse bonding in the cores of the dislocations they would define partial dislocations of higher energy, which do not occur.

Osipiyan and Smirnova (1971) carried out a complete ball-and-wire model analysis of the basal and prismatic plane partial dislocations in the wurtzite structure and published drawings of the models of typical partial dislocations of all these types. They also analysed the splitting or dissociation of unit dislocations into partials and stacking faults, as well as the types of reactions that could occur between partial dislocations. (For details see Osipiyan and Smirnova 1971.)

TEM studies of stacking faults in wurtzite-structure materials

No HRTEM evidence on the atomic core structures of defects in wurtzite-structure materials has as yet appeared. Indirect evidence bearing on these matters is of two kinds. Evidence from studies of the orientation dependence of the photoplastic effect, to be reviewed in Section 4.3, provides support for one of the features exhibited by these models. This is that basal plane dislocations are polar and so charged and exhibit the photo- and electro-plastic effects whereas prismatic plane dislocations are not and do not, as noted above. TEM studies provided information on several features of defects in, especially, II-VI compounds, with the wurtzite structure models

In Section 4.1.2, the necessity for the weak-beam TEM technique to resolve the structure of dissociated dislocations in diamond-structure semiconductors, especially Si, with their high-energy and, therefore, narrow stacking faults, was discussed. The faults in wurtzite structure materials are wide so ordinary diffraction contrast TEM could be used. Consequently, many early studies of stacking faults in vapour-phase grown thin platelets of AlN and of II-VI compounds with the wurtzite structure were reported. These showed that wide stacking faults were common on both basal planes and type II prismatic planes, i.e. those of the form $\{11\bar{2}0\}$, shown in Fig. 4.26.

Faults seen by TEM frequently bend from a plane of one type to one of another type. For example, the fault in the closed loop of stacking fault in wurtzite structure ZnS in Fig. 4.35 (Blank *et al.* 1962) bends repeatedly from the basal- to a prism-plane and back.

In general, the independent model analyses of Blank *et al.* (1962) and Delavignette *et al.* (1961), and the more complete work of Osipiyan and Smirnova (1971) are in good agreement with each other and with the findings of TEM studies like those of Drum (1965) for AlN.

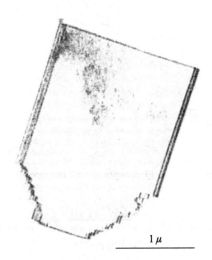

1 μ

Figure 4.35 A closed loop of stacking fault in wurtzite structure ZnS which is composed of alternating basal plane and prismatic plane faults. (After Blank *et al.* 1962.)

The mechanisms of the wurtzite-sphalerite and polytype transformations

Evidence that the stacking fault energy in II-VI compounds is low is provided by the numerous observations of wide stacking faults to be outlined below and by the observation of martensitic (i.e. not diffusion controlled) phase transformations from the wurtzite to the sphalerite structure at a critical temperature in certain compounds. These transformations occur by the formation and widening of regular arrays of stacking faults to transform the stacking sequence from that for the wurtzite structure to that for the sphalerite structure. Finally the common occurrence of polytypes and stacking disorder are evidence of low fault energies.

Polytypes are long repeat-distance stacking sequence structures. The sphalerite and wurtzite structures are produced by stacking double atom (A and B) planes in particular stacking sequences, as shown in Fig. 4.36. The double atom plane positions are represented by corresponding letters of the Latin and Greek alphabets thus: a α, b β, c γ. It is convenient now to omit the Greek letters when dealing with complex polytype structures, but remember that the second planes are implied in all cases: $a(\alpha)$, $b(\beta)$, and $c(\gamma)$.

Polytype sequences can be represented in several ways. One of these is to use stacking sequence diagrams as shown in Fig. 4.36. The possible successive stacking

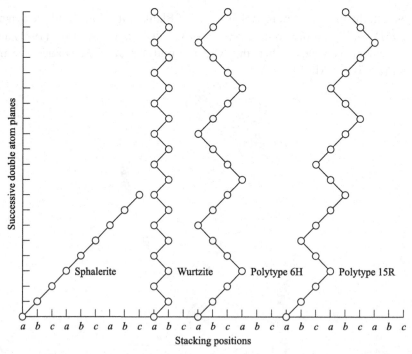

Figure 4.36   Ramsdell stacking sequence diagrams for four common polytypes. Sphalerite is polytype 3C and wurtzite is polytype 2H.

positions *a b c a b c* . . . are marked along the abscissa and the positions of successive (double) planes are represented by successive positions up the ordinate. The stacking sequence, *a b c a b c* . . . , of the sphalerite structure which is referred to as the 3C (three layer high repeat sequence, cubic) structure in polytype notation, is then represented by a 45° straight line as shown. The sequence, *a b a b* . . . , of the wurtzite (or 2H for two layer repeat sequence, hexagonal) structure is represented by a vertical two step zig-zag as shown. Another common polytype in ZnS and SiC is the 6H structure. Others have rhombohedral symmetry. An example of this type, the 15R structure, is also shown. Such polytypes occur most commonly in such semi-conducting compounds as SiC and ZnS, as well as in CdI_2. Each polytype has unique properties. This phenomenon, which was first observed in SiC about a century ago, was found subsequently to be fairly common in other materials as well (see e.g. review by Trigunayat 1991 and references therein). The polytypism has been since systematically investigated and several theories were proposed to explain it but a comprehensive explanation for its occurrence has not been found yet (e.g. Trigunayat 1991).

An early comprehensive review of polytypism in crystals was provided by Verma and Krishna (1966). Pandey and Krishna (1983) reviewed the origin of polytype structures in crystals, Jepps and Page (1983) provided a review of polytypic transformations in SiC, and Steinberger (1983) and Mardix (1986) outlined polytypism in ZnS. Later, Trigunayat (1991) provided an extensive overview of the theoretical and experimental developments in this field. He reviewed different theories of polytypism, as well as various factors affecting polytype formation (e.g. the effect of impurities and the role of stacking faults and solid-state phase transformations). There are equilibrium theories of the formation of polytypes that assume they are thermodynamically stable phases, i.e. they are formed to minimize the energy. Non-equilibrium theories propose their creation by kinetic crystal growth mechanisms involving dislocations, or impurities. Hence the latter are also known as growth kinetic theories (see e.g. review by Trigunayat 1991 and references therein). Fig. 4.38 (see below) shows the growth kinetic mechanism for polytype formation in ZnS.

SiC wafers with the 4H and 6H structures are commercially available for various electronic device applications. Another promising polytype being developed for commercial applications is 3C SiC, which is more difficult to grow reproducibly without high densities of defects. However, it has some advantages over hexagonal SiC polytypes. These include isotropic properties and a potential for a greater wafer size by heteroepitaxial growth on Si substrates. These SiC polytypes have a unique combination of properties (wide energy gap, high thermal conductivity, high saturation electron drift velocity, high breakdown electric field and superior stability). Therefore, they are employed in high-temperature and high-power devices, and in blue and green light-emitting devices. SiC substrates are also employed for III-N epitaxy. A major obstacle to these applications is the presence of defects such as micropipes, tilt boundaries, and inclusions in SiC (see Section 4.8), which may also generate defects in the epitaxial layers.

Known polytype structures include some 200 in ZnS (Alexander *et al.* 1970, Steinberger *et al.* 1973), about 150 in SiC and about 200 in CdI_2 (e.g. Trigunayat 1991), and a few in other II-VI compounds (Schlossberger 1955) as well as other various compounds (e.g. Trigunayat 1991), and can all be regarded as regular arrays of stacking faults. Irregular arrays of faults can also occur. This is known as stacking disorder and also occurs commonly in ZnS (Jagodzinski 1949, Ebina and Takahashi 1967, Wilkes 1969, Holt and Culpan 1970 and Holt and Brada 1997).

Polytype structures occur profusely as narrow bands in striated ZnS platelets as shown in Fig. 4.37a. These platelets grow by sublimation in a stream of a carrier gas, usually H_2S. The ZnS, sublimed from the hot zone of the furnace, deposits in the

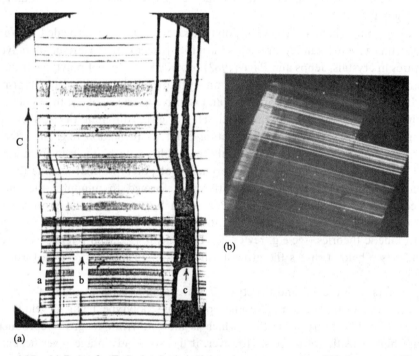

(a)

(b)

Figure 4.37   (a) Part of a ZnS platelet in a light microscope, seen at 100 times magnification under crossed polarizers. The striated regions perpendicular to the C axis, shown to the left of the crystal, are polytypes. Lines like those marked b and c, in the general direction of the c-axis, are prismatic growth facets, referred to as linear markings. The originally straight linear markings and the edge of the plate, marked a, have become kinked as a result of the periodic slip processes responsible for the polytype transformations. (After Mardix *et al.* 1968.) (b) SEM CL micrograph showing localized light emission from some polytype bands. The full width of the crystal was 350 μm. (After Holt and Culpan 1970.)

cooler zone with various characteristic morphologies in particular regions. The platelets are one of these forms. The platelets received much attention because they were for a long time the largest and best crystals of ZnS available and they were found to exhibit the 'anomalous' photovoltaic effect. That is, hundreds of volts per cm can be generated along the common c-axis under monochromatic illumination (Merz 1958, Brafman *et al.* 1964, Shachar and Brada 1968). These structures exhibit other interesting properties such as the strongly varying cathodoluminescence shown in Fig. 4.37b.

The formation of polytypes in vapour phase ZnS platelets was shown to take place in a series of steps as follows (Alexander *et al.* 1970). Firstly, a hexagonal 2H (wurtzite) needle, with the c-axis along the axis of the needle, grows around a screw dislocation with a large Burgers vector as shown in Fig. 4.38. These Burgers vectors are many times greater than the height of the wurtzite unit cell (e.g. Mardix 1984, 1991) so the structure is a thick-slab spiral ramp. In many cases the needles widen to form platelets. Secondly, a stacking fault or a set of stacking faults is introduced in a basal plane or a set of arbitrarily spaced parallel basal planes. Thirdly, at lower temperatures, when cooled after growth, the hexagonal structure becomes unstable. The arbitrary set of parallel faults, therefore, expands around and up the spiral ramp layer as shown at the bottom of Fig. 4.38. That is, the partial dislocations leading the stacking faults glide round climbing the spiral ramp. This constitutes 'periodic slip' and sets up a regularly repeated arbitrary stacking sequence as indicated in Fig. 4.38. This establishes the stacking sequence of some particular polytype and also results in a particular tilt of the prismatic facets, i.e. of the linear markings as seen in Fig. 4.37. It was shown by the Jerusalem group that the observed polytype sequences could all be accounted for in detail by this mechanism (Alexander *et al.* 1970) and that the measured tilts of the prismatic facets were equal to those calculated from the numbers, spacings and Burgers vectors of the partial dislocations in the periodic slip for the particular polytype structure. It was found that platelets of ZnS could be mechanically stressed, e.g. by indentation with a pin, in the direction of the polytype kink shear and so made to transform (Steinberger and Mardix 1967) as also happened frequently due to thermal stresses in cooled samples (Mardix and Steinberger 1970).

These polytype transformations in ZnS are the best-understood martensitic transformation by faulting in the II-VI compounds (Mardix 1991).

The energy of stacking faults in wurtzite and sphalerite structure compounds

The fault energy depends on the difference in energy between the wurtzite and sphalerite structures. This cannot be calculated with sufficient accuracy for the small temperature-dependent difference between them, which determines, for example, which structure is stable.

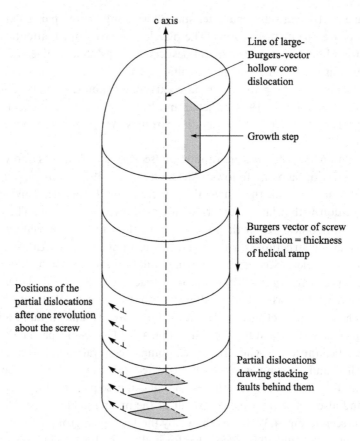

Figure 4.38 The periodic slip mechanism of polytype transformation. The initial structure of the needle is wurtzite. From such needles, platelets like those in Figure 4.37, grow out sideways like flags from flagpoles. A grown-in screw dislocation of large Burgers vector turns the whole needle (and platelet) into a spiral or helical ramp of pitch equal to the large Burgers vector. An arbitrary set of partial dislocations drawing stacking faults behind them is shown in the lowest turn of the ramp. After advancing through one rotation they will pass through the second turn of the ramp and so on. The Burgers vector of the screw dislocation determines the thickness of the helical ramp. This becomes the repeat distance, i.e. the unit cell height of the stacking sequence for the polytype.

The development of the weak-beam TEM imaging technique (see Sections 3.6.3 and 4.1.2) enables the narrow width of extended dislocations to be resolved in high stacking fault energy materials. From the observed separations of the partial dislocations in such cases the fault energy can be obtained. This was done for the industrially important crystalline semiconductors. Takeuchi and Suzuki (1999) presented the compiled results of the stacking fault energies for various semiconductors (see Table 4.3; see also Alexander 1976, Gai and Howie 1974, Gottschalk *et al.* 1978).

Table 4.3. *Stacking fault energies in various semiconductors. Stable crystal structure (D: diamond, Z: zincblende, and W: wurtzite); stacking fault energy γ; reduced stacking fault energy γ'; and ionicity f_i (after Takeuchi and Suzuki 1999; see also references therein)*

Semiconductor	Structure	γ mJ/m^2	γ' meV/bond	f_i
Si	D	55 ± 7	44 ± 5.5	0
Ge	D	60 ± 10	52 ± 8.5	0
GaP	Z	43 ± 5	34 ± 4	0.327
GaAs	Z	45 ± 7	39 ± 6	0.310
GaSb	Z	53 ± 7	53 ± 7	0.261
InP	Z	18 ± 3	17 ± 3	0.421
InAs	Z	30 ± 3	30 ± 3	0.357
InSb	Z	38 ± 4	43 ± 5	0.321
ZnS	Z	≤6	≤5	0.623
ZnSe	Z	13 ± 1	11 ± 1	0.630
CdTe	Z	9 ± 1	10 ± 1	0.717
CdS	W	8.7 ± 1.5	8.0 ± 1.4	0.685
CdSe	W	14 ± 5	14 ± 5	0.699

The table also lists the fraction of ionic character, f_i, of the bonds in the compounds according to the Phillips theory (see Section 1.6.3). The Madelung constant of the wurtzite structure is larger than that of sphalerite. Hence it is customarily assumed that increasing ionicity favours the wurtzite over the sphalerite structure. Since the stacking fault is a lamina of wurtzite-structure material in sphalerite it is reasonable to expect the fault energy to decrease with increase in the fractional ionicity of the bonding, f_i, as the table shows that it does.

Note that the Suzuki effect, discussed below, lowers the fault energy in the presence of impurities and results in more widely dissociated dislocations than in pure specimens of the same material. Elastic stresses, applied or internal, can drive the bounding partials further apart and also result in lower values for the fault energy calculated from the width of the fault ribbon.

The Suzuki effect

Suzuki (1952) pointed out that impurity atoms may have a different solubility in the structure of the stacking faults in dissociated dislocations than in the structure of the matrix crystal. For example, if an impurity dissolved more readily in the structure of the stacking faults it would segregate to the stacking faults, lowering the energy of the faults, and hence their effective surface tension. This results in the widening of the extended dislocation due to the mutual repulsion of the partial dislocations which is unaffected. It is thought the very widely extended dislocations seen in

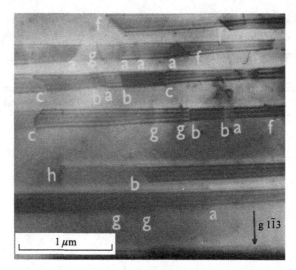

Figure 4.39 Transmission electron microscope image of stacking faults in vapour-phase grown and Te-doped GaP. Dislocations marked a, b and c are Shockley partials with different Burgers vectors of the form $a/6\langle112\rangle$. Frank partials with **b** $= a/3\langle111\rangle$ are marked f. Unit (undissociated) dislocations are marked g, that marked h is a partial of undetermined type. (After Chase and Holt 1972.)

many III-V semiconductors in the early days, when the purity was low, were due to this Suzuki effect (see e.g. Fig. 4.39). The Suzuki effect also makes the dislocations less mobile, since to glide they have to leave the energy-lowering cloud of dissolved impurity atoms behind.

What can be interpreted as evidence for a Suzuki effect was found by Caveney (1968). He examined by TEM CdS platelets into which Hg had been diffused. He found numerous basal plane stacking faults whereas no faults were present in pure specimens. Caveney himself interpreted his observations in terms of the change in the degree of ionicity of the bonding produced by the Hg. However, he recognized that this effect could be greatly enhanced by segregation of the Hg atoms to the vicinity of the dislocations where they would be more effective than elsewhere.

Crystallography of stacking faults in wurtzite-structure compounds

Most III-V compounds have the cubic sphalerite structure but AlN and GaN, like many of the II-VI compounds, grow with the hexagonal wurtzite structure. These materials can generally be grown from the vapour as needles or platelets, which are thin enough for TEM examination. Hence, extensive early defect studies were made on them, e.g. Delavignette *et al.* (1961), Blank *et al.* (1962, 1964), Chadderton *et al.* (1963, 1964), Chikawa (1964), Fitzgerald *et al.* (1966), Secco D'Aragona and Delavignette (1966), Fitzgerald and Mannami (1966), Fitzgerald *et al.* (1967), Caveney (1968).

It was found that numerous wide faults occurred on both basal (b) planes (0001) and on the prism (p) planes of the form {11$\bar{2}$0} in AlN and ZnS and that the faults frequently fold from b- to p-planes and back (Blank *et al.* 1962, 1964). The example shown in Fig. 4.35 is a fault loop in wurtzite structure ZnS, some portions of which are on basal- and others on prism-planes. Faults on p-planes were always connected to b-plane faults.

Delavignette *et al.* (1961) and Blank *et al.* (1962) considered the crystallography of defects in the wurtzite structure. They pointed out that, as shown in Fig. 4.40a, the possible partial dislocation Burgers vectors in the basal plane are of the type of Aσ, σB, etc. whereas unit Burgers vectors are of the type AB. Thus partial dislocations shear atoms from one stacking position to another of a different kind, while unit dislocations shear them to a neighbouring site of the same stacking position.

They also considered the core structure of dislocations in the wurtzite structure as shown in Fig. 4.41. Their conclusion was that dislocations in glide planes of type I could only be perfect, but that dislocations in type II planes can dissociate. The case of the 60° dislocations in a type II plane is shown in Fig. 4.41(a). This can take up either of the extended forms shown in Figs. 4.41b or 4.41c. These two forms arise in the cases that the dissociation, involving the atomic movements indicated by the arrows, occurs in the next type I glide plane above (case b) or below (case c) the type II glide plane of the original perfect dislocation (shown in a).

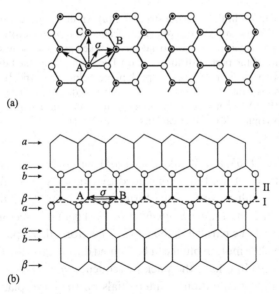

(a)

(b)

Figure 4.40 Diagrams of the wurtzite structure (after Blank *et al.* 1962). (a) Projection on the basal plane, with the possible shear vectors of partial dislocations marked Aσ, Bσ, Cσ and their negatives. (b) Side view of the wurtzite structure shown in section through AC in (a), projected on a {1$\bar{1}$00} plane. The two possible types of slip plane are indicated by I and II. Glide planes of type II can only contain perfect dislocations according to Blank *et al.* (1962).

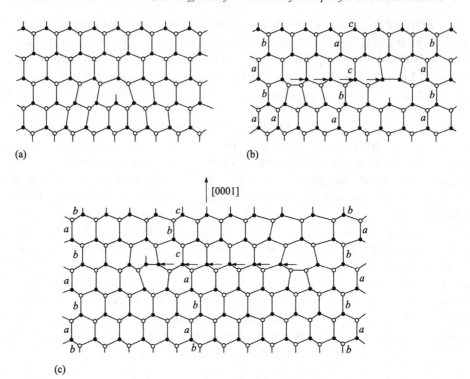

Figure 4.41 Dislocations in the wurtzite structure. (a) An undissociated 60° dislocation with type II glide plane. (b) The dislocations of (a) dissociated into partials with the formation of a fault of the form *b a b c a c a*. . . The dissociation has taken place in the next type I glide plane above the glide plane of the perfect dislocation. (c) The form of extended dislocation obtained by dissociation into partials in the next type I plane below the type II glide plane of the original dislocation. A fault *b a b a c b c b*. . . is formed. (After Delavignette *et al.* 1961 and Blank *et al.* 1962.) (Reprinted with permission from *Journal of Applied Physics*, **32**, pp. 1098–1100. Copyright 1961, American Institute of Physics.)

Chadderton *et al.* (1963) and Blank *et al.* (1964) considered the crystallographically possible distinct types of fault and concluded that three types of fault could occur. Type I faults have fault vectors of the form $\mathbf{t} = 1/3\langle 1\bar{1}00\rangle + 1/2\langle 0001\rangle$ and can be described by the stacking sequence *abab/cbcb*. . . This fault vector is the sum of a basal plane shear and one parallel to the c-axis. Therefore, this type of fault cannot be formed by a single glide motion but it can be formed during growth. Type II faults with fault vectors $\mathbf{t} = 1/3\langle 1\bar{1}00\rangle$ have the stacking sequence *abab/caca*. . . Type III faults are produced by the precipitation of interstitials on the basal plane with no offset. They have the fault vector $\mathbf{t} = 1/2\langle 0001\rangle$. The precipitation of interstitials with an associated basal plane shear produces type I faults. In the case of the precipitation of vacancies to remove a portion of a double atom plane, i.e. to produce an intrinsic fault, a basal plane shear of the form Aσ will take place giving a type I fault. A model of a type I fault lying partly on basal planes (type I b-faults) and partly on a prism plane (p-fault) is shown in Fig. 4.42.

The shear faults (type II) are those that arise in extended dislocations and which are responsible for the martensitic transformation of the wurtzite structure to sphalerite as discussed in Section 4.1.5 above. Closed loops of b- and p-faults, as in the case shown in the TEM micrograph in Fig. 4.35, involve folding from b- to p-planes and back again as shown diagrammatically in Fig. 4.42. This was explained by Blank *et al.* (1964) in terms of a growth mechanism since it would be very difficult to form such fault loops by means of the movement of partial dislocations, i.e. by a post-growth glide mechanism. They put forward detailed arguments to explain why the non-basal portions lay accurately on prismatic planes and why the basal faults are always planar whereas the prismatic faults connected to them are often folded, as in Fig. 4.35 where the narrow, roughly horizontal lines are basal faults while the broad, fringed, nearly vertical bands are prismatic faults. This detailed analysis by Blank *et al.* (1964) was applied to wurtzite-structure ZnS.

Similar observations of basal and prism plane faults were made on CdS (Chadderton *et al.* 1964, Chikawa 1964), on AlN (Drum 1965) and on ZnSe (Fitzgerald *et al.* 1966). The role of stacking faults in the growth of platelets from

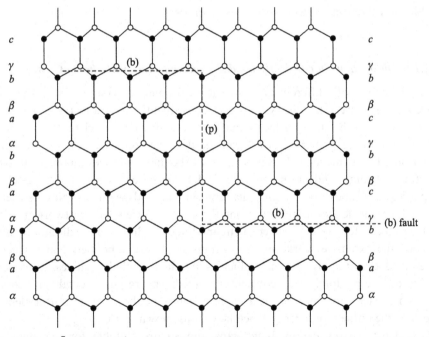

Figure 4.42 (1$\bar{1}$00) plane projection of the wurtzite structure containing a growth domain boundary (dotted lines) consisting of basal and prism plane portions (marked b and p respectively) of a type I stacking fault. (After Blank *et al.* 1964.)

initial whisker crystals have been considered for CdS by Chikawa and Nakayama (1964), for AlN by Drum (1965) and for ZnS and ZnSe by Fitzgerald *et al.* (1967).

Fitzgerald and Mannami (1966) used a transmission electron microscope moiré fringe method to determine the Burgers vectors of b- and p-faults in ZnS. The basal plane faults were found to be of type II while the prismatic plane faults were of type I or II. The partial dislocations terminating faults on basal planes were found to have Burgers vectors like those of the faults, i.e. of the form $1/6\langle0\bar{2}23\rangle$. Caveney (1968) used TEM to show that, in the case of the basal plane faults produced in CdS by the in-diffusion of Hg, type II faults predominated over type III faults and that no type I faults occurred. Drum's (1965) TEM study of basal and prism plane faults in AlN showed that the faults on the $\{1\bar{2}10\}$ prism planes had fault vectors $\mathbf{t_p} = (3/2)\mathbf{p} + (1/2)\mathbf{c}$ where $\mathbf{p} = 1/3[10\bar{1}0]$ and $\mathbf{c} = [0001]$ i.e. $\mathbf{t_p} = 1/6[20\bar{2}3]$. Examples of faults of the types $\mathbf{t} = 1/2\,\mathbf{a_2}$ on (0001), $\mathbf{t} = 1/2\,\mathbf{c}$ on (1010) and a more complex type were also found. Drum presented ball-and-wire models of the $(1\bar{2}10)$ faults. He also found that the expected stair-rod dislocations occurred at the intersections of the faults and that they had screw components to their Burgers vectors.

A common polytype in SiC has hexagonal symmetry like the wurtzite structure. A number of TEM studies of dislocations and stacking faults in the '6H' polytype structure were published (Amelinckx *et al.* 1960, Van Landuyt and Amelinckx 1971, Stevens 1972a, 1972b, 1972c). This work showed similarities between the defects in 6H SiC and those in wurtzite-structure materials.

4.1.6 Dislocations in covalently bonded semiconductors with the NaCl structure

Many materials with different types of chemical bonding crystallize with the NaCl (rocksalt) structure. The slip systems and consequently the geometry of the dislocation cores and the stacking faults change also with changing chemical bonding.

Three groups of materials, which can be described as predominantly ionically, or covalently or metallically bonded, must be distinguished among the NaCl structure AB compounds, as noted in Table 4.4. The core structure and the electrical properties of dislocations in ionically bonded materials with the NaCl structure received a great deal of attention at one time. For a review see Whitworth (1975).

Due to the difference between covalent and ionic bonding and the difference in slip systems, the extensive work on dislocations in ionically bonded NaCl-structure materials does not apply to semiconductors like the lead chalcogenides. These semiconductors are predominantly covalently bonded. This results in octahedral coordination, i.e. each atom has six nearest neighbours, of the other chemical element, placed at the corners of a regular octahedron as shown in Fig. 1.17.

They have narrow energy band gaps and so are suitable for detecting and emitting infrared radiation. Epitaxial lead chalcogenide films are of considerable interest for applications in infrared detector arrays (see Zimin *et al.* 2002, and

Table 4.4. *Slip Systems in Materials with the NaCl Structure but with different types of chemical bonding*

Predominant chemical bonding	Examples of such materials	Slip plane	Slip direction	References
Ionic	NaCl, LiF	{110}	⟨110⟩	Gilman (1959)
Covalent	PbS, PbSe, PbTe	{100}	⟨110⟩	Matthews and Isebeck (1963), Levine and Tauber (1968)
Metallic	UC, TiC	(111)	⟨110⟩	van der Walt and Sole (1967)

references therein). They were of interest for use in infrared lasers at one time but they have been displaced in this application by quantum cascade lasers (QCL) in recent years (see Section 1.7.4). They also have interesting fundamental properties that are unusual (at least among semiconductors) like ferroelectricity and useful thermoelectric properties (see Delin *et al.* 1998 and references therein).

Frank and Nicholas (1953) analysed the possible dislocations in NaCl-structure materials in terms of their elastic energy. They concluded that the only possible Burgers vectors for stable dislocations were $a/2\langle 110\rangle$ and $a\langle 100\rangle$. All other possible Burgers vectors would result in such large energies that it would be favourable for them to dissociate into others with Burgers vectors of the two forms just mentioned. The fact that the slip direction in all types of materials with the NaCl structure is ⟨110⟩, means that dislocations with $a/2\langle 110\rangle$ Burgers vectors should be by far the more common. In fact there is evidence for some slip in ⟨100⟩ directions also in the case of mineralogical PbS (galena) (Mugge 1914, as quoted by Matthews and Isebeck 1963).

Unfortunately no detailed geometrical analysis of dislocations in covalently bonded NaCl-structure materials has been carried out. There has, however, been some experimental work in this field as we now discuss.

Observations of dislocations in lead chalcogenides

Matthews and Isebeck (1963) carried out a TEM study of the dislocations in epitaxial films of PbS grown by evaporation onto cleavage faces of NaCl. The films grew in (100) orientation and long screw dislocations were found aligned in ⟨110⟩ directions in the plane of the films and other dislocations were seen to lie in inclined {110} slip planes. The majority of the dislocations had $a/2\langle 110\rangle$ Burgers vectors but some were found to have $a\langle 100\rangle$ Burgers vectors. Dislocations were seen to move under observation in the microscope. Some of these movements corresponded to the operation of the ⟨110⟩{100} slip system but others to the operation of the ⟨110⟩{1$\bar{1}$0} slip system.

Yagi *et al.* (1971) studied the epitaxial growth of PbSe evaporated onto (100) films of PbS in the TEM. They found layer-by-layer growth (Frank and van der Merwe mode) to occur, not three-dimensional nucleation (Volmer-Weber mode, see Section 2.3.11). The misfit dislocations formed in the PbSe/PbS interface constituted a square network made up of two sets of dislocations. These ran in the two orthogonal ⟨110⟩ directions in the (100) plane. The dislocations of each set were pure edges with Burgers vectors $a/2\langle 110\rangle$ in the direction perpendicular to the dislocation lines and lying in the (100) plane. Thus, the Frank and van der Merwe model for the relief of misfit in this type of growth, by the introduction of classic, pure edge misfit dislocations (see Section 2.3.11), applied.

4.2 Semiconductor dislocation dynamics

The dynamics of dislocations in semiconductors is essentially represented by the velocity-stress relation for these materials, as discussed in Section 2.5.4 and Fig. 2.26, which compared the velocity-stress curves for crystals of materials with different types of bonding. It also includes the study of the effects of doping, electrical fields, etc. on the mobility of semiconductor dislocations, as initially discussed in Section 2.5.4.

Dislocation mobility in semiconductors occurs through the kink mechanism for moving a dislocation from one Peierls potential trough in the slip plane to the next (see Section 2.5.2 and Fig. 2.24).

4.2.1 Dislocation motion

The movement of dislocations in semiconductors can, in principle, be considered in atomistic terms using ball-and-wire models like those of Hornstra (1958) as was shown in Fig. 2.23. The development of quantitative treatments of dislocation motion started from the continuum approach of elasticity theory through the Peierls model.

As more experimental and theoretical evidence on the atomic core structures of dislocations in semiconductors appeared, it became possible to interpret atomistically, and in more detail, the effects of doping, electrical fields, etc. on dislocation mobility and so on the mechanical properties of these materials.

The Peierls stress and the double-kink mechanism

As initially discussed in Section 2.5.2, since Peierls (1940), it has been recognized that the atomic configuration and so the core energy of a dislocation will vary periodically over the slip plane. This 'Peierls potential', $U_p(x)$, varies strongly with a repeat distance, a ($\approx b$), in the direction of motion in semiconductors. Semiconductor dislocations align strongly with the Peierls troughs, which, by symmetry, run in low index directions, as shown in Fig. 4.2. The Peierls stress (to push a straight dislocation over the slip plane) is the maximum slope of the Peierls potential:

$$\tau_p = 1/b \ (dU_p/dx)_{max} \qquad (4.5)$$

where b is the length of the Burgers vector.

Early calculations of τ_p gave values that exceeded the critical resolved shear stress for dislocations found experimentally (Haasen and Alexander 1968). It was therefore proposed that dislocations do not move forward as straight lines but by the thermal nucleation and subsequent widening of double-kink loops (Fig. 2.24).

Dislocation mobility

Dislocation velocity is measured as a function of stress by repeated etching. The first etch marks the starting position of the dislocation by a surface pit. The sample is stressed to a value τ for a time Δt at a temperature T and re-etching marks the new dislocation etch pit site a distance Δx further on (Chaudhuri et al. 1962, Kabler 1963, Schäfer 1967). This works well because dislocation etchants produce pits at the points where the dislocation line emerges at present, as the rate of dissolution is higher at the high-energy core. Re-etching, after moving the dislocations, produces new pointed-bottom pits at the new dislocation sites. The pits at the former sites of the dislocations are widened but not deepened (the faster etching, normal to the surface at the dislocation, no longer occurs here) and so are readily distinguishable, as shown diagrammatically in Fig. 4.43a.

The form of the dislocation velocity stress relation varies characteristically from materials of one type of bonding to another (Fig. 2.26). These relationships can be approximated by power laws of the form:

$$v = B_1\tau^m \qquad (4.6)$$

where τ is the stress, v the velocity and B_1 is a constant for a given temperature and doping concentration. The covalently bonded semiconductors have by far the lowest power dependence of dislocation velocity on stress of any class of materials with m close to 1.

Lang x-ray topography showed the ends of the loops in the experimental arrangement of Fig. 4.43b to be of 60°-type (Patel and Freeland 1970) so Schäfer's data applies to dislocations of this type (Schaumburg 1970). Schaumburg (1972) showed that the exponent m varies slightly with dislocation type and stress (Table 4.5a) and the activation energy, E_d, with stress.

The outcome of many dislocation velocity-stress measurements on Ge, Si (Chaudhuri et al. 1962, 1963, Patel and Freeland 1967, Kannan and Washburn 1970, Erofeev and Nikitenko 1971, George et al. 1972, Alexander and Haasen 1972), InSb (Chaudhuri et al. 1962, Mihara and Ninomiya 1968, Steinhardt and Schäfer 1971, Erofeeva and Osipyan 1973, 1975) and GaAs (Choi and Mihara 1972,

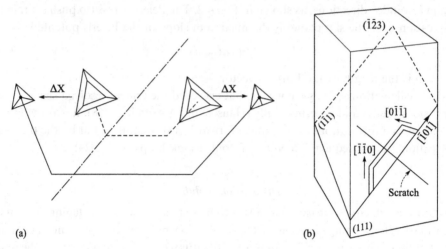

(a)   (b)

Figure 4.43   The measurement of dislocation velocity by the etch, stress and re-etch method in a diamond structure semiconductor. (a) A {111} surface is scratched along a ⟨112⟩ direction, the dash-dot line, to produce a half loop shown by the dotted line and etched. The crystal is then stressed and re-etched. It will then have the surface emergence points of the initial position of the loop, marked by wide, flat-bottomed pits. The end points of the new position of the half-loop, indicated by the solid line, will be marked by small sharp-bottomed pits. (b) The form and orientation of the compression specimens used by Schäfer (1967) and Patel (1970). The scratch was made in the ⟨112⟩ direction perpendicular to the [$\bar{1}$01] trace of the main (1$\bar{1}$1) slip plane. Dislocation loops then result as shown and their Burgers vector was shown by Patel and Freeland to be [$\bar{1}$01]. (After Patel and Freeland 1970. Reprinted with permission from *Journal of Applied Physics*, **41**, p. 2814. Copyright 1970, American Institute of Physics.)

Table 4.5a. *Values of the stress exponent m in Equation (4.6) for the velocity of dislocations in Ge (Schaumburg 1972).*

T (°C)	Stress range (kg/mm^2)	60° dislocations	Screw dislocations
500	>2	1.6 ± 0.2	1.6 ± 0.2
447	”	”	”
”	<2	3.5 ± 0.15	
400	>2	1.7 ± 0.2	1.3 ± 0.3
”	<2	4.6 ± 0.3	

Osvenskii and Kholodnyi 1972, Erofeeva and Osipyan 1973, Osvenskii *et al.* 1973) was that, generally, m ≅ 1.5.

   Chaudhuri *et al.* (1962) and Schäfer (1967) found that the dislocation velocity in Ge could be expressed as:

$$v = B_0 \tau^m \exp(-E_d/kT) = B_1 \tau^m \qquad (4.7)$$

Table 4.5b. *Values of the parameters in the velocity-stress Equation (4.7) for dislocations in semiconductors*

Material and reference	At τ in kg/mm^2	E_d in eV	m	T (°C)
Si 60° dislocations	0.3	2.6	0.9	800
(Alexander and Haasen 1972)				
"	1	2.3	1.3	650
"	3	2.1	-	-
Si screw dislocations	all	2.0	1.1 ± 0.1	-
(Alexander and Haasen 1972)				
InSb and GaAs (Möller 1978)	-	-	1 to 1.5	
GaAs α dislocations		0.93	"	
(Osvenskii and Kholodnyi1972)				
GaAs screw dislocations (ibid.)		1.11	"	
GaAs β dislocations (ibid.)		1.57	"	

(Note that in some figures of earlier works the activation energy is denoted U).

Applying the microdynamical theory of plasticity to macroscopic deformation confirmed that for Ge, m ≈ 1.5, and suggested the same approximate power law for Si, InSb and GaAs also (Alexander and Haasen 1968), in agreement with all the results listed above and those summarized in Table 4.5b.

4.2.2 Modification of dislocation behaviour by the electronic environment

An interesting aspect of the movement of dislocations in semiconductors is the occurrence of several large effects, which have no analogues in the plastic properties of metals.

As we shall see in Chapter 5, the dislocation core contains broken bonds that can act as acceptors (or donors) so the dislocation line will generally be charged negatively. The broken bonds are chemically reactive as well. Thus we can expect that there will be electrostatic interactions between the charged dislocation lines and ionized point defects and short-range chemical bonding effects may occur. These reactions tend to reduce (neutralize or passivate) the electrical and luminescence effects of dislocations especially in the more ionically bonded II-VI compounds than in Si, Ge or the III-Vs.

The long-range electrostatic interaction strengthens the attraction of the dislocation to its Cottrell impurity atmosphere and strengthens the pinning effect opposing dislocation motion. The charge states of the dislocation and point defects can be altered by light of appropriate photon energies. These effects are believed to be responsible for the photoplastic and electroplastic effects in the more ionic II-VI compounds and similar effects in ionic crystals (see Section 4.3 and Ossipiyan *et al.* 1986). For example, applying an electric field across e.g. a ZnS crystal under compressive deformation exerts a force on the charged dislocations and can reversibly alter the stress required to keep the crystal deforming at the same rate. This is known as the electroplastic effect and is illustrated in Fig. 4.44. The effect is a large one. The flow stress was increased and decreased by about 15% by the field in Fig. 4.44.

Doping can increase or decrease the flow stress (hardness) of semiconductors as we shall see. In metallic materials only hardening occurs due to elastic interactions between impurity atoms (solution hardening) or impurity precipitate particles (precipitation hardening) and dislocations. It is thought that the effects of doping of semiconductors result from changes in the core bond structure due to changes in the Fermi level which increase or decrease the mobility of the dislocation (Hirsch 1985).

The effects of liquid environments on surface charges and band bending underlie the chemomechanical effects (see Section 4.2.3) on the deformation of semiconductors.

Effects of doping on dislocation mobility

The effects of doping, i.e. the effects of the charge carrier sign and density, on dislocation mobility were discovered by Patel and Chaudhuri (1966). Patel and Chaudhuri found that heavy doping of Ge (to 10^{19} atoms cm^{-3}) with As, a donor,

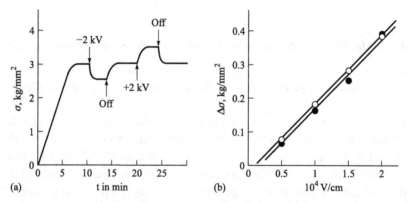

Figure 4.44 (a) The electroplastic effect is the change of hardness that occurs on applying an external potential difference across, e.g. a ZnS specimen undergoing compression testing. The specimen was 4×3×1 mm with the potential applied between the large faces and compression parallel to the long axis. This is the stress strain curve for the material. When an electrostatic potential difference was applied across the contacts the material softened by 15% (showed a fall in flow stress of this amount). (b) The dependence of the change in flow stress on the electrical field. The open circles are for hardening. The solid circles represent softening observations. The deformation rate for both (a) and (b) was 10 μm min^{-1}. (After Osipiyan and Petrenko 1976b. Copyright 1978, American Institute of Physics.)

raised the dislocation velocity for a given stress. Doping with Ga, an acceptor, (to 2×10^{19} atoms cm^{-3}), in contrast, lowers it compared with intrinsic, i.e. pure Ge (Fig. 4.45, Patel and Chaudhuri 1966). As is to be expected from the microdynamical theory of the yield point (see Section 4.4.4), the changes in the velocity-stress relation affect the upper yield stress (see Fig. 4.92 below). That is, since Ga doping makes the dislocations less mobile, a higher upper yield stress is needed to produce the velocity

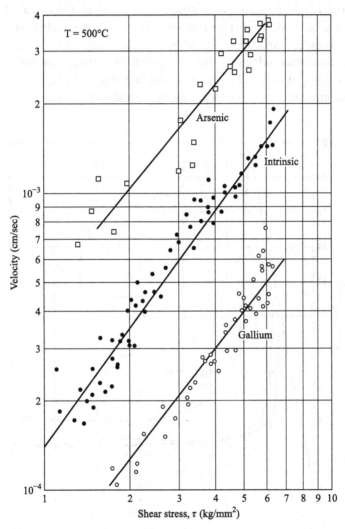

Figure 4.45 The velocity of dislocations in Ge at $500°$ C plotted against the resolved shear stress. The effect of 10^{19} atoms cm^{-3} of As is to increase the velocity whereas 2×10^{19} atoms cm^{-3} of Ga reduces it. [After Patel and Chaudhuri 1966. Reprinted with permission from *Physical Review*, **143**, pp. 601–8, 1966 (http://publish.aps.org/linkfaq.html). Copyright 1966 by the American Physical Society.]

to accommodate the imposed strain rate. Conversely, As-doping increases the dislocation mobility and so lowers the upper yield point as Patel and Chaudhuri found (Fig. 4.46). These observations attracted considerable attention because the occurrence of 'solution softening' by n-type As-doping was unexpected.

The measurements of the dislocation velocity as a function of the concentration of the impurity shown in Fig. 4.47, demonstrated that the impurities had no effect until the concentration reached the intrinsic carrier density for the test temperature.

In addition, Patel and Chaudhuri doped Ge with electrically neutral impurities such as Sn and Si in Ge to concentrations in the range in which electrically active (donor or acceptor) impurities were effective. However, such impurities had no effect on the dislocation velocity or the yield point. Thus these effects were not due to elastic interactions, as solid solution hardening in metals is, but due to alterations in the electronic environment of dislocations. Specifically, they were due to changes in

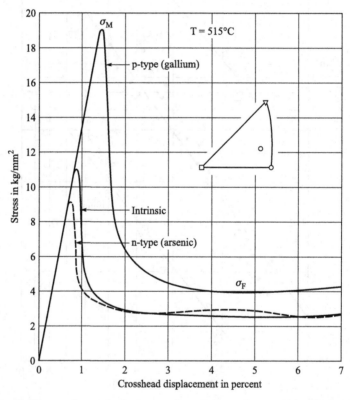

Figure 4.46 Gallium and arsenic doping to the concentrations used in Figure 4.45 has the effects shown on the yield point in dislocation-free Ge. This was tested at 515°C and a crosshead velocity of 0.005 cm min^{-1}. [After Patel and Chaudhuri 1966. Reprinted with permission from *Physical Review*, **143**, pp. 601–8, 1966 (http://publish.aps.org/linkfaq.html). Copyright 1966 by the American Physical Society.]

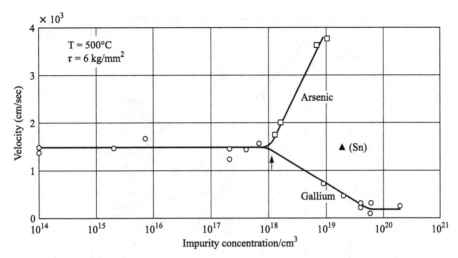

Figure 4.47   Dislocation velocity in Ge at 500° C plotted against the impurity concentration for a constant shear stress 6 kg mm^{-2}. The arrow indicates the intrinsic carrier concentration in Ge. [After Patel and Chaudhuri 1966. Reprinted with permission from *Physical Review*, **143**, pp. 601–8, 1966 (http://publish.aps.org/linkfaq.html). Copyright 1966 by the American Physical Society.]

the sign and density of the majority free charge carriers. Moreover, Patel and Chaudhuri found that the velocity was altered through changes in the activation energy for dislocation movement, E_d, in Equation (4.7) for the thermally activated movement of dislocations, as can be seen in Fig. 4.48. In Equation (4.7), B_0 and m are constants, B_1 is a constant for constant temperature and τ is the resolved shear stress on the slip plane. This is the form of expression of the dependence used in the early literature. In the recent literature the dislocation velocity is written as:

$$v = v_0(\tau/\tau_0)^m \exp(-E_d/kT) = B_1\tau^m \qquad (4.8)$$

where v_0, m, and E_d, the activation energy for dislocation movement, are material constants, B_1 is a constant for constant temperature, and τ_0 is 1 MPa (see e.g. Sumino and Yonenaga 2002).

In silicon, doping with donors like arsenic and antimony in concentrations greater than 10^{18} cm^{-3} increased the dislocation velocity and lowered the upper yield stress just as in germanium (Patel and Freeland 1967, Erofeev *et al.* 1969). The effect of acceptors like boron is temperature dependent. At the lowest test temperatures (450° C) B increased the velocity of 60° dislocations (Erofeev *et al.* 1969) while at 600° C it has little or no effect on the velocity of screw dislocations (Erofeev *et al.* 1969, Patel *et al.* 1976) as can be seen in Fig. 4.49.

It is difficult to account for this behaviour of silicon, so it is regarded as anomalous. Siethoff (1970) measured the lower yield stress in Si crystals doped with P and B. Application of the microdynamical equation for the lower yield stresses to

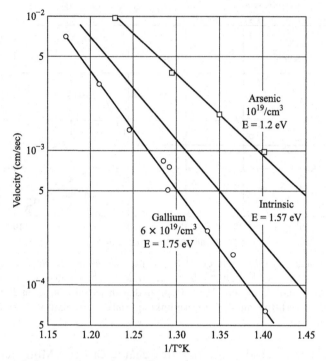

Figure 4.48   Plot of the log of the dislocation velocity against 1/T for As- and Ga-doped Ge crystals at a constant shear stress of 6 kg mm^{-2}. The activation energy is higher for gallium-doping and lower for arsenic-doping than for intrinsic crystals. [After Patel and Chaudhuri 1966. Reprinted with permission from *Physical Review*, **143**, pp. 601–8, 1966 (http://publish.aps.org/linkfaq.html). Copyright 1966 by the American Physical Society.]

low and high strain rate results led him to an interpretation in terms of a strong short-range electrostatic interaction between charged dislocations and oppositely charged solute atoms leading to impurity clouds around the dislocations. This seems to be different from the activation-energy-change mechanism found by Patel and Chaudhuri (1966) in Ge and by Patel and Freeland (1967) and Erofeev et al. (1969) in Si. Evidence for an effect other than a change in the activation energy for dislocation movement was also provided by Erofeev et al. (1969). For stresses less than 4 kg mm^{-2} at 600° C dislocations were found not to move. That is they found evidence of a step in the velocity-stress relation at a 'starting stress' in silicon, which depended somewhat on the doping.

Sumino and co-workers (Sumino and Harada 1981, Imai and Sumino 1983, Sumino 1997) employed *in situ* x-ray topography to study the generation, motion and impurity pinning of dislocations in silicon, as well to elucidate the influence of impurities on dislocation movement (Sumino 1997).

Sumino and Yonenaga (2002) provided an extensive review of the present understanding of the effect of impurities on (i) the variation of the dislocation mobility in

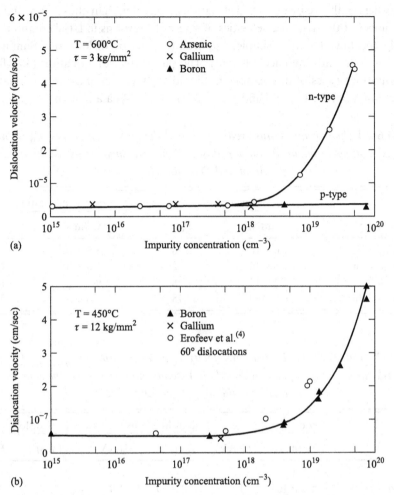

Figure 4.49 Velocity versus impurity concentration for screw dislocations in silicon at the temperatures and resolved shear stresses marked on the graphs. In (b) the results of Erofeev *et al.* (1969) for 60° dislocations are plotted for comparison. [After Patel *et al.* 1976. Reprinted with permission from *Physical Review*, **B13**, pp. 3548–57, 1976 (http://publish.aps.org/linkfaq.html). Copyright 1976 by the American Physical Society.]

glide motion, (ii) the immobilization of dislocations, and (iii) the macroscopic mechanical properties of these crystals. They also presented extensive up-to-date experimental data related to the effects of impurities on those properties in Si and III-V compounds. In numerous studies it was observed that in a doped semiconductor the dislocation velocity could be either increased or decreased depending on the electrical activity of the impurity and the dislocation type. Moreover, the results vary from one semiconductor to another. For instance, it was shown that the velocity of α (A(s)) dislocations is greater than those of β (B(s)) and screw dislocations in InAs and

GaAs, whereas the velocity of β dislocations is greater than those of α and screw dislocations in InP. Also, the velocities of screw dislocations in III-V compounds are typically the lowest of the simple, important forms of dislocation (Sumino and Yonenaga 2002 and references therein). The parameters of Equation (4.8) for the most important types of dislocations in Si and III-V compounds were tabulated by Sumino and Yonenaga (2002) and are listed in Tables 4.6a and 4.6b.

Table 4.6a. *Values of dislocation velocity* v_0 *and the activation energy* E_d *of 60° and screw dislocations in undoped and doped Si and under a shear stress of 20 MPa (after Imai and Sumino 1983)*

Silicon doping	60° dislocations		Screw dislocations	
	v_0 (m/s)	E_d (eV)	v_0 (m/s)	E_d (eV)
High purity	1.0×10^4	2.20	3.5×10^4	2.35
P: $6.2 \times 10^{18}\,\mathrm{cm}^{-3}$	3.3×10^1	1.66	7.0×10^1	1.74
P: $1.5 \times 10^{19}\,\mathrm{cm}^{-3}$	1.7×10^1	1.58	9.5	1.57
B: $1.4 \times 10^{19}\,\mathrm{cm}^{-3}$	6.0×10^3	2.18	7.0×10^3	2.23

Table 4.6b. *Values of the dislocation velocity* v_0 *and activation energy* E_d *of α, β, and screw dislocations in undoped and doped III-V semiconductors. (For the sources of this data see Sumino and Yonenaga 2002.)*

	α dislocations		β dislocations		Screw dislocations	
	v_0 (m/s)	E_d (eV)	v_0 (m/s)	E_d (eV)	v_0 (m/s)	E_d (eV)
InAs						
undoped	5.6×10^4	1.4	2.8×10^4	1.4		
S: $1.3 \times 10^{18}\,\mathrm{cm}^{-3}$	5.3×10^3	1.30	1.4×10^3	1.35		
Zn: $3.7 \times 10^{17}\,\mathrm{cm}^{-3}$	7.8×10^3	1.22	5.2×10^2	1.35		
InP						
undoped	4×10^4	1.6	5×10^5	1.7	4×10^4	1.7
S: $8 \times 10^{18}\,\mathrm{cm}^{-3}$	1×10^{-1}	0.8	2×10^4	1.7	8×10^1	1.4
Zn: $6 \times 10^{18}\,\mathrm{cm}^{-3}$	7×10^{-2}	1.3	1.6	1.2	1.3×10^{-1}	1.4
GaAs						
undoped	1.9×10^3	1.30	5.9×10^1	1.30	1.2×10^2	1.4
Si: $4 \times 10^{18}\,\mathrm{cm}^{-3}$	8.0	1.10	2.7×10^2	1.60	5×10^1	1.7
Te: $6 \times 10^{18}\,\mathrm{cm}^{-3}$		1.10	8.8×10^1	1.60		
Zn: $2 \times 10^{19}\,\mathrm{cm}^{-3}$	6.7×10^1	1.40		1.15		1.6
In: $2 \times 10^{20}\,\mathrm{cm}^{-3}$		1.40	8.3×10^1	1.40		1.3
Cr: $2 \times 10^{17}\,\mathrm{cm}^{-3}$	1.1×10^3	1.25	1.9×10^2	1.35		

Yonenaga (2003) studied the dislocation-impurity interactions in Si doped with acceptor (B), donor (P) and neutral (Ge) elements and compared them with material containing O. In these experiments, the specimens underwent high temperature three-point bending in a vacuum. Dislocation generation and motion from a scratch were followed by the etch pits, produced by a modified Sirtl etchant at 20° C (Yonenaga 2003). He found that the dislocation velocity increased with increasing B and P concentrations. However, Ge had only small effect on dislocation velocity increase. With increasing B and P concentration, the critical stress for dislocation generation increased. This was attributed to dislocation immobilization by impurity segregation. It was also found that dislocation generation from a surface scratch in heavily-doped Si (B and P concentrations $>1 \times 10^{19} \, cm^{-3}$) was suppressed, indicating complete pinning of the sources. Fig. 4.50 plots the velocities of 60° dislocation (under a shear stress of 30 MPa) in Si against the electrical type and concentration of the impurities.

Fig. 4.51 depicts the effects of doping on the temperature dependence of the velocities of 60° dislocations, under a shear stress of 30 MPa, in Cz Si. It can be seen that the dislocations move much faster in P- and B-doped Si than in undoped and O-containing Si at temperatures below about 800° C. Yonenaga (2003) attributed the velocity increase, i.e. the activation energy for dislocation motion decrease, to the electrical effects of donor and acceptor impurities on the basic formation and/or migration of kinks to produce dislocation motion. As also noted by Yonenaga (2003), the results above 800° C, showing that the dislocation velocities in heavily B and P doped Si are still greater than in undoped Si, may not be due to the same mechanism. This is because of the breaks in the temperature dependence of

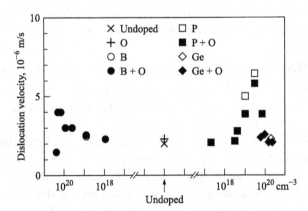

Figure 4.50 Dependence of the velocities of 60° dislocations (under a shear stress of 30 MPa at 800° C) in Si on the electrical type and concentration of the main impurities. (After Yonenaga 2003. Reprinted from *Materials Science in Semiconductor Processing*, **6**, Dislocation–impurity interaction in Si, pages 355–8. Copyright 2003, with permission from Elsevier.)

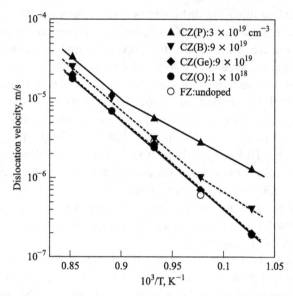

Figure 4.51   Dependence of the velocities of 60° dislocation, under a shear stress of 30 MPa, in Cz Si doped with B, P and Ge, on the reciprocal temperature. The results for undoped and O-containing Cz Si are also shown. (After Yonenaga 2003. Reprinted from *Materials Science in Semiconductor Processing*, **6**, Dislocation–impurity interaction in Si, pages 355–8. Copyright 2003, with permission from Elsevier.)

dislocation velocity in B and P doped Si. Yonenaga (2003) related this to the kink configuration of a dislocation in heavily doped semiconductors. For the case of such a high doping level, the sample was treated as a dilute solid solution. (High doping densities, by semiconductor standards, correspond to dilute solutions in metals and other materials, which are generally much less pure.)

Theories of the effects of doping on dislocation velocities

Two theories were proposed to account for the effects of doping on dislocation mobilities in semiconductors. Both ascribe this to the influence of the line charge on the creation or motion of double kinks.

The first was that of Patel and his co-workers (Frisch and Patel 1967, Patel *et al.* 1976, Patel and Testardi 1977a). This postulated that the kinks responsible for dislocation movement are associated mainly with charged dislocation acceptor sites (dangling bonds). It also assumed that the enhancement of the dislocation velocity due to doping is proportional to the fraction of charged sites, i.e. that $v \propto f$. Originally Frisch and Patel (1967) considered the charge balance around an isolated dislocation assuming that the concentrations of the various species entering the charge-balance equation obey Boltzmann statistics. Patel *et al.* (1976) presented a revised form of the theory using Fermi-Dirac statistics and applied it to their

Si observations but the interpretation was not self-consistent. Patel and Testardi (1977a) applied the Fermi-Dirac statistics theory to Ge and showed that a self-consistent interpretation could be given for all the observations on the doping and temperature dependence of the dislocation velocity in germanium. The values for the dislocation energy levels in Si and Ge obtained from their theory were consistent with values obtained from early studies of the effect of plastic deformation on the resistance and Hall constant of Ge using the Read theory (see Section 5.2).

These values, however, disagreed with later results obtained by the Göttingen school (see Section 5.2.6), members of which, therefore, criticized the theories of Patel *et al.* (Schröter *et al.* 1977) and put forward an alternative theory (Haasen 1975). Haasen pointed out that the electrostatic self-energy of a system of charges on a straight line, such as a dislocation with a number of dangling bonds containing two or zero electrons, is lowered by any deviation from straightness. He calculated the difference, ΔU, in the electrostatic part of the self-energy between a straight dislocation and one with a double kink in the saddle point configuration and found it to be proportional to the square of the line charge, Q^2. He showed that this effect could roughly account for the experimentally observed variation of the activation energy, E, with doping (Schröter *et al.* 1977). Later, Haasen (1979) put forward a treatment of kink formation and migration as dependent on the Fermi level.

Effects of doping on dislocation motion in semiconducting compounds

Effects similar to those found in Ge and Si also occur in semiconducting compounds where dislocations of opposite polarity are affected differently.

Choi and Mihara (1972) measured the velocities of individual dislocations by the etch-stress-re-etch method and identified the dislocations by x-ray topography. They reported, as shown in Fig. 4.52, that in GaAs α-dislocations, as they were then known, i.e. Ga(s) dislocations are much faster in *n*-doped crystals than are β (As(s)) dislocations. In *p*-doped crystals the velocities of the two types of dislocations are similar with the order perhaps reversed.

Osvenskii *et al.* (1973) measured dislocation velocities in GaAs over the temperature range 200 to 500° C and the stress range 0.25 to 5 kg mm^{-2} as affected by Zn and Te doping. They found that Zn (*p*-type) doping increased the activation energy for movement of screw dislocations, E_d, in the expression for dislocation velocity

$$v = v_0(\tau/\tau_0)^m \exp(-E_d/kT) \tag{4.9}$$

i.e. of the same form as Equation (4.8), as shown in Fig. 4.53. This appears to agree with Choi and Mihara who observed that *p*-doping reduced the velocity of dislocations of both polarities. However, Osvenskii *et al.* reported that due to an increase in the pre-exponential factor of Equation (4.9), the velocity of screw dislocations in fact increases in Zn-doped crystals despite the rise in the activation energy.

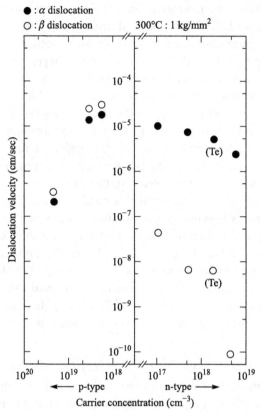

Figure 4.52   Velocities of α (Ga(s)) and β (As(s)) dislocations in GaAs for 1 Kg mm⁻² at 300° C. The n-type crystals were doped with Si except in the case marked, which was doped with Te. The p-type crystals were doped with Zn. (After Choi and Mihara 1972.)

Osvenskii *et al.* (1973) also reported the influence of doping on the temperature dependence of the velocity of 60° α-(Ga(s)) and β-(As(s)) dislocations under a stress of 2 kg mm⁻² as shown in Fig. 4.54. That is, they found that doping with both Zn and Te reduced the difference between the activation energies of the motion of the Ga(s) and As(s) dislocations, because E_d for the Ga(s) dislocations increases, whereas E_d for the As(s) dislocations decreases. They also reported that a preliminary heat treatment of the Te-doped crystals at 900° C made the dislocations immobile throughout the range of stresses that were investigated. This they related to the formation of complexes composed of the Te atoms and intrinsic point defects, which then interfere with the kink nucleation of movement.

In addition to these papers on dislocation velocities there are a number of papers on the effects of doping on the plastic properties of GaAs. Sazhin *et al.* (1966) investigated the indentation hardness of Zn- and Te-doped GaAs and Weiss and Hartnagel (1977) extended this to Cr-doped (semi-insulating) GaAs with a view to

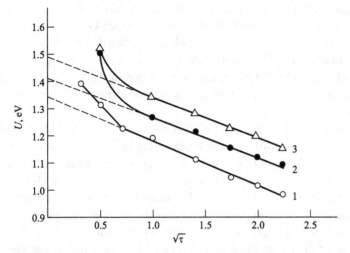

Figure 4.53 The dependence of the activation energy of the movement of screw dislocations in GaAs on the square root of the applied stress. Curve 1 is for undoped crystals, curve 2 for crystals doped with 5×10^{18} Zn atoms cm^{-3}, and curve 3 for Zn doping to a concentration of 5×10^{19} cm^{-3}. (After Osvenskii *et al.* 1973. Reprinted with permission from *Soviet Physics Solid State*, **15**, pp. 661–2. Copyright 1973, American Institute of Physics.)

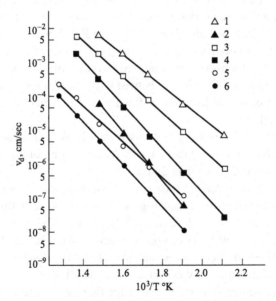

Figure 4.54 The temperature dependence of the velocity of Ga(s) and As(s) 60° dislocations in GaAs for the following impurity concentrations: curves 1 and 2 undoped crystals, curves 3 and 4 Zn-doped to $p_0 = 5 \times 10^{18}$ cm^{-3}, and curves 5 and 6 Te-doped crystals with $n_o = 8 \times 10^{18}$ cm^{-3}. Curves 1, 3 and 5 are for Ga(s) dislocations. Curves 2, 4 and 6 are for As(s) dislocations. (After Osvenskii *et al.* 1973. Reprinted with permission from *Soviet Physics Solid State*, **15**, pp. 661–2. Copyright 1973, American Institute of Physics.)

determining the relative sensitivities of these materials to damage around thermo-compression bonds. They found that Cr- and Zn-doped GaAs were harder than Te-doped, which in turn was harder than undoped material. This Weiss and Hartnagel ascribed to an atomic-misfit, elastic hardening effect since the Cr atoms would not be ionized. Osvenskii *et al.* (1969) studied the creep in three-point bending of GaAs crystals oriented so that either mainly Ga(s) or mainly As(s) dislocations were introduced.

The observations of Yonenaga and Sumino (1993) on the variation of velocity of α-, β- and screw-dislocations with temperature in variously doped InP crystals is shown in Fig. 4.55. Isovalent Ga and As dopants do not influence the velocity of any type of dislocations, whereas Zn acceptors reduce the velocities of all three types and S donors increase the velocity of α dislocations but reduce the velocities of β and screw dislocations. These impurity effects on dislocation dynamics also correlated with the mechanical strength of InP (Yonenaga and Sumino 1993). A reduction in the dislocation velocity in Zn-doped InP caused an increase in the yield strength. Dislocation immobilization (due to impurity locking) in S- and Zn-doped material also results in an increase in the yield strength at high temperatures (Yonenaga and Sumino 1993).

The dynamic behavior of α and β dislocations in undoped and S- and Zn-doped InAs crystals was studied as a function of stress and temperature by employing the etch pit technique (Yonenaga 1998). The α dislocations moved faster than β dis-locations in both the undoped and doped InAs (see Fig. 4.56). It was also found that S donors reduced the velocities of both α and β dislocations, whereas Zn acceptors increased the velocity of α dislocations but reduced that of β dislocations (Yonenaga 1998). These observations, which vary between semiconductors, were attributed to the electronic state of the dislocations influencing the elementary processes of the motion (Yonenaga 1998).

The parameters of Equation (4.8) for selected III-V compounds are listed in Table 4.6b. The values of m vary from 1.4 to 2.1 depending on the material and the doping type (Sumino and Yonenaga 2002).

Yonenaga (1998) also compared the dynamics of dislocations in InAs with that of those in other semiconductors. From Fig. 4.57, it can be seen that the activation energies for dislocation motion depend linearly on the band-gap with an apparent distinction between different types of semiconductors. That is, the slope of the linear increase in the activation energy as a function of the band-gap is different for the elemental, IV-IV compound, III-V compound, and II-VI compound semiconductors. Furthermore, these slopes increase in an order that, as pointed out by Yonenaga (1998), suggests it may depend on the ionicity of the material. Yonenaga (1998) also indicated that, from the dependencies of the activation energy on the band-gap, one could deduce information for materials in which dislocation velocities have not been measured. For instance, in semiconductors like ZnSe and GaN, which may be used for blue lasers, dislocation motion is expected to influence the device degradation,

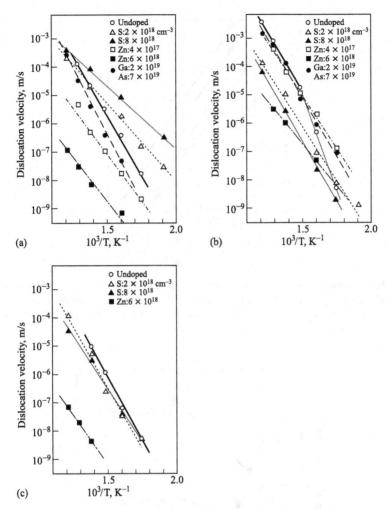

Figure 4.55 Dependence of the dislocation velocities (under a resolved shear stress of 20 MPa) in InP crystals doped as indicated of (a) α dislocations, (b) β dislocations, and (c) screw dislocations on the reciprocal temperature. (After Yonenaga and Sumino 1993. Reprinted with permission from *Journal of Applied Physics*, **74**, pp. 917–24. Copyright 1993, American Institute of Physics.)

e.g. by the DLD (dark line defect) failure mechanism (see e.g. Section 5.11.5). In Fig. 4.57 one can interpolate that the activation energy of dislocation motion is about 1 eV for ZnSe (which has a band-gap energy of 2.3 eV). In GaN (with band-gap energy of 3 eV) it should be about 2 eV. Thus, dislocations would be expected to be more mobile in ZnSe than in GaN. Hence, ZnSe-based devices are more likely to experience degradation due to dislocation motion and multiplication, than GaN devices (Yonenaga 1998).

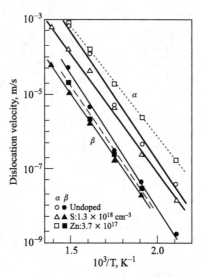

Figure 4.56 Dependence of dislocation velocities (under a resolved shear stress of 20 MPa) of α- and β-dislocations on the temperature in InAs crystals doped as marked. (After Yonenaga 1998. Reprinted with permission from *Journal of Applied Physics*, **84**, pp. 4209–13. Copyright 1998, American Institute of Physics.)

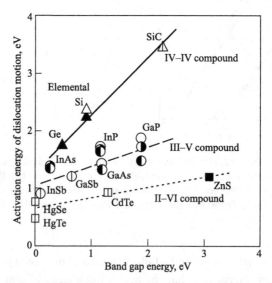

Figure 4.57 Plots of the activation energy of dislocation motion versus the band-gap energy for the semiconductors marked. (After Yonenaga 1998, who included data from the references he quotes. Reprinted with permission from *Journal of Applied Physics*, **84**, pp. 4209–13. Copyright 1998, American Institute of Physics.)

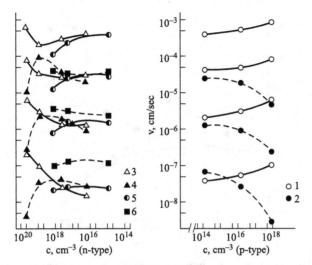

Figure 4.58   The variation in the velocities of In(s) dislocations (continuous curves) and Sb(s) dislocations (dashed curves) with the concentrations of Te donors and of Zn or Ge acceptors in InSb for a stress of $1.5\,\mathrm{kg\,mm^{-2}}$. Curves 1 and 2 are for Te-doped, curves 3 and 4 for Zn-doped and curves 5 and 6 for Ge-doped crystals. (After Erofeeva and Osipiyan 1973. Reprinted with permission from *Soviet Physics Solid State*, **15**, pp. 538–40. Copyright 1973, American Institute of Physics.)

Erofeeva and Osipiyan (1973) studied the effect of Te, Zn and Ge doping on the mobility of dislocations in InSb, and found large differences in the effects on In(s) and Sb(s) dislocations. Their results are shown in Fig. 4.58. It can be seen that the velocity of the In(s) dislocations increased with the Te-doping concentration, decreased with the concentration of Ge acceptors and increased with the concentration of Zn acceptors. The variation of the velocities of the Sb(s) dislocations was generally of the opposite sign to those of the In(s) dislocations.

Markedly different changes in the activation energies for dislocation movement of the In(s) and Sb(s) dislocations with the concentration of electrically active impurities were also observed by Erofeeva and Osipiyan (1973) as shown in Fig. 4.59. They pointed out that the velocity variations of Fig. 4.58 couldn't be accounted for by the changes in activation energy of Fig. 4.59 alone. (For example, the dashed velocity curve no. 4 in Fig. 4.58 exhibits a marked maximum at a concentration of about $10^{19}\,\mathrm{cm^{-3}}$ whereas the corresponding activation energy curve (dashed no. 4 in Fig. 4.59) shows a much smaller fall at higher concentrations.) This they interpreted as evidence that charged impurities influence not only the nucleation of double kinks which alters E but also the velocity of the kinks which would alter the pre-exponential factor in the dislocation velocity-stress relation. They also ascribed the opposite effects of doping on In(s) and Sb(s) dislocations to the differences between the electrical properties of these dislocations.

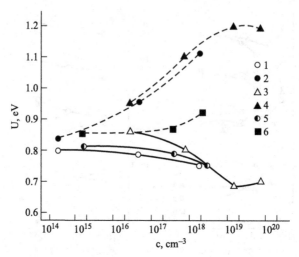

Figure 4.59 The variations with impurity concentration in the activation energy for movement of In(s) dislocations (continuous curves) and for Sb(s) dislocations (dashed curves) in InSb at a stress of $1.4\,\mathrm{kg\,mm^{-2}}$. Curves 1 and 2 are for Te doping, curves 3 and 4 are for Zn-doping and curves 5 and 6 for Ge-doping. (After Erofeeva and Osipiyan 1973. Reprinted with permission from *Soviet Physics Solid State*, **15**, pp. 538–40. Copyright 1973, American Institute of Physics.)

Seltzer (1966) studied the movement of dislocations in PbS as a function of non-stoichiometry (see Section 6.3.2). He measured the indentation etch rosette 'wing' length (see Section 4.1.3), i.e. the maximum distance dislocations moved from the indentation, and the hardness (indentation size) on specimens whose charge-carrier concentration had been fixed by high temperature equilibration in controlled sulphur pressures as shown in Fig. 4.60. He found that both the hardness and the rosette wing length were nearly independent of the concentration of free electrons in the range from 2×10^{17} to $3 \times 10^{19}\,\mathrm{cm^{-3}}$ whether the n-type material were prepared by the introduction of an excess of Pb or by the addition of bismuth donors. Similar measurements on p-type, sulphur-excess crystals, whether pure or doped with Ag acceptors, showed a marked increase in the hardness and decrease in rosette wing length with increasing hole concentrations. Seltzer suggested that this behaviour results from an electrostatic interaction between charged dislocations and acceptor point defects.

Much further work will be required to clarify obscure and contradictory results in this field and to elucidate the mechanisms involved.

4.2.3 Chemomechanical effects in semiconductors

Related to the doping-dependence of dislocation velocities in semiconductors are chemomechanical effects on the deformation and fracture of semiconductors

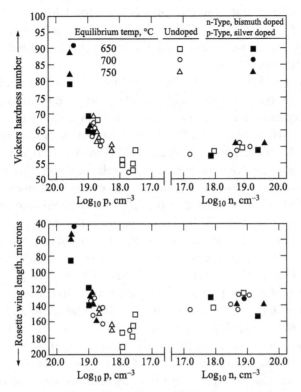

Figure 4.60 The variation of the diamond pyramid hardness (upper graph) and the etch rosette wing length (lower graph) for room temperature indentation of PbS crystals of various p- and n-type carrier concentrations. (After Seltzer 1966. Reprinted with permission from *Journal of Applied Physics*, **37**, pp. 4780–4. Copyright 1966, American Institute of Physics.)

(Schröter and Haasen 1977) and other non-metals (Westwood 1974, Westwood and Latanision 1976, Westwood *et al.* 1981). Such effects were first reported by Rebinder in 1928 in the then Soviet Union and long studied by him and his school (Rebinder 1928, Likhtman *et al.* 1958, Rebinder *et al.* 1944) but there were difficulties in reproducing some of their results (for a historical review see Westwood *et al.* 1981). Later chemomechanical effects gained scientific interest (see e.g. the Proceedings of the Nato Advanced Study Institute on Surface Effects in Crystal Plasticity (Latanision and Fourie 1977)) since they could be ascribed to the interaction between the line charges on dislocations and the charge on the surface of a crystal, which can be widely varied by chemical adsorption. This can alter the dislocation mobility and thereby change the plastic properties within some sub-surface region. This is well confirmed in ionically bonded materials and minerals.

Schröter and Haasen (1977) considered such interactions in terms of their models for the electrical state of dislocations in Si and Ge (see Section 5.2.5) and derived a possible mechanism for such a change of the plastic properties in the surface region.

They pointed out that data for their model were available only for Si and Ge, which are not plastically deformable at room temperature. They suggested that some II-VI compounds were more suitable for experimental observation of such chemomechanical effects.

Ahearn *et al.* (1978a), following this suggestion, set out to study the relation between surface charge and hardness in ZnO. They adopted this as a model material because the electronic effects in ZnO resulting from altering the surface charge were reasonably well understood. (ZnO has the wurtzite structure. It is a wide-gap, normally *n*-type II-VI compound in which the bonding has a large ionic component. It can be deformed at room temperature and is relatively soft.) To alter the surface charge a ZnO single crystal was made the working electrode in an electrolytic cell, as shown in Fig. 4.61. The bias voltage of the cell, V, was varied relative to a standard Calomel electrode. Under anodic bias (ZnO positive) the surface charge is negative and the sub-surface space-charge region is depleted, forming a Schottky barrier as shown in Fig. 4.62a. (There are no ZnO surface states to complicate this picture, Dewald 1960.) The surface charge can be increased to zero (Fig. 4.62b) and then made positive (Fig. 4.62c) by reducing the anodic bias. Thus biasing allows control of the band bending and so of carrier density in the conduction and valence bands and the occupancy of the dislocation and defect levels in the space charge layer. The influence of these changes on the micro-hardness can thus be studied.

Ahearn *et al.* (1978a) measured the dependence of the Vickers micro-hardness of (0001) surfaces on bias voltage. Basal plane dislocations are involved in this case. The Vickers hardness is obtained from the observed size of the indentation produced in the surface of the material by a pyramidal diamond under a fixed load. The harder the material the smaller the indentation produced. They found, unexpectedly, that

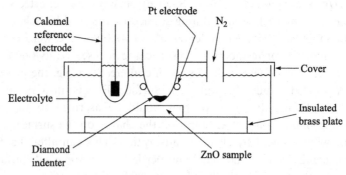

Figure 4.61 Diagram of the set-up allowing hardness indentations to be made in ZnO at a controlled bias voltage in an electrolytic cell. (After Ahearn *et al.* 1978a. Reprinted with permission from *Journal of Applied Physics*, **49**, pp. 96–102. Copyright 1978, American Institute of Physics.)

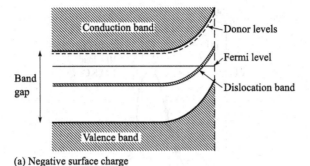

(a) Negative surface charge

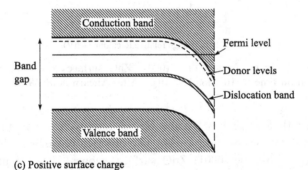

(b) Zero surface charge

(c) Positive surface charge

Figure 4.62 Diagram of the band bending in n-type ZnO with (a) negative, (b) zero and (c) positive surface charges. (After Ahearn *et al.* 1978a. Reprinted with permission from *Journal of Applied Physics*, **49**, pp. 96–102. Copyright 1978, American Institute of Physics.)

the hardness maximum (see Fig. 4.63) did not occur for the zero surface charge, flat band condition, but for a slight positive surface charge.

Ahearn *et al.* (1978b) made similar observations on the {10$\bar{1}$0} surfaces of ZnO. In this case it is prismatic dislocations that are forced to move to produce the micro-hardness indentations. [Osipiyan and Smirnova (1968) had carried out a ball-and-wire study of the core structures of dislocations in the wurtzite structure.

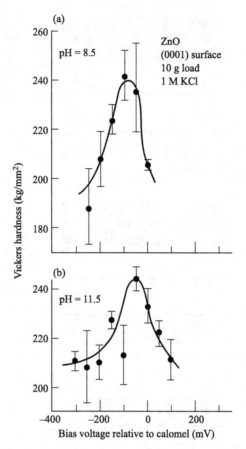

Figure 4.63 The Vickers hardness of a (0001) ZnO surface versus the bias voltage in 1 M KCl. Indentation time = 10 s, load = 10 g. (After Ahearn *et al.* 1979.)

They concluded that while basal plane dislocations were polar, prismatic plane dislocations were not (see Section 4.1.5).] The observations of Fig. 4.64, on {10$\bar{1}$0} surfaces, like those for the (0001) ZnO surface, showed that the micro-hardness varies with surface charge. Again, however, the maximum did not occur at zero surface charge, i.e. the flat band potential bias, as hitherto assumed, but for a slight positive charge, i.e. downward band bending.

To account for the hardness maximum occurring for a small positive surface charge they suggested a mechanism involving charge exchange between donor levels and the conduction band near moving dislocations (Ahearn *et al.* 1978a, 1979, Westwood *et al.* 1981). As shown in Fig. 4.65, both the conduction and valence bands are distorted near a dislocation. (This can be due to (i) the dislocation having a positive or negative line charge and a surrounding screening cloud of conduction electrons of interstitial zinc ions, or (ii) coupling of the strain field around the

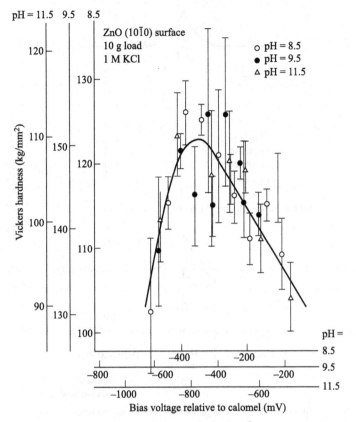

Figure 4.64 Master curve for the Vickers hardness of {10$\bar{1}$0} surfaces of ZnO versus bias voltage in 1 M KCl at three values of pH. The indentation load of 10 g was applied for 10 seconds. (After Ahearn *et al.* 1979.)

dislocation and the energy bands through the deformation potential, or (iii) the piezoelectric effect.) Now if, for example, the dislocation induces an upward band distortion, as in Fig. 4.65, its motion will require the movement of the band distortion. Under flat band conditions (zero surface charge), Fig. 4.65a, only rearrangement of the conduction electrons would be required, so the motion of the dislocations would be relatively easy and the crystal relatively soft. As the bands are bent down (positive surface charge) and E_c and E_d approach E_F, the Zn^+ donors away from the dislocation become neutralized. Hence dislocation motion in this case requires excitation of electrons from the neutral donor levels to unoccupied levels in the conduction band as shown in Fig. 4.65b. This excitation requires about 50 meV/donor so there will be a decrease in dislocation mobility and an increase in hardness because this additional energy must supplied to allow motion. Further downward band bending results in the donors in the space charge layer all becoming neutralized, as shown in Fig. 4.65c. In this case the dislocation can again move

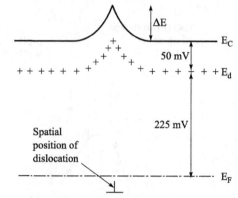

(a) Flat bands: All donors ionized

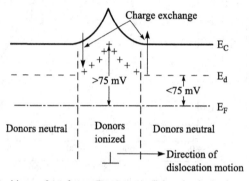

(b) Positive surface charge: Donors near dislocation ionized

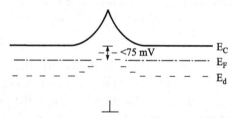

(c) Increased positive surface charge: All donors neutralized

Figure 4.65   Schematic diagram of the distortion in electronic energy levels in ZnO in the neighbourhood of a basal dislocation near the edge of the conduction band. (After Ahearn *et al.* 1978b. Reprinted with permission from *Journal of Applied Physics*, **49**, pp. 614–17. Copyright 1978, American Institute of Physics.)

without exciting donors. Hence the dislocation mobility will be less in the situation of Fig. 4.65b than in either of the others resulting in a maximum in the micro-hardness. A similar argument applies in the case of a dislocation that results in downward band bending locally (a dislocation with a positive line charge).

The amount of environmentally induced band bending required to produce a maximum in hardness depends on the magnitude of ΔE (Fig. 4.65a). It is necessary to assume a value for ΔE to account for the position of maximum hardness. Agreement between theory and experiment was obtained for a value $\Delta E = 200 \, \text{meV}$ for the presumed negatively charged basal O(s) dislocations controlling the hardness of the $(10\bar{1}0)$ surface. The best-fit value in the case of the presumed uncharged prismatic dislocations controlling the hardness of the (0001) surface was $\Delta E = -50 \, \text{meV}$ (Ahearn *et al.* 1978a and b, 1979).

Czernuska (1989) used the experimental set-up of Fig. 4.61 to indent $\{1\bar{1}00\}$ faces of ZnO crystals, producing arrays of polar basal plane dislocations. The applied voltage was initially set at its most negative value and indentations were made at successively more positive values. Etching revealed rosettes of pits, with arms of unequal length. The crystals were cleaved along the basal plane and etched to determine the polarity following Mariano and Hanneman (1963). Czernuska concluded that the longer arms consisted of Zn(g) dislocations and the shorter of O(g) so the relative mobilities of polar dislocations could be studied. His results are plotted as the length of the Zn(g) minus that of the O(g) rosette arms against the applied voltage at two values of pH in Fig. 4.66. It can be seen that the Zn(g) dislocations were the more mobile for pH = 9.5 for positive voltages and the O(g) dislocations for negative voltages. For pH = 9.0 the O(g) dislocations become the more mobile only for negative biases $> -250 \, \text{meV}$. This influence of the bias on the relative mobilities constitutes an electromechanical effect. It can be seen in Fig. 4.66 that a reversal of the sign of the relative mobilities can be produced for negative voltages $< 250 \, \text{meV}$ by changing the pH which is a chemomechanical effect. The observed effect extended to a depth of the order of the Debye length. In air, with no applied voltage the Zn(g) basal plane dislocations were the more mobile.

Czernuska (1989) interpreted his observations in terms of the model of Hirsch (1980). This postulates that the dislocation velocity in semiconductors depends on the energy difference between the Fermi level and a dislocation kink energy level in such a way that a greater energy difference increases the concentration of charged kinks and thus the dislocation velocity. The kink level will move with the conduction and valence band edges while the Fermi level remains constant as shown in Fig. 4.62. Thus an applied voltage can change the kink to Fermi energy difference, thereby changing the dislocation mobility. This influence will extend into the crystal only as far as the band bending does, i.e. of the order of the Debye length as observed by Czernuska. This difference from the more ionic approach of Ahearn *et al.* (1979) suggests that for ZnO the change in velocity is due to a change in the interaction of a charged dislocation with charged Zn ions through the applied voltage altering the charge on these ions.

Czernuska (1989) checked for any chemomechanical effect in GaAs. GaAs was much harder so the indentations produced much smaller dislocation arrays (etch pit rosette arms). Consequently, only the hardness could be measured and there was no

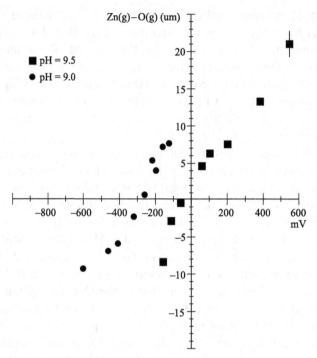

Figure 4.66 The difference in rosette length of Zn(g) and O(g) dislocations versus the applied voltage at two values of pH. [After Czernuszka 1989, In *International Symposium on Structural Properties of Dislocations in Semiconductors*, Oxford Conference Series No. 104, 1989. (Bristol: Institute of Physics), pp. 315–19.]

detectable variation with applied voltage or pH at loads of 5–50 g. This is to be expected since it is well known that in GaAs the energy levels are pinned at the surface by surface states so that applied voltages can produce little or no band bending.

Pugh *et al.* (1966) studied the effects of several liquid metals, from groups II to V of the periodic table, on the fracture strength of germanium single crystals tested in three-point bending at temperatures from 30 to 600° C. The effect of gallium from 30 to 650° C, which they divided into ranges I to IV, is shown in Fig. 4.67.

In range I, 30 to 100° C, Ge was purely brittle and showed the large scatter of the fracture stress that is usual for damage-free specimens. Neither the presence of Ga nor the temperature in this range altered the fracture stress. In range II, 100 to 370° C, the behaviour was still purely brittle and the fracture stress in air remained unaffected by the temperature but that for samples tested in Ga fell, over a narrow range of temperatures, from 120 to about 10 kg/mm². Examination of fractured Ga-coated specimens showed large etch-pits in the Ge surface which acted as notches, facilitating the initiation of cracking. This is consistent with equilibrium

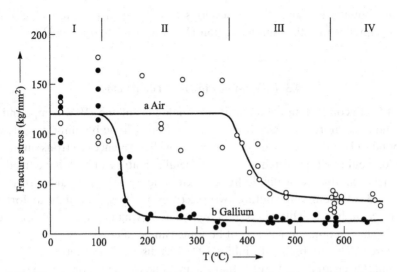

Figure 4.67   Fracture stress versus testing temperature (a) in air and (b) in gallium. (After Pugh *et al.* 1966.)

data for the Ga-Ge system, which indicates that the onset of solution of Ge in Ga occurs at about 100° C. The Ge specimens became ductile in air from about 370° C and exhibited well-defined yield drops (see Section 4.4.4) over the whole of range III (to 565° C). Ga-coated specimens, however, remained completely brittle and the fracture stress was temperature independent. In range IV, 565 to 650° C, both Ga-coated specimens and those tested in air deformed plastically.

The liquid metals studied were zinc and cadmium from group II, gallium, indium and thallium from group III, tin and lead from group IV and antimony and bismuth from group V. Limited tests carried out at temperatures in the brittle range showed that large reductions in fracture stresses were produced and in each case these were associated with etching effects. At higher temperatures, when the significant increase in ductility reduces the notch effect of the etch pits, true liquid metal embattlement phenomena were observed.

Another semiconductor chemomechanical effect was observed in chemomechanical polishing (e.g. McHugo and Sawyer 1993, and references therein). During such polishing, impurities may be introduced in single and polycrystalline silicon. McHugo and Sawyer (1993) demonstrated (using EBIC) that chemomechanical polishing of boron-doped float-zone silicon introduces impurities that decorate and electrically activate swirl defects. In boron-doped polycrystalline silicon, chemomechanical polishing and preferential etching generates etch pits that increased in density with decreasing resistivity; this suggests an interaction between the polishing impurity and a dopant (McHugo and Sawyer 1993). They also demonstrated that in this case, the presence of the impurity increases the electrical activity of dislocations and grain boundaries, and it also develops a uniform distribution of electrically active precipitates in a thin surface region (McHugo and Sawyer 1993).

4.3 Dislocations in II-VI compounds

The II-VI semiconducting compounds are analogues of the III-V compounds, but their bonding is more ionic (see Section 1.3.2). Purely ionic bonding, $f_i = 1$, would correspond to $II^{2+}VI^{2-}$, e.g. $Cd^{2+}S^{2-}$. That is, the two valence electrons on the metal atom, Cd, would be transferred to the non-metallic S atom so both have filled octets and attract each other ionically. Fully covalent bonding, $f_i = 0$, would correspond to each II and VI atom contributing two electrons to its sp^3 orbitals to form four shared covalent bonds. This can be represented as leaving the filled-electron-shells as $Cd^{2-}S^{2+}$. The fractional ionicity, f_i, of bonding can be calculated in several ways that do not agree. (The approach of Phillips 1973 was discussed in Section 1.2.3) but the conclusion (Osipiyan *et al.* 1986) is that the metallic element atoms have some positive charge in II-VI compounds, i.e. there is partial ionic bonding. Many II-VI compounds occur in both the sphalerite and wurtzite structures in different temperature ranges and since they change structure on cooling after growth, they tend to contain many wide stacking faults in extended dislocations. Complex polytype (long stacking sequence) structures also occur, especially in ZnS, as discussed in Section 4.1.5.

Many of the II-VIs have wide band-gaps and so are transparent and semi-insulating at room temperature and below. They are used in optoelectronics as phosphors (light-emitting powders for fluorescent lights and CRT (including TV) screens and photoconductors). The II-VIs were reviewed by Aven and Prener (1967) and Hartmann *et al.* (1982).

Unlike other ionic crystals, in the II-VIs the electronic sub-system dominates the electrical properties, i.e. they are semiconductors. Moreover, unlike most other (covalent) semiconductors, the more ionically bonded II-VIs can easily be plastically deformed at room temperature, and during plastic deformation, the moving dislocations transport charge. This charge is due to both the ions in the core and charge carriers in localized energy states. The wide band-gap II-VIs contain few free charge carriers at the temperatures at which they can be plastically deformed. Hence large changes can be produced by illumination and other means of carrier injection and this can change dislocation line charges and mobilities. Interest in dislocations in II-VI materials was first aroused by the discovery by Osipiyan and Savchenko (1968) of an effect caused in this way, i.e. a large photoplastic effect in CdS. This is one of a group of effects of the types illustrated in Fig. 4.68. Others are the electroplastic effect caused by the application of electrical fields and deformation-induced luminescence, a form of triboluminescence, which are discussed below. These unique effects of the interaction of the defect and charge carrier systems in the II-Vs

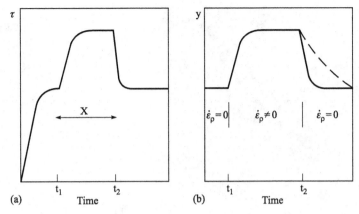

Figure 4.68 Plastic-physical effects in crystals of II-VI semiconductors. (a) If a crystal is deformed at a constant strain rate the application of an external influence from times t_1 to t_2 may produce a change in the flow stress τ. An example is the large photoplastic effect, produced by light of photon energy $h\nu > E_g$, discovered in CdS by Osipiyan and Savchenko (1968). There are similar electro- cathodo- and injection-plastic effects. (b) Conversely, if a physical property y is measured as a function of time, plastic deformation during the interval t_1 to t_2 can cause a substantial but temporary change in y, which may then decay abruptly or gradually (the solid or dashed curves). Examples are deformation-induced-conductivity, -luminescence, -electron emission (exoemission) and -dislocation current flow. (After Osipiyan *et al.* 1986.)

were systematically studied by Osipiyan and co-workers and others and are reviewed in Osipiyan *et al.* (1986) and Osipiyan (1989) which accounts are largely followed here. For a review of dislocations in other ionic crystals see Whitworth (1975).

4.3.1 *The core form and mobilities of dislocations in II-VI compounds*

As in the III-V compounds, observations on etch rosettes (see Section 4.1.3) around indentations in II-VI compounds make it possible to observe how far different dislocations have moved from the indentation and so obtain their relative mobilities. A striking observation of this kind on the $(1\bar{2}10)$ surface of a wurtzite (hexagonal) structure crystal of CdS is shown in Fig. 4.69. Clearly the mobility of the dislocations that moved over the $(000\bar{1})$ slip plane, identified as Cd(g) here, is greater than that of the S(g) polar type.

Osipiyan and his co-workers found the relative mobilities of dislocations in six II-VI materials to be as listed in Table 4.7. Osipiyan *et al.* (1986) point out that the mobility ratios given here show that the metal(g) dislocations are the more mobile in the wurtzite structure crystals but the less mobile in the sphalerite ones. The mobility ratio also increases with the flow stress of the material.

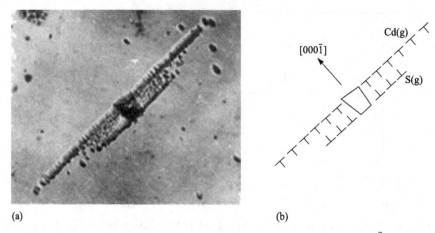

(a)   (b)

Figure 4.69   Rosette of etch pits of dislocations due to an indentation on a $(1\bar{2}10)$ surface of a wurtzite structure crystal of CdS. (b) Schematic interpretation of (a) on the assumption that the (polar) dislocations have structures of the glide form. If they belong in fact to the shuffle set, the pits of the upper, inclined, more mobile row would be due to S(s) and those of the lower row to Cd(s) dislocations. (After Strukova 1977.)

Table 4.7. *The more mobile of the two types of polar dislocations and the factor by which their mobility is the greater in some II-VI compounds*

Compound	Structure	More mobile dislocation	Factor by which they are more mobile	Reference
ZnO	Wurtzite	Zn(g)	> 100	Petrenko *et al.* (1982)
ZnSe	Sphalerite	Se(g)	"	Kirichenko *et al.* (1978)
CdS	Wurtzite	Cd(g)	> 20	Osipiyan *et al.* (1973)
CdSe	Wurtzite	"	> 10	Petrenko and Whitworth (1980)
ZnTe	Sphalerite	Te(g)	> 5	"
CdTe	Sphalerite	"	> 2	"

Early TEM studies of dislocations in II-VI materials confirmed that they have Burgers vectors of $\langle 110 \rangle$ form in the sphalerite and $\langle 11\bar{2}0 \rangle$ type in the wurtzite structure. They were also found to tend to align with low index crystallographic directions in the slip plane as screw, edge, 30° or 60° dislocations. This suggests that the Peierls potential troughs are relatively deep. Weak beam TEM observations, like those in Fig. 4.70, showed that dislocations in the II-VI compounds are extended even when moving, just like those in Si, Ge and the III-Vs (see Section 4.1.2).

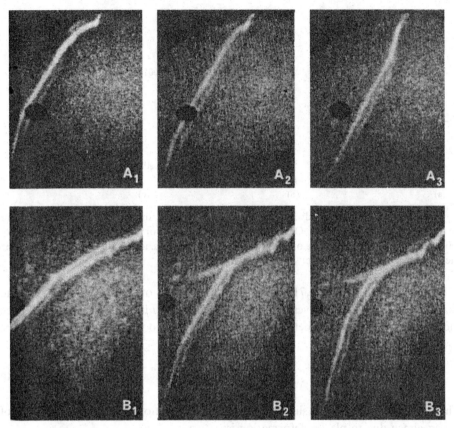

Figure 4.70 Still photographs from a video record of weak-beam TEM observations of the (0001) plane of CdS showing two extended dislocations, in sequences in A and B, moving past a dark spot. (After Cockayne *et al.* 1980, Lu and Cockayne 1983. Reprinted with permission from *Philosophical Magazine*, **A42** (1980), pp. 773–81; Taylor & Francis Ltd., the Journal's web site: http://www.tandf.co.uk/journals.)

4.3.2 *Dislocation current flow and the determination of dislocation line charges*

Suppose that a II-VI crystal, such as ZnSe, is compressed as in Fig. 4.71. In II-VI compounds one polar type of dislocation is much more mobile (the Se dislocations in ZnSe, see Table 4.7) so the deformation will be due essentially to them alone. In the mid 1970s it was found that when in motion dislocations in the II-VIs carried large line charges that were not in thermal equilibrium with the conduction and valence bands of the semiconductor. Consequently, during deformation the movement of the dislocations (of one polar type) constituted a current flow that is characteristic of those polar dislocations. These dislocation currents can be measured in an experimental set up like that in Fig. 4.71a and it was hoped that this would provide information about dislocation cores as we now discuss (Osipiyan *et al.* 1986).

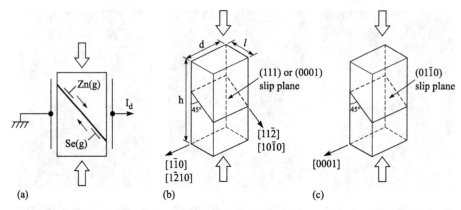

Figure 4.71 (a) Zn and Se dislocations move in opposite directions during slip but the Se ones are much more mobile. Orientation of II-VI crystals for single slip on (b) the (111) or (0001) basal planes in sphalerite or wurtzite structure materials, and (c) on a prismatic plane in the latter structure. (After Osipiyan *et al.* 1986).

In the notation of Fig. 4.71b, suppose a dislocation of length d along $[1\bar{1}0]$ and charge q per unit length moves a distance Δx in time Δt in the $[11\bar{2}]$ direction. Then the displacement of charge between electrodes applied to the faces of area dh can be written as:

$$\Delta Q = \frac{qd\Delta x}{l_1} \qquad (4.10)$$

where l_1 is the length of the slip plane in the $[11\bar{2}]$ direction. The change in length of the specimen due to plastic deformation is:

$$\Delta h_P = \frac{b_z \Delta x}{l_1} \qquad (4.11)$$

where b_z is the component of the Burgers vector along the compression axis. Hence the dislocation current is given in terms of the rate of deformation, dh_P/dt by:

$$I_d = \frac{dQ}{dt} = q\frac{d}{b_z}\frac{dh_P}{dt} \qquad (4.12)$$

Hence the line-charge, q, can be obtained from measurements of I_d in experiments as shown in Fig. 4.73 (see below).

The line charge on dislocations in the II-VIs due to trapped carriers is in thermal equilibrium with the crystal, but not that arising from ions in the core. That is, only the electronic portion of the charge is altered by interaction with any changes in the local charge-carrier density due to injection or to applied electric fields.

The influence of the electronic component of line charge, however, suffices to render ineffective the interesting method of Speake *et al.* (1978) for determining the glide or shuffle character of the dislocations that will now be described. They pointed out that, if the material can be treated as ionic, the direction of dislocation current flow during plastic deformation depends on the shuffle or glide form of the core. They observed the current flow due to slip during compression of sphalerite-structure ZnS crystals, oriented as shown in Fig. 4.72a and 4.72b. In the case of shuffle set dislocations (the upper pair in Fig. 4.72a) the positive dislocation has an S-ion core while the negative one has a Zn-core. The Zn ions are positive and the S ions negative. Under compression the dislocations must move apart as shown, in order to accommodate the compressive strain. Hence the Zn dislocations carry positive charge to the surface electrode connected to the electrometer and current flows from the crystal through the electrometer to earth. For the glide set dislocations (the lower pair in Fig. 4.72a), it is the S-core dislocations that move toward the electrometer, carrying negative charges so the current flow would be the reverse of that in the shuffle set case.

Thus, as illustrated in Fig. 4.72, if the charge on the dislocations in ZnS were purely ionic, the sign of charge reaching the electrometer would unambiguously determine whether the dislocations were glide or shuffle, as mentioned in Section 4.1.2. The dislocation charge, even in ZnS, one of the most ionic of the II-VI compounds, however, is not entirely ionic. Although the II-VIs are partially ionically bonded, they are electronic semiconductors. A dislocation model for II-VI materials, in which the ionic line charge is modified by the capture and emission of carriers by dislocation energy states, was, therefore, developed by Kirichenko *et al.* (1978).

Petrenko and Whitworth (1980) set out to determine the core form of dislocations in II-VI compounds by the method of Fig. 4.72a, but interpreting their results using the model of Kirichenko *et al.* (1978). They plastically deformed crystals of three sphalerite-structure compounds and four with the wurtzite-structure, oriented for slip on a single plane, but with two slip directions equally favoured. The polar character of the dislocations was deduced, in all cases, as shown in Fig. 4.72 for ZnS.

The nature of the mobile dislocations, α or β-type as shown in Fig. 4.72b, was determined by etching in reagents that distinguish polarity (see Section 3.3.1 and Section 4.1.3). They found that in the three sphalerite structure materials studied, ZnSe, ZnTe and CdTe, the mobile dislocations were α-type i.e. either Zn(s) or S(g), as shown in Fig. 4.72b. In the three wurtzite-structure materials, ZnO, CdS and CdSe they were β-type. Five of these materials were *n*-type and in all cases the charge carried by the mobile dislocations was negative. In the *p*-type ZnTe it was positive. This strongly suggests that the charge on the mobile dislocations was essentially electronic rather than ionic and Petrenko and Whitworth concluded their observations did not give evidence whether the dislocations were shuffle or glide. (Petrenko and Whitworth (1980) also studied samples of ZnS but these were structurally complex so the significance of the results was even less clear.) The magnitudes of the charges on moving dislocations in II-VI compounds (Osipiyan *et al.* 1986) and their dependence on the band gap is discussed in Section 5.2.7.

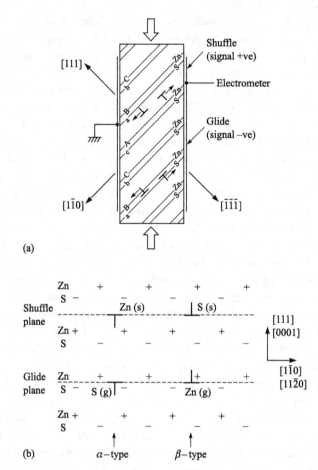

(a)

(b)   α–type   β–type

Figure 4.72   (a) A sphalerite-structure ZnS crystal undergoing compression. (111) planes of Zn and S with cubic stacking *aα bβ cγ aα*... are shown. The upper pair of dislocations, on the widely spaced *a* and α (111) planes of atoms, are of the shuffle set and must move outward to accommodate the compressive strain. Thus the Zn(s) dislocation moves to the right to give an assumed positive current to the electrometer. For the lower pair of glide set dislocations, moving between α and *b* planes, the S(g) dislocation moves to the right to give a presumed negative current. (After Speake *et al.* 1978.) (b) Diagram of the positions of Zn (+) and S (−) ions, in either simple crystal structure of ZnS, since neither cubic nor hexagonal stacking sequence is specified. Also shown are the positions of the shuffle and glide planes and the positions of the four types of edge dislocations. Note that the positive and negative dislocations of both sets are interchanged, left to right, compared to (a). (After Petrenko and Whitworth 1980. Reprinted with permission from *Philosophical Magazine*, **41** (1980), pp. 681–99; Taylor & Francis Ltd., the Journal's web site: http://www.tandf.co.uk/journals.)

The large, non-equilibrium line charges on dislocations in II-VI compounds also result in (i) the electroplastic effects (effects of externally applied electrical fields on the plastic flow stress, Osipiyan and Petrenko 1976a and 1976b), which will be discussed in Section 4.3.4, and (ii) observable non-uniform charge distributions after deformation as shown in Fig. 4.73.

Kirichenko *et al.* (1978) deformed ZnSe crystals, as shown in Fig. 4.73, between non-conducting compression plates and with no electrodes applied. At low deformation rates the charge density varied roughly linearly across the crystal but at higher rates there was an accumulation of negative charge in the crystal under the surface to which the dislocations were moving. This was taken to be evidence that dislocation charge in the crystal corresponded to a density of electrons of about $3 \times 10^{18}\,\mathrm{m}^{-3}$. This was much higher than the density of free electrons (10^{12} to $10^{13}\,\mathrm{m}^{-3}$) but lower than that of electrons in point defect states near the Fermi level (10^{21} to $10^{22}\,\mathrm{m}^{-3}$). This did not make possible the determination of the line charge on individual dislocations but confirmed the transport of charge by moving dislocations.

4.3.3 The photoplastic effect

The photoplastic effect, discovered by Osipiyan and Savchenko (1968), is further evidence that dislocations in II-VI materials are electrically charged and that charge is largely electronically determined. Osipiyan and Savchenko tested CdS crystals in

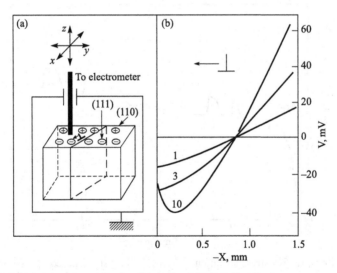

Figure 4.73 (a) Arrangement for probing the charge distribution over a ZnSe crystal deformed by compression along the y axis. (b) Variations of the electrometer signal for x-scans over the top face during deformations at 1, 3 and 10 μm min^{-1} and a temperature of 293 K. (After Kirichenko *et al.* 1978. Reprinted with permission from *Soviet Physics JETP*, **47**, pp. 389–94. Copyright 1978, American Institute of Physics.)

compression at a constant strain rate and found that the flow stress could be reversibly altered by a large percentage by turning a visible light source on and off as shown in Fig. 4.74.

Light from a cinema projection lamp increased the flow stress, i.e. it strengthened the crystal. This occurred for all stages of the deformation and in crystals of widely different resistivities and degrees of non-stoichiometry. The relative hardening, $\Delta\sigma/\sigma$, due to the light was greater at low test temperatures (75° C) than at higher ones (200°). It increased with the light intensity at first but saturated at intensities above about 1000 lux as shown in Fig. 4.75.

The effect was also strongly wavelength dependent as shown in Fig. 4.76. The peak occurs close to the peak absorption for CdS at 530 nm where $hv \cong E_g$, the forbidden gap energy.

Carlsson and Svensson (1969) reported a similar increase in the flow stress of ZnO crystals under illumination. They compressed ZnO crystals at a constant strain-rate that were oriented for basal plane only, as in the work of Osipiyan and Savchenko (1968). Up to $\Delta\sigma/\sigma = 30\%$ increase could be produced by illumination. Again $\Delta\sigma/\sigma$ first rose and then saturated with increasing intensity of illumination and there was a relatively sharp peak in the spectral response curve with the $\Delta\sigma/\sigma$ maximum occurring at about the forbidden gap energy of 3.2 eV. Carlsson and Svensson suggested that the effect was due to light produced electron-hole pairs. Holes are

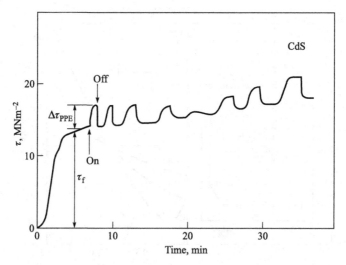

Figure 4.74 Compressive stress-strain curve for a CdS crystal with compression axis [11$\bar{2}$3], slipping on the basal system, (0001) [11$\bar{2}$0]. The reproducible photoplastic effect (PPE) increase in flow stress $\Delta\tau_{PPE}$ occurs only while the crystal is under illumination. Deformation occurred at 75° C and a strain rate of 10^{-5} cm sec^{-1}. (After Osipiyan and Savchenko 1968. Reprinted with permission from *JETP Letters*, **7**, pp. 100–2. Copyright 1968, American Institute of Physics.)

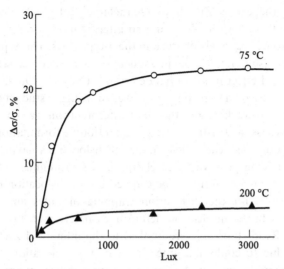

Figure 4.75 The dependence of the relative increase of the flow stress on the intensity of illumination and the test temperature. (After Osipiyan and Savchenko 1968. Reprinted with permission from *JETP Letters*, **7**, pp. 100–2. Copyright 1968, American Institute of Physics.)

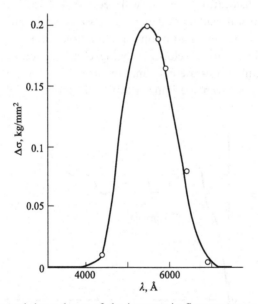

Figure 4.76 The spectral dependence of the increase in flow stress on illumination of CdS crystals in compression. These measurements were performed at 75° C and a constant intensity of 150 lux. (After Osipiyan and Savchenko 1968. Reprinted with permission from *JETP Letters*, **7**, pp. 100–2. Copyright 1968, American Institute of Physics.)

then trapped by the excess Zn atoms characteristic of ZnO. These are thereby indirectly ionized from Zn^+ to Zn^{2+} and so interact more strongly with the dislocations assumed to be negatively charged in the n-type ZnO. This type of explanation makes the photoplastic effect in semiconducting compounds closely analogous to the photomechanical effect in alkali halide crystals (Nadeau 1964).

The photoplastic effect in semiconductors was thoroughly studied by Osipiyan and Carlsson and their co-workers and the latter interpretation was confirmed.

Carlsson and Svensson (1970) presented a quantitative formulation of their model basing themselves on the Read (1954) model of dislocations having a negative line charge and a balancing positive space charge in a surrounding cylindrical region. Holes produced by illumination may be trapped by the dislocation or by interstitial Zn ions. Expressions for the dislocation trapping of holes and the consequent photoconductivity, due to the electrons left as free carriers, were taken from Figielski (1965). Shockley-Read statistics were applied to the interstitial Zn^+ trapping. This second process, due to coulombic interaction of the dislocations and Zn^{2+} ions, produces pinning and hardening. A number of expressions were obtained and found to be in reasonable quantitative agreement with the observations.

Three other types of evidence for the physical model were produced. Firstly, the pinning effect implies that the photo-ionization occurs essentially in the space-charge region round the dislocations and it will require a certain time to take place. This provided a natural explanation for the repeated yield points, under constant strain-rate deformation and continuous illumination, in Fig. 4.77. Stationary or slowly moving dislocations develop a pinning cloud of ionized Zn^{2+} traps. This increases the constant-strain-rate deformation stress to an upper yield value. Dislocations then pull away from the ionized traps and move rapidly at a lower stress.

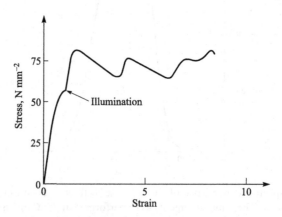

Figure 4.77 Compression stress-strain curve for a crystal of ZnO under continuous illumination from the point marked onwards, showing repeated yielding. (After Carlsson and Svensson 1970. Reprinted with permission from *Journal of Applied Physics*, **41**, pp. 1652–6. Copyright 1970, American Institute of Physics.)

The flow stress falls to a lower yield value, the dislocations slow and again become pinned. Hence, the effect should depend on strain rate. It was found that no photoplastic hardening occurred on illumination for strain rates that exceeded about $5 \times 10^{-2} \sec^{-1}$. Secondly, the spectral response curve for the photoconductivity had a peak at the same photon energy as the photoplastic hardening. Finally, the concentration of interstitial Zn atom traps, c, is a few ppm (parts per million) so the strength of the coulombic interaction should be proportional to $c^{1/2}$. A number of crystals of different c, i.e. differing non-stroichiometry, were prepared by heating in different partial pressures of Zn vapour. The flow stresses were measured, both in the dark and in the light, and found to be indeed proportional to $c^{1/2}$.

A similar model was proposed independently by Osipiyan and Petrenko (1973a). Their model arose directly from evidence on the photoplastic effect in CdS. The trap effectiveness was shown by the fact that the photoplastic strengthening was approximately the same in CdS with free electron densities in the dark from 10^{12} to $10^{18} \, cm^{-3}$ although the light intensity used produced no more than about 10^{13} holes cm^{-3}. Thus the holes are very rapidly trapped, within 10^{-9} to 10^{-8} sec, while the electrons have longer lifetimes. This interpretation was confirmed by the observation of infrared quenching (the decrease of the photo-induced conductivity increment, $\Delta \sigma$, on illuminating with infrared in addition to the visible light producing the photoplastic effect) (Osipiyan *et al.* 1971). The infrared photons provided the activation energy to remove the holes from the traps and thus decrease the pinning of the dislocations by the traps. The spectral dependence of the infrared quenching showed that the energy level of the traps was 1.35 eV above the top of the valence band in CdS. This identified the traps as one of the well-known photosensitization centres of CdS photoconductivity. The decrease in photoplastic hardening with increasing temperature (Fig. 4.75) can be explained in terms of the thermal activation of the holes from the traps at higher temperatures.

Using a model which characterized the strength of the interaction in terms of the angle of bowing of the dislocations pinned at the traps, Osipiyan and Petrenko (1973a) obtained the dependence of the photo-strengthening on the density of photoelectrons. This was in agreement with experimental results.

The photoplastic effect is strongly orientation dependent. Carlsson (1971) showed that ZnO crystals deformed by basal plane slip exhibited the photoplastic effect as described above but those oriented for ($\bar{1}100$) prismatic slip were not affected by illumination. This he explained in terms of the ball-and-wire models of Osipiyan and Smirnova (1968) (see Section 4.1.5). These models showed that dislocations with the basal slip plane were polar, i.e. were either A- or B-edged in AB compounds and so should have markedly different electrical effects. However, dislocations on prismatic slip planes were non-polar, i.e. their core atoms with dangling bonds alternated between A and B, so their electrical effects should tend to cancel. This orientation dependence of the photoplastic effect shows the value of the ball-and-wire models of dislocations in semiconducting compounds. Fig. 4.78 is further evidence of the

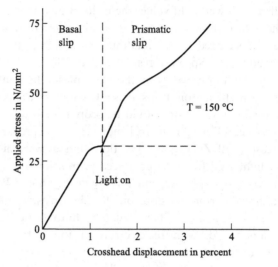

Figure 4.78 Stress-strain curve for compression of ZnO crystals, oriented for single slip on the basal plane (0001), at 150° C. The first part of the curve involves basal slip in darkness. During illumination the basal plane dislocations are pinned and glide starts on prismatic planes leading to rapid work-hardening and the steeply rising curve. A stress-strain curve obtained in darkness is shown dashed. (After Carlsson 1971. Reprinted with permission from *Journal of Applied Physics*, **42**, pp. 676–80. Copyright 1971, American Institute of Physics.)

radical difference between basal and prismatic plane dislocations. Switching the light on pinned the basal dislocations. This stopped single slip on the basal planes, which would result in easy glide, i.e. deformation without hardening that occurs in the dark, along the dashed curve. With light pinning the basal glide dislocations, only prismatic ones are left to produce slip on intersecting prismatic planes to accommodate the imposed strain rate. The intersection of the prismatic slip with the previously introduced basal dislocations results in rapid hardening as exhibited by the steeply rising stress-strain curve. Osipiyan and Savchenko (1973) showed that there is no effect of light on the flow stress for prismatic slip in CdS either.

Carlsson and Ahlquist (1972) studied the photoplastic effect in CdTe. The flow stresses in the dark and light both changed with variations from stoichiometry as shown in Fig. 4.79a. The variations in stress in (a) are similar to those of the free hole and electron concentrations shown in Fig. 4.79b. Thus the photoplastic effect occurs in both *p*- and *n*-type samples of this, the only II-VI compound that can be made highly conducting in both forms. The occurrence of the minima in both light and dark flow stresses at the near-stoichiometric composition, with its minimum carrier and vacancy or interstitial densities, is consistent with the model of the effect as due to charged point defect pinning. The spectral peak in the photoplastic hardening occurred as expected for photon energies close to the band gap value.

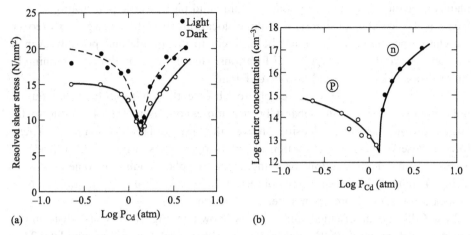

Figure 4.79 (a) The yield stresses in darkness and under illumination with white light of CdTe crystals after heating under various partial pressures of cadmium at 1200° C to alter the stoichiometry and hence the concentration of charge carriers either holes, p, or electrons, n, as shown in (b). (After Carlsson and Ahlquist 1972. Reprinted with permission from *Journal of Applied Physics*, **43**, pp. 2529–36. Copyright 1972, American Institute of Physics.)

Gutmanas *et al.* (1979a and b) also studied *n*- and *p*-type CdTe and found both positive and negative PPEs. They found that rapid cooling of plastically deformed samples from high temperatures and an increase in the light intensity produced a change from a positive to a negative PPE. They also studied the spectral and thermal dependences of the PPE in CdTe. They put forward a detailed model to account for their results and distinguished two cases, i.e. $\sigma > \sigma_P$ and $\sigma < \sigma_P$. Here σ is the flow stress and σ_P the Peierls stress. When $\sigma > \sigma_P$ the interaction of dislocations with obstacles is important and the PPE has to be explained in terms of changes in those interactions, e.g. due to changes in charge state of the defects as the result of the illumination. However, when $\sigma < \sigma_P$ these interactions are unimportant and the PPE is due to photon-induced changes in the ease of formation or of motion of kinks from one Peierls potential trough to the next (Gutmanas *et al.* 1979b, see Section 4.2.1).

Ahlquist *et al.* (1972) reported photoplastic effects in CdS and ZnO with the wurtzite structure and CdS, CdSe, CdTe, ZnSe and ZnTe with the sphalerite structure. ZnS, the hardest of these materials, fractured without deforming under illumination although it plastically deformed in the dark. The flow stress in the dark and the maximum increase in flow stress on illumination both increase with the band gap energy from one compound to another for both structures. They interpreted their observations in terms of a microdynamical theory (see Section 4.4).

Osipiyan *et al.* (1973) investigated the photoplastic effect for polar dislocations in CdS. The chemical difference between the dangling bonds in Cd- and S-edged basal plane dislocations, the partially ionic bonding and piezoelectric polarization by dislocation elastic fields all suggest that the dislocations should be charged. Moreover, the charge per unit length should be different in magnitude and possibly even in sign on the two different types of dislocations. Hence the strength of the pinning by charged point defects should also be different.

Osipiyan *et al.* (1973) were unable to observe the photoplastic behaviour of polar-opposite dislocations separately in compression, as in Fig. 4.68, due to the large grown-in dislocation densities in even the best available CdS crystals. Therefore, to observe the separate motion of polar-opposite dislocations they introduced excesses of mobile dislocations of a single type by polar bending of some crystals as in Fig. 4.80 and in other cases they observed the separate polar 'wings' or 'petals' in dislocation rosettes from an indenter (see Section 4.1.3).

In a CdS crystal oriented and bent as shown in Fig. 4.80, Cd(s) dislocations move inward on basal (0001) planes, toward the neutral axis and accumulate. They come from sources near the surfaces, as shown. The S(s) dislocations, however, move outward and many are lost through the surface. In a large central volume, cross-hatched in the figure, slip is by the movement of dislocations of one sign only and the magnitude of the photoplastic hardening here will be that for these dislocations. Crystals were bent at 100°C with the bending supports moving at $5\,\mu\text{m}\,\text{min}^{-1}$. The cross-hatched area was illuminated by a helium neon laser. Fig. 4.81 gives results for Cd(s) and S(s) bending in two orientations. Photoplastic hardening, $\Delta\sigma$, was greater for S(s) than for Cd(s) dislocations and greater for $\langle 1\bar{2}10 \rangle$- than for $\langle 10\bar{1}0 \rangle$-axis bending for both polarities. Measurements on rosette wing lengths also indicated that S(s) dislocations were more strongly retarded by illumination than

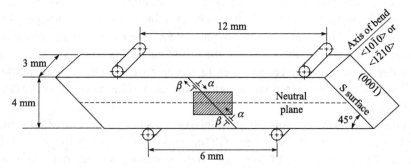

Figure 4.80 The orientation of a wurtzite-structure CdS specimen for four-point polar bending. To introduce an excess of Cd(s) dislocations on the basal slip plane, the more-closely spaced supports are driven upward. To introduce an excess of S(s) dislocations the crystal would have to be bent with the opposite curvature, i.e. with the closely and widely spaced supports interchanged between the top and bottom surfaces of the crystal, and the closely spaced pair, driven downward. (After Osipiyan *et al.* 1973. Reprinted with permission from *Soviet Physics Solid State*, **15**, pp. 1172–4. Copyright 1973, American Institute of Physics.)

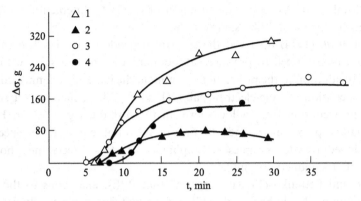

Figure 4.81 The dependence of the photoplastic increase in flow stress, $\Delta\sigma$, on the deformation time, i.e. on the strain. Curves 1 and 3 are for crystals with an excess of S(s) dislocations and curves 2 and 4 are for an excess of Cd(s) dislocations. The axis of bend in the case of curves 1 and 3 was $\langle 1\bar{2}10 \rangle$ and of curves 2 and 4 it was $\langle 10\bar{1}0 \rangle$. (After Osipiyan *et al.* 1973. Reprinted with permission from *Soviet Physics Solid State*, **15**, pp. 1172–4. Copyright 1973, American Institute of Physics.)

were Cd(s) dislocations. This is strong, direct evidence for a difference in line charge between the two signs of dislocation in CdS.

As we discussed above, Osipiyan and Petrenko (1973a) had established that the photoplastic effect is due to trapped positive charges. The above mentioned evidence that S(s) dislocations were more strongly retarded than Cd(s) dislocations, but that both are retarded, therefore means that both have negative line charges but that on S(s) dislocations is the greater. CdS crystals are always *n*-type. This means that the Fermi level is high in the forbidden gap in all cases. The energy levels of both Cd(s) and S(s) dislocations can therefore be expected to fall below the Fermi level and trap electrons to make the line charges of both types of polar dislocation negative.

Photoplastic hardening occurs for basal slip in wurtzite ZnO and CdS but, in general, prismatic plane slip is unaffected by illumination. Later more careful measurements by Osipiyan and Shiksaidov (1973) showed that illumination of some CdS crystals deformed mainly by prismatic plane slip led to significant softening, i.e. to a 'negative' photoplastic effect (PPE). The maximum reduction in flow stress obtainable in the negative PPE was about one-half the largest increase obtainable in the positive PPE. Spectral observations showed that maximum photoconductivity coincided with the negative and positive PPE maxima at a wavelength close to the fundamental absorption edge (band-gap photon energy) for CdS. Negative PPE only occurred in specimens that exhibited red luminescence with $\lambda \cong 720\,\text{nm}$ which is known to arise from complexes of the types $(V_{Cd}^{-2} + V_S^{+2})^{\circ}$ and $(Cu_{Cd}^{-1} + V_S^{+1})^{\circ}$. A possible mechanism for the observed softening effect is a change in the charge state of such centres under illumination, which might partly compensate the elastic field around a local centre to reduce the strength of its elastic interaction with moving dislocations (Osipiyan and Shiksaidov 1973).

Osipiyan *et al.* (1974) observed the spectral dependence of the PPE in ZnS:Al, i.e. ZnS activated (doped to produce cathodoluminescence emission) with Al as a phosphor. In addition to the principal maximum in the hardening connected with the interband excitation of hole-electron pairs, (1) in Fig. 4.82, there was a subsidiary peak due to excitation from a local level in the forbidden gap, (2) in Fig. 4.82. This subsidiary peak corresponds to one in the photoconductivity spectrum of similarly doped powder samples so doping of crystals affects the photoplastic effect and photoconductivity similarly.

Osipiyan and Petrenko (1973a) suggested that PPEs, analogous to those in the II-VI compounds, should be found in all crystals that can be plastically deformed in a temperature range in which they also exhibit photoconductivity, i.e. in which the carrier density can be increased by a significant percentage by light. Maeda *et al.* (1977) looked for such an effect in III-V compounds, choosing GaP for their experiments because its wide band gap, $E_g = 2.24\,\text{eV}$, means that the thermal equilibrium carrier density is relatively low for a given temperature. They did not observe actual changes in flow stress due to illumination, as in the II-VI compounds, but changes in dislocation mobility due to light. Similar effects were found in CdTe by Gutmanas *et al.* (1979b) and in Si by Küsters and Alexander (1983).

Since GaP crystals are brittle below 500° C, Maeda *et al.* (1977) used dislocation etch rosette patterns around indentations (see Section 4.1.3), to study

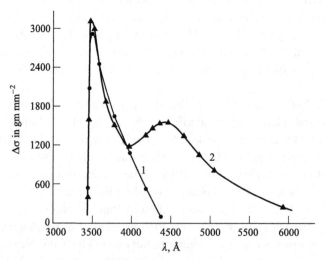

Figure 4.82 The spectral dependence of the photoplastic effect for undoped ZnS crystals (curve 1) and for ZnS:Al crystals (curve 2). The impurity peak occurs at about 450 nm. (After Osipiyan *et al.* 1974. Reprinted with permission from *JETP Letters*, **20**, pp. 163–4. Copyright 1974, American Institute of Physics.)

low-temperature plasticity. They indented the surfaces of GaP crystals at temperatures from 100 to 300° C. They found that illumination with light of photon energies near the band gap value increased the mobility of dislocations, i.e. the effect was equivalent to a fall in the flow stress in a stress-strain test and constituted a negative PPE. They interpreted their results in terms of illumination-induced changes in the dislocation line charge and the effect of this on mobility.

Küsters and Alexander (1983) used the double etching technique on Si crystals. In this, the first etch gives the initial position of dislocations. After stressing at a known level for a fixed time, a second etch gives their final position. The distance travelled gives the velocity. They used light of photon energy $hv = 1.17\,\text{eV}$ (the band gap value) and intensity $1\,\text{W cm}^{-2}$ on specimens at temperatures in the range between 325 and 422° C. They reported that screw dislocations and those 60° dislocations that dissociated into a 30 and a 90° partial dislocation, had higher velocities when illuminated. Thus the Si PPE was equivalent to a fall in flow stress during deformation, i.e. a negative PPE. The activation energy required for motion of the 30/90 extended dislocations, $E = 1.82\,\text{eV}$, was reduced by $\Delta E = 0.68\,\text{eV}$ to $1.14\,\text{eV}$ under illumination. They concluded that the difference in behaviour between the two types of 60° dislocation suggested that recombination enhanced movement was responsible rather than the effect of a changed charge state. They suggested that the 'phonon-kick' mechanism of Kimerling (1978) might account for the PPE in Si.

4.3.4 The electroplastic effect

Due to their large charge per unit length, dislocations in II-VI compounds experience a force due to any external electric field and this will produce an electroplastic effect (EPE). Thus an electric field in the direction of motion adds to the force exerted by the stress on a moving dislocation so the total force acting can be written as:

$$f = \phi_1 qE + \phi_2 \tau b \tag{4.13}$$

where q is the charge per unit length of dislocation, E is the electric field, τ the shear stress, b is the Burgers vector and ϕ_1 and ϕ_2 are the appropriate geometrical orientation factors.

We shall see in discussing the microdynamical theory of plastic deformation in semiconductors (Section 4.4) that the rate of deformation can be written as:

$$\frac{d\varepsilon}{dt} = \rho_m b v_{av} \tag{4.14}$$

where ρ_m is the density of moving dislocations and v_{av} is their average velocity which depends on the stress (see Section 2.5.4) as:

$$v = B_0 \tau^m \exp(-E_d/kT) \tag{4.15}$$

where B_o and m ($\cong$1) are constants, τ is the stress and E_d is the activation energy for dislocation movement.

Microdynamical theory thus predicts that to keep the dislocation velocity constant, which is necessary to accommodate the constant imposed strain rate, any change in the field **E** must be compensated by a change of opposite sign in the flow stress τ. Thus the stress required to maintain the constant strain rate will be reduced by the application of an electric field in such a 'forward' direction. The flow stress will be restored to its original value if the field is turned off and an increase of flow stress will be observed if an equal and opposite field is applied. Experiment showed that, in fact, in ZnS, application of an external bias voltage between the surfaces as shown in Fig. 4.83a did result in a hardening or a softening of the specimen depending on the polarity of the voltage as can be seen in Fig. 4.83b so this is a polar EPE. Moreover, the flow stress changed linearly with applied voltage for both polarities as shown in Fig. 4.84.

Later Osipiyan and Petrenko (1976b) found an EPE in studying the effects of electrical boundary conditions on the plastic properties of ZnS. Their undoped sphalerite structure crystals had a high density of grown-in stacking faults. The specimens were compressed and slipped on the {111} planes parallel to the faults. The electrical boundary conditions could be varied in two ways. In the first, specimens were immersed in a liquid, either a conducting one such as Hg to equalize the potentials on all the side surfaces, or an insulating one to allow maximum potential differences between faces to develop. In the second, metal contacts to the

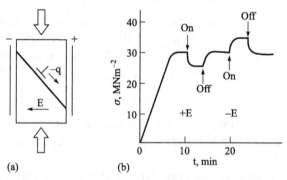

(a) (b)

Figure 4.83   (a) The application of an electric field as shown during a stress-strain test at constant strain rate will apply an additional force to move the dislocations of the type assumed here in the direction of motion. (b) Consequently, the observed flow stress decreased on the application of such a field and increased equally on the application of a reverse field. The specimen was a crystal of ZnS, of dimensions $4 \times 3 \times 1$ mm, with the potential applied between the large faces and compression parallel to the long axis and the deformation rate was $10 \, \mu\text{m min}^{-1}$. The applied potential difference was $2 \, \text{kV}$. (After Osipiyan and Petrenko 1976b. Reprinted with permission from *Soviet Physics JETP*, **42**, pp. 695–9; and *Soviet Physics Doklady*, **21**, pp. 87–8. Copyright 1976, American Institute of Physics.)

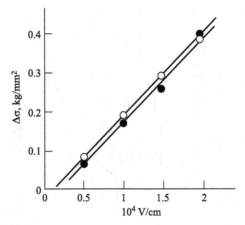

Figure 4.84   The change in flow stress with the electrical field in the electroplastic effect in ZnS. The open circles are for hardening and the solid are for softening (reverse field) observations. The experimental conditions were as quoted in Fig. 4.83. (After Osipiyan and Petrenko 1976b. Reprinted with permission from *Soviet Physics JETP*, **42**, pp. 695–9; and *Soviet Physics Doklady*, **21**, pp. 87–8. Copyright 1976, American Institute of Physics.)

larger of the two sets of opposite side faces could be short circuited by a switch outside the deformation chamber, so the potentials of the two large faces become equal. All measurements were made at room temperature. Using Hg, when the level of the liquid was raised to short-circuit the faces, the flow stress was reduced by about $\Delta\sigma/\sigma = 40\%$. Lowering the Hg level to allow interfacial potential differences to develop resulted in recovery of the initial flow stress. These changes could be observed repeatedly in a sample. The same changes occurred on short- or open-circuiting the large faces by means of an externally switched circuit. The flow stress change, $\Delta\sigma$, increased with the electrical resistance of the samples. Potential differences of 2 to 2.5 kV were developed between the wide faces which were 1 to 1.5 mm apart. This potential difference rose rapidly with the rate of plastic deformation. When the potential difference exceeded 2 to 2.5 kV, periodic discharges took place through the air. This resulted in a sawtooth variation of V and of the flow stress with time.

The occurrence of substantial electrical forces on the highly charged dislocations in, e.g. ZnS (Equation 4.13) and the consequent effect on the flow stress (via Equations 4.14 and 4.15) is clearly the explanation for the effects of externally applied voltages as in Fig. 4.83a and for the effects of shorting the internally generated field by immersion in Hg or via electrical contacts. Thus this electroplastic effect is clear and direct evidence that the dislocations in ZnS are charged.

The polarity of the EPE shown in Figs. 4.83 and 4.84 is the strongest evidence that Zn(s) and S(s) dislocations are differently charged. That this is so follows from the elementary geometrical considerations of Fig. 4.85. It can be seen that under

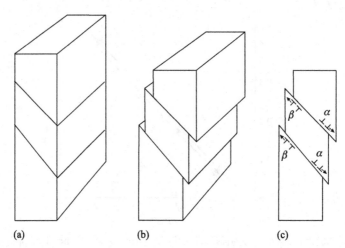

(a)   (b)   (c)

Figure 4.85   (a) Geometry of a sphalerite-structure crystal undergoing single slip or of a wurtzite-structure crystal undergoing basal plane slip in compression. The offsets on slip planes after compression must be as shown in (b) to accommodate the reduction in the height of the crystal. To produce these offsets, positive and negative dislocations of opposite signs must have moved across the planes as shown in (c). In semiconducting compounds the opposite sign dislocations must be of opposite polarity, i.e α and β as marked, these might be e.g. A(s) and B(s). Alternatively, they might be B(g) and A(g) dislocations, respectively.

compression, the sign of the shears or offsets to be produced across the slip planes by dislocation movement is unambiguously determined (Fig. 4.85b). This in turn unambiguously requires the specific opposite directions of movement of positive and negative edge dislocations that are shown in Fig. 4.85c, so the dislocations emerging at opposite faces must be of opposite sign. In the sphalerite structure, positive and negative edges are of different chemical type so that depending on the polar orientation of the crystal, either A or B dislocations (in the case of ZnS, these are Zn(s) (or S(g)) and S(s) (or Zn(g)) types) only will move to the large left-hand side face, for example.

The orientation of the ZnS specimens of Osipiyan and Petrenko (1976a) was such that the dislocations moving to either side were actually of 60° type. Since the two wide faces developed opposite charges during deformation in air as their experiments showed, the charges on the 60° A(s) and B(s) dislocations, say, must be different. They need not be of opposite sign although Osipiyan and Petrenko assumed that they were in interpreting these observations. We saw above in relation to Figs. 4.68 and 4.69 that there is clear evidence in CdS that the charges on both are negative. The charges on these two types of dislocation must, however, both by that evidence and this, be substantially different in magnitude at least. Osipiyan and Petrenko produced a quantitative theory of this polar electroplastic effect, which enabled them to obtain the line charges on the 60° dislocations from three different experimental quantities. They first calculated it from the deformation current, carried by the moving dislocations, and secondly from $\Delta\sigma$ arising from changes in the externally applied field, via Equation (4.13). The third method derived the line charge from $\Delta\sigma$ due to short-circuiting of the surface charges. The latter charges repel the later-approaching, similarly charged dislocations, so their elimination reduces the back stress on the dislocations and allows the flow stress to fall for a constant velocity, i.e. for a constant strain rate. The values obtained by the three methods were in good agreement and gave the line charge, assumed to be of opposite sign for Zn(s) and S(s) dislocations, as 1.3 ± 0.15 electron charges per Burgers vector length, i.e. per 38.2 nm along the dislocation. This value is 1.5 to 3 times the value calculated from observations of the numbers of deformation-induced luminescence flashes produced, as will be discussed next (in Section 4.3.5). It was found in this work that the values of the line charge obtained varied with the time of storage in darkness, due to a photoconductive effect with a long relaxation time characteristic of this material, so this agreement can be regarded as relatively satisfying.

If the photoplastic effect is due to trapped holes, similar changes in flow stress should result from the injection of holes by other means. Osipiyan and Petrenko (1973b) showed that there is indeed such an injection-plastic effect, using sphalerite structure crystals of n-type ZnSe. Electrical contacts were applied to the large top and bottom faces of these platelets and they were deformed at a constant rate in three-point bending. Voltages were applied producing fields of up to $100\,\text{kV cm}^{-1}$. These voltages increased the flow stress up to 100%, and the increase disappeared on turning off the voltage. The magnitude of the effect varied with the contact metals used, and the magnitude of the flow-stress increase fell with increasing temperature and disappeared at a somewhat lower temperature than that for the disappearance of the photoplastic effect. This electroplastic effect (EPE) is non-polar since hardening occurred regardless of the sign of the potential difference between the top and bottom of the platelets. The non-polar EPE results from bipolar injection, i.e. the injection of holes and electrons since the minimum voltage for hardening is that corresponding to the start of the $I \propto V^3$ portion of the current-voltage characteristic of these crystals (Osipiyan, private communication).

A related cathodo-plastic effect caused by electron beam bombardment was observed in CdS when irradiated with $25\,\text{keV}$ electrons at a beam current of about $10^{-11}\,\text{A m}^{-2}$ (Maeda and Sakamoto 1977 and Maeda *et al.* 1981, see Osipiyan *et al.* 1986).

Osipiyan and Petrenko (1976b) then studied the effect of illumination on the deformation currents in ZnS. They could thus determine, via the theory mentioned above, the effect of illumination on the dislocation line charge. They used similar specimens and deformation conditions to those of Osipiyan and Petrenko (1976a), described above.

Shielding of the dislocation line charges by free photocarriers could be ignored as measurements and calculations showed dislocation spacings were two orders of magnitude less than the radius of the carrier screening volumes, and the concentrations of the photocarriers were many orders of magnitude less than the total charge on the dislocation lines. They found that the dislocation line charge, calculated from the deformation current, approximately doubled when the crystals were illuminated with light of wavelength 415 nm. It decreased, but not fully down to the original dark value, when the light was switched off. The value of the deformation current in the light and in the dark was found to be proportional to the rate of plastic deformation confirming the deformation current in both cases was due to the movement of dislocations (see Equation 4.12).

The higher dislocation line charge in the illuminated specimens was confirmed by the observation that when an external electric field was applied across the ZnS specimens it caused a greater softening in the illuminated specimens. Line charges calculated from field softening values agreed well with those derived from the deformation currents for both illuminated and dark conditions. Both the photo-plastic hardening and the increase in dislocation line charge on illumination were smaller in specimens cut from purer ingots than in those from less pure ones. Osipiyan and Petrenko suggested this could be because the charged dislocations exchange electrons and holes with point defects as they move, to establish a steady-state value of the line charge. This value would then depend on two things. Firstly, on the concentration of point defects, hence the difference in line charge change between more and less pure specimens. Secondly, on the charge state of the point defects, hence the difference between the increases in the line charge on illumination between the more and less pure specimens.

Before leaving the genuine, bulk photo- and electro-plastic effects, a warning should be given concerning the spurious so-called photomechanical and electro-mechanical effects in semiconductors. Some hundreds of papers appeared concerning these in the 1950s and 1960s. The (non-existent) photo- and electro-mechanical effects must not be confused with the real photo- and electro-plastic effects. The photo- and electro-mechanical (non)effects were alleged changes, due to illumination and applied potentials, in surface hardness, measured by very-low-load indentation testing. These proved to be illusory as shown by Hanneman and Jorgenson (1967) and Hall 1968. For a review of this embarrassing topic and much else see Westbrook (1968).

In addition to the photoplastic and electroplastic effects represented schematically in Fig. 4.68a, there are a number of plastic-physical effects. These are represented in Fig. 4.68b. In these, during plastic deformation, the value of some physical property is altered or a physical effect is produced. These are deformation-induced conductivity, luminescence, electron emission (exoemission) and the dislocation current flow that was discussed above.

4.3.5 Triboluminescence

Triboluminescence (TL) is the emission of light by a material under mechanical stress. In some cases TL is due to plastic deformation and in others to fracture.

The energy for deformation-induced TL comes from the mechanical work done in deforming the crystal. Triboluminescence in ZnS, which can be seen on rubbing the mineral sphalerite with one's fingernails, has probably been known for centuries. (For a review see Walton 1977.)

Triboluminescence in ZnS crystals was shown to be due to dislocations moving during deformation (Bredikhin and Shmurak 1974). Their wurtzite-structure single-crystal platelets were oriented for single slip on their basal planes as shown in Fig. 4.86, and loaded with square pulses of mechanical stress of duration 0.2 to 30 secs. There were two forms of light emission. The brighter consisted of flashes, lasting 0.05 to 0.2 μs, detected by photomultiplier, and all of approximately equal intensity. No flashes were emitted for stresses below the elastic limit. Above that, flashes were observed by oscilloscope and the number per unit time, N, determined. Stress birefringence observations showed that a slip band appeared at the same time as each flash of light. The cumulative total number of flashes emitted increased linearly with the total deformation.

Bredikhin *et al.* (1975) and Bredikhin and Shmurak (1975a,c, 1977) showed that the pulses occur when the field, built up during deformation, discharges. This field can be detected directly or through the birefringence it produces and which decreases suddenly with the emission of each pulse of light. Pulsed luminescence is eliminated by immersing the crystal in a transparent conducting liquid (Zaretskii and Petrenko 1982, unpublished but quoted by Osipiyan *et al.* 1986).

The pulsed triboluminescence can be avoided not only by immersing the specimen in a conducting liquid but also by using an insufficiently high-resistivity crystal. A second form of less intense, continuous emission can then be seen. Bredikhin and Shmurak (1977, 1979) studied this in crystals doped with Mn or Cu and containing

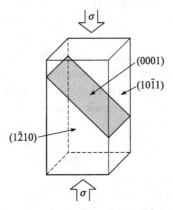

Figure 4.86   Orientation of the ZnS samples under uniaxial compressive deformation. (After Bredikhin and Shmurak 1975a. Reprinted with permission from *JETP Letters*, **21**, pp. 156–7. Copyright 1975, American Institute of Physics.)

Al as co-activator. The continuous triboluminescence was generated only during plastic deformation and decreased rapidly after it ceased. The emission spectra depended on the doping and were closely related to those of photoluminescence excited in the same crystals. This is evidence that the triboluminescence is produced at the same impurity centres they suggested, but the differences between the tribo- and photo-luminescence spectra mean the emission mechanisms differ in detail.

In ZnS crystals, illuminated with band-gap radiation, electrons are excited into deep traps, stable at room temperature. Illumination or heating releases them to produce photo- or thermo-luminescence, respectively. Plastic deformation, between the irradiation and the heating, reduces the thermo-luminescence. Hence, apparently, moving dislocations in II-VIs can empty the traps giving rise to thermo-luminescence.

The occurrence of deformation-induced luminescence and the photoplastic effect in the same type of ZnS crystals suggested the likelihood of cross-effects so Bredikhin *et al.* (1975) studied the effect of light on the pulsed triboluminescence emission. N/N_L was measured where N and N_L are the numbers of flashes emitted per unit time under stress in the dark and under illumination respectively. Illuminating the specimens with light of wavelengths in the absorption bands of the activator impurities (Cu, Mn) decreased the number of flashes emitted, N_L, to give values of N/N_L greater than unity. (Activators are impurities added to enable ZnS to emit visible light.) In contrast, illumination with light of photon energy equal to the band-gap, i.e. of wavelength $\lambda \cong 340\,\text{nm}$, produced values of N/N_L less than unity.

This was shown to be due to the light reducing or increasing the number of discrete slip events occurring. That is, each flash arises from a large number of dislocations moving over a particular slip plane or a few neighbouring planes at a particular time. Reductions or increases in N_L compared to N are due to photoplastic changes in the number of slip pulses. Light of wavelengths absorbed by point defects such as impurity atoms or complexes, change their charge state and lead to photoplastic changes in N_L. Thermally activated surmounting of the photobarriers, i.e. the pinning interactions with point defects then account for the temperature dependence of N_L. The fact that $\lambda \cong 340\,\text{nm}$ light softens the crystals and increases N_L was explained in terms of an assumed easier operation of surface sources of dislocations. The absorption coefficient of $\lambda \cong 340\,\text{nm}$ light in ZnS is large so it is absorbed within a few microns of the surface and no bulk negative photoplastic effect is observed. This type of mechanism is supported by the observation that, while the minima in N/N_L occurred at different wavelengths characteristic of the different impurities in Cu- and Mn-activated crystals, the maxima occurred at the same wavelength, 340 nm, in all the specimens.

The effect of electrical fields on the relative rate of emission of flashes of TL was found by Bredikhin and Shmurak (1975a) to be as shown in Fig. 4.87. Again they used wurtzite-structure ZnS specimens oriented for single slip as in Fig. 4.86. Electrical fields were applied either between indium contacts soldered to the left

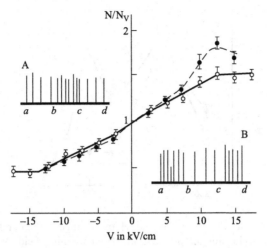

Figure 4.87 Dependence of the ratio N/N_V on the magnitude and polarity of the electric field in ZnS; the dark circles represent measurements with the field applied to soldered contacts, the open circles: field applied to a capacitor containing the specimen. The insets A and B are oscillograms illustrating the influence of fields of opposite polarity on the light flashes emitted. The times at which the crystals were loaded and unloaded are labelled a and d and the times when the external voltage was applied and removed are b and c. (After Bredikhin and Shmurak 1975a. Reprinted with permission from *JETP Letters*, **21**, pp. 156–7. Copyright 1975, American Institute of Physics.)

and right faces in Fig. 4.86 (the solid circle experimental points in Fig. 4.87) or between the electrodes of a capacitor in which the specimen was placed (the open circle points in Fig. 4.87). The specimens were again loaded with square stress pulses. N/N_V was measured where N_V is the number of flashes emitted per unit time with the field applied. Fields of electroplastic hardening polarity decreased N_V relative to N while fields of softening polarity led to values of N/N_V less than one as shown in Fig. 4.87. These effects did not depend on whether the fields were applied to contacts or to capacitor plates adjacent to the specimens. This can be understood in terms of the polar electroplastic effect, which we saw (Section 4.3.4) could be explained as follows. Fields that oppose the motion of the charged dislocations of opposite Zn(s) and S(s) polarity in opposite directions increase the flow stress necessary for deformation at a constant imposed strain rate. Such an electrical back stress, for constant stress pulses, as applied by Bredikhin and Shmurak, results in fewer slip events and flashes per unit time, N_V. Conversely, a softening field applies an additional forward stress (Equation 4.13) to the dislocations and under constant-stress pulses this results in more instances of slip and an increase in N_V.

The variation with temperature of the change of the rate of light flashes induced by hardening and softening electric fields is shown in Fig. 4.88.

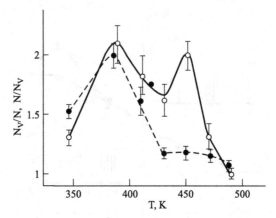

Figure 4.88   The temperature variation of the relative change in the number of luminescence flashes per unit time following the application of a hardening field, V (solid circles) and a softening field, –V (open circles). (After Bredikhin and Shmurak 1975b. Reprinted with permission from *Soviet Physics Solid State*, **17**, pp. 1628–9. Copyright 1975, American Institute of Physics.)

The other plastic-physical effects represented in Fig. 4.68, deformation-induced conductivity and electron emission (exoemission) will not be discussed here. Accounts of these phenomena can be found in Osipiyan *et al.* (1986).

Clearly, the physical-plastic effects, especially the photoplastic and electroplastic effects and plastic-physical effects such as deformation triboluminescence, provide unique and powerful tools for the study of line charges and the kinetics of dislocations in the II-VI compounds.

4.4 Plastic deformation and the microdynamical theory of plasticity

This is of considerable research interest because it was early found possible in semiconductors to compute stress-strain and creep curves, i.e. to predict deformation behaviour from basic, if semi-empirical, dislocation velocity-stress and generation rate relations. This is referred to as the microdynamical theory of plasticity. The study of dislocation movement (velocity-stress relations) is called dislocation dynamics and this is another advanced branch of dislocation theory that is perhaps better understood in semiconductors than in metals, ionic crystals or ceramics. As a result, the conditions for deformation and its effects are well understood quantitatively in semiconductors. This knowledge is available for the diagnosis and avoidance of problems arising in crystal growth and processing.

Knowledge of the generation, multiplication, movement, and interactions of dislocations is also of practical importance. It is essential for the identification and elimination of the defects found in new materials and devices (see Section 5.11).

Specific and characteristic configurations of dislocations and other defects tend to appear each time a new material or growth or processing technique is introduced (see Section 4.8).

Plastic deformation introduces large numbers of dislocations and point defect debris, either deliberately or accidentally. Macroscopic plastic deformation results from the movement of large numbers of dislocations. The microdynamical theory deduces plastic behaviour from the velocity-stress relation for individual dislocations and expressions for the numbers and interactions of dislocations during deformation. For a review see Gilman (1961, 1969). Here we will limit ourselves to the principles, following these accounts.

The starting point for microdynamical theories of plasticity is the relation between the macroscopic strain rate and the velocity of the moving dislocations. This may be derived (Gilman 1961, 1969) by considering a unit cube of a crystal undergoing shear strain through the movement of a number of dislocations (Fig. 4.89). The displacement of the top of the cube relative to the bottom is Δ, the sum of the contributions is due to individual dislocations. A dislocation sweeping over an area A of a slip plane of unit area produces a displacement $\delta_i = bA$ (the displacement is b when $A = 1$, i.e. the dislocation has swept the whole area of the slip plane) so the shear strain is:

$$\varepsilon = \Delta/h = \Delta = \sum_1^N bA = NbA \qquad (4.16)$$

assuming $h = 1$, where N is the number of moving dislocations. The strain rate is:

$$d\varepsilon/dt = Nb\ dA/dt \qquad (4.17)$$

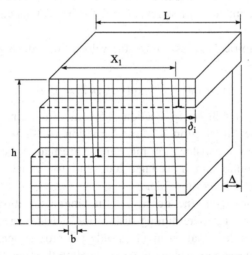

Figure 4.89 Diagram showing the shear strain, Δ, produced by dislocations moving across a cube of unit dimensions and each producing an offset, δ_i. (After Gilman 1961.)

dA/dt, the area swept per unit time is Lv where L is the length of the dislocation and v is its velocity. Hence:

$$d\varepsilon/dt = \rho_m b \, v_{av} \qquad (4.18)$$

where the dislocation density, ρ_m, is the length of dislocation per unit volume *moving* with an average velocity, v_{av} for those in all orientations, i.e. edge, screw or 60°.

Microdynamical theories introduce expressions for ρ as a function of ε and v as a function of τ and T into Equation (4.18). For our subject, the relations required are those specific to covalent semiconductors leading to treatments specifically for these materials (for reviews see Haasen and Alexander (1968) whose account we now follow). Relations can thus be found between the macroscopic variables for the conditions normally used in mechanical testing. For example, expressions for $\tau = f(\varepsilon)$ derived in this way with T and $d\varepsilon/dt$ constant represent what are called stress-strain curves. These relations exhibit the phenomenon of work hardening, i.e. the increasing stress for continued deformation (the flow stress), at the constant strain rate, $d\varepsilon/dt$. Such deformation measurements at constant temperature, T, and strain rate, $d\varepsilon/dt$, are referred to as dynamic testing. Measurements of the strain, ε, as a function of time, $f(t)$, with τ (stress) and T constant (static testing) yield 'creep' curves. Thus, these standard ways of determining the mechanical strength of materials can be correctly simulated by the microdynamical theory, as we shall see.

4.4.1 Dislocation multiplication in semiconductors

In the remainder of Section 4.4 we shall follow the account given by Haasen and Alexander (1968, 1972), as the treatment from here on is specific to semiconductors. Reference should be made to these reviews for further details and references to the original literature.

A simple semi-empirical expression for the velocity of a dislocation in a semiconductor, v, is:

$$v = v_0(\tau/\tau_0)^m \exp(-E_d/kT) = B_1 \tau^m \qquad (4.19)$$

[this relation was introduced in Section 2.5.4 as Equations (2.20) and (2.21) and further discussed in Section 4.2.2 as Equation (4.8)], where v_0, m, and E_d (the activation energy for dislocation movement) are material constants, and τ_0 is 1 MPa (see e.g. Sumino and Yonenaga 2002 and for values of v_0 and E_d, see Tables 4.6a and 4.6b in Section 4.2.2).

An expression for $\rho_m(\varepsilon)$, the density of moving dislocations is also needed. Experimental methods for determining dislocation densities generally give ρ_t, the total density whereas deformation involves only the mobile density, ρ_m. To determine the density of mobile dislocations in the initial stages of deformation in semiconductors was possible, however, because the initial, grown-in density was low and such dislocations tended to be decorated and pinned, i.e. rendered immobile by impurities. Other dislocations may be immobile because their Burgers vector corresponds to slip on a system that is inoperative because of the low resolved component of the applied stress or to a large back stress arising internally from defects, etc. Hence, to a good approximation, in slightly deformed semiconducting crystals, the observed total density is equal to the required mobile density $\rho_t = \rho_m$. The validity of this identification is confirmed by the good agreement between the experimental results and calculations based on this assumption.

It was thus found experimentally that the rate of generation of dislocations, $d\rho/d\varepsilon$, during strain is of the simple linear form:

$$d\rho/d\varepsilon = K\tau_{\text{eff}} \tag{4.20}$$

where ε is the strain, τ_{eff} is the effective stress experienced by the moving dislocations, K is a rate constant = 8 mm/kg for Si in the range $950° < T < 1300°$ C and K = 10.5 mm/kg for Ge (Berner and Alexander 1967).

A widely used expression is that the dislocations accumulating during deformation exert an 'internal' elastic back-stress proportional to the square root of the dislocation density, that is:

$$\tau_{\text{int}} = C\rho^{1/2} \tag{4.21}$$

so $$\tau_{\text{eff}} = \tau_{\text{appl}} - C\rho^{1/2} \tag{4.22}$$

where τ_{appl} is the stress due to the externally applied load and $C = \mu b/\{2(1-v)\} \approx \mu b/3$.

Measurements of ρ and ε must be made in the same microscopic area of the crystal, because crystals with low initial dislocation densities deform inhomogeneously. Direct experimental checks then confirm relation (Equation 4.18) for such microscopic areas.

4.4.2 Microdynamical equations for the initial stages of deformation

To treat the macroscopic stress-strain-time relations for the initial stages of the deformation of semiconductors, two simultaneous equations are obtained. The first is obtained by substituting for the velocity from the short form of Equation (4.19) and for τ (Equation 4.22) in Equation (4.18):

$$d\varepsilon/dt = \rho_m b \, v_{\text{av}} = b\rho_m B(T)\{\tau - C\rho^{1/2}\} \tag{4.23}$$

An expression for $\rho_m(\varepsilon)$, the density of moving dislocations is also needed. This is obtained from Equations (4.18), (4.19) and (4.20) as follows:

$$\frac{d\rho}{dt} = \frac{d\rho}{d\varepsilon}\frac{d\varepsilon}{dt} \tag{4.24}$$

$$\text{i.e.} \qquad \frac{d\rho}{dt} = K\tau_{\text{eff}} \, b\rho_{\text{m}} v_{\text{av}} \tag{4.25}$$

and therefore

$$\frac{d\rho}{dt} = K \, b\rho \, B(T) \, [\tau - C \, \rho^{1/2}]^2 \tag{4.26}$$

where we will from now on drop the descriptive suffixes.

Plastic properties are determined at constant temperature in (i) static or creep tests in which a constant stress τ is applied and $\varepsilon(t)$ is measured, and in (ii) dynamic tests which impose a constant strain rate $d\varepsilon/dt$ and in which $\tau(\varepsilon)$ is measured. Note that it is $d\varepsilon_{\text{total}}/dt = d(\varepsilon + \varepsilon_{\text{el}})/dt$ that is constant where ε_{el} is the elastic strain.

4.4.3 Microdynamical theory of creep

Creep is slow deformation at a constant, relatively low load and hence at approximately constant resolved shear stress, τ. The strain rate, $d\varepsilon/dt$, and the strain, ε, are observed as functions of time.

The normalized creep curve of Fig. 4.90a was obtained (Peissker *et al.* 1961) by numerical integration of Equations (4.23) and (4.25), using the functions ρ and v in Fig. 4.90b with m = 1.5 in Equation (4.19), though the result is little different if m = 1 as assumed above, as generally accepted now.

The calculated curve in Fig. 4.90 is of the form found experimentally and shown in Fig. 4.91. Similar results for GaAs and InSb were reported by Laister and Jenkins (1973) and Walter (1977). S-shaped creep curves occur because in stage I there is exponential dislocation multiplication and accelerating strain. The back stress due to the rising density of dislocations then reduces the dislocation velocity and rate of multiplication, giving a constant strain rate in stage II. The curvature of the strain-time relation reverses sign at the point of inflexion at time t_{w}. The slope tends to zero in stage III as the back stress $C\rho^{1/2}$ rises toward equality with the applied stress.

4.4.4 Microdynamical theory of the yield point

The important phenomenon to be explained in the initial stages of stress-strain curves determined in dynamic testing is a large yield drop.

In the dynamic deformation of semiconductor crystals with low initial dislocation densities, macroscopic strain begins at the constant applied strain rate at an upper yield stress τ_{uy}. The stress falls with continuing strain to a lower yield stress, τ_{ly}, at which it continues for some time as shown in Fig. 4.92. Such yield points occur not only in elongation but also in constant strain rate compression, torsion and bending tests (Alexander and Haasen 1968).

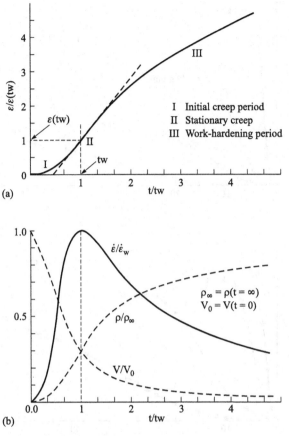

Figure 4.90 (a) The strain-time (creep) curve in normalized coordinates calculated from Eqs. (4.23) and (4.25). (b) The factors determining the plastic strain rate and underlying the creep curve in (a). (After Peissker *et al.* 1961. Reprinted with permission from *Philosophical Magazine*, **7** (1961), pp. 1279–1303; Taylor & Francis Ltd., the Journal's web site: http://www.tandf.co.uk/journals.)

Microdynamical theory readily accounts for this quantitatively. The substantial rate of applied strain requires large initial dislocation velocities because the dislocation density is low (see Equation 4.23). The crystal strains elastically until, at τ_{uy}, v becomes sufficiently large to give a significant plastic contribution. At these high stresses (τ_{eff} in Equation 4.25) dislocation multiplication will be rapid and lower values of v will come to suffice. Hence by Equation (4.19) lower stresses will be needed until the minimum value τ_{ly} is reached. Strain continues at this stress until $\tau_{int} = C\rho^{1/2}$ rises to reduce τ_{eff} significantly. The total, observed value of τ then rises to overcome the effect of this rising internal (work-hardening) back stress.

When silicon crystals are deformed at the same strain rate but a series of increasing temperatures, the initial yield drop decreases as shown by the curves marked

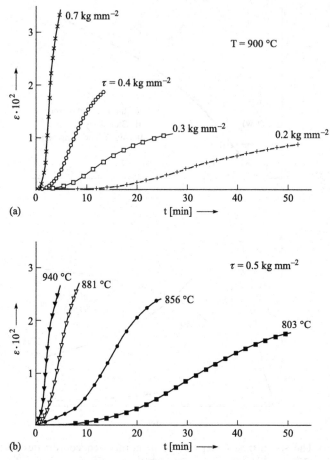

(a)

(b)

Figure 4.91 Experimental creep curves for silicon in compression at the temperatures and stresses indicated. (After Reppich *et al.* 1964. Reprinted from *Acta Metallurgica*, **12**, Kreichen von silizium-einkristallen, pages 1283–8. Copyright 1964, with permission from Elsevier.)

(1) to (4) in Fig. 4.93. This, too, can readily be understood because the greater thermal activation energy available at higher temperatures enables the few dislocations present initially to begin moving at lower stresses.

4.4.5 Microdynamical theory of the polar bending of GaAs

Booyens *et al.* (1978) subjected GaAs single crystals to three-point bending in opposite polar senses (see Section 4.1.3) but about a ⟨110⟩ axis chosen so slip would occur on two intersecting {111} planes rather one chosen to produce single slip like the ⟨112⟩ axis in Fig. 4.21. They treated the deformation on the assumption that ρ is large enough to be treated as a continuous function and employed

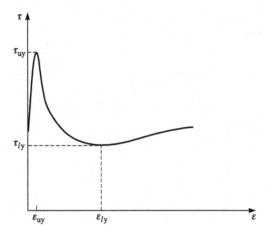

Figure 4.92   The initial yield-point region of a typical stress-strain curve of an adamantine semiconductor.

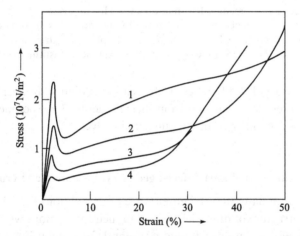

Figure 4.93   Tensile stress-strain curves for silicon specimens with an initial dislocation density of $N_D = 2 \times 10^4\,\mathrm{cm}^{-2}$ that were deformed at a rate of $\dot{\varepsilon} = 1.2 \times 10^{-4}$ at temperatures of (1) 800°, (2) 850°, (3) 900° and (4) 950° C. (Modified from Yonenaga and Sumino 1978.)

expressions for dislocation multiplication analogous to Equation (4.25) and for dislocation velocity analogous to Equation (4.19), etc. They derived expressions to make it possible to calculate the stress-strain curves for polar bending. They assumed that either α [A(s)] or β [B(s)] dislocations were introduced and labelled their experimental curves accordingly.

   Their results in Fig. 4.94 showed that 'α' bending occurred at lower stresses than those required for 'β' bending. Thus, for example, the upper yield stress (that at the initial peak) was greater for β than for α bending and since these stresses are those to get the few initial mobile dislocations moving as rapidly as is necessary, this is fairly

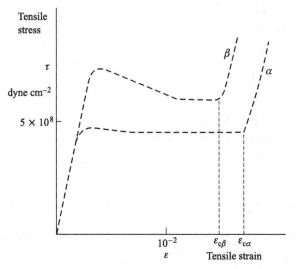

Figure 4.94   Average experimental deformation curves for opposite polar 'α' and 'β' bending of GaAs single crystals, assumed to have occurred by the motion exclusively of A(s) and B(s) dislocations, respectively. (After Booyens *et al.* 1978. Reprinted with permission from *Journal of Applied Physics*, **49**, pp. 5435–40. Copyright 1978, American Institute of Physics.)

direct evidence of a polar difference in dislocation mobility. Booyens *et al.* (1978) used the fitting of curves calculated from their model to the experimental results to extract values for a number of the parameters involved.

4.5 Dislocations and area defects: geometry, formation and properties

It was pointed out initially that the importance of dislocations lies partly in the role they play in the structure of other defects and in their interactions with other defects. The interactions of dislocations with native point defects, impurity atoms, vacancies and interstitials, were introduced in Section 2.3.12. They will be further discussed in Section 4.7.

The role of dislocations in relation to area defects is also important. For example, partial dislocations are the boundaries of stacking faults and provide a mechanism for several structure transformations and for deformation twinning. Moreover, certain types of interface can be resolved into sets of dislocations and these are dealt with here.

4.5.1 Dislocations and stacking faults: twinning and phase transformations

The dissociation of unit dislocations into a stacking fault, bounded by two Shockley partial dislocations in materials with an f.c.c. lattice based structure, was treated in Section 2.3.7 and Section 2.3.8. Such extended dislocations occur in both the semiconducting elements with the diamond structure (i.e. Ge, Si and diamond) and in semiconducting compounds with the sphalerite structure. Stacking faults constitute a localized change in stacking sequence and so they play an important role in the processes of phase transformation from the sphalerite to the wurtzite structure and vice versa and of the formation of polytypes in ZnS (see Section 4.1.5). The movement of Shockley partial dislocations, drawing stacking faults behind, is also the mechanism of formation of deformation twins and other phases in the deformation zones of indentation hardness impressions (see Section 4.5.1).

The {111} double-atom plane stacking sequence in the diamond and sphalerite structures is . . . $A\alpha\ B\beta\ C\gamma\ A\alpha\ B\beta\ C\gamma$. . . An intrinsic fault is produced by the passage of a Shockley partial dislocation with an $a/6\langle112\rangle$ Burgers vector like $\mathbf{b_2}$ or $\mathbf{b_3}$ in Fig. 4.95. This alters this sequence to one of the form . . . $A\alpha\ B\beta\ C\gamma/B\beta\ C\gamma$ $A\alpha\ B\beta$. . . (as in the central section, reading upward, of Fig. 4.95b, where for simplicity only the Latin letters for the stacking positions are shown). The four double-atom planes flanking the fault, $B\beta\ C\gamma/B\beta\ C\gamma$, have the wurtzite stacking sequence. Wurtzite structure materials have the stacking sequence . . . $A\alpha\ B\beta\ A\alpha\ B\beta\ A\alpha\ B\beta$ $A\alpha\ B\beta$. . . and the passage of a Shockley partial dislocation will shear this into the sequence: . . . $A\alpha\ B\beta\ A\alpha\ B\beta/C\gamma\ A\alpha$. . . Again a group of four double atom planes, two on either side of the fault plane, constitute a thin slab of the other, sphalerite, structure in the wurtzite matrix.

The volume of transformed-structure material can be altered by the glide of the Shockley partial dislocations bounding the fault area. This movement can be produced by stresses, e.g. due to large temperature gradients in cooling after growth or during processing. Impurity reduction of the fault energy is known as the Suzuki effect (Suzuki 1952). The unchanged mutual elastic repulsion of the two partial dislocations is opposed by the reduced 'surface tension' of the stacking fault so

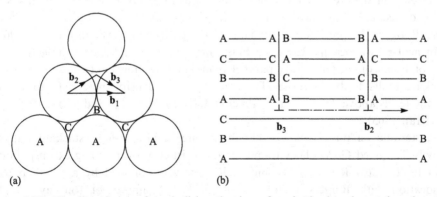

(a)   (b)

Figure 4.95   The dissociation of a unit dislocation in an f.c.c. lattice-based crystal, such as the diamond- and sphalerite-structure semiconductors. (a) The Burgers vectors, $\mathbf{b_1}$ of a unit dislocation and $\mathbf{b_2}$ and $\mathbf{b_3}$ of the Shockley partial dislocations into which the unit dislocation can dissociate. (b) The area between the partial dislocations (viewed end-on and indicated by the dash-dot line) is a stacking fault.

extended dislocations widen. Wide dislocations in heavily doped semiconductors were commonly observed in the early years, e.g. in Te doped GaAs and GaP as will be discussed below (Laister and Jenkins 1968 and Chase and Holt 1972).

Transmission electron microscope studies show that extended dislocation widths generally increase from the IVB elements through the III-V and II-VI compounds, suggesting fault energies fall in this order. This accords with the fact that Si and Ge have no hexagonal phase (a hexagonal crystalline phase of carbon, known as Lonsdaleite, occurs rarely in minerals) and the sphalerite structure greatly predominates among the III-Vs, whereas the II-VI compounds all occur in both sphalerite and wurtzite phases. Several II-VI compounds transform from the one structure to the other at a critical temperature so their relative energies are temperature dependent. Widening of stacking faults by partial dislocation movement is the mechanism of this transformation (Blank *et al.* 1964, Secco D'Aragona and Delavignette 1966) and of transformation to 'polytype' stacking sequences in ZnS (Mardix *et al.* 1968, and Steinberger *et al.* 1973) and other phase changes in CdS (Holt and Wilcox 1971).

Many TEM studies have been carried out on stacking faults in semiconducting compounds, especially GaAs, GaP and a number of II-VI compounds, and these will be discussed now.

Abrahams and Buiocchi (1967) studied eptitaxial GaAs grown by CVD (chemical vapour deposition) on (111) and (100) GaAs substrates and found both stacking faults and microtwins. [Microtwins are regularly faulted structures (see Section 4.5.1) that are several (double) atom planes thick.] In the (100) material almost all the defects were microtwins. These only occurred, however, when mechanical damage was not completely removed from the substrates by chemical polishing. Tetrahedral stacking faults were found in GaP by Chase and Holt (1972) but they were a rare occurrence in III-V compounds, not common as in epitaxial Si (see Section 4.8.1). In Se-doped epitaxial GaAs both stacking faults and coherent twin boundaries were decorated by precipitates, probably of Ga_2Se_3 (Abrahams and Buiocchi 1967).

In melt-grown heavily Te-doped bulk GaAs, Laister and Jenkins (1968) found sets of faults on neighbouring parallel planes. Diffraction contrast showed these stacking faults to be extrinsic. The stacking fault density increased rapidly for doping levels above $2 \times 10^{18}\,cm^{-3}$. Precipitates were also observed above $10^{18}\,cm^{-3}$ (Meieran 1965).

Laister and Jenkins (1969, 1971) analysed three- and four-layer stacking faults in heavily Te-doped GaAs. They found these consisted of overlapping extrinsic faults with fault vector $\mathbf{R} = 1/3\langle 111\rangle$ and each layer was bounded by a Frank partial dislocation with Burgers vector $\mathbf{b} = +1/3\langle 111\rangle$. Examples of four-layer faults are shown in Fig. 4.96. These four-layer faults were found in thicker regions of TEM specimens from heavily Te-doped GaAs that had been annealed for long times (24.4 h at 879° C in a stoichiometric atmosphere of arsenic). This suggests that the Suzuki effect was involved (see Section 4.1.5).

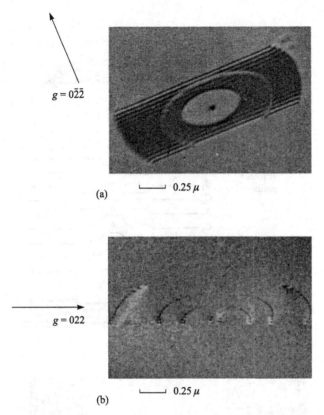

$g = 0\bar{2}\bar{2}$

$\longmapsto$ 0.25 μ

(a)

$g = 022$

$\longmapsto$ 0.25 μ

(b)

Figure 4.96   TEM micrographs of two different four-layer extrinsic stacking faults in heavily Te-doped and annealed GaAs. (a) Stacking fault fringe contrast does not occur where three faults overlap. The fringe contrast in the small, central fourth fault cannot be seen partly due to the absorption in the relatively thick specimen. (b) Portions of the four concentric Frank partial dislocation loops bounding the four faults can be seen in dotted contrast. The remaining sectors of the loops were removed in thinning the specimen. (After Laister and Jenkins 1971. Reprinted with permission from *Philosophical Magazine*, **24** (1971), p. 705; Taylor & Francis Ltd., the Journal's web site: http://www.tandf.co.uk/journals.)

Abrahams and Buiocchi (1970) examined epitaxially grown, Zn-doped GaAs and found that the faults which nucleated at the mechanically polished substrate surfaces were intrinsic and were bounded by Shockley partial dislocations. Near each end of most of these grown-in faults there were three overlapping faults as can be seen in Fig. 4.97a. The overlapping faults in these three-layer fault structures were shown to be intrinsic as shown schematically in Fig. 4.97b. The two faults that overlap the grown-in fault form by vacancy migration and are bounded by Frank partials. Unfaulting of these three-layer defects was observed in annealed specimens. The mechanism of unfaulting was the nucleation and motion of Shockley partials. This eliminated the three-layer faults as in Fig. 4.97c or, eventually, the entire fault.

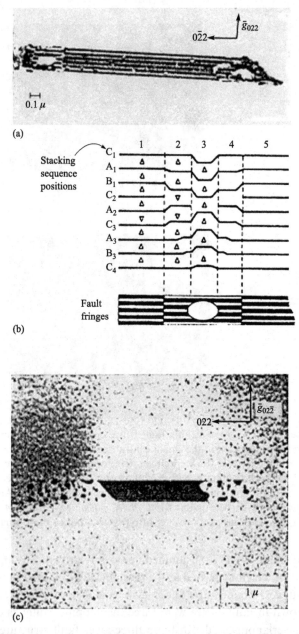

(a)

(b)

(c)

Figure 4.97   TEM images of grown-in intrinsic stacking faults in GaAs. (a) There are two additional fault loops on adjacent planes at either end of the fault. (b) Schematic diagram of the stacking sequence for the three-layer defects seen in (a) and (below) the appearance of the faults when viewed in the [100] direction. (c) Partial annihilation of the faulted structure in an annealed specimen. Here the three-layer defects had unfaulted leaving only the central section of the initial intrinsic fault. (After Abrahams and Buiocchi 1970. Reprinted with permission from *Journal of Applied Physics*, **41**, pp. 2358–65. Copyright 1970, American Institute of Physics.)

Chase and Holt (1972) studied Czochralski-grown bulk and liquid- and vapour-phase epitaxial (i.e. LPE and VPE) GaP using TEM. The VPE GaP was found to contain many intrinsic stacking faults. Most were planar but a few were tetrahedral, and both Frank and Shockley bounding partials were found. Tetrahedral stacking faults consist of faults on two or more of the four orientations of {111} planes in the diamond and sphalerite structures. These four {111} planes are oriented as the faces of a tetrahedron (see Thompson's tetrahedron shown in Fig. 2.2). An area containing a number of sets of overlapping intrinsic stacking faults is shown in Fig. 4.98. The fault density increased with increase in the Te dopant concentration in the VPE material. The Czochralski GaP also contained intrinsic stacking faults and its defect content was higher than that of layers grown by LPE.

Stacking faults are generally a more prominent feature of the defect structure of II-VI than of III-V compounds. This indicates that the fault energy in the II-VI compounds is relatively low.

Wurtzite-sphalerite phase transformation (the crystallography of stacking faults in wurtzite structure compounds)

The mechanisms of the wurtzite-sphalerite and polytype transformations, as well as the crystallography of stacking faults in wurtzite-structure compounds, were outlined in Section 4.1.5. As noted in Section 4.1.5, such transformations occur by the

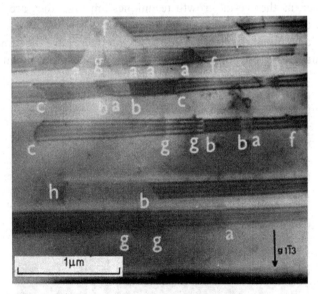

Figure 4.98   TEM micrograph of a number of sets of overlapping intrinsic stacking faults in Te-doped VPE GaP. The dislocations marked a, b and c are Shockley partials with different Burgers vectors of the form $a/6\langle112\rangle$, those marked f are Frank partials with $\mathbf{b} = a/3\langle111\rangle$, g marks unit perfect dislocations and h is a partial of undetermined type. (After Chase and Holt 1972.)

formation and widening of regular arrays of stacking faults to transform the stacking sequence from that for the wurtzite structure to that for the sphalerite structure.

4.5.2 Coherent interfaces and dislocations

Interfaces separate distinguishable volumes of crystalline material. If the volumes have the same crystal structure and differ only in orientation, the interface is a grain boundary. Polycrystalline material is entirely divided up by such boundaries. Polycrystals, in which certain orientations predominate, are said to have a 'preferred orientation', otherwise they are randomly oriented. If the material all has the same orientation, it is described as a single crystal or monocrystalline. Thin films grown on monocrystalline substrates are called epitaxial if monocrystalline or predominantly so. If material contains only regions having misorientations less than 1°, it is said to have a mosaic structure of subgrains, separated by small-angle or subgrain boundaries. If mainly monocrystalline material contains a few misoriented regions, these are referred to as included grains.

Grains with certain simple low energy orientations relative to the monocrystalline matrix are called twins. Their coherent interface boundaries are of such low energy that they tend to be planar and extend over large distances and twins with such boundaries are of such low energies that they occur profusely in the early stages of development of the crystal growth of new or difficult-to-grow materials (see Fig. 4.99). Later, as the crystal growth techniques improve, they are seen as many (see Fig. 4.99) or a few included grains (see Fig. 4.103 below).

If regions in a compound, on either side of an interface, have the same orientation and crystal structure, but differ in the occupation of the sublattices by the

Figure 4.99   Secondary electron SEM micrograph of a multiply twinned area in Cr-doped semi-insulating liquid encapsulated (LEC) Czochralski-grown InP. (After Holt and Salviati 1990.)

constituent elements of the compound, the interface is an antiphase boundary (APB). In sphalerite and wurtzite structure materials, it can also be described as a polarity-reversal boundary.

Interfaces between regions with different crystal structures or compositions are phase boundaries. The two phases may also differ in orientation. The phase boundaries of importance for semiconductor devices are heterojunctions and contacts. Heterojunctions are interfaces between two different semiconducting materials and, electrically, are generalizations from *p-n* junctions. Contacts are metal/semiconductor interfaces. The structurally simple contacts are metal-silicide/semiconductor interfaces. (For a review of heterojunction and metal-silicide/semiconductors see Ourmazd *et al.* 1991.) Grain boundaries, contacts and heterojunctions may consist of dislocation arrays or can contain them, as we shall discuss below.

Coherent interfaces are those in which the structures and orientations of the two regions are so related that good atomic matching occurs across the interface. Such interfaces have low energies. Coherent interfaces, especially twins and APBs, predominate when the growth conditions approach those for single crystal growth. This is because, then, defect formation energies are important, e.g. in the epitaxial growth of one material on another. Because of the good matching and chemical bonding at coherent interfaces in semiconductors they have relatively low densities of interfacial recombination or trapping states. An ideally coherent interface is one at which there is perfect atomic matching so that there are no strained, wrong or broken chemical bonds. This is the case for strained-layer or pseudomorphic epitaxial interfaces. It is also the case for coherent first order twin boundaries in the diamond- and sphalerite structures (see Fig. 4.100), which alter only next-nearest neighbour configurations

It has been found experimentally that many interfaces tend to approximate locally to some nearly equivalent semi-coherent, and therefore low-energy interface structure. Mathematical tools have been developed to deal with the crystallographic structure and energy of coherent grain boundaries (GBs). To define the crystallography of interfaces between grains of different orientations, materials or crystal structures the methods of the coincidence site lattice (CSL), of the 'O' lattice and of coloured symmetry bicrystallography have been developed (for reviews see e.g. the book by Bollmann 1970 and the first chapter of Sutton and Baluffi 1995). Methods of calculation, developed for finding the energies of interfaces, include that of grain boundary and misfit dislocations and the variational method, dealing with the problem in reciprocal space. We will now briefly introduce the most widely used CSL approach and just mention, with references, the other methods which have thus far been less important for the study of interfaces in semiconductors.

Subsymmetry and coherence: The coincidence lattice

The ideas of the coincidence- and O-lattices arose from consideration of twins and other low-energy grain boundaries. Twins are grains with crystallographic

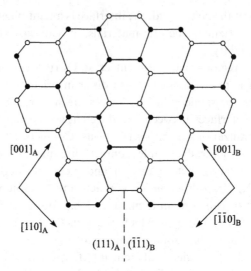

Figure 4.100 The first order (111) twin, i.e. the 70°32′ ⟨110⟩ symmetrical tilt boundary. The figure is a (1$\bar{1}$0) projection and corresponding planes and directions in the two grains A and B are labelled. Open and solid circles represent atoms at heights 0 and z/2 above the plane of the figure. (After Hornstra 1959. Reprinted from *Physica*, **25**, Models of grain boundaries in the diamond lattice I. Tilt about ⟨110⟩, pp. 409–22. Copyright 1959, with permission from Elsevier.)

orientations relative to the single crystal matrix such that the twin/matrix interfaces are of particularly low energy. These orientations can be derived from that of the matrix either by a rotation θ about a particular twinning axis [*hkl*] or by reflection in a mirror plane (*pqr*). Twins occur very commonly in 'ingots', produced by freezing large volumes of the molten material under conditions that are getting near those required for single crystal growth, as first reported in InSb by Haasen (1957a). Coherent twin boundaries are readily recognizable because they are strikingly planar and crystallographic and generally extend over long distances, as shown in Fig. 4.99.

It was also early reported that twins were produced in adamantine semiconducting materials under the impressions made with pyramidal diamond indenters used to determine hardness (Churchman *et al.* 1956). Such deformation twinning and related phase transformation was later studied by Pirouz (1987, 1989a, 1989b). He showed them to be due to shears over certain planes, produced by the passage of twinning or transformation dislocations.

Friedel (1926) pointed out that if the lattices of the twin and matrix were continued across the interface (composition plane) so as to interpenetrate, a certain fraction of sites of the twin and matrix would be found to coincide to constitute a 'coincidence lattice'.

Friedel defined a unit cell of the coincidence lattice, of volume *V*, as having as base the smallest mesh of coincidence sites in the mirror plane (*pqr*) and height equal to the coincidence site spacing in the direction normal to this plane. Let the unit cell volume of the crystal structure be v and the multiplicity:

$$V/v = \sum \tag{4.27}$$

Friedel proved that, for all Bravais lattices and hence for all crystal structures, Σ, the Friedel index, is odd. That is, for all crystal structures and all possible orientation relationships, the fraction of coincidence sites, $1/\Sigma$, can only be the reciprocal of an odd integer.

Friedel used Σ to classify twins and called it the twin index. For all crystals of the cubic system in which $[hkl]$ can be taken as normal to (pqr):

$$\sum = p^2 + q^2 + r^2. \tag{4.28}$$

Twins in cubic system crystals can now be specified by Friedel indices of the form:

$$\sum = 3^n \tag{4.29}$$

where n is an integer, the order of the twin.

Ellis and Treuting (1951) constructed the atom positions in the first order twin in the diamond cubic structure. This corresponds to an 180° (twofold) rotation about the [111] (threefold rotation symmetry) axis and has the coherent interface twin plane (111). They found that all the sites, of every third double-atom plane parallel to the interface, coincide. This twin has index $\Sigma = 1^2 + 1^2 + 1^2 = 3$ and the fraction of coincidence sites is $1/3$ in accordance with Friedel. The higher order twins are produced by repeated twinning about the different $\langle 111 \rangle$ axes in the structure. Thus, for example, the second order twin has an orientation, relative to the matrix, that is produced by a 180° rotation about the [111] axis followed by a second 180° rotation about [1$\bar{1}$1] say, and the fraction of coincidence sites for second-order twins is $1/9$.

Since $\Sigma = p^2 + q^2 + r^2$, the possible forms of twin planes, $\{pqr\}$ corresponding to any twin index, can be found. Thus for first order twins, $\Sigma = 3$ so there are $\{111\}$ and $\{112\}$ twins. Next, methods were needed for finding the twinning (coincidence) rotations and twin indices for particular twinning axes. F. C. Frank developed a procedure applicable to rotations about [100] and [111] axes. C. G. Dunn and H. Brandhorst then developed one for rotations about [110]. This work was never published but is referred to by Brandon *et al.* (1964) and an account of these procedures is given by Ranganathan (1966). Ranganathan also gave a method for finding the possible values of the Friedel index, Σ, in the structures of the cubic system for rotations about the $[hkl]$ axis and listed values of Σ and θ for [111] and [210] axes.

Rotation matrices, twins and semicoherent grain boundaries

The rotational symmetry operations of a crystal structure can be represented by means of matrices. The transformation matrix, corresponding to any rotation through an angle θ about an axis oriented at angles α_1, α_2 and α_3 to the crystal axes, will be found in Hornstra (1960).

In his pioneering work on grain boundaries in diamond-structure semiconductors Hornstra (1960) proved that if a grain boundary in the diamond structure is regular, i.e. if it has a periodically repeating structure and so is resolvable into a regular array of dislocations, it must be a plane of a coincidence lattice. [As mentioned in Section 2.3.10, Frank (1949) showed that, formally, the Burgers vector content of any grain boundary could be found. Only in the case of those with periodically repeated structures, however, can separate dislocations, in principle, be recognized.] Hornstra also proved that, in this case, the axis of rotation must be a rational direction and $\cos \theta$ must be a rational fraction, i.e. equal to the ratio of two integers. He used this rule to select particular tilt angles, about the rotation axes $\langle 110 \rangle$ and $\langle 100 \rangle$ in the diamond structure, for which to construct ball-and-wire crystal models. These represented grain boundary core structures in the adamantine (diamond like) semiconductors that have a recognizable dislocation-related content (Hornstra 1959, 1960).

Hornstra (1960) further proved that, for regular grain boundaries, all the elements of the transformation matrix must be rational fractions. That is, when the matrix is multiplied by n, the least common multiple of all the denominators, all the elements of the matrix become whole numbers. Hornstra pointed out that since n = Σ, the calculation of the transformation matrix is a general means for finding the Friedel index. He also pointed out that a low index means a short period along the boundary, i.e. a small dislocation spacing in the boundary. This, in turn, means a rapid fall-off of the elastic strains from the boundary plane and, therefore, a low elastic energy of the boundary.

Coherent interfaces can now be described as containing a high density of coincidence sites so atomic matching across the interface is good. Coherent interfaces thus occur with planar orientations corresponding to high-site-density planes in the coincidence lattice, and have orientation relations between the grains which correspond to small values of the Friedel index.

In polycrystalline materials, the grain boundaries (GBs) can be random in misorientation and structure. Some GBs, however, will correspond to planes of high densities of coincidence lattice sites, and have relatively simple, regular structures. Small Σ boundaries, especially coherent twin boundaries, have low energy structures. They, therefore, tend to occur more frequently than by random chance.

At present, software is available for the analysis of EBSD (electron back scattering diffraction) patterns from polycrystalline materials in SEMs. This rapidly gives the orientations of very large numbers of grains and hence the axis and angle of misorientation of the GBs between pairs of them. (For introductory accounts of this technique see Section 3.7.4 and the books by Randle 1993, by Dingley *et al.* 1993 and by Randle and Engler 2000.) In many cases these misorientations correspond to coincidence site lattices with values of Σ, the Friedel index, which can then also be found. The latest EBSD software thus can rapidly produce orientation maps. These are emissive mode SEM images of polycrystalline material distinguishing, for example, $\Sigma = 3$ boundaries as broad white lines, $\Sigma = 9$ boundaries as thin white lines and all other GBs as black lines. They also give read-outs of the percentage of each in the total GB projected length in the field of view. An example, showing how common twin boundaries can be, was published by Randle (2001). This showed an area of a nickel-based superalloy of the type used for gas turbine blades. In that field of view, 47% of the total length consisted of $\Sigma3$, first order twin boundaries and 9% of $\Sigma9$, second order twin boundaries. The ease with which such data can be obtained has led to much use of this technique especially for the study of polycrystalline metals and ceramics. There is some question as to the significance of these data, however. Randle (2001) presented the arguments to emphasize that in addition to the Σ value, the orientation of the boundary plane in the coincidence lattice and structural periodicity are important in ensuring low boundary energy and 'special' GB properties.

Raza (1994, Ph.D. Thesis, and Raza and Holt 1991, 1995) carried out an early, extensive study of Wacker SILSO. This is commercial polycrystalline silicon, produced as large square wafers for fabricating low-cost Si solar cells for terrestrial power generation. It is grown by a proprietary method of directional solidification, from the bottom up, in large, cooled ingot moulds. The structure, therefore, consists of relatively large (mm to cm) columnar grains. By observing corresponding areas on the top and bottom surfaces of slices of SILSO Si it was confirmed that the parttern of grains matched. Thus the GBs run vertically through the slices over comparatively long distances with constant character. Hydrogen passivation (see Section 5.5.8) is normally used to render the GBs electrically inactive and optimize the efficiency of the solar cells made from the material.

Relatively large areas of this material were examined by EBSP (electron backscattering pattern) analysis and EBIC microscopy in an SEM. The EBSP software then available gave the orientations for both grains, as well as the rotation axis and misorientation angle characterizing the GB between them. By consulting tables, in many cases, the Friedel index, Σ, of the boundary could also be found (Raza and Holt 1995). The structures of the GBs could then be related to their electrical properties as determined by EBIC (see Section 5.10.4).

Raza and Holt (1995) reported that there were numerous boundaries that produced no EBIC contrast at room termperature but appeared in low-T EBIC images, like those marked 29 and 30 in Fig. 4.101. Of these, the great majority were first order, $\Sigma = 3$ coherent twin boundaries and the rest other low Σ GBs or small angle boundaries (walls of well-separated dislocations). Low Σ interfaces generally tend to have little electrical effect, e.g. in this case they were electrically inactive in EBIC at room temperature. The GBs that produced strong EBIC contrast at all temperatures were large Σ boundaries or did not correspond to coincidence site lattice misorientations. Such GBs are relatively electrically active and generally

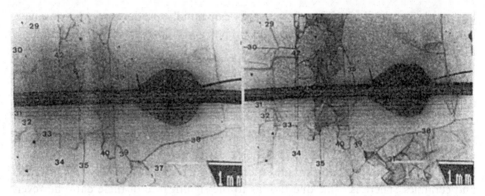

Figure 4.101 An area of unpassivated polycrystalline SILSO Si shown in SEM EBIC micrographs at (a) room temperature and (b) liquid nitrogen temperature. In (b) in addition to the grain boundaries seen in (a) many additional subgrain boundaries (e.g. those marked 29 and 30, both of which are coherent first order, $\Sigma = 3$ twin boundaries) are in contrast. The numbers identify GBs that were structurally characterized by EBSP and electrically analysed by EBIC. (After Raza and Holt 1995.)

harmful to device operation. Many grain boundaries in randomly oriented polycrystals, i.e. those with no 'preferred orientation', are not crystallographically oriented or planar and so do not have large areas of constant character.

The O-Lattice treatment of crystalline interfaces

Grimmer, Bollmann and Warrington (1974) generalized the coincidence site lattice (CSL), which they termed the displacement shift complete lattice (DSCL) or 'complete pattern-shift lattice'. For a rational rotation or a twin and matrix, a CSL exists as Hornstra proved. It is obtained when one site of each lattice is made to coincide and taken as origin. However, other periodic patterns with the same period but without containing a CSL can be produced for the same rotation by a translation of the second lattice with respect to the first. The CSL is then the finest (smallest unit cell) common superlattice of the two crystal lattices. The coarsest lattice that contains the two crystal lattices as sublattices is the DSCL. Grimmer *et al.* treated the cubic lattices: simple, i.e. primitive cubic, face centred cubic (f.c.c.) and body centred cubic (b.c.c.). They showed that the ratio of the volume of the DSCL unit cell to that of the original lattice is $1/\Sigma$, i.e. it is the reciprocal of the Friedel index of the CSL. For the f.c.c. lattice, the CSL and the DSCL have to be face-centred. Grimmer *et al.* gave explicit methods for determining the CSL, the DSCL and the planar density of coincidence sites in all three cubic lattices and tabulated the rotation, the CSL and DSCLs for all Friedel indices up to 49.

Bollmann (1967a, 1967b, 1970) further generalized the coincidence site lattice to the 'O-lattice'. The coincidence lattice is the set of sites at which points of the two lattices coincide. The O lattice consists of the set of points at which any corresponding points relative to lattices 1 and 2, whether lattice sites or not, coincide. Thus the O points are coincidences of points which are in equivalent positions, that is have the same coordinates, not necessarily rational fractions of the unit cell edges, relative to the unit cells of the two lattices. The O points may constitute an arrangement of points, of lines or of planes. Each O element (point, line or plane) is enclosed by a cell which is composed of planes which are the bisectors of the vectors joining the O elements. The cell walls are closed polyhedra, open tubes, and parallel planes for O-point, -line and -plane lattices respectively. The O elements lying in the boundary (interface) are positions of best matching, i.e. coherence sites while the intersections of the boundary surface and the cell walls are lines of worst matching and are dislocations. These are described as primary dislocations and separate areas of good matching (coherence). Deviations from these optimal boundary structures can be accommodated by superposing additional 'secondary' dislocations. Their DSCL defines the possible Burgers vectors of secondary dislocations.

This approach was extensively developed and reviewed a number of times (Bollmann 1970, 1972, 1974, Smith and Pond 1976, Balluffi *et al.* 1982). It can be used to predict the structures of grain boundaries and phase interfaces on the assumption that they consist of some optimal structure with, if necessary, a super-imposed 'secondary' dislocation network. The latter accommodates any difference between the actual crystallographic parameters of the interface and those of the optimal structure. The optimal structure may be an interface of the coincidence lattice type discussed above, or, for a low angle grain boundary or epitaxial interface, that of the perfect crystals. Experimental evidence from both field ion microscopy and transmission electron microscopy (Smith and Pond 1976, Balluffi *et al.* 1982) indicated that high angle grain boundaries in metals do consist of areas of good fit separated by line defects of dislocation or boundary step character.

Symmetry and bicrystallography

A still more general and abstract treatment of crystalline interfaces is 'bicrystallog-raphy', developed by Pond and Bollmann (1979), Pond and Vlachavas (1983) and Pond and Bastaweesy (1984). Bicrystals are two single crystal grains meeting in a planar grain boundary or phase interface. Bicrystallography derives the spacegroup symmetry of the bicrystal in a series of steps from the most symmetric starting group possible. For our purposes it is enough to know that such a powerful branch of crystallography is available to deal with interfaces and that it can be used to predict new types of interface line defects in epitaxial metal/semiconductor contacts (Pond 1989a) and crystal variants like antiphase domains (Pond *et al.* 1984) in particular cases. It also led to improved understanding of the importance of surface and interface steps. For a clear and complete account of all the concepts of grain boundary and interface geometry introduced above see Chapter 1 in Sutton and Balluffi (1995).

Even when the geometry of possible interfaces is understood, as in principle it now is, it is necessary to calculate the energy of the boundary. This is important for determining the atomic relaxations that may occur to minimize the energy relative to the structures obtained crystallographically. These atomic details help to determine the physical and chemical properties of the interface.

Interface energy is also important for crystal growth and defect formation. For example, epitaxy occurs because the depositing atoms add on to the substrate so as to match it in symmetry and lattice parameter as well as possible. That is, the atoms on either side of the interface are in relatively good registry, so the interfacial energy per unit area is minimized. Moreover, the discussion in Section 2.3.11 and Fig. 2.13 showed that the energy of the film/substrate interface also plays an important role in determining which mode of epitaxial growth (the Frank and van der Merwe two dimensional 2D monolayer growth mode, the Volmer-Weber 3D nucleation mode or the Stranski-Krastanow mode of first 2D and then 3D growth) occurs.

Interface energy calculations

The equilibrium configuration of any interface is determined by minimizing the free energy. This is not an absolute minimization, since no defects larger than point defects occur in thermodynamic equilibrium. An unconstrained crystal energy minimization would, therefore, require the boundary to migrate out of the sample. The minimization is subject to crystallographic constraints appropriate to the particular interface under consideration. For example, in the case of epitaxial deposition, the crystal structure and the surface orientation of the substrate are fixed and the deposit grows with that (or those) crystal structure(s) and orientation(s) that minimize the interfacial energy.

There are several methods for calculating the energy and/or predicting the structure of coherent interfaces (Fletcher 1969) like low-energy grain boundaries and epitaxial heterojunctions. These are the geometrical or crystallographic, the Volterra dislocation (elasticity theory), the Frank and van der Merwe (van der Merwe 2001, Matthews 1979) and the variational methods.

The geometrical methods are those of the coincidence lattice, its generalized form the 'O' lattice and bicrystallography discussed above. The coincidence- and O-lattices yield the low energy orientations for interface planes between two crystals of known structures and relative orientations. They suggest low energy interfaces contain high area densities of coincidence sites.

The Volterra (elastic singularity) dislocation method is that used by Shockley and Read (1949, 1950) to treat small-angle grain boundaries (see Section 2.3.10). The Frank and van der Merwe method treated epitaxial heterostructures as two semi-infinite elastic solids meeting in the interface, which was represented by a periodic interaction potential. It was used to treat the growth of epitaxial films and predicted the occurrence of misfit dislocations (see van der Merwe 2002, and

Section 2.3.11). It was also used to deal with interfacial energies (van der Merwe and Ball 1975). Since misfit dislocations are known to occur, they can simply be treated using elasticity theory. That is, the interface can be regarded as a dislocation boundary analogous to a grain boundary. This approach was applied to epitaxial heterojunctions by Igarashi (1971).

The energy of the interface is expressed so its variation with relative orientation, alterations in the positions of individual atoms, etc. may be calculated using the variational method. For this type of calculation, Fletcher and co-workers (Fletcher and Adamson 1966, Fletcher 1967, Darling and Field 1973, Fletcher and Lodge 1975) pioneered the use of periodic analytical expressions for the interatomic potentials. They manipulated these expressions mathematically in reciprocal space by computer. They found certain relative orientations to correspond to local energy minima. Such orientations would then be expected to be e.g. the preferred low-interfacial-energy orientations for epitaxial nucleation. However, reliable predictions of epitaxial orientation relations, or realistic misfit dislocation geometries, would require the use of anisotropic elasticity theory and realistic interfacial interaction potential expressions.

4.5.3 Grain boundaries

Hitherto we have discussed grain boundaries in the abstract as examples of crystalline interfaces. Now we must discuss them as they occur in polycrystalline semiconducting materials with a view to understanding their effects on the operation of polycrystalline devices.

Polycrystalline semiconductors, especially polycrystalline thin-films, are extensively employed in devices. Both the GB structure and impurity segregation to it affect GB electrical properties. In microelectronic applications, individual grains and grain boundaries can produce variations in the material's electrical properties and influence device operation. Devices with dimensions comparable to the polycrystalline grain size are particularly affected. Thus, control of GB properties is especially necessary in nm-scale devices with active regions containing a few grains only.

Grain boundaries, whose core structures contain dangling bonds or contaminating impurities, can be expected to introduce deep levels in the band gap. Such GBs can be expected to be electrically active, i.e. they are likely to contain many minority carrier recombination centres.

In addition, grain boundaries in crystalline materials can provide paths for enhanced atomic diffusion. Impurities are also likely to segregate at grain boundaries (GB), which consequently affects many properties of a polycrystalline semiconductor. (For reviews, see e.g., Grovenor 1985, Seager 1985, Pizzini *et al.* 1989, Möller 1993). This can, sometimes, be beneficial. As discussed in Section 5.12.1, an example is the use of grain boundaries to 'getter' (collect) unwanted impurity atoms making them inactive. There are a group of important 'electroceramic' (sintered oxide semiconductor) devices including thermistors, varistors, and boundary layer capacitors. These have interfacially trapped charges at grain boundaries producing potential barriers which may be switched by applied voltage (varistors) or temperature (positive temperature coefficient of resistivity devices). Another form of 'defect engineering' has been developed to control properties like the critical voltage for switching, by controlled doping (see Section 5.12.4).

Thus it is important to understand GB structures and properties.

Core structures of grain boundaries and the Hornstra models

Hornstra (1958) showed that it was useful to construct ball-and-wire models of dislocations in covalently bonded diamond cubic structure semiconductors, to clarify the possible core structures (see Section 4.1.1). He then built ball-and-wire crystal models of grain boundary core structures to find how the bonds in the core could be accommodated with the least strain at the grain boundary (GB) interfaces (Hornstra 1959, 1960). He built models of structures that he had shown, using rotation matrices, would correspond to regular periodic GB structures which were also coincidence lattice planes (see Section 4.5.2) and constructed models for them. That is, he selected particularly significant tilt angles, about the rotation axes $\langle 110 \rangle$ and $\langle 100 \rangle$, for his models.

Particularly significant misorientations relative to the matrix are those of twins, related to that of the matrix by sub-symmetry operations. In the diamond structure, for example, the first order twin orientation is obtained by rotating the structure through 180° (a twofold rotation) about a $\langle 111 \rangle$ (threefold symmetry) axis. It can also be described as the 70°32′ $\langle 110 \rangle$ symmetrical tilt boundary. The atom sites in this twin are mirror images in the twin plane of those in the matrix (Fig. 4.100). It can be seen that this model could be constructed without dangling or strained bonds indicating that it is a particularly low-energy interface without interface energy states in the forbidden band gap. This conclusion is confirmed by the general lack of strong electrical or gettering effects in first order twin boundaries. The twinning operation can be repeated by rotating first about one $\langle 111 \rangle$ axis and then about another to produce second and higher order twins. The second order twin boundary is shown in two forms in Fig. 4.102.

As remarked previously, twin boundaries are of particularly low energies so twins occur profusely as included grains in the largely single crystal matrices of ingots produced from the melt when the growth conditions are not yet fully optimised. The coherent interface boundaries of twins, especially first order twins, are of such low energy that they tend to be planar over large areas. Haasen (1957a) first reported profuse twinning in an otherwise monocrystalline ingot of InSb (such twinning was shown in Fig. 4.99). Later similar twins were found in most III-V compounds as well as e.g. in bulk CdTe as shown in Fig. 4.103.

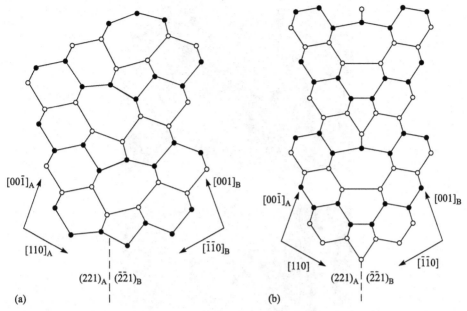

(a)   (b)

Figure 4.102   The second order (111) twin, i.e. the symmetrical boundary for a rotation
$\theta = 38°57' = \arccos 7/9$ tilt about [110]. Two alternative possible low strain ball-and-wire
models are shown. (a) has the necessary density of dislocations accommodated in a zig-zag
arrangement while (b) has overlapping 'double dislocations' in the boundary plane. Open
and solid circles represent atoms at heights 0 and $z/2$ above the plane of the drawing.
(After Hornstra 1959. Reprinted from *Physica*, **25**, Models of grain boundaries in the diamond
lattice I. Tilt about ⟨110⟩, pages 409–22. Copyright 1959, with permission from Elsevier.)

Similarly, bulk LEC (liquid encapsulated Czochralski) semi-insulating InP is
prone to twinning (Fig. 4.99). To produce semi-insulating, i.e. very high resistivity
InP, the material can be grown with a relatively heavy doping concentration of either
Cr (as in the example shown in Fig. 4.99) or Fe. Fig. 4.104 contains two scanning
electron microscope images taken with a photodiode set to near the specimen so as to
receive a large fraction of the cathodoluminescence emitted under electron beam.
When this is video displayed as signal, an emissive cathodoluminescence (ECL)
picture is obtained. It can be seen in Fig. 4.104a that, as in the case of CdTe shown in
Fig. 4.103, the short, lateral twin boundaries, which contain many twinning dislocations, are much brighter than the long, vertical, low energy, interface-state-free
coherent first order twin boundaries.

Small- and large-angle grain boundaries

Grain boundaries (GBs) in general can be classed as small- or large misorientation
angle types. Small angle (<10°) GBs were shown to consist of walls of distinct
dislocation lines (Section 2.3.10). The earliest quantitative confirmation of any

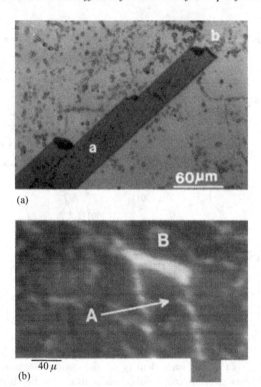

(a)

(b)

Figure 4.103 (a) First order twin included grain in CdTe, marked (a), intersecting a {111} Cd surface and terminated by a lateral twin boundary, marked (b) lying on the planes {110}–{114}. Etched with Nakagawa's reagent (Nakagawa, K., Maeda, K. and Takeuchi, S. *Appl. Phys. Letters* **34** (1983) 691). (After Durose and Russell 1990.) (b) SEM EBIC image of another twin band in CdTe (marked A) which is terminated by a lateral twin boundary (marked B). B appears in much brighter EBIC contrast than do the side, coherent boundaries of the twin. [After Durose and Russell 1987. In *Microscopy of Semiconducting Materials 1987*. Conference Series No. 87 (Bristol: Institute of Physics), pp. 327–32.]

prediction of dislocation theory came through observations on such grain boundaries in Ge. While the Burgers vector content of a large angle GB can be found, individual dislocations cannot be distinguished in such interfaces, as was discussed in Section 2.3.10.

Hornstra (1959, 1960) showed that the atomic core structures of large angle grain boundaries in the diamond structure could be usefully represented by building ball-and-wire models as well as by applying crystallographic matrix methods as was discussed above. His first grain boundary paper (1959) concerned ⟨110⟩ tilt boundaries. It can be shown, by consideration of the rotation matrices, that, due to the symmetry of the f.c.c. lattice-based diamond structure, a single orientation relation can be represented by a number of different rotations. In particular, the first order twin in the diamond structure, which can be described as related to the

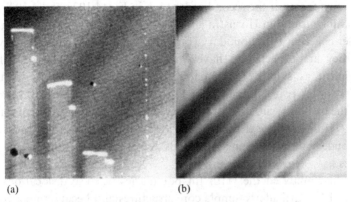

(a)   (b)

Figure 4.104 Emissive CL (ECL) images of first order twin included grains in semi-insulating LEC Czochralski grown InP:Fe. (a) The vertical bright bands are twins in bright contrast due to channelling and the coherent vertical twin boundaries are much less visible than are the lateral twin boundaries (the roughly horizontal bright lines across the tops of the twins, compare Figure 4.103b). A vertical row of bright dislocations seen end-on appears at the right of the image as well as a few bright twinning dislocations along the vertical coherent boundaries. (b) Another region showing numerous alternating parallel bands of twin and matrix. (After Holt and Salviati 1993.)

matrix crystal by a 180° rotation about a $\langle 111 \rangle$ direction, can also be represented by a 70°32′ rotation about a $\langle 110 \rangle$ direction (see Fig. 4.100 and the discussion thereof). Similarly, second order twinning (see Fig. 4.102), corresponds to a rotation of $180° - 2 \times 70°32′ = 38°56′$ and third order twinning to a rotation of $3 \times 70°32′ - 180° = 31°36′$ about a $\langle 110 \rangle$ axis. Thus the $\langle 110 \rangle$ symmetrical tilt boundaries include the coherent interfaces of the integral order twins in the diamond structure. The wide measure of agreement, at least on general features, concerning the twin interfaces reached by independent geometrical analyses by Ellis and Treuting (1951), Kohn (1958) and Hornstra (1959) led to considerable confidence in this approach.

Hornstra constructed ball-and-wire models of these grain boundaries in order to find how the bonds could be accommodated with least strain at the interface and published drawings giving projected views of the structures so produced. He made use in this work of the results of his previous ball-and-wire model analysis of the structures of the cores of dislocations in the diamond structure (Hornstra 1958, see Section 4.1.1).

For tilt angles less than $\theta = 26°32′ = \arccos 17/19$ the dislocations in the boundaries are sufficiently widely separated so that, in the ball-and-wire models, the cores do not interfere, but for larger tilts, the dislocations touch. For larger tilts the increased dislocation density can be accommodated with low strain in the models either by arranging them in a zig-zag array or by allowing core overlap and rearranging the bonding into a different 'double dislocation' arrangement (see e.g. the two forms of the second order twin in Fig. 4.102a and 4.102b) of which the details will be found in Hornstra (1959). Neither the first- nor either form of the second-order coherent twin boundary (Figs. 4.100 and 4.102) contains any dangling bonds so they are expected to have little or no electrical activity. This agrees with experiment as shown, for example, by the low contrast .(visibility) of the coherent first order twin boundaries in Figs. 4.103b and 4.104a. Hornstra gave an account of the ball-and-wire models for the regular tilt boundaries in all ranges of tilt angle. The ⟨110⟩ symmetrical tilt boundaries could not all be constructed without dangling bonds, but when dangling bonds were necessary, they were present in low densities.

Hornstra first analysed the ⟨110⟩ tilt boundaries because dislocations with this line direction had particularly simple core structures and because most accidentally grown grain boundaries had been observed to be of this type, especially the first order twins (Figs. 4.99, 4.103 and 4.104). Most work on intentionally grown bicrystals had concerned ⟨100⟩ tilt boundaries, however, so Hornstra (1960) analysed these GBs also. A typical structure obtained in this work is that in Fig. 4.105. In this case again, no dangling bonds were found to be needed to construct the boundary. It was found that for some ⟨100⟩ tilt boundaries also, alternative structures were equally possible mechanically. Thus the boundary with [001] tilt axis, (010) median plane and a tilt angle of $\theta = 22°37'$ could be constructed using either edge dislocations or 45° (angle between the line direction and Burgers vector) dislocations.

Impurities and grain boundaries

Early studies of GB diffusion were enlightening in the present context, since they were interpreted in terms of crystal models. Diffusion is often more rapid along grain boundaries than through defect free material so the impurity or dopant atoms penetrate more deeply along grain boundaries (GBs) than elsewhere. Early studies of this type were reviewed by Amelinckx and Dekeyser (1959). This deeper penetration along GBs was observed in early Si and Ge (Karstensen 1957, Queisser *et al.* 1961). A consequence is that the presence of even small-angle 'mosaic' boundaries leads to the formation of non-planar junctions with generally disastrous results for device operation.

This deeper penetration along grain boundaries was observed by diffusing donors into uniformly doped *p*-type material. Staining techniques revealed *p-n* junctions to give the positions of the isoconcentration boundaries where the donor density equals the original acceptor concentration. Queisser *et al.* (1961) used this technique to measure the velocity of advance of the 'spike' in the diffusion front at small angle grain boundaries due to the faster diffusion along the dislocations of which such boundaries consist, as shown in Fig. 4.106.

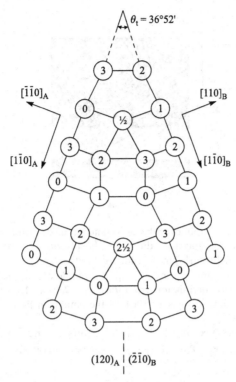

Figure 4.105   Ball-and-wire model of the symmetrical grain boundary with tilt axis [001] and an angular misorientation between the grains of 36°52'. The height of the atoms above the plane of the diagram is given by the numbers in the circles in multiples of $a/4$ where a is the lattice parameter. (After Hornstra 1960. Reprinted from *Physica*, **26**, Models of grain boundaries in the diamond lattice. II. Tilt about ⟨001⟩   and theory, pages 198–208. Copyright 1960, with permission from Elsevier.)

Queisser *et al.* (1961) based their analysis of such GB diffusion on the proposition that it adds a total flux, F, of solute atoms through a plane, y = constant, perpendicular to the dislocations given by

$$F = -D'\, W_D \partial c/\partial y \qquad (4.30)$$

where D' is the (bulk) diffusion coefficient and $\partial c/\partial y$ is the concentration gradient. The grain boundary diffusion enhancement factor, W_D, was considered to be due to the higher concentration of impurity atoms in the vicinity of the boundary times an additional enhancement factor for a higher diffusion coefficient within the boundary (Queisser *et al.* 1961). It can be written as:

$$W_D = b\,\theta\, \exp(U_D/kT) \qquad (4.31)$$

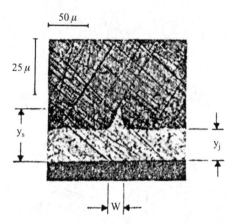

Figure 4.106 Light micrograph of a bevelled and stained surface of a silicon bicrystal showing the *p-n* junction produced by diffusing phosphorus upward from the lower edge. The diffusion front (the light-dark junction line) exhibits a 'spike' due to enhanced diffusion along the grain boundary running vertically up the middle of the specimen. [After Queisser *et al.* 1961. Reprinted with permission from *Physical Review*, **123**, 1245–54, 1961 (http://publish.aps.org/linkfaq.html). Copyright 1961 by the American Physical Society.]

where b is the Burgers vector of the dislocations in the boundary and θ is the tilt angle of the boundary. The dislocation activation or 'favouring' energy U_D is given by:

$$U_D = U_i + U_v + U_j \tag{4.32}$$

where U_i is the maximum value of the Cottrell first order elastic interaction energy of the impurity and the dislocations, i.e. the binding energy, which is responsible for the enhanced local grain boundary concentration. $\exp(U_v/kT)$ is the factor by which vacancies are concentrated in the dislocation core and U_j is the amount by which the potential energy barrier for an atom jumping into a vacancy is lowered in the dislocation core.

Queisser *et al.* (1961) derived the spike velocity equation:

$$(v \tan \varphi)^{-1} = (W_D \sin^2 \varphi - W_0)/2D' \tag{4.33}$$

where v is the velocity of advance of the *p-n* junction spike along the boundary, φ is the semi-angle of the spike and W_0 is the equivalent width of the grain boundary core. That is, the tilt boundary contains extra impurity atoms per unit area equal to those in a slab of bulk material of thickness:

$$W_0 = [b^2 \exp(U_i/kT)]/D \tag{4.34}$$

where $D = b/\theta$ is the spacing of the dislocations in the boundary. Experimental measurements of v, φ and D' yield values of W_0 and W_D, using the spike velocity Equation (4.33).

Queisser *et al.* (1961) measured the diffusion of phosphorus into Si at 1050° and 1200° C and obtained values of U_D of 1.5 and 1.6 eV, respectively. The central problem of interpreting data on diffusion down dislocations or, as in this case, down grain boundaries, is constructing a model for the core of the defect so that theoretical values for the impurity, vacancy and jump favouring energies U_i, U_v and U_j can be obtained for comparison with the experimental value of their sum U_D (Equation 4.32). For this, Queisser *et al.* considered Hornstra's ball-and-wire models. Values of W_0, W_D and U_i were estimated using Cottrell's model of impurity atmospheres around dislocations. Their estimates of several factors involved led to a value of $U_i(P) = 1.1$ eV. This is so large that the impurity concentration enhancement factor $\exp(U_i/kT)$ for the site at the centres of the dislocations in the GB is about 80 times as great as for the next nearest position. Thus, practically all the contribution to W_0 and to W_D will come from the column of sites nearest to the centre line of the dislocation core. This approach led to a value of W_0 in agreement with experiment. The value of U_D was more difficult to determine because it depends also on the unknown quantities U_v and U_j. However, their highest estimate made $U_i = 1.5$ eV which is equal to the experimental value of U_D.

The treatment of Queisser *et al.* (1961) made it clear that the strong segregation of impurities to the dislocation lines in small-angle GBs means that unless the most stringent measures be taken to avoid or control such segregation, the electrical properties that might be observed will not be the intrinsic properties of such boundaries.

Grovenor (1985) reviewed the theoretical concepts and experimental information on the structure and electronic properties of semiconductor grain boundaries and the segregation of impurities and dopants to the boundaries. Impurity segregation is thought to influence strongly the electrical properties. Grovenor (1985) suggested the possibility that all the observed potential barrier heights, and associated properties, are solely determined by segregated impurities, not by the intrinsic structure of the boundaries. Grovenor (1985) emphasized that in order to elucidate the effect of contamination, strictly uncontaminated grain boundaries must be obtained. They can then be contaminated in a controlled manner. Grain boundaries typically provide paths for enhanced atomic diffusion as discussed above. Such diffusion along GBs crossing a semiconductor device junction can cause changes in the junction. Leamy *et al.* (1982) employed EBIC to study the effect of GBs on *n*-channel (polycrystalline Si) thin-film transistors. They concluded that GBs impede channel current flow and facilitate enhanced diffusion paths for source and drain dopants. Swaminathan *et al.* (1982) studied As diffusion in polycrystalline Si films by using RBS to measure the As concentration profiles. They found that the latter is controlled by both the lower As diffusion in the grains and substantially greater diffusion in GBs (Swaminathan *et al.* 1982). Grovenor (1985) tabulated the measured activation energies for GB diffusion of various dopants in polycrystalline Si. In the temperature range around 1000° C, the activation energy (for e.g. P and As) is about 2 eV. Grovenor (1985) also emphasized that in thin-film structures (e.g. solar cells), incorporating different polycrystalline compound semiconductors in contact, possible substantial interdiffusion along GBs may modify their electrical properties.

Wong *et al.* (1985) studied equilibrium As segregation to grain boundaries in polycrystalline Si. They found that increasing As segregation results in increasing resistivity. They proposed a model of segregation to grain boundary sites that produce trapping states, like dangling bonds, and their consequent removal, as well as segregation to other boundary sites to form a degenerately doped interfacial layer. Wong *et al.* (1985) suggested that the segregation results in a very high density of charge trapped in GBs, and that the main cause of the increased resistivity is due to carrier scattering by this sheet of positive charge.

Maurice and Colliex (1989) showed that the increased electrical activity of Si grain boundaries after heat treatment is due to Cu and Ni precipitates. This was demonstrated on a $\Sigma25$ bicrystal, and the role of these metal impurities was compared to that often ascribed to oxygen (Maurice and Colliex 1989).

Seifert *et al.* (1993) employed EBIC and TEM to study the recombination at grain boundaries in polycrystalline silicon. They observed very weak EBIC contrast in grain boundaries with no dislocations or precipitates. They concluded that the recombination depends on the dislocation density in the boundary. Decoration by precipitates contributes to that activity (Seifert *et al.* 1993).

Kittler *et al.* (1995) used EBIC to study the interaction of iron with a grain boundary in boron-doped polycrystalline Si. They found substantial iron depletion around the GB (within 20 μm of both sides of the GB). The iron concentration was reduced from 1.4×10^{14} cm^{-3} to 3.5×10^{13} cm^{-3}, implying gettering of about 4×10^{11} cm^{-2} iron at the GB.

An example of the use of grain boundaries for impurity gettering and oxygen precipitation in polycrystalline sheet silicon was presented by Lu *et al.* (2003). By employing EBIC, DLTS, FTIR, and optical microscopy, they confirmed the preferential precipitation of transition metals and nitrogen along the grain boundary (Lu *et al.* 2003). For instance, the GB gettering of transition metals produced the higher collection efficiency (brighter) halos seen along the GBs in the EBIC image in Fig. 4.107. This EBIC image also reveals bright contrast in the smaller grains, indicating that the GBs act as gettering sites and that the carrier lifetimes are greatly enhanced thereby (Lu *et al.* 2003).

GB electrical properties may depend on both the GB structure and impurity contamination. Chen *et al.* (2004) studied the influence of GB type and impurity contamination on the recombination at GBs in polycrystalline Si. For this, EBIC was employed to assess GB recombination. The GB types were found by electron back-scattered diffraction (EBSD) analysis (Chen *et al.* 2004). They concluded that the recombination strength of uncontaminated GBs was weak. The GB type had no substantial influence on recombination activity, whereas with increasing Fe contamination the recombination activity of the GBs increased significantly. In addition,

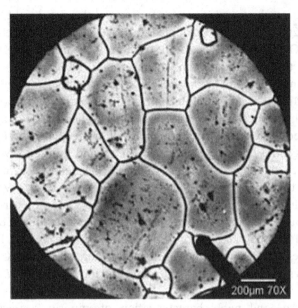

Figure 4.107   A room temperature EBIC image showing higher collection efficiency, and, therefore, bright halos along the grain boundaries and bright contrast throughout the smaller grains. (After Lu *et al.* 2003. Reprinted with permission from *Journal of Applied Physics*, **94**, pp. 140–4. Copyright 2003, American Institute of Physics.)

for the same degree of the contamination, the recombination strength was greater for the high-Σ than for low-Σ GBs (Chen *et al.* 2004). These observations were explained by the gettering properties of GBs, i.e. random GBs getter the most strongly, and $\Sigma 3$ {111} GBs, little if at all (Chen *et al.* 2004).

As mentioned above, grain boundaries generally provide paths for enhanced diffusion. However, Stolwijk *et al.* (1996) studied gold diffusion into silicon and observed retarded impurity incorporation in polycrystalline material, compared to that in monocrystals. Stolwijk *et al.* (1996) suggested a model, of high impurity mobility in the lattice and strong segregation to grain boundaries. According to Stolwijk *et al.* (1996) this model would apply in cases of (i) high impurity diffusivity through the lattice, (ii) insignificant diffusion along the grain boundaries, and (iii) strong impurity segregation to the grain boundaries. Yan *et al.* (2001) employed TEM and EDS to investigate the effect of O on the diffusion of S in polycrystalline CdS/CdTe heterojunctions. They observed that S diffusion along GBs was enhanced in the absence of oxygen, but reduced in the presence of oxygen. In the latter case, it was suggested that Cd-O bond formed in the presence of O during growth of CdTe, prevented GBs enhancing S diffusion along GBs (Yan *et al.* 2001).

To obtain reproducible GB properties, Kamiya *et al.* (2002) employed oxidation and annealing to control the GB potential barriers in polycrystalline Si. In this case, selective GB oxidation at around 700° C, followed by annealing at about 1000°C, results in an increased potential barrier height and resistance (Kamiya *et al.* 2002). This they attributed to structural changes (at the grain boundaries) in the Si-O network and the competition between surface oxygen diffusion and oxidation from the GBs in the grains (Kamiya *et al.* 2002).

Dangling bonds in grain boundaries in semiconductors introduce deep levels, which act as efficient recombination centres, in the energy band gap. As discussed in Section 5.5.8, hydrogen passivation, i.e. neutralization of the dangling bonds by atomic hydrogen attachment, is often employed for reducing the number of such recombination centres (e.g., see reviews by Pankove and Johnson 1991, Pearton *et al.* 1992, and Nickel 1999). Hydrogen passivation of extended defects in semiconductor devices substantially improves their performance. Seager *et al.* (1979, 1980, 1981) demonstrated that hydrogenation substantially reduces minority carrier recombination at grain boundaries and improves diode characteristics. SEM EBIC and CL are often employed to study passivation. For instance, cross-sectional EBIC examinations of samples, treated with hydrogen plasma for 1 h at 310°C, revealed that grain boundaries were passivated to a depth of 100 μm (Krüger *et al.* 2000, Vyvenko *et al.* 2000).

For the micro- and nano-characterization of impurity segregation at grain boundaries in semiconductors, there are techniques that provide complementary information on electronic properties, composition, structure and chemistry. These include electron microscopy techniques (Section 3.6–Section 3.8) including SEM-EBIC (Section 5.12.4) and REBIC (Section 5.10.4), scanning Auger electron spectroscopy (AES) and secondary ion mass spectrometry (SIMS) techniques (e.g. Kazmerski 1982, 1993, 1994, Kazmerski *et al.* 1980, 1984), and specialized scanning probe microscopy techniques (Kazmerski 1993).

A combination of EBIC (for assessing the GB recombination strength) and EBSD, the electron backscattered diffraction technique, for establishing GB type, is a particularly useful SEM technique for investigating grain boundaries in semiconductors (e.g. Chen *et al.* 2004). An additional technique for determining the bonding state of the segregated species is x-ray photoelectron spectroscopy (XPS). Electrical measurements (e.g. DLTS and resistivity) can be used to elucidate the effect of impurity segregation on the GB electrical properties. Some of these techniques (i.e. AES and SIMS) also allow obtaining three-dimensional compositional and chemical information during a single analytical run (for details, see Kazmerski 1994, and references therein). Such volume mapping for a grain boundary is schematically illustrated in Fig. 4.108. This acquires specific digital (indexed) information, which is stored in a computer, at each point of a microvolume, during a single depth-resolved measurement, employing control of the beam and analyser (Kazmerski 1994). This subsequently allows the display of the desired information for any point, or along any line, or on any plane, within the analysed microvolume (Kazmerski 1994). Such AES maps for a Si sample with localized impurity segregation on grain boundaries are shown in Fig. 4.109.

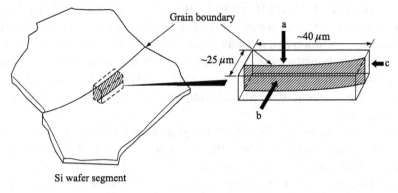

Figure 4.108 Illustration of three-dimensional (volume-indexed) surface analysis mapping method showing three possible viewing directions. (After Kazmerski 1994.)

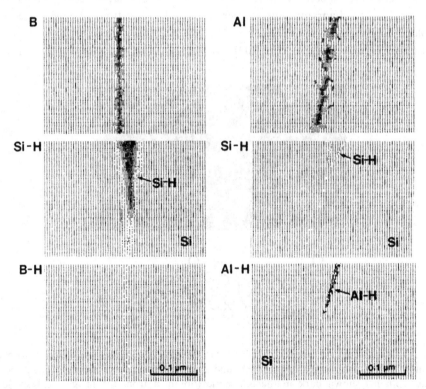

Figure 4.109 Volume-indexed AES maps for a Si grain boundary with various segregated impurities. (After Kazmerski 1994.)

Kazmerski and Dick (1984), using EBIC and SIMS, studied the relationship between the chemistry and composition of polycrystalline Si grain boundaries and charge carrier generation and recombination. They demonstrated that oxygen segregated to Si grain boundaries during thermal processing of the materials or devices degraded solar cell characteristics (Kazmerski and Dick 1984). They also showed that the device characteristics could be improved by hydrogen passivation of the grain boundaries.

Using atomic-resolution Z-contrast imaging in a STEM (see Pennycook and Jesson 1991, Pennycook *et al.* 1999), the atomic arrangement of extended defects in materials, and specifically impurities segregated in a grain boundary, can be imaged directly. Chisholm *et al.* (1998) employed this technique to directly image As segregated in specific atomic columns in a Si GB (see Fig. 4.110). By combining experimental image intensity analysis and theoretical calculations, they demonstrated that As atoms are segregated in the form of isolated dimers (Chisholm *et al.* 1998).

Chisholm *et al.* (1998) interpreted the observed structure as follows. The GB core structure comprises a continuous sequence of dislocations: a perfect edge dislocation (labelled 1 in the diagram) and two perfect mixed dislocations (2,3) arranged as a dipole, followed by the same sequence (1′, 2′, 3′) mirrored across the boundary plane (Chisholm *et al.* 1998). In the ⟨001⟩ projection, the dislocations appear as a

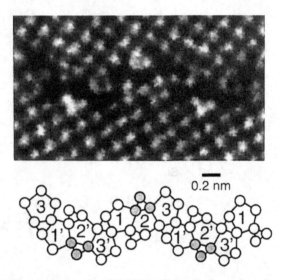

Figure 4.110 Z-contrast image of a 23°⟨001⟩ tilt grain boundary in Si doped with As. The more intense periodic features in the boundary correspond to As segregates. The projected atomic columns shown dark in the schematic (obtained from the image) correspond to brighter (periodically repeated every 0.69 nm) features in the image due to segregated As. [After Chisholm *et al.* 1998. Reprinted with permission from *Physical Review Letters*, **81**, pp. 132–5, 1998 (http://publish.aps.org/linkfaq.html). Copyright 1998 by the American Physical Society.]

connected array of pentagonal and triangular arrangements of atomic columns (Chisholm *et al.* 1998).

HRTEM observations of GB core structures

When TEMs first became capable of resolving atomic structure, by what is now called high resolution transmission electron microscopy (HRTEM, see Section 3.6.1), it was immediately applied to observe the structure of a GB in germanium by Krivanek *et al.* (1977). Under their experimental conditions the dark areas in the images correspond to columns of atoms and the light areas to channels through the crystal. Krivanek *et al.* analysed a grain boundary with a ⟨110⟩ tilt axis and a misorientation of 38°56.6′, i.e. a second order twin boundary with $\Sigma = 9$. However, its structure was complicated by intersections with a number of first order twin planes. Thus the structure of this boundary is comparable with Hornstra's models (see above) only over a short length. Fig. 4.111 is a micrograph of the boundary.

The rows of white dots leading up to the boundary in the grains above and below it all represent the normal tunnels of six-atom rings in Ge. At the boundary there are 15 distinctly larger white dots, which Krivanek *et al.* interpreted as 7-atom rings. On this assumption they built the model of tubular connecting rods photographed in [110] projection in Fig. 4.112a. Its agreement with the corresponding section of the HRTEM image in Fig. 4.112b is clarified by removing the irrelevant sharp detail from the model by photographic blurring as shown in 4.112c.

The model of Fig. 4.112a is shown more clearly in Fig. 4.113. A 38°57′ rotation about a ⟨110⟩ axis, resulted in a $\Sigma = 9$ coincidence lattice so 1/9th of the sites of the two orientations coincide. The boundary in the model of Fig. 4.113 runs along a direction that enables it to contain a maximum density of coincidence lattice sites.

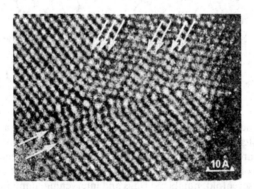

Figure 4.111 Transmission electron micrograph of an area of a thin Ge film containing a complex ⟨110⟩ 39° tilt boundary running almost horizontally across the middle of the picture. The structure is complicated by intersections with a number of first order twin planes, which are indicated by arrows. This structure is interpreted in Fig. 4.112. [After Krivanek *et al.* 1977. Reprinted with permission from *Philosophical Magazine*, **36** (1977), pp. 931–40; Taylor & Francis Ltd., the Journal's web site: http://www.tandf.co.uk/journals.]

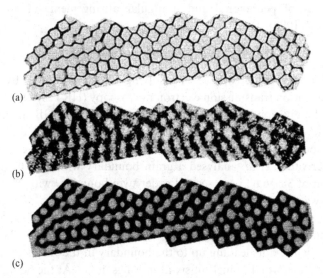

(a)

(b)

(c)

Figure 4.112 (a) A model of the structure of the boundary in Fig. 4.111, assuming the presence of 7-m and fivefold rings. (b) The portion of Fig. 4.111, containing the grain boundary. Compare (b) with (c) a photographically blurred image of the model of Fig. 4.111. [After Krivanek *et al.* 1977. Reprinted with permission from *Philosophical Magazine*, **36** (1977), pp. 931–40; Taylor & Francis Ltd., the Journal's web site: http://www.tandf.co.uk/journals.]

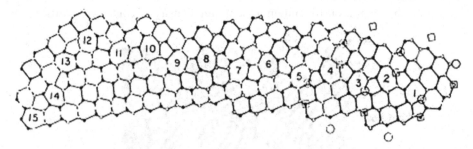

Figure 4.113 The model for the boundary of Fig. 4.112a. The empty and full circles indicate atoms occurring in alternate (022) planes, i.e. above and below the plane of the diagram. The larger circles and squares at the right-hand end of the diagram mark the two interpenetrating f.c.c lattices. The line of intersection of the boundary with the plane of the diagram is marked by a row of distorted sevenfold, numbered rings and intervening, unnumbered fivefold rings of atoms. The diamond structure in undeformed regions is seen as a set of interconnected sixfold rings in this [011] projection. [After Krivanek *et al.* 1977. Reprinted with permission from *Philosophical Magazine*, **36** (1977), pp. 931–40; Taylor & Francis Ltd., the Journal's web site: http://www.tandf.co.uk/journals.]

A pair of one sevenfold ring (numbered in the figure) and one fivefold ring separate neighbouring coincidence sites. However, the boundary only runs in this plane from the sevenfold rings 1 to 5 before changing alignment due to interaction with other twin planes. Complete details of this boundary and its interpretation will be found in the paper of Krivanek *et al.* (1977). The model of Fig. 4.113, from ring 1 to 5, should be compared with Hornstra's model for the boundary for a 38°57′ tilt about a ⟨110⟩ axis in Fig. 4.102a, which predicted the alternating seven- and five-fold rings of atoms in the boundary core. This is strong confirmation of the value of Hornstra's approach.

This work was followed by very many other HRTEM and Z-contrast FEGSTEM studies resolving the atomic structures of grain boundaries in Ge and Si. For reviews see Thibault *et al.* (1991) for the HRTEM literature and Chisholm *et al.* (1999) for the Z-contrast work.

Crystallographic polarity and GBs in the sphalerite structure

As discussed in Section 4.1.3, the non-centrosymmetry of the sphalerite and wurtzite structures of many important semiconducting compounds, doubles the number of core types of defects possible as compared with the diamond cubic structure of the important semiconducting elements. That is, to each type of dislocation in the elements there are two polar opposite types of dislocation in the sphalerite structure (Section 4.1.3). Similarly, to each type of GB in the elements there are two physically distinct polar opposite types in the sphalerite structure (Holt 1964).

Although the electrical and optical properties of GBs in sphalerite structure compounds were studied with an eye to possible device applications (see Section 5.10.1) no such application ever became practical. However, the results of ball-and-wire modelling will be briefly reviewed in the interests of completeness and because polarity makes possible two distinct types of GB which should have markedly different properties.

Hornstra (see Section 4.5.3) analysed the commonly observed grain boundaries with tilt axes ⟨100⟩ and ⟨110⟩. These are both rotational symmetry axes in the diamond structure but are roto-inversion axes in the sphalerite structure. Thus symmetry in the diamond structure makes grain boundaries with two different tilt angles equivalent for each of these axes. That is, for tilts about the fourfold-rotation ⟨100⟩ axis, grain boundaries with tilts of θ and $\Theta = \theta + 90°$ are indistinguishable, but in the lower-symmetry sphalerite structure, the two tilts result in physically distinct boundaries, differing by the polar inversion of one of the grains. Similarly, in the diamond structure, θ and $\Theta = \theta + 180°$ tilts about the twofold-rotation ⟨110⟩ axis, produce indistinguishable grain boundaries. However, in the sphalerite structure these boundaries differ by a polar inversion of one grain.

The simplest example is the first order {111} twin in the sphalerite structure, shown in Fig. 4.114. A symmetric boundary tilted about a ⟨110⟩ axis to a

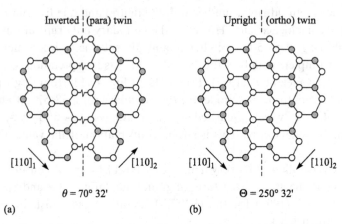

Figure 4.114 The [110] tilt boundaries with tilts of (a) $\theta = 70°32'$ and (b) $\Theta = \theta + 180°$. The zig-zag lines indicate wrong A-A or B-B bonds in the AB semiconducting compound. The Θ boundary is identical with the (111) boundary formed by a 60° twist about the $[1\bar{1}1]$ axis. It is the commonly observed coherent, upright first-order twin with Friedel index $\Sigma = 3$. The θ boundary is the inverted first-order twin with Friedel index $\Sigma = -3$. The rotation of 70°32′ is the first-order inversion twinning operation and the rotation of 180° about [110] is a polarity-reversal (antiphase) operation. (After Holt 1964.)

misorientation of $\theta = 70°32'$ is the inverted or para twin. A misorientation of $\Theta = \theta + 180° = 250°32'$ produces the upright or ortho twin. In the diamond structure the atoms are all the same, the two tilts are equivalent and the two boundaries are indistinguishable. In the sphalerite structure the upright twin, with tilt $\Theta = 250°32'$ and Friedel index $\Sigma = 3$, in Fig. 4.114b, has all correct (A-B) and unstrained cross-boundary bonds and so is of low energy like the (only) first order twin in the diamond structure. This can be transformed into the inverted, polarity-reversal or 'para' twin with tilt $\theta = 70°32'$ and index $\Sigma = -3$, by interchanging the occupation of all the sites in one of the grains, that on the right in the diagrams of Fig. 4.114. This inverted boundary has all wrong (A-A) or (B-B) cross-boundary bonds and is of higher energy. The interface of the inverted (upright) twin is a mirror (mirror inversion) plane.

There is some confusion in terminology in the literature. The first report of the observation of 'inversion twins' in wurtzite structure BeO by Austerman and Gehman (1966) described them as involving 'exact inversion of the sense of polarity across a . . . boundary' apparently following Aminoff and Bromé (1931, Z. Krist. 80, p. 355, as quoted by Austerman and Gehman). Such a boundary would now be better described as a (pure) antiphase boundary (APB) as will be discussed in the next section. Similarly, Heiland and Kunstmann (1969) reported 'inversion twins' in wurtzite structure ZnO. Their experimental evidence is the inverse polar etching of areas on both the facing surfaces created by (0001) cleavage shown in Fig. 4.115. Only the polarity, not the orientation, is changed between these domains.

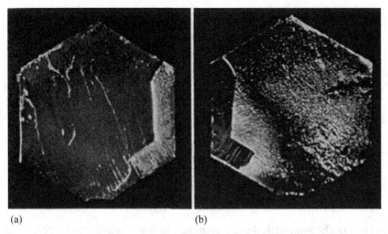

(a)   (b)

Figure 4.115 Optical reflection micrographs of two matching cleavage surfaces showing polarity reversal domains in wurtzite-structure ZnO revealed by the different etching of the (0001) Zn and (000$\bar{1}$) O regions of the surface in 6% HCl. In (a) the greater part (matrix) exhibits the smooth etching behaviour of the Zn surface, the smaller the pitting of the O surface. (b) The matching cleavage face exhibits the reverse etching behaviour. Observations of the sign of the piezoelectric voltage confirmed that the polar c-axis perpendicular to these surfaces was reversed between the matrix and the domain. (After Heiland and Kunstmann 1969. Reprinted from *Surface Science*, **13**, Polar surfaces of ZnO crystals, pages 72–84. Copyright 1969, with permission from Elsevier.)

Polar pairs of grain boundaries in the sphalerite structure can be of strongly polar character like that in Fig. 4.115. Such boundaries can occur in an all A atom form as shown in this diagram or in an all B form, i.e. one in which only B atoms are connected by wrong bonds and/or have dangling bonds. That is, they can occur in two polar forms that are the equivalent of the α- and β-forms, i.e. the A(s) or B(s) forms of dislocations in the sphalerite structure (see Section 4.1.3). These α- and β-like forms of GB are to be expected to be relatively strongly electrically active and distinctly different in their effects.

The form of boundary shown in Fig. 4.116b, the Θ boundary, is non-polar as it contains equal numbers of A and of B (black) dangling bonds and it is possible that the two types of dangling bonds will tend to compensate to some extent, e.g. acting as donor and acceptor centres. Its electrical effects should be much less than those of the θ boundaries shown in Fig. 4.116a. There are not two distinct polar forms of such boundaries.

Only a few early studies of the electrical effects of grain boundaries in sphalerite structure materials have been published and these will be discussed in Section 5.10.

Interchanging the occupation of the sites, black-for-white and white-for-black, in one grain transforms one polar form of a boundary into the other (e.g. the boundary type shown in Fig. 4.116a into that in Fig. 4.116b). The interchange of site

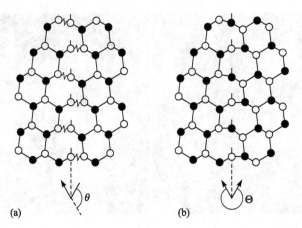

(a)   (b)

Figure 4.116   (a) The symmetrical grain boundary in the sphalerite structure formed by a tilt of θ = 129°31′ about [110]. The zig-zag lines indicate the wrong like-like bonds. All the wrong and dangling bonds (short vertical lines) involve one kind of atom (white, A atom, in this diagram) only. (b) The symmetrical boundary formed by a tilt Θ = θ + 180° = 309°31′ about [110]. There are no wrong bonds and equal numbers of dangling bonds from A and B (white and black) atoms. (After Holt 1964.)

occupation can be described as the polarity-reversal or antiphase operation. The term antiphase relates to a supposed site-occupation function whose sign (phase) indicates the type of atom occupying the site. Thus +ve sites, say, would be those occupied by A atoms and −ve sites by B atoms of the AB compound. These signs would be those of the net charges in the Wigner-Seitz cells containing the atoms, A ion positive, B ion negative, if the bonding were ionic. The antiphase operation would reverse the sign of the function at all sites and thus indicate the interchange of the A and B atoms.

Antiphase boundaries

If the polarity of one grain in a sphalerite-structure material is reversed this transforms one polar form of grain boundary to the other. This suggests the possibility of reversing the polarity in one region relative to the other without any misorientation between them. Such volumes can be described as (pure) polarity reversal or antiphase domains. Between domains in antiphase relation, the A and B (lower and higher valence) atoms interchange occupation from one f.c.c. sublattice (or h.c.p. sublattice, in the case of the wurtzite structure) to the other. In other words, in the sphalerite structure, the two atoms of the basis unit interchange occupation between the 000 and the 1/4,1/4,1/4 sites. Antiphase is a metallurgical (materials) term. Many physicists (fiercely) prefer the 'polarity reversal' description. The term 'antiphase' can be understood by reference to a fictious occupation function. This is of opposite sign for A and B atom occupation of a site. Thus a phase reversal of this function represents interchange of the site occupation.

Evidence of surface and bulk (pure) polarity reversal appeared early. There were observations of reversed polar etching behaviour (see Section 4.1.3). That is, some samples exhibited patches of B-type surface on a mainly A-type surface as in Fig. 4.115 [Bobb *et al.* 1966 (GaAs), Wilson 1966 (CdS), Heiland and Kunstmann 1969 (ZnO) and Kubo and Tomiyama 1971 (ZnO)]. The observation that the piezoelectric properties of CdS single crystals were structure dependent, i.e. that they varied widely from one sample to another, was initially surprising (Wilson 1966). This was also true for polycrystalline films (Rozgonyi and Foster 1967) where it is due to the occurrence of grains with polar directions differing by arbitrary angles. It had been assumed at the start of crystal growth work on CdS for use in acoustoelectric devices that the piezoelectric constants were invariant physical properties of the material. The variation of the average, macroscopic piezoelectric constant was found to be due to the development of charge of reverse sign over parts of polar faces in the areas of reverse polarity (Heiland and Kunstmann 1969, Wilson 1966). Wilson showed that the piezoelectric constant increased towards its ideal, perfect-crystal value for CdS as the reverse polarity areas were reduced by annealing, although he did not realise the antiphase character of the areas of reversed polarity.

The grain boundary work led to the later realisation (Holt 1969) that this was evidence of the occurrence of antiphase domains (APDs i.e. volumes of opposite polarity). Holt (1969) considered the crystallography of APBs in the sphalerite structure using ball-and-wire modelling.

Since antiphase domains are volumes between which only the polarity is (180°) reversed while the orientation is unaltered, cleavage cracks generally propagate undeviated through APBs. This was observed in the case of the ZnO crystal in Fig. 4.115 and the epitaxial GaP film in Fig. 4.117.

Inversion, polarity-reversal or para-twins differ both in site occupation and in orientation from the matrix as shown in Fig. 4.114a. This difference can be expressed in terms of Friedel indices by extending the definition of the latter to negative values as follows (see Fig. 4.118). The positive odd integers 3, 5, 7... represent the coincidence lattice orientation relations for grain boundaries and the subset $\Sigma = 3^n$ represents the twin orientation relations. These coincidence lattice orientations correspond to particular rotation/axis combinations, which can be described as partial symmetry operations.

Partial symmetry operations produce coincidence of a fraction of finish sites with sites of the start structure. The fraction is the reciprocal of the Friedel index. The symmetry operations of a space group produce complete coincidence between the start and finish sites of a space lattice. This can be represented by $\Sigma = 1$ (complete site coincidence). The index $\Sigma = -1$ corresponds to negative (antiphase) coincidence, i.e. every atom has been carried to a wrong site. In dichromatic coloured symmetry this corresponds to an antisymmetry operation, which carries every lattice

Figure 4.117 Light micrograph of an etched surface of a film of GaP epitaxially grown on a (100) surface of NaCl and floated off in water. The APBs appear as irregular closed curves. The straight vertical and horizontal black lines are cracks. The cracks run straight across the APBs because the orientations of both domains are the same, i.e. the Friedel index of the APBs is $\Sigma = -1$. (After Morizane 1977. Reprinted from *Journal of Crystal Growth*, **38**, Antiphase domain-structures in GaP and GaAs epitaxial layers grown on Si and Ge, pages 249–54. Copyright 1977, with permission from Elsevier.)

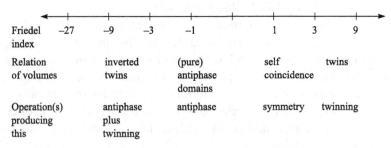

Friedel index	−27	−9	−3	−1		1	3	9
Relation of volumes		inverted twins		(pure) antiphase domains		self coincidence	twins	
Operation(s) producing this		antiphase plus twinning		antiphase		symmetry	twinning	

Figure 4.118 Extended range of values of the Friedel index, showing the defect represented and the corresponding symmetry or partial symmetry operations.

point to one of the other colour. The indices −3, −9 . . . are those for twins like the 3, 9 . . . twins but with the occupation of the sites in the twin reversed. Thus, $\Sigma = 3$ indicates that 1/3 of the sites of the two lattices coincide correctly. $\Sigma = -3$ means that the A atom sites of the first lattice coincide wrongly with B atom sites of the second while B atom sites of the first lattice coincide with A atom sites of the second.

Across APBs the site occupation reverses. Therefore, geometrically, all bonds that cross APBs must of necessity be 'wrong' (A-A or B-B) higher-energy bonds, not 'right' A-B bonds. Two types of APBs are crystallographically possible in the sphalerite structure. Type I antiphase planes (Fig. 4.119a) have equal numbers of A and B atom wrong bonds, while type IIB planar APBs involve all B (Fig. 4.119b) 'wrong' bonds while type IIAs are crossed only by A-A wrong bonds (Fig. 4.119c). It can readily be shown that an APB on any plane (uvw) will be of type I if $[uvw] \langle 111 \rangle = 0$ and will be of type II if $[uvw] \langle 111 \rangle \neq 0$ for any $\langle 111 \rangle$ direction (Holt 1969). Clearly, a type II APB, like that in Figs. 4.119b and 4.119c, can, by making frequent 'demi-steps', i.e. those from a 000 site to a ¼¼¼ one of the basis unit, change from the IIB to IIA form and vice versa. This will provide both an increase in entropy and some balancing of any net charges associated with A-A and B-B bonds. Hence, such atomic-scale demi-step raggedness is likely to occur in type II APBs.

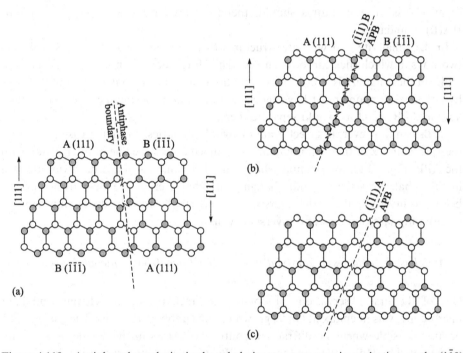

Figure 4.119 Antiphase boundaries in the sphalerite structure seen in projection on the $(1\bar{1}0)$ plane. Zig-zag lines denote wrong bonds. (a) $(\bar{1}1\bar{3})$ antiphase boundary. This is a type I APB. There are equal numbers of wrong A-A and B-B bonds crossing the boundary and the relative numbers of A and B atoms are unaffected by the presence of this APB. (b) $(\bar{1}\bar{1}1)$B APB which is of type II and involves wrong B-B bonds only. More B atoms than A atoms are incorporated in material containing type IIB APBs. (c) $(\bar{1}\bar{1}1)$A antiphase boundaries are also of type II but incorporate a non-stoichiometric excess of A atoms into the material. (After Holt 1969.)

The lower energy of A-B bonds compared to A-A or B-B bonds is what is responsible for ordering the compound and APBs constitute local disorder. Thermal disordering in binary semiconducting compounds has not been observed. APBs in sphalerite and wurtzite structure compounds occur, therefore, only as growth defects.

Thermal disordering from the chalcopyrite to the sphalerite structure does occur in ternary II-IV-V$_2$ and I-III-VI$_2$ compounds, however, as well as in some semiconductor alloy systems (see Shay and Wernick 1975, Binsma *et al.* 1980, 1981, Madelung *et al.* 1985, and Zunger 1987). In metal alloys, order to disorder transformations occur through the introduction of increasing areas of APBs, where the order is locally disturbed, into ordered material. Whether APBs play a role in order-disorder transformations in semiconducting materials does not seem to have been investigated, however.

Type I APBs do not affect the relative numbers of A and B atoms in the material, but a preponderance of area of type IIA or of IIB configuration APBs would alter the stoichiometry. Growth of crystals under conditions favouring the incorporation of an excess of A(B) atoms should, therefore, favour the incorporation of type IIA(B) boundaries.

In the sphalerite and wurtzite structures there are just two sublattices and only two ways in which they can be occupied. Therefore, there cannot be intersections of APBs in these materials, for, if there were three domains in contact, two of the three hypothetical domains must have the same occupation scheme, i.e. the same phase, and no APB therefore would separate them.

APBs can, however, intersect grain or twin boundaries. In such a case the sign of the Friedel index of the twin boundary is changed across the line of intersection with the APB. Fig. 4.120, for example, shows an APB joining a first order twin interface in the sphalerite structure and changing the character of the twin from upright, below, to inverted, above the intersection. This emphasises the distinction between (pure) antiphase domains and inversion twins.

Antiphase boundaries in epitaxial films of II-VI and III-V compounds grown on Ge and Si

One of the first observations of polarity reversal, by surface etching and light microscopy, was in epitaxial GaAs grown on Ge (Bobb *et al.* 1966). The use of TEM became possible when the diffraction contrast of APBs in the sphalerite structure were treated by Holt (1969) and that of APBs in the wurtzite structure by Blank *et al.* (1964), who referred to them as antistructure boundaries.

Holt *et al.* (1976) studied (110) epitaxial films of II-VI compounds that grew by the three-dimensional-nucleation, Volmer-Weber mode (see Section 2.3.11). Their films contained a 'domain-form structure' formed by the well-known double-position twinning mechanism (see Fig. 4.121). In double positioning nuclei deposit in two orientations, related by a 180° rotation about the substrate normal. Both have the

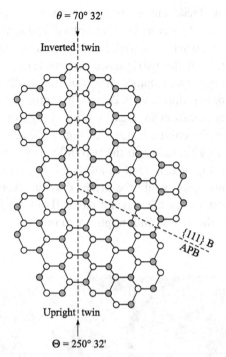

Figure 4.120   Intersection of a {111}B antiphase boundary with {111} twin boundaries. (After Holt 1969.)

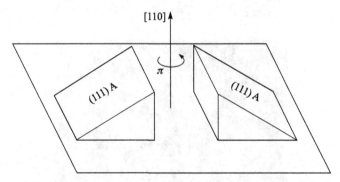

Figure 4.121   Epitaxial double positioning arises through growth from nuclei in two azimuthal orientations related by a 180° rotation about the [110] normal to the substrate surface as shown. This puts the nuclei in antiphase and twinned relation to each other. Consequently, the differently oriented {111} facets shown are of the same (A) polar type. The two types of antiphase domains, grown from these nuclei, fault on these differently oriented planes. This gives rise to the strong contrast seen in Fig. 4.122. (After Holt *et al.* 1976.)

same film/substrate interfacial energy so they form in equal numbers and grow equally fast. The rotation, however, is an antiphase sub-symmmetry operation; so one nucleus has reverse polarity, as well as a different orientation, relative to the other. Hence equal areas of the two 'phases' are nucleated and grow to have the characteristic jigsaw puzzle piece shapes shown in Fig. 4.122.

Such double-positioning domains occurred in epitaxial films of CdS, CdTe and CdSe grown by evaporation in vacuum onto (110) surfaces of NaCl and Ge. The strong black and white contrast shown in Fig. 4.122 arose from the diffraction due to the numerous stacking faults on the two differently oriented, but equivalent, {111} planes, shown in Fig. 4.121. That is, all the black domains in Fig. 4.122 were faulted on the one plane only, while the bright domains were faulted only on the other plane. For full details see the original paper (Holt *et al.* 1976). This contrast evidence showed that the black and bright domains were double-position-twin, antiphase related.

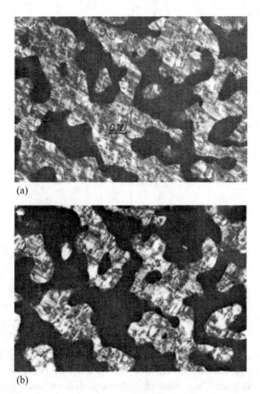

(a)

(b)

Figure 4.122   The domain-form structure in an epitaxial film of CdSe evaporated in vacuum onto a (110) face of Ge. The two types of domain are in strong contrast due to {111} faulting, with (a) the TEM image formed by electrons diffracted by the faults in the one type of domain and (b) the image formed by electrons diffracted by faults in the domains of the 'other phase'. (After Holt *et al.* 1976.)

When the (110) sphalerite-structure films of CdSe were annealed to eliminate the profuse stacking faulting responsible for the strong contrast in Fig. 4.122, the APBs could be resolved in TEM as shown in Fig. 4.123. These boundaries showed the predicted diffraction behaviour (Holt 1969), confirming that the film consisted of antiphase domains.

Striking and conclusive evidence of antiphase domains in epitaxial films of III-V semiconductors and of the orientation dependence of their occurrence was obtained by Morizane (1977). He grew GaP on Si and Ge and GaAs on Ge by chemical vapour deposition. He found that in GaP layers grown on Si and Ge there were usually evenly scattered antiphase domains as shown in Fig. 4.124. In these observations as well as those of Fig. 4.123, the APBs form closed loops or run continuously across the field of view. These are the only possibilities since, in single crystals, as was discussed above, APBs can not intersect or meet in triple points in the sphalerite or wurtzite structures. Hence metastable foam-like morphologies like those characteristic of grain boundaries in polycrystalline materials are not possible.

The occurrence of antiphase domain structures in these films could be seen by eye, without special specimen preparation. This was because the surface growth morphology was polarity dependent. The typical surface morphology due to the presence of antiphase domains in (100) heteroepitaxial GaP films was as shown in Fig. 4.125. Surfaces with such morphologies scatter the light strongly and, therefore, appear matt or milky, not smooth and shiny (specularly reflecting). This makes possible a striking macroscopic demonstration of the orientation dependence of the growth morphology.

Morizane grew GaP on monocrystalline hemispheres of Ge in order to observe the whole range of possible orientation dependence of the domain morphology.

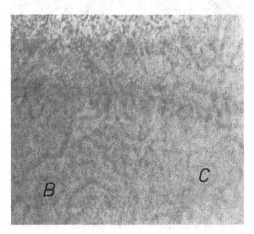

Figure 4.123   TEM image of an annealed domain-form structure film of CdSe/Ge (110), showing APBs in dark diffraction contrast. The width of the field of view is 1 µm. (After Holt *et al.* 1976.)

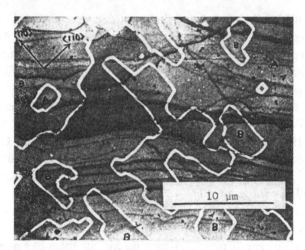

Figure 4.124   A TEM image showing antiphase boundaries, in bright contrast, in films of GaP epitaxially grown on a (100) surface of Si. A and B are regions in antiphase (i.e. of reverse polarity). (After Morizane 1977. Reprinted from *Journal of Crystal Growth*, **38**, Antiphase domain-structures in GaP and GaAs epitaxial layers grown on Si and Ge, pages 249–54. Copyright 1977, with permission from Elsevier.)

Figure 4.125   Light micrograph of the as-grown surface of a film of GaP grown on (100) oriented Si. (After Morizane 1977. Reprinted from *Journal of Crystal Growth*, **38**, Antiphase domain-structures in GaP and GaAs epitaxial layers grown on Si and Ge, pages 249–54. Copyright 1977, with permission from Elsevier.)

The result is shown in Fig. 4.126a and is stereographically indexed in Fig. 4.126b. Around the {100} poles the domain morphology of Fig. 4.125 occurred and extended over an area corresponding to deviations of a few degrees from the {100} orientation. The boundary line forming the sides of the triangle appeared straight at low magnifications but at higher magnifications was seen to be randomly curved. Isolated domains were sometimes found on either side of the boundary. Thus it appears that, on surfaces of any orientation between {100} and {110}, i.e. those of the form {hk0}, antiphase domains can be formed.

In GaAs films grown on Ge, the occurrence of antiphase domains was the exception rather than the rule as in GaP, and domain structures were confined to areas of substrate surface irregularity such as grooves or other regions of surface curvature.

Preliminary analysis of the diffraction contrast fringes in TEM images of these APBs showed that they were not of the simple or pure type illustrated in Fig. 4.117. Morizane (1977) suggested that his diffraction contrast results could be accounted for by assuming a small displacement of the lattice in one domain relative to that on the other side of the boundary. Such a displacement had been found to occur at antistructure or inversion (i.e. antiphase) boundaries in BeO by Chikawa and Austerman (1968).

In recent years, (100) GaAs/Ge and GaAs/Si have become important for their applications. For example, optoelectronic devices in the GaAs can be combined with microelectronic circuits in the Si. For GaAs solar cells for space applications,

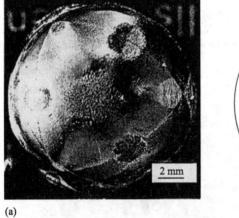

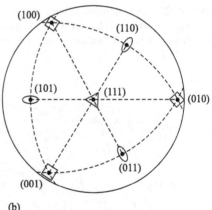

(a)   (b)

Figure 4.126   (a) Surface morphology of an epitaxial GaP film grown on a hemisphere of Ge. A spherical triangle with its apices at the {100} poles can be seen due to domain surface growth morphology (roughening) effects. (b) A stereogram on which the points on the surface with the orientations of the three singular-surface forms: {100}, {110} and {111} are plotted. (After Morizane 1977. Reprinted from *Journal of Crystal Growth*, **38**, Antiphase domain-structures in GaP and GaAs epitaxial layers grown on Si and Ge, pages 249–54. Copyright 1977, with permission from Elsevier.)

Ge and Si wafers provide larger and, in the case of Si, stronger epitaxial substrates than any III-V wafers. However, antiphase boundaries tend to form in profusion in both GaAs/Ge and GaAs/Si (Kroemer 1987).

Holt *et al.* (1996) found that APBs occurred only in a naked-eye-visible, narrow band near the periphery of two inch squares of (100) GaAs/Ge material grown for solar cell fabrication as shown in Fig. 4.127. This was only realized, however, after the presence of APBs in the material had been found in EBIC images (see Fig. 4.128a). The reason for the visibility of the band was the presence of surface

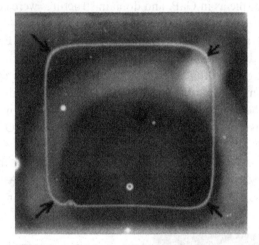

Figure 4.127   Demagnified camera photograph of a two inch square (508×508 mm) wafer of GaAs/Ge, grown for solar cell fabrication. The light band marked by arrows is due to light scattering by a band of surface roughness due to APBs as shown in Fig. 4.128b. (After Holt *et al.* 1996.)

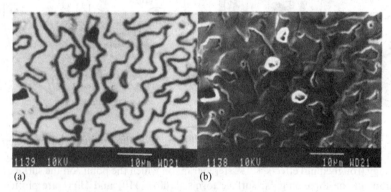

(a)   (b)

Figure 4.128   SEM (a) EBIC and (b) secondary electron images of APBs in (100) GaAs/Ge. The three surface dirt particles appearing black in (a) and white in (b) show the coincidence of the two fields of view. (After Holt *et al.* 1996.)

roughness due to the profuse APB structure, which produced surface markings as shown in Fig. 4.128b.

Another important case in which APBs can be a problem is that of (100) multi-quantum well (MQW) structures of e.g. GaAs/Ge (Neave *et al.* 1983). Growth of an epitaxial layer of any polar, sphalerite-structure material on a non-polar, diamond-structure (100) substrate containing monatomic ('demi')steps can result in APBs forming (Pond *et al.* 1984, Kroemer 1987).

4.6 Epitaxial interfaces and misfit dislocations

Major efforts have been and are being devoted to the growth of materials having low-dimensional structures. This became practical with the development of such techniques as molecular beam epitaxy (MBE) and metal-organic chemical vapour deposition (MOCVD) (see Section 1.1). Low MBE growth rates, about a monolayer per second, for example, allow high-quality crystalline materials to be produced, with atomically abrupt interfaces and excellent uniformity, stoichiometry, thickness and impurity control. Thus, structures in which layers of different materials and thicknesses of a few monolayers alternate, can be grown. Such thin layers have the properties of quantum mechanical two-dimensional (2D) systems. These epitaxial techniques are used to grow the energy band-gap engineered heterostructures (see Section 1.7) essential for the fabrication of many devices.

As discussed in Section 2.3.11, selection of deposited- and substrate-materials closely matched in symmetry and lattice parameter favours the epitaxial growth of heterostructures of high crystalline perfection. Lattice parameter misfit can lead to the introduction of interfacial misfit dislocations to relax long-range elastic strain as illustrated in Figs. 2.12 and 2.15. These dislocations are not unavoidable, however. Small misfit heterostructures can be produced as pseudomorphic or 'strained-layer structures' if the epilayers are sufficiently thin. Instead of misfit dislocations the lattice mismatch is accommodated by long-range elastic strain.

Moreover (Section 2.3.11), it was once believed that there was a '15% rule' i.e. that for a misfit greater than this value, epitaxy was not possible, and this motivated Frank and van der Merwe 1949a, 1949b) to develop their model and introduce the idea of misfit dislocations. It was later shown by the accumulating evidence (Pashley 1956) that virtually every material will grow epitaxially on any other once the best growth technique and conditions are found. However, it is still the rule that close matching in symmetry and lattice parameter favours the growth of epitaxial material of high perfection. It is also now clear that misfit dislocations only begin to appear when the growing epitaxial film thickness exceeds a first critical threshold. The final equilibrium number is only present after a larger second thickness value is reached (Section 4.6.2).

In practice, heterojunctions are epitaxial interfaces between semiconductors with the same or related structures, in parallel alignment. Oldham and Milnes (1964) first pointed out that such Si/Ge (111) heterojunctions would be expected to contain edge misfit dislocations (Section 4.6.2). Experimentally, the number and type of dislocations depend on the conditions of growth and the mechanism, e.g. of slip, by which they are introduced. Moreover, the alignment of the materials may not be accurately parallel, implying that the dislocations of the corresponding misorientation grain boundary will also be present (Igarashi 1971). Moreover, any dislocations in the initial epitaxial substrate material must either thread through the heterojunction and epilayer or turn and run along the interface (Section 4.6.2) since dislocations cannot end in the interior of a crystal. Thermal stresses can result in slip and introduce additional dislocations.

4.6.1 Materials selection for epitaxy

For optimum epitaxy then the epilayer material should match the substrate in symmetry and lattice parameter as closely as possible. Semiconductors from the family of diamond-structure elements and sphalerite-structure compounds have related symmetries and grow in parallel alignment. Hence, symmetry matching is not a great problem. However, antiphase domains can appear when one of these compounds is grown on a substrate of Ge or Si (see Section 4.5.3).

To select the materials for use in epitaxial heterostructure devices requires consideration of both their band gaps (to satisfy the energy band engineering requirements of the device) and their lattice parameters (to minimize the probability that epitaxial misfit dislocations will appear). To do this for semiconductors of importance for optoelectronics, the III-V compounds and alloys, the III-nitrides and alloys, and the II-VI compounds (ZnS and ZnSe), Fig. 4.129 is used.

Fig. 4.129 shows that GaAs is closely lattice-matched to AlAs and the ternary alloys of $Al_xGa_{1-x}As$ which have compositions falling on the line between the two compounds. In this alloy system, the energy gap varies from about 1.4 to 2.2 eV so the energy gap of these $Al_xGa_{1-x}As$ alloys can be tuned across the photon-energy range from the visible to the near infrared. For efficient luminescence, a direct band gap is needed, so light emitting devices are only made of alloys with the range of compositions and band gaps corresponding to the solid section of the GaAs–AlAs line above. In contrast, lattice matching of $Ga_xIn_{1-x}As$ on InP substrates only occurs for one value of x, and hence of E_g, i.e. photon energy. Moreover, even if a heterojunction is lattice-matched at the growth temperature, it may develop a significant mismatch due to the difference in thermal expansion coefficients of the two materials, on cooling to room temperature. Hence, a large difference in this property between epilayer and substrate is undesirable. The relevant data for selecting semiconductors for heterojunctions are given in Table 4.8.

A double heterostructure in which a few-nm thick narrower energy-gap semiconductor is sandwiched between thicker layers of a wider gap semiconductor,

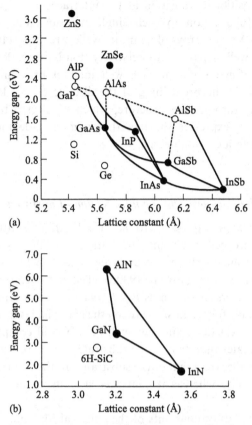

(a)

(b)

Figure 4.129 Plots of the energy gap versus lattice constant for direct- (solid line and filled circle) or indirect-gap (dotted line and open circle) semiconductors showing (a) the III-V compounds and alloys first used in red-, yellow- and green-emitting LEDs and lasers, and two wider energy-gap II-VI compounds, and (b) the wider energy-gap III-nitrides used for blue and ultra-violet emitting diodes. The lattice constants for favoured substrates, Si and Ge in (a) and 6H-SiC in (b), are also plotted.

Table 4.8. *Properties of selected semiconductors at room temperature (300 K).*
Structure: D = diamond, S = sphalerite

Material/Property	Ge	Si	GaAs	InP	InAs
Energy gap (eV)	0.67	1.12	1.42	1.35	0.36
Lattice constant a (Å)	5.646	5.431	5.653	5.869	6.058
Structure	D	D	S	S	S
Thermal expansion coefficient (10^{-6} K^{-1})	5.75	2.56	5.73	4.60	4.52

with a sharp compositional transition at the interfaces, is a quantum well (QW) (see Section 1.7). In addition to such single-quantum wells (SQWs), multiple-quantum wells (MQWs, i.e. arrays of quantum wells) are important (see Section 1.7). In MQWs with the wells separated by relatively thin barriers (less than 10 nm), the electrons in the confined states in each well interact by tunnelling through the barriers. This results in the broadening of the discrete energy levels into so-called minibands extending through the structure, which is referred to as a superlattice. The term superlattice expresses the fact that a man-made longer periodicity is present, in addition to a crystalline one.

4.6.2 *Film thickness and the introduction of misfit dislocations*

Strained layer epitaxial growth occurs for sufficiently thin misfitting layers. Then no misfit dislocations are generated and the lattice mismatch is accommodated by long-range elastic strain in the epitaxial layer. This conclusion came out of the Frank and van der Merwe (1949a, 1949b) elastic model (see Section 2.3.11). It showed that, for layers of thickness $h < h_c$, where h_c is a critical thickness, it is energetically favourable for the misfit between the epitaxial layer and the substrate to be accommodated by elastic strain. This will be a biaxial compression if the layer lattice constant is greater than that of the substrate, or a biaxial tension (if the layer lattice constant is smaller than that of the substrate). In the former case, the epitaxial layer elongates laterally, whereas it contracts laterally in the case of the biaxial tension.

If a_f and a_s are the lattice constants of the epitaxial film and substrate materials, respectively, one can express the lattice mismatch by the so-called misfit parameter f_m related to the substrate reference lattice:

$$f_m = \frac{|a_f - a_s|}{a_s} \qquad (4.35)$$

(for more details, see Section 2.3.11) Pseudomorphic or strained-layer structures can be formed if the film (epilayer) is sufficiently thin, so the lattice mismatch is accommodated by strain. But, with the increasing layer thickness, it becomes energetically favourable for misfit dislocations to be generated to accommodate the misfit strain. The critical thickness for the onset of dislocation generation, h_c, varies inversely with the misfit parameter. As the layer thickness increases above the critical value, i.e. for $h > h_c$, an increasingly large part of the misfit strain is accommodated by increasing numbers of misfit dislocations. That is, at the critical thickness it begins to be energetically favourable to relax the long-range stress by introducing a network of interfacial dislocations. There are altogether four mechanisms for accommodating lattice mismatch strain. They are (i) elastic distortion of the epitaxial layer (pseudomorphism), (ii) plastic relaxation through misfit dislocations,

(iii) epitaxial layer roughening, and (iv) interdiffusion (for details and discussion on the competition between these mechanisms, see Hull 1999, and references therein).

The usual type of the misfit stress relaxation is through misfit (interfacial) dislocation formation which reduces the coherence between the substrate and epitaxial layer. The concept of misfit dislocations and critical thickness were introduced by Frank and van der Merwe (1949b), in an energy balance model. That is, the critical thickness was first calculated by minimizing the sum of the interfacial energy and the strain energy of the epitaxial layer. Subsequently, this was developed into a generalized relaxation mechanism by Matthews *et al.* (1970, 1974, 1975) who calculated the critical thickness by considering the forces acting on the dislocations as will now be discussed.

TEM studies led Matthews to recognize the important mechanism for the formation of misfit dislocations from pre-existing threading dislocations as shown in Fig. 4.130. A threading dislocation bends over and runs along the interface, to form a misfit dislocation segment, under the influence of the misfit stress.

The critical thickness can now be calculated by considering the forces acting on the dislocations, e.g. the force acting to elongate the threading dislocation in the interface in Fig. 4.130, due to the misfit strain, and the force from the dislocation line tension that resists this glide. This is an equilibrium theory that does not consider the energy required for the threading dislocation motion. The Matthews model (Matthews *et al.* 1970, 1974, 1975) for the balance of forces acting on a threading dislocation leads to an expression for the equilibrium critical thickness as:

$$h_c = \frac{b(1 - v\cos^2\theta)}{8\pi f_m(1 + v)\cos\lambda}\left(\ln\frac{h_c}{b} + 1\right) \tag{4.36}$$

where b is the Burgers vector, v is the Poisson coefficient, θ is the angle between the dislocation line and its Burgers vector, and λ is the angle between the slip direction and the direction perpendicular to the line of intersection of the slip plane and the

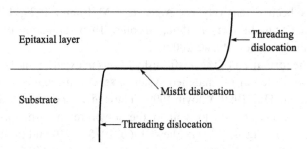

Figure 4.130   A threading dislocation from the substrate grew up through the epilayer. Under the misfit stress it bends over and runs along the interface leaving a segment of misfit dislocation. (Adapted from Matthews 1975.)

interface. (It should be noted that in his model, Matthews defined the misfit f_m as related to the epilayer reference lattice, not to the substrate as in Equation (4.35)).

The Matthews model has certain limitations. First, it is for pre-existing threading dislocations moving over the interface to form misfit dislocations. Also, this model does not take into consideration the influence of growth conditions, and therefore it has been continuously refined (for reviews on other critical thickness models, see e.g. Dunstan 1997, Hull 1999, and the references theren). In some cases, like $In_xGa_{1-x}As/GaAs$ strained-layer heterostructures, the generation of misfit dislocations from threading dislocations is not the primary relaxation mechanism (see, e.g. Liu *et al.* 1999). Nevertheless, the Matthews model is considered to be a fairly good approximation with accuracy from 20 to 50% (Hull 1999). The critical thickness expression has been published in several modifications for different cases and assumptions (e.g. Freund 1990, Willis *et al.* 1990; for a review of other models, see Dunstan 1997, and the references therein).

Numerous observations showed that, generally, misfit strain relaxation is partial, and full relaxation is never reached. It may take place in several stages, and is influenced by the dislocation kinetics (see e.g. Matthews *et al.* 1970, Dodson and Tsao 1987, Hull *et al.* 1989, Houghton 1991, Perovic and Houghton 1995, Hull 1999, Hull *et al.* 2002). Mechanisms such as (i) nucleation of half loops, (ii) dislocation multiplication, (iii) blocking of dislocations during glide over intersecting dislocations, and (iv) thermally activated dislocation glide result in the presence of kinetic barriers for dislocation formation. This implies that epitaxial layers may be grown, thicker than predicted by the equilibium theory, prior to the introduction of significant numbers of dislocations. However, although discrepancies between theory and experiment appear, caution must be exercized, since some microscopical methods, like TEM, cannot examine large enough volumes to detect the first few misfit dislocations indicating the onset of relaxation.

4.6.3 Mechanisms of introduction and types of misfit dislocations

This topic has been extensively reviewed (e.g. Matthews 1979, Fitzgerald 1991, Beanland *et al.* 1996, Jain *et al.* 1996, Mooney 1996, Dunstan 1997, Hull 1999, Vdovin 1999, and van der Merwe 2002).

To explain the formation of the misfit dislocations in experimental observations on a wide variety of different mismatched systems, several mechanisms were proposed (for reviews, see e.g. Hull 1999, Vdovin 1999). These include (i) the bending over of the threading dislocations already present in the substrate to form misfit dislocation segments as shown in Fig. 4.130 (Matthews 1966, 1975, 1979) and (ii) mechanisms for the formation of new dislocations to be discussed below. Although no universal mechanism can explain all the cases of nucleation of misfit dislocations, some general categorization schemes have been suggested. These are (i) homogeneous nucleation of dislocation loops and half-loops, (ii) heterogeneous nucleation at specific strain concentrations, e.g. at substrate defects or growth artifacts, and (iii) multiplication mechanisms involving dislocation pinning and interaction processes (for a review see e.g. Hull 1999). These mechanisms of initial misfit dislocation generation have been categorized by Vdovin (1999) for (i) the homogeneous case as 'bending' (Matthews 1966), and 'surface generation' (Matthews *et al.* 1970) and (ii) the heterogeneous case as 'interface generation' (Perovic and Houghton 1995), 'substrate surface precipitation' (Perovic *et al.* 1989, Fitzgerald *et al.* 1989), and 'diamond defect generation' (Eaglesham *et al.* 1989, Humphreys *et al.* 1991). Dislocation nucleation may also be related to strained epitaxial layer surface roughening (Cullis *et al.* 1994, Tersoff and LeGoues 1994, Cullis *et al.* 1995, Jesson *et al.* 1995). For example, it was demonstrated that ripple troughs are sources for misfit dislocations in InGaAs/GaAs structures (Cullis *et al.* 1995). The various mechanisms of secondary generation (i.e. multiplication) of misfit dislocations have been summarized by Vdovin (1999).

During strain relaxation in heteroepitaxial layers, mutual interactions between dislocations are of great importance (e.g. Pichaud *et al.* 1999). Gosling *et al.* (1994) developed a model for the kinetics of strain relaxation in lattice-mismatched semiconductor layers, including nucleation, propagation and interaction of misfit dislocations, with emphasis on interaction processes. Beanland (1995) described specific multiplication mechanisms and compared the characteristics of dislocation generation mechanisms (i.e. homogeneous- or heterogeneous- nucleation and multiplication) with experimental data in the InGaAs/GaAs system. Hence, he concluded that dislocation multiplication is the dominant nucleation mechanism. Liu *et al.* (2003) demonstrated that, in GaAs/In$_x$Ga$_{1-x}$As/GaAs double heterostructures with strained layer thicknesses less than the critical value predicted by the Matthews model, misfit dislocations can be introduced during postgrowth treatment at a temperature higher than the growth temperature. Thus, they concluded that the formation of misfit dislocations depends, in addition to strained-layer thickness and composition, also on temperature through the temperature-dependent Peierls force (Liu *et al.* 2003).

In GaAs/Si(100) thin-film heterostructures grown by MOCVD, Narayan and Oktyabrsky (2002) found by high-resolution transmission electron microscopy that the misfit dislocation network in the interface consisted of two types of misfit dislocations, i.e. 60° and 90°. About 60% were 90° dislocations and 40% were the closely spaced pairs of 60° dislocations with intersecting glide planes. They observed that these percentages did not change much with rapid thermal annealing at 800° C although some of the 60° dislocation pairs split to form stacking faults. They suggested that the 60° dislocation pairs, having parallel screw components, couldn't combine to form a 90° dislocation. On the basis of their observations, Narayan and Oktyabrsky (2002) proposed a model for the generation of these dislocations. According to it, some 60° dislocations are generated at the undulated surface above a critical thickness, whereas others are nucleated at a greater thickness at the smoother surface (Narayan and Oktyabrsky 2002). From atomic-resolution Z-contrast imaging, Lopatin *et al.* (2002) also confirmed the presence of both 60° and 90° dislocations at the GaAs/Si interface. Dangling bonds were observed in both 60° and unreconstructed 90° dislocations and thus both of these are expected to introduce energy levels in the energy gap.

The density of threading dislocations in the epitaxial layer is related to the density of misfit dislocations and the velocity of dislocation motion. From experimental determination of dislocation nucleation rates and activation energies, a semiempirical expression for the generation rate of misfit dislocations was derived by Houghton (1991):

$$\frac{dN(t)}{dt} = BN_0(\tau_e/G)^n \exp(-E_n/kT) \tag{4.37}$$

where B and n are constants, N_0 is related to the initial density of heterogeneous nucleation sites, G is the shear modulus, and E_n is the activation energy for dislocation nucleation. The derived values for these constants are $B \approx 10^{18}\,s^{-1}$, $n \approx 2.5$, and from the dependence of $N(t)$ as a function of temperature, E_n was found to be $2.5 \pm 0.5\,eV$ (Houghton 1991).

Dislocation motion typically accompanies the strain relief. Dislocations typically move by the formation of kinks (see Section 4.2.1). The dislocation propagation can be described by an expression for the velocity of dislocation motion as:

$$v_d = v_o\tau_e^m \exp(-E_v/kT) \tag{4.38}$$

where v_o is a constant, E_v is the activation energy for dislocation glide, and τ_e is the excess stress providing the driving force for the motion (the values of m vary between 1 and 2, see Section 4.2.1). Dislocations slip over glide planes, and can also move by climb (i.e. move out of the glide plane). By comparing the glide and climb velocities of dislocations at various temperatures, it was determined that the relaxation of misfit strain proceeds mainly by glide mechanisms (Jesser and van der Merwe 1994). However, climb processes may be important in bypassing obstacles and redistributing an irregular array of misfit dislocations into a regular array of lower energy (Jesser and van der Merwe 1994).

It is necessary to avoid high densities of dislocations to obtain high-quality epitaxial heterostructures. Various methods have been employed to prevent the nucleation and propagation of dislocations. These techniques include (i) incorporation of buffer layers, (ii) promotion of dislocation interaction and annihilation, (iii) incorporation of compositionally graded layers, (iv) selected area growth, and (v) insertion of strained-layer superlattices (see the review by Hull 1999, and the references therein). Alternatively, so-called compliant substrates, i.e. thin, free-standing ones (Lo 1991, Powell *et al.* 1994) can be employed. This results in strain partitioning in coherent compliant heterostructures, i.e. the film and thin substrate both undergo long-range elastic strain to match at the interface.

The information available on specific heterostructures, GeSi/Si, GaAs/Si, and III-nitride systems, is outlined below.

4.6.4 Misfit dislocation anisotropy in {100} sphalerite stucture films

As discussed in Section 4.1.3, the anisotropic ⟨110⟩ directions in the {100} faces of sphalerite structure semiconducting compounds can be identified by etching. When one such material is grown on a {100} face of another, under such conditions that neither anisotropic cracking nor bending results, the long-range elastic stress due to lattice mismatch may lead to the introduction of misfit dislocations (MDs).

The densities of MDs in the two sets, which run in the two orthogonal ⟨110⟩ directions in the interface, differ greatly, as in the example shown in Fig. 4.131. The dislocations in these two orthogonal grids differ in polarity and, therefore, in mobility. Introduction of the more mobile type of MD begins by glide, therefore, at a lower stress than that required for the other type and this accounts for the observed differences in the numbers in the two orthogonal sets.

4.6.5 Misfit dislocations in the GeSi/Si system

An important heterostructure system is GeSi/Si, which combines highly developed Si technology with the benefits of the introduction of Ge in Si-based devices

Figure 4.131   X-ray topographs showing the two orthogonal sets of misfit dislocations at the interface of a wafer of GaAs on which had been grown a triple epitaxial layer structure of GaAlAsP/GaAs/GaAlAsP. (a) The micrograph taken with the 220 reflection shows a high density of vertical dislocations while (b) the image taken with the 2̄20 Bragg reflection reveals that there are only a few horizontal misfit dislocations. X-ray topographs are printed at a magnification of 50×, so this sample has undergone only a small relaxation introducing relatively few dislocations. (After Bartels and Nijman 1977. Reprinted from *Journal of Crystal Growth*, **37**, Asymmetry of misfit dislocations in heteroepitaxial layers on (001) GaAs substrates, pages 204–14. Copyright 1977, with permission from Elsevier.)

(for reviews see Mooney 1996, Hull 1999, and Mooney and Chu 2000). Introducing Ge significantly improves the performance, especially the operating frequency, of Si devices since the carrier mobility in GeSi is higher than in Si. The smaller energy gap and larger refractive index of GeSi suggest that these heterostructures should also be suitable for optoelectronic applications, e.g. detectors in the 1.3 to 1.6 μm wavelength range, which can be employed in optical fibre communication. The critical thickness values in such systems are between about 1 nm for pure Ge (there is about a 4% lattice mismatch between Si and Ge) and about 100 nm for $Si_{0.8}Ge_{0.2}$ (with a 1% mismatch).

Several mechanisms for the nucleation of misfit dislocations have been proposed for this system. These typically involve growth defects, such as oxide particles (Tuppen *et al.* 1989), SiC precipitate plates (Perovic *et al.* 1989), diamond defects (Eaglesham *et al.* 1989, Humphreys *et al.* 1991), or Ge-rich platelets (Perovic and Houghton 1993).

Cullis *et al.* (1994) demonstrated from TEM, AFM, and x-ray diffraction studies of surface morphology that misfit dislocation nucleation in Ge_xSi_{1-x} epitaxial layers grown on Si is related to compressive lattice distortion at the ripple troughs. Jesson *et al.* (1995) observed from AFM and TEM studies of the morphologies of strained $Ge_{0.5}Si_{0.5}/Si(100)$ that the nucleation of edge dislocations is associated with cusped surface morphologies related to surface roughening.

As mentioned in the previous section, there are techniques for preventing the nucleation and propagation of misfit dislocations (see the review by Hull 1999). In the case of Ge_xSi_{1-x}/Si heterostructures it was also demonstrated that misfit dislocation generation could be controlled by point defect injection (Stirpe *et al.* 1997). A report on the growth of relaxed SiGe on compliant silicon-on-insulator (SOI) substrates was presented by Powell *et al.* (1994), i.e. by deposition on an ultrathin SOI substrate with a superficial silicon thickness less than that of the SiGe layer. Such a substrate relaxes without the generation of threading dislocations in the SiGe layer, due to the fact that the dislocations are produced and glide in the thin Si layer.

4.6.6 Misfit dislocations in the GaAs/Ge and GaAs/Si systems

III-V compounds grown on Si substrates attracted great interest in recent years. Such heterostructures are useful for integrating a wide range of optoelectronic applications in the III-V compounds with Si circuitry in a monolithic chip. Moreover, Si provides a mechanically stronger (than GaAs), large area substrate that also has a greater thermal conductivity.

Problems in obtaining high-quality III-V epitaxial heterostructures like GaAs/Si are (i) the possibility of high dislocation densities due to the lattice constant mismatch between the epilayer and substrate at the growth temperature,

(ii) additional stresses due to the difference in thermal expansion coefficients of the epilayer and substrate, and (iii) the formation of structural defects (i.e. antiphase boundaries) due to the epitaxial growth of a polar crystal like GaAs on a non-polar substrate like Si. To overcome these problems, one can employ (i) thermal cycle annealing, (ii) growth interrupts, (iii) selected area growth, and (iv) insertion of strained-layer superlattices.

An example of the lattice-mismatch-induced generation and propagation of misfit and threading dislocations in GaAs/Ge layers grown on Si substrates is presented in Fig. 4.132. It includes both cross-sectional and plan-view TEM micrographs of GaAs/Ge layers grown on Si substrate and demonstrates that the dislocation density falls with increasing GaAs thickness.

Fig. 4.133 compares the densities of threading dislocations as a function of film thickness in four heterostructures. It demonstrates that the dislocation density falls with the same power of the thickness of the epitaxial layer in all four cases. This implies that similar mechanisms of density reduction by dislocation annihilation interactions and the formation of closed loops are responsible.

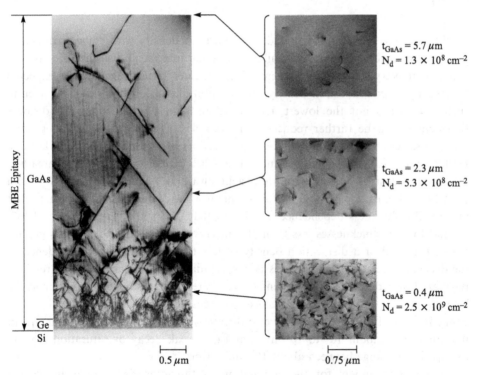

Figure 4.132   Cross-sectional and corresponding plan-view TEM micrographs of GaAs/Ge layers grown on a Si substrate. [After Sheldon *et al.*, *Journal of Applied Physics*, **58**, p. 4186 (1985).]

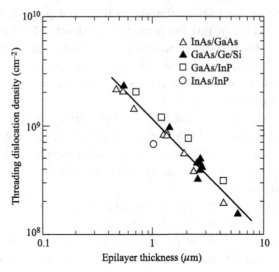

Figure 4.133   A plot of the density of threading dislocations versus thickness of the epitaxial layer for various heterostructures. [After Sheldon *et al.*, *Journal of Applied Physics*, **63**, p. 5609 (1988).]

Although the reduction of threading dislocation density with epilayer thickness for different heterostructures is similar, the actual interfacial dislocation density is expected to vary. (A heterostructure with a greater lattice mismatch is expected to have a greater dislocation density.) In addition, the dislocation densities given in such figures are not the lowest achievable in these systems, since the dislocation density can be further reduced, for example, by thermal annealing, growth interrupts, selective area growth or the insertion of strained-layer superlattices. Furthermore, for thicker films and lower dislocation densities, the threading dislocation densities tend to saturate (Tachikawa and Yamaguchi 1990, Beanland *et al.* 1996). For example, for GaAs/Si, Tachikawa and Yamaguchi (1990) found that inverse film thickness dependence, similar to that of Sheldon *et al.* in Fig. 4.133, is valid for film thicknesses less than 10 μm. However, for film thicknesses greater than 50 μm and/or a dislocation density of less than $10^7\,\mathrm{cm}^{-2}$, the dependence of the dislocation density on thickness is exponential. To summarize, there are three regimes of threading dislocations density. They are: (i) an entanglement region at the heterointerface, with threading dislocation densities ρ_{TD} in the range between about 10^{10} and $10^{12}\,\mathrm{cm}^{-2}$, (ii) an inverse thickness (h) dependence region, for ρ_{TD} in the range from about 10^7 to $10^9\,\mathrm{cm}^{-2}$, and (iii) weak decay or saturation regime, for ρ_{TD} in the range between about 10^6 and $10^7\,\mathrm{cm}^{-2}$.

An analytical model for the elimination of threading dislocations in hetero-epitaxial systems was proposed, based on reactions between dislocations that are within a characteristic interaction distance (Speck *et al.* 1996, Romanov *et al.* 1996,

1997, 1999, and Mathis *et al.* 1999). It successfully accounts for both the experimentally observed inverse thickness dependence for higher dislocation densities and the saturation behaviour for lower dislocation densities, i.e, thicker films. Sufficiently close threading dislocations may annihilate or fuse. The reaction radius is related to the shear modulus μ of the materials, the Burger's vector b, and the type of the reaction (annihilation or fusion). The expression for the reaction radius in the case of annihilation is (Speck *et al.* 1996):

$$r_A \approx \frac{\mu b}{2\pi\sigma_p} \qquad (4.39)$$

where σ_P is the Peierls stress. For cubic semiconductors, the effective annihilation radius was determined to be in the range between 50 and 500 nm, while by fitting the experimental data for selected III-V semiconductors, the reaction radius was found to be from about 50 to 100 nm (Speck *et al.* 1996).

The threading dislocation density can be expressed as (Speck *et al.* 1996):

$$\rho_{TD} = \frac{1}{K(h + \hat{h})} \qquad (4.40)$$

where K is the dislocation reaction cross-section that is proportional to r_A, and $\hat{h}$ is a parameter related to the initial dislocation density. For h much greater than $\hat{h}$, the inverse dependence $1/Kh$ is obtained.

4.6.7 Defects in III-Nitride systems

Another important heterostructure system is that of GaN and InN, which is used to fabricate ultraviolet detectors, light emitting diodes and blue lasers (see the reviews by Pankove 1999, and Jain *et al.* 2000, and references therein). GaN epilayers are typically grown on sapphire and SiC substrates. There is a large lattice mismatch, as well as a thermal mismatch, between GaN and these substrates, so growth of GaN layers on, e.g. sapphire substrates is facilitated by incorporation of a thin AlN buffer layer. For devices in previously used III-V semiconductors, dislocation densities in excess of about $10^3 \, \text{cm}^{-2}$ significantly degrade device performance, and densities greater than $10^6 \, \text{cm}^{-2}$ were too high for the operation of light emitting devices based on these semiconductors (see Section 5.1.2 and Fig. 5.2). This is because dislocations act as non-radiative recombination centres that significantly reduce the light emission. In addition, the recombination at the dislocations may generate heat, resulting in motion and multiplication of defects so that eventually the devices no longer emit light.

The unique characteristic of the III-nitride devices is that they can emit bright luminescence in spite of the high dislocation density (see Section 5.1.2). The dislocation concentration, due mainly to the lattice mismatch between the

III-nitrides and sapphire substrates, in the active layers of high efficiency GaN LEDs can be as high as 10^{10} cm^{-2} (for a review, see Jain *et al.* 2000). To explain this, it was first proposed that dislocations in fact did not act as recombination centres in these materials. However, it was later demonstrated (by TEM and CL) that dislocations do act as non-radiative recombination centres (e.g. Jain *et al.* 2000 and references therein). This was demonstrated by a substantial reduction in the density of the photo-generated minority carriers near dislocation lines (Rosner *et al.* 1997). From these experiments, a carrier diffusion length for holes of 250 nm was also derived. This conclusion was confirmed in other observations, using TEM and CL, which found a lower diffusion length for holes of 50 nm in *n*-type GaN (Sugahara *et al.* 1998). It was also found that the CL efficiency is high as long as the minority diffusion length is shorter than the averaged dislocation spacing. Thus, the explanation of the remarkable insensitivity of GaN-based devices to very high densities of dislocation may be that the dislocations are clustered in cellular structures while large regions are free of them. Thus, if the diffusion length of the injected carriers is sufficiently small, most carriers can recombine radiatively prior to reaching the dislocations. The cross-sectional and plan-view TEM studies of dislocations in epitaxial GaN clearly demonstrate the distribution of dislocations, which are clustered with the separation of about 300 nm (Ponce 1997). Such cellular arrangement of dislocations corresponds to a crystallite columnar structure with slight angular distribution in orientation (Ponce 1997).

The dislocation mobility in GaN-related materials at room temperature is substantially lower (by 10^{-10}–10^{-16}) than in GaAs (Sugiura 1997). Thus, in practice the dislocations do not propagate into the active region of the device, where they could be harmful. Neither can they multiply after growth.

The dislocation reduction with layer thickness differs between GaAs and GaN. In GaAs the elimination of threading dislocations typically occurs in a relatively short distance from the substrate surface because the line directions are inclined in this case. The dislocation density in GaAs falls roughly as $\sim 1/h$ as described by Equation (4.40). A slower rate of fall in density, i.e. as $\sim 1/h^{2/3}$, was obtained for GaN (Mathis *et al.* 2000). This was explained by lower probabilities of reaction between threading dislocations in GaN since line directions in this case are nearly normal to the layer thickness (Mathis *et al.* 2000).

Although LEDs based on such GaN material operate satisfactorily, with sufficient lifetime, in the case of laser diodes it was recognized that, due to the generation of excessive heat during operation, the dislocations cause device degradation. Thus, to optimize the performance of devices fabricated in films grown on lattice-mismatched substrates, an efficient extended-defect-reduction method of *lateral epitaxial overgrowth* (LEO) was developed (Sakai *et al.* 1997, Zheleva *et al.* 1997). This requires interrupting the initial growth of the GaN layer on a foreign substrate (e.g. sapphire, Si, or SiC) in order to deposit and pattern a dielectric film (SiO$_2$ or SiN$_x$) shown in black in Fig. 4.134. Further GaN growth begins in the uncovered

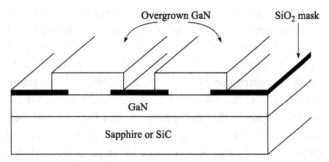

Figure 4.134 Diagram of the lateral epitaxial overgrowth (LEO) method which involves three steps: (i) interrupting the initial GaN growth on a foreign substrate (sapphire or SiC), (ii) depositing and patterning a dielectric layer (e.g. SiO_2), and (iii) continuing GaN growth resulting in GaN laterally grown over the dielectric stripes.

areas and spreads laterally over the dielectric film. Dislocations cannot move into these overgrown areas resulting in significant reductions in the threading dislocation density in the material over the dielectric stripes (see Fig. 4.134). Lasers made from this material have improved performance. For recent reviews on this topic, see Jain *et al.* (2000) and Beaumont *et al.* (2001).

Carrier scattering at threading dislocations and the effect of dislocations on several properties of the III-nitrides were discussed by Weimann *et al.* (1998), Look and Sizelove (1999), Speck and Rosner (1999), Cherns (2000), and Farvacque *et al.* (2001). A comprehensive review of the growth, characterization and properties of III-nitrides was provided by Jain *et al.* (2000), and a paper on the growth, doping, physical properties and device applications of GaN films grown by plasma-assisted MBE on sapphire substrates was presented by Moustakas *et al.* (2001).

4.7 Dislocations and point defects

Dislocations can emit or absorb native point defects, i.e. vacancies and interstitial atoms as well as impurities, as discussed in Section 2.3.12. They are important as sources and sinks of native point defects and make possible rapid attainment of the new thermodynamic equilibrium density when the temperature changes. Point defects are incorporated at the edge of the extra half-plane of edge dislocations, creating jogs so the dislocation becomes ragged and climbs, up or down, out of the slip plane. Point defect absorption by screw dislocations results in climb out onto the spiral ramp into which the crystal is turned by the screw. Screw dislocations subjected to heat treatment in the presence of high densities of point defects are turned into helical dislocations. Their presence indicates that such an episode has occurred in the history of the sample.

When jogs of height, h, greater than one Burgers vector length, b, are dragged by a moving screw dislocation, dipoles are left behind them (Section 2.3.12) as illustrated in Fig. 4.135b. Dipoles produced in this way are often seen in TEM studies of plastic deformation as, for example, in the case of the GaAs shown in Fig. 4.135a.

Leipner *et al.* (2000) deformed doped and undoped Si and GaAs samples in compression at temperatures from 900 to 1100 K by up to 27%. They employed positron annihilation spectroscopy to determine lifetimes at temperatures from 15 to 600 K. They considered the movement of screw dislocations with dissociated jogs as shown in Fig. 4.136 and noted that movement of the Frank partial jog could produce either vacancies or interstitials in the elemental semiconductors while in GaAs vacancies or interstitials of either Ga or As can be generated. They calculated the

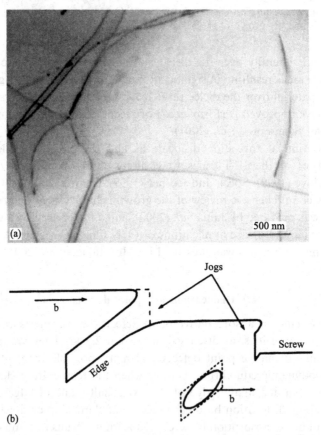

Figure 4.135 (a) TEM image of GaAs deformed at 625°C and containing dipoles. (b) Diagram of dipole formation by the dragging of jogs with $h > b$ by a dislocation moving into the plane of the figure. Segments of the two dislocations may climb and annihilate each other at points along the dipole to pinch off the dipole into separate prismatic loops like that shown behind the dislocation. Dragging jogs with $h = b$ produces vacancies or interstitials depending on the sign of the jog. (After Leipner *et al.* 2000, *Journal of Physics: Condensed Matter*, **12**, pp. 10071–8, 2000.)

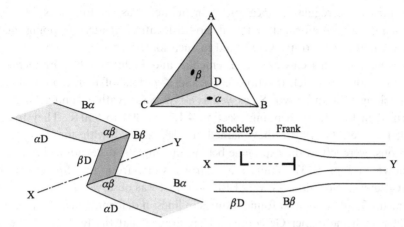

Figure 4.136 Diagram of an acute-angle extended jog on a screw dislocation with Burgers vector BD in Thompson's tetrahedron. The cross-section along X–Y shows the dissociation of the jog into a Shockley (mobile, the upright L) and a Frank (sessile, horizontal T) partial dislocation with the Burgers vectors indicated. The Frank partial can only move in the direction of slip by emitting or absorbing point defects. (After Leipner *et al.* 2000, *Journal of Physics: Condensed Matter*, **12**, pp. 10071–8, 2000.)

formation energies of strings of point defects such as would be immediately left behind dragged unit jogs and of point-defect clusters of increasing sizes for both Si and GaAs. The positron lifetime in relaxed three-dimensional clusters increases monotonically with the number of vacancies. Thus cluster sizes could be determined. In GaAs two defect related lifetimes, τ_{d1} and τ_{d2}, were found. These were ascribed to divacancies and V_{14} clusters. In Si the lifetime found was that for divacancies.

Antisites are atoms on sites that should be occupied by an atom of the other element in the compound, like Ga_{As}, which represents an atom of Ga on an As site. Leipner *et al.* (2000) point out that antisites can be formed from native point defects produced by jog dragging followed by the reactions $Ga_i + V_{As} \rightarrow Ga_{As}$ or $As_i + V_{Ga} \rightarrow As_{Ga}$. They observed shallow positron traps to be present in the deformed material, at an energy depth corresponding to that which had been found for Ga_{As}^- in irradiated GaAs.

Silicon is brittle at normal temperatures and pressures, but under high pressures it can be strained in compression at room temperature. Leipner *et al.* (2004) compared Si deformed both at 5 GPa and room temperature (which samples, however, contained microcracks with associated dislocation loops) and at 800° C using positron spectroscopy to study the trapping defects present. In the high-T deformed material there were two types of such centres. Large vacancy clusters had a positron-trapping lifetime of 500 ps while dislocation-bound vacancies had one of 280 ps. After additional annealing at 800° C the vacancy cluster traps disappeared but the vacancy centres bound to dislocations remained. The dislocation-bound vacancies acted as deep traps while regular dislocation segments acted as shallow ones. However, in room-temperature-deformed material no dislocation related trapping occurred. Only vacancy cluster traps were found in such material.

Eremenko and his co-workers (Eremenko and Fedorov 1995, Eremenko *et al.* 1999) used the Sirtl etch (Section 3.3.1) and a variant of it, for both photo- and dark-etching of *n*- and *p*-type Si, as well as *n*-type SiGe (with 0.8 and 4.8 at % of Ge) deformed in four-point bending (Section 4.1.3) at 800 to 950 K. They found that needle-like defects, which they named 'trails', aligned with the ⟨110⟩ or ⟨112⟩ directions, were left in the slip plane behind moving 60° dislocations in Si and SiGe as shown in Fig. 4.137. Trails are clearly distinct from the dipoles left behind moving screw dislocations. Strong EBIC contrast was observed at slip planes in such specimens. Trails were also found along slip lines in deformed $Si_{99.2}Ge_{0.8}$ but not in the SiGe with the higher Ge content. They argued that the trails might be a form of interstitial aggregate aligned with the ⟨110⟩ and ⟨112⟩ directions.

The electrical properties of the point-defect related debris left by dislocations moving through semiconductors during deformation are discussed in Section 5.1.4 and Section 5.9.1.

It was realised very early that impurities are important in determining the electrical and luminescence effects of dislocations and grain boundaries as is discussed repeatedly throughout this book. Later evidence made it clear that native point defect debris left behind moving dislocations also had electrical effects. Jones (2000) review, for example, again emphasized the need to distinguish the intrinsic properties of the dislocation core from those due to point defects or impurities. In summary, it is now well established that the interactions of dislocations with point defects and impurities affect (i) dislocation generation and multiplication, and (ii) dislocation dynamics (see Section 2.5.4 and Section 4.2), i.e. electrically active impurities

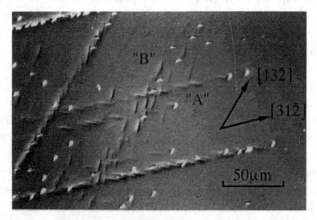

Figure 4.137 Etched (112) surface of deformed Si. The short needle-like defects lie in inclined rows along the slip lines of two glide systems. (After Eremenko *et al.* 1999.)

influence dislocation mobility and pinning, as well as (iii) control dislocation elec-trical and optical properties, and (iv) make possible important impurity gettering processes.

There are two general types of interaction of dislocations with point defects and impurities (e.g. Pizzini 1999, 2002): (i) direct, by the formation of chemical bonds between point defects and dangling bonds at dislocations, and (ii) indirect, i.e. via electrical or strain field interactions, or through the dislocations acting as sinks or sources of point defects.

4.7.1 Dislocation interactions with impurities

Impurity atoms, whether larger or smaller than the atoms they replace, can lower the elastic energy of the crystal by segregating to edge dislocations to reduce the hydrostatic stress. This energetic attraction results in the concentration of impurities around dislocation lines in 'Cottrell atmospheres' (Cottrell and Bilby 1949). The Suzuki effect (Suzuki 1952) is the preferential solution of an impurity in the crystal structure of the stacking fault in extended dislocations, lowering the fault energy and increasing the fault width (see Section 2.3.12). In some more ionic semiconductors there will be additional long-range electrostatic interactions between charged dislocations and ionized impurities and short-range reactions of some impurities with any broken bonds in dislocation cores.

Types of long-range or indirect interaction between point defects and dislocations that were treated early include (e.g. Bullough and Newman 1963) the first-order elastic interaction (due to the size misfit), the second-order size interaction (due to the difference in elastic constants), and electrical interactions (e.g. due to the dangling bonds).

The first-order elastic interaction, due to the size misfit, is the most important. The interaction energy $E_i(r)$ for a straight dislocation can be expressed as (Bullough and Newman 1963, 1970):

$$E_i(r) = \frac{4}{3} \frac{\mu b \delta r_0^3 \sin \theta}{r} \left(\frac{1+v}{1-v} \right) \qquad (4.41)$$

where μ is the shear modulus, v is Poisson's ratio, r is the distance between a point defect and the edge dislocation, b is the magnitude of the Burgers vector, and δ is related to ε (the size misfit between the host atoms and defects) as $\delta = 3\varepsilon (1-v)/(1+v)$. In this case, r_0 and $r_0(1+\varepsilon)$ are the radii of host atoms and defects, respectively. Thus, the interaction energy is inversely proportional to the dislocation-impurity separation. It is a maximum about one atomic distance from the dislocation core centre and it is typically a fraction of an eV (e.g. Bullough and Newman 1963, Pizzini 2002). The second-order size interaction is proportional to the inverse square of the defect separation (Bullough and Newman 1963, 1970).

Dangling bonds in the cores of dislocations with edge components act as acceptor states. In n-type semiconductors, therefore, on Read's model (see Section 5.2), dislocations have a negative line charge. This is screened by a positive space charge in a surrounding cylindrical region (see Fig. 5.6). The negative line and positive space charges are responsible for the electrical interactions with point defects and impurities.

Point defects in the presence of dislocations can move by diffusion and drift. Diffusion is the thermally activated transport of point defects down a concentration gradient, while drift is thermally activated movement in the direction of an elastic or electrical attraction to e.g. dislocations.

The total diffusion-drift flow is (e.g. Bullough and Newman 1963, 1970):

$$F(r, t) = -D\left[\nabla C(r, t) + \frac{C(r, t)}{kT} \nabla E \right] \qquad (4.42)$$

where E is the total interaction energy, D is the diffusion coefficient and $C(r,t)$ is the defect concentration at a distance r from a long straight dislocation at time t. (The first term expresses Fick's Law that the rate of diffusion is proportional to the concentration gradient. The second term expresses the effect of the gradient in the interaction energy, E.)

Diffusion generally follows Fick's Law. A number of cases of 'anomalous', non-Fickean diffusion in semiconductors are known, however. One of these involves dislocations and will be discussed in Section 4.7.2.

The concentration, C, in the stress field, if Fick's Law is followed, must satisfy the equation (e.g. Bullough and Newman 1963, 1970):

$$\frac{\partial C(r, t)}{\partial t} = D\left[\nabla^2 C(r, t) + \frac{\nabla [C(r, t)\nabla E]}{kT} \right] \qquad (4.43)$$

(Again the first term is for the diffusion and the second for the drift.)

The interaction of dislocations with impurities strongly affects semiconductor dislocation dynamics (see Section 4.2) and the mechanical properties of semiconductors as outlined by Sumino and Yonenaga (2002).

The interactions of dislocations with point defects and impurities make possible gettering (see Section 5.12.1), which renders unwanted impurities electrically inactive by employing defects as sinks, to which the impurities 'get'. In gettering, the material is heated to activate the diffusion of impurities to the defects where they precipitate and become electrically inactive. Segregation of impurities to an extended defect is defined as the process of formation of an impurity-rich region around that defect. This contrasts with early stage precipitation in which the impurity condenses to form a new phase (Sumino 2003). Which of the two, precipitation or impurity segregation, occurs at an extended defect depends on whether a short-range reaction takes place between the impurity atoms and extended defect. Such a possible reaction depends on the core structure of the dislocation, the type and concentration of impurities, and the cooling rate of the crystal (e.g. Sumino 1999, 2003).

If the impurity concentration does not exceed the solubility limit and remains below the lattice site density, the impurity distribution in thermal equilibrium around a dislocation can be described by Boltzmann statistics and is a Cottrell atmosphere. Then the impurity concentration around the dislocation can be expressed as:

$$C(r) = C_0 \exp(-E_i/kT) \tag{4.44}$$

where C_0 is the mean concentration of impurities in a dislocation-free lattice and $C(r)$ is the concentration at the point where the impurity-dislocation binding energy is E_i.

4.7.2 The role of dislocations and grain boundaries in diffusion

As mentioned in relation to Equation (4.42), the rate of diffusion of impurities is usually directly proportional to the concentration gradient for that type of atom. The constant of proportionality is D, the diffusion coefficient. This is the well-known Fick's Law. When this 'Law' is not obeyed, the diffusion is said to be anomalous or non-Fickean.

The first example of diffusion in semiconductors that did not correspond to a single constant value of D, to be discovered and explained theoretically, was the rapid, structure-sensitive diffusion of Cu in Ge. It was observed that when copper was diffused into n-type Ge, converting it to p-type, the p-n junction occurred at a shallow depth in defect free regions, with long protuberances of approximately constant radius, along the dislocation lines (Tweet and Gallagher 1956). This could not be explained as due to rapid diffusion down the dislocations as this would require impossibly rapid diffusion in these regions. Frank and Turnbull (1956) suggested that Cu dissolves in Ge both substitutionally and interstitially. Substitutional Cu atoms are those occupying Ge lattice sites and are more numerous, by about two orders of magnitude, than the interstitial Cu atoms. The latter sit in the relatively large spaces between Si atoms, called interstices. However, the substitutional atoms diffuse much the more slowly. Frank and Turnbull (1956) suggested that for substitutional diffusion D was of the order of $6 \times 10^{-15} \, \mathrm{cm}^2 \, \mathrm{sec}^{-1}$, while for interstitial diffusion it was of the order of $6 \times 10^{-8} \, \mathrm{cm}^2 \, \mathrm{sec}^{-1}$. The fast diffusing interstitial atoms drop into vacancies and become virtually fixed in substitutional positions. Since dislocations can act as sources of vacancies this occurs preferentially near dislocations lines, where more vacancies are available. Hence the observation of Cu along the dislocations at much greater depths than the slow moving substitutional diffusion front, i.e. the depth of the junction in dislocation-free material. When there are many sources of vacancies, i.e. in highly imperfect crystal, the interstitials are rapidly absorbed into substitutional sites everywhere, so no anomalous behaviour occurs. However, in highly perfect crystals, there are often not enough dislocations to provide the necessary density of vacancies throughout the bulk.

The type of fast diffusion involving both substitutional and interstitial atoms is referred to as dissociative because the inter-conversion of the two forms can be thought of in terms of a dissociation:

substitutional impurity ↔ vacancy + interstitial impurity

This phenomenon is thought to operate in such important cases as the diffusion of Au in Si and of Zn in GaAs (Willoughby 1968, Lee 1974, Warburton and Turnbull 1975).

Another example of non-Fickean diffusion is the more rapid diffusion along grain boundaries, which will be discussed below in this section (see below the subsection *Rapid diffusion along grain boundaries and 'dislocation pipes'*).

Diffusion-induced misfit dislocations

In some important cases, the in-diffusion of dopants can result in slip, introducing many misfit dislocations into devices. Misfit dislocations accommodate 'misfit', that is, a change of lattice parameter between adjacent regions in a crystal. This may be lattice parameter misfit between an epitaxially grown layer and the substrate (see Section 2.3.11, Section 4.5.2 and Section 4.6). Alternatively, the change of lattice parameter may occur at a boundary between a region that is heavily doped with an impurity that has a very different atomic size from the host atom(s) and a less heavily doped or undoped region. Then the boundary is a diffusion or growth interface. When a high concentration of an element with a large misfit diffuses, in the steepest portion of the concentration profile, there will be a steep lattice parameter gradient. This constitutes a localized strain corresponding to an elastic stress. At the high temperatures required for diffusion, semiconductors are plastic. The diffusion-induced stress can, therefore, suffice to cause plastic deformation so many dislocations glide toward the diffusion front, to relax the strain.

Prussin (1961) and Czaja (1966a) treated this by elasticity theory. Prussin assumed that the strain components ε are proportional to the concentration, C, i.e. $\varepsilon = -\beta C$, where β is a constant. Let the concentration gradient run in the z-direction and consider an element of the material lying between z and $z + dz$ in which the dislocation density is ρ. The dimensional change in an orthogonal direction, dx, is given by $\rho \alpha dz$ where α is the component of the Burgers vector along x. If the lattice strain due to the solute is entirely compensated by dislocations then

$$dx = \beta(\delta C/\delta z)dz = \rho \alpha dz \qquad (4.45)$$

so that

$$\rho = (\beta/\alpha)\, \delta C/\delta z \qquad (4.46)$$

Prussin and Czaja discussed diffused layers in detail using this relation between dislocation density and concentration gradient. The dislocations generally glide in on several different slip systems and undergo complex interactions to produce misfit dislocation networks.

The details of the slip and interaction processes vary with (i) the diffusion conditions used, (ii) the orientation of the crystal and (iii) the material concerned (Czaja 1966a). Diffusion-induced misfit dislocations were found to be introduced by the in-diffusion of high densities of B or P into Si and of Zn into GaAs. Misfit dislocations are found especially near the edges of areas diffused through masks. The best-known case is that of 'emitter edge' dislocations at the periphery of diffused emitters in planar Si devices (Schwuttke 1970) as seen in Fig. 4.138. Diffusion-induced misfit dislocations were studied (i) by x-ray topography (e.g. Schwuttke and Queisser 1962, Blech *et al.* 1965, Black and Jungbluth 1967, Schwuttke 1970 and Ghezzi and Sevidori 1974), (ii) in two of the first SEM-EBIC papers (Czaja and Wheatley 1964 and Czaja 1966b), and (iii) by etching (Czaja 1966a). However, the resolutions of these techniques (about 1 µm) cannot resolve the individual dislocations at the high densities often produced by diffusion-induced slip. For this TEM was used by Washburn *et al.* (1964), Joshi (1965), Joshi and Wilhelm (1965), and Levine *et al.* (1967a and b). It was confirmed that the nature of the misfit dislocation networks in the (111) orientation in Si was determined by the glide and interaction processes.

Rapid diffusion along grain boundaries and 'dislocation pipes'

Another form of deviation from simple Fick's Law diffusion arises from the more rapid diffusion along extended defects especially grain boundaries. Grain boundaries

Figure 4.138   X-ray topograph of 'emitter edge' dislocations due to diffusion-induced stresses at the edge of the triangular emitter area of a planar silicon transistor. (After Schwuttke 1970. Reprinted from *Microelectronics and Reliability*, **9**, Silicon material problems in semiconductor device technology, pages 397–412. Copyright 1970, with permission from Elsevier.)

can be the locale of numerous bonds and small atomic-scale voids and in a given time the impurity penetrates farther along such boundaries than elsewhere. This was found to occur in Ge and Si (Karstensen 1957, Queisser *et al.* 1961). A consequence is that the presence of even small-angle, so called 'mosaic' boundaries, leads to the formation of non-planar *p-n* junctions in, for example, Zn diffused GaAs as shown in Fig. 5.78 and discussed in Section 5.10.1, generally with disastrous results for device operation.

Rapid 'pipe' diffusion along dislocation lines was often suspected of causing problems in device production. This is based on experimental observations indicating that dislocations can modify the diffusion coefficient. For instance, Braga *et al.* (1994) demonstrated (using SEM EBIC) greatly enhanced diffusion of arsenic along individual misfit dislocations in SiGe/Si heterostructures. It was estimated that the dislocation diffusion coefficient for arsenic was several orders of magnitude greater than that in the host crystal (Braga *et al.* 1994). (Braga *et al.* 1994 list earlier references to the literature on this topic.) In general, pipe diffusion may cause problems especially in lattice-mismatched epitaxy (e.g. GaAs/Si and GaN/Al$_2$O$_3$) generating high dislocation densities in the interface region, which may become conductive as a result of donor decoration of dislocations (e.g. Look *et al.* 2001). Dislocations are known to correlate with microplasma breakdown sites in reverse-biased Si *p-n* junctions. This, however, is now known to be the effect of metal precipitates decorating heavily contaminated dislocations.

The best evidence pointing to 'pipe' diffusion along dislocations is the clear observation of rapid diffusion along grain boundaries, especially in the case of small-angle boundaries, which can be resolved into walls of dislocations (see Section 2.3.10 and Fig. 2.10). Karstensen (1957) reported not only the preferential diffusion of Sb along small-angle grain boundaries in Ge but also a dependence of this effect on the direction of the dislocation lines in the boundary.

4.7.3 Dislocations and volume defects

Dislocations also interact with volume defects in important ways. However, because of the high purity of semiconductors, many of the volume defects commonly encountered in metals, such as second phase material and relatively large precipitate particles, have seldom been seen. Antiphase domains occur commonly in the growth of sphalerite structure compounds on diamond structure elements and small precipitates decorating dislocations can occur under special circumstances and these will be dealt with here.

Decoration

Impurities are attracted to dislocation lines. These, therefore, act as important sites for the precipitation of concentrations of impurities that come to exceed the solid solubility limit as the temperature is lowered. Thus dislocations are sometimes introduced into the backs of Si slices by mechanical polishing to 'getter' unwanted impurities (see Section 5.12.1). An early technique for rendering dislocations visible was to heat treat a crystal containing an impurity, so precipitates formed along the dislocation lines. The historically important work of Dash was based on precipitating Cu on dislocations and using transmission infrared microscopy to observe them (Dash 1957, 1958). The picture of a Frank-Read source in Fig. 2.29 is such a decoration micrograph. The formation of long needle-like precipitates containing Fe and other impurities on dislocations in Si processing can be fatal, i.e. lead to short-circuiting and rejection of the device or circuit on test (see Section 5.11.2).

Yonenaga and Sumino (1996) studied the effect of the precipitation of oxygen impurities along dislocations on the mechanical strength of Czochralski-grown Si. They found that the yield strength of Si crystals increased through the early precipitation stage, but fell during the later stage (Yonenaga and Sumino 1996). The increase of yield strength during early precipitation was attributed to the locking of the dislocations by small closely spaced precipitates. However, during the later stage, the locking effect is reduced as the larger precipitates grow at the expense of the smaller ones. Then the longer dislocation segments between the larger, more widely spaced precipitates can move at lower stresses (Yonenaga and Sumino 1996). [Equation (2.26) gave the stress to operate a Frank-Read source, i.e. a length of dislocation pinned at its ends, by precipitate particles in this case, as inversely proportional to its length L. Thus the longer segments between the larger precipitates will 'blow out' and generate dislocations to produce deformation at lower stresses.]

This phenomenon is well known as 'overaging' in precipitation hardened metallic alloys because a reduction of hardness results from allowing precipitation to continue overlong.

Fig. 4.139 presents TEM images of dislocations in samples after precipitation at 900°C for times of (a) zero, (b) 0.75, (c) 1.5, and (d) 3.0 hours (Yonenaga and Sumino 1996) and shows that both the particle size and separation increase with time. The size d and the average spacing l of precipitates are plotted against the time at 900°C in Fig. 4.140. The increasing deviation of dislocations from straight lines (see Fig. 4.139) for longer heat treatment times was attributed to dislocation climb (Yonenaga and Sumino 1996).

Precipitates and dislocation sources

Equation (2.26) ($\tau_o = \alpha Gb/R$) gave the stress, τ_o, needed to operate a Frank-Read source of length R. The surface of a precipitate (or void or crack) with a suitably large dimension, R, can also act under stress to generate large numbers of dislocations in a similar way. This is well known from TEM observations on plastically deformed semiconductors. Moreover, if a dislocation runs from one precipitate to another so they act as obstacles to its motion, that 'pinned' segment can act as a

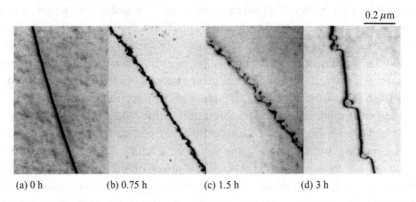

0.2 μm

(a) 0 h   (b) 0.75 h   (c) 1.5 h   (d) 3 h

Figure 4.139   TEM images showing the variation in size and separation of precipitate particles along decorated dislocations for the increasing precipitation heat treatment times at 900° C marked. (After Yonenaga and Sumino 1996. Reprinted with permission from *Journal of Applied Physics*, **80**, pp. 734–8. Copyright 1996, American Institute of Physics.)

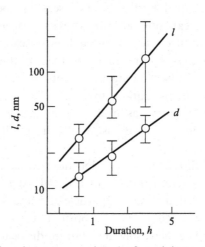

Figure 4.140   The size, *d*, and average spacing, *l*, of precipitates as a function of increasing precipitation times at 900° C. (After Yonenaga and Sumino 1996. Reprinted with permission from *Journal of Applied Physics*, **80**, pp. 734–8. Copyright 1996, American Institute of Physics.)

Frank-Read source. Such segments bowing out between the precipitates generate dislocation loops at the stress given by Equation (2.26).

Precipitation hardening, due to precipitates acting as obstacles to dislocation glide, is important in metallic alloys. This can also occur in semiconducting alloys and even in heavily doped elements and compounds. It is known that the addition of certain impurities to InP, etc. produces harder, more damaged and deformation resistant material for substrate. The role of precipitates changes with their spacing. At small spacing, *R*, the dislocations cannot bow between the precipitates but must cut through them if plastic deformation is to occur. At some larger spacing, τ_o becomes less than the cutting stress. In the case of duralumin, the first important aeronautical alloy, it was discovered that the growth of the precipitates, by the diffusion of the Cu through the Al, could occur at temperatures only slightly above ambient. The result was that, during storage, the material first got harder (more resistant to deformation) with time, as the precipitates nucleated closely spaced obstacles. The phenomenon is, therefore, known as age hardening. Under continued storage in such conditions, however, the larger precipitates grow at the expense of the smaller ones. The spacing of the precipitates, R, then grows, so τ_o and, consequently, the hardness falls with time. This is known as 'overaging'. It is not known whether such phenomena occur in semiconductors.

4.8 Growth and processing induced defects

Most bulk crystals are produced by Czochralski pulling as circular-cross-section boules and sliced into wafers. These are etch-polished for further processing into devices or for use as substrates for epitaxial growth. It is thus useful to distinguish between (i) growth defects in bulk crystals (wafers) and epitaxial layers, and (ii) process induced defects.

Defects in as-grown crystals fall into four broad categories: (i) point defects (Section 6.1 and Section 6.3), (ii) dopant (or impurity) striations (Section 6.5.4), (iii) microdefects (point defect clusters, Section 6.5.5), and (iv) dislocations (Mahajan 1989). To complete the picture, we will briefly discuss striations and microdefects below before turning to dislocations and growth defects that are often characteristic of the growth of a particular material or by a particular technique of epitaxy.

Impurity growth striations appear as regular roughly parallel sets of lines or stripes. In Czochralski growth, a rotating seed crystal is pulled up out of the melt. Striations in Cz crystals are spiral ramps with pitch equal to the pull velocity divided by the number of rotations per unit time (see Fig. 6.6). Growth striations in melt-grown crystals arise from convective fluctuations in the melt. These cause growth-rate fluctuations and hence periodic variations in the grown-in impurity concentrations, i.e. striations (for more detail, see Section 6.5.4). Impurity growth striations in melt-grown crystals can be suppressed by magnetic quenching of the convection currents in electrically conducting melts.

During cooling, after crystal growth, point defects can condense to produce microdefects. These are distributed in typical swirl patterns observed in cross-sectional slices (see Section 6.5.5 and Fig. 6.7).

Grown-in defects occur either in the bulk or in interfaces. Bulk defects include precipitates, dislocations and dislocation loops. Interface defects include dislocations, dislocation clusters, stacking faults, microtwins, and misfit dislocations. As discussed in Section 4.6, defects such as dislocations present in the substrate can also propagate into the epitaxial layer.

To obtain a high yield of reliable semiconductor devices, grown-in and process-induced defects must be minimized and passivated or eliminated. In practice this requires (i) the identification of these defects and their sources, as well as (ii) their structural and electronic properties and device effects, and (iii) the discovery of means for either eliminating, passivating or using these defects.

Much of semiconductor materials development effort in the early days was devoted to minimizing and, eventually in a few cases, eliminating extended defects from as-grown crystals [see Section 1.1.2 for the story of Dash's (1958) success in developing a Czochralski method of growing large dislocation-free semiconductor crystals]. However, each new material and growth technology raises new challenges. Often new forms of grown-in defect come to the fore as discussed in this section. For further details of this large subject, reference should be made to books, conference proceedings and journals on crystal growth especially in the *Journal of Crystal Growth*.

Similarly a large continuous effort is devoted, especially in industry, to minimizing the defects introduced by new production processes. Unfortunately, often considerations of commercial advantage, in having eliminated a problem without telling competitors, and of prestige delay or prevent publication. (Sales people do not like the idea of publishing pictures of defects in the company's products.) However, some of this work does get published and reviewed. Growth and processing-induced defects and their role in, especially silicon, devices were reviewed by Queisser (1969), Schwuttke (1970), Jowett (1979), Ravi (1981), Mahajan (1989, 2000), Claeys and Vanhellemont (1993), Holt (1996), Queisser and Haller (1998), and Cerva *et al.* (2001) and the role of defects in devices is discussed in Section 5.11 and Section 5.12.

Early work on the effect of process-introduced structural defects on Si transistors clearly correlated poor reverse junction characteristics with dislocations in junction areas (Jungbluth and Wang 1965). These authors employed x-ray topography at several stages in the processing of electrochemically and mechanically polished wafers to trace the origin of device defects. They also concluded that electropolished wafers yield more acceptable junctions than mechanically polished slices (Jungbluth and Wang 1965).

4.8.1 Grown-in defects

Mahajan (1989) lists three distinct sources of dislocations incorporated during crystal growth. These are (i) dislocations in seed crystals that propagate into a growing material, (ii) dislocations formed at the edges of the growing crystal, that propagate into the interior due to stresses caused by thermal gradients (so-called quenching stresses), and (iii) point defect supersaturation, during post-growth cooling, that result in clustering to form dislocation loops. The density of dislocations is greatly reduced by high concentrations of some dopants mainly due to impurity-induced lattice strengthening. This reduces the probability of dislocation formation as a result of thermal gradient stresses. In addition, impurity atom interactions with point defects inhibit clustering into dislocation sources (e.g. Mahajan 1989, 2000).

Reductions in grown-in dislocation densities in doped InP and GaAs were reported by Seki *et al.* (1976, 1978) who proposed a model for cases of strong dopant-host lattice atom bonding. Then dislocations will be pinned to the impurities, since glide requires breaking crystal bonds near the dislocation. Ehrenreich and Hirth (1985) proposed a model for In-doped GaAs crystals in which InAs$_4$ tetrahedral units, acting like solute atoms in metals, cause solution-hardening and reduce the dislocation density. Guruswamy *et al.* (1989) showed that adding In, Sb or B raised the critical resolved shear stress resulting in reduced dislocation densities in as-grown GaAs crystals. Lattice strengthening by dopants could be due to the fact that some impurities reduce the mobility of dislocations electronically (see Section 4.2.2) rather than to solution hardening. Yonenaga *et al.* (1984) showed for Si that strengthening by oxygen was related to locking of dislocations rather than impurity-reduction of the dislocation mobility (see also Yonenaga and Sumino 1992 for the case of GaAs crystals doped with In, Zn, and Si, and Sumino and Yonenaga (2002) for a review of the interactions of impurities and dislocations, especially of their influence on dislocation electrical properties and the mechanical strength of semiconductor crystals).

Grown-in defects in epitaxial layers include misfit dislocations (see Section 4.6), threading dislocations, and stacking faults (see e.g. Mahajan 1997). Threading dislocations in homoepitaxial and lattice-matched heteroepitaxial layers originate mainly from dislocations in the substrate. The mechanisms for the formation of misfit dislocations in many different mismatched systems are known. These involve either the propagation of substrate threading dislocations into the epitaxial layer, or the formation of new dislocations. There are methods for preventing such nucleation and propagation of dislocations. These include incorporation of graded buffer layers, promotion of dislocation interaction and annihilation, selected area growth, and insertion of strained-layer superlattices (see Section 4.6).

The frequent appearance of an otherwise rare type of defect in a particular material or in material produced by a particular technique is typical of semiconductor technology. These include tetrahedral stacking faults in epitaxial silicon, hillocks in MOVPE layers, oval-shaped defects in MBE-grown III-V compound layers, grappes in Czochralski InP and micropipes and nanopipes in SiC and GaN (see Mahajan 1997).

Tetrahedral stacking faults in epitaxial silicon

At one time epitaxial Si always grew containing large numbers of tetrahedral stacking faults. These were nucleated at the substrate surface as the result of carbon contamination, e.g. from the gases used for vapour phase epitaxy. Stacking faults occurred on the {111} planes corresponding to the faces of Thompson's tetrahedron inclined to the growth direction. Hence the tetrahedra grew upwards and outwards from a point as shown in Fig. 4.141 to intersect the epilayer surface in squares on (100) films (Charig *et al.* 1962, Finch *et al.* 1963) or triangles on (111) films, with sizes that increased with epilayer thickness (Booker 1964). In growth on (111) substrates, the fault tetrahedron consists of three faults lying in the {111} planes other than (111), connected by stair-rod dislocations lying along the edges of the tetrahedron as in Fig. 4.141. Etching reveals many triangles corresponding to the edges of the faults (Queisser *et al.* 1962, Booker and Stickler 1962, Charig *et al.* 1962, Miller *et al.* 1963 and Unvala and Booker 1964) as shown in Fig. 4.142.

In addition to complete stacking fault tetrahedral, a variety of other open and closed stacking fault figures can be formed. These produced the various linear etch figures seen in Fig. 4.142 and this has been confirmed by TEM (Booker and Howie 1963). Detailed accounts of these defects and of dislocation mechanisms for their formation were given by Booker (1964, 1966) and Mendelson (1964a, 1964b, 1967), and for a review of mechanisms for the nucleation of stacking faults in epitaxial films see Stowell (1975).

Tetrahedral stacking faults often occurred together with the more serious tripyramid defects, to be discussed next (Booker 1964, 1965). Elimination of carbon contamination later eliminated these defects.

Tetrahedral stacking faults occur in other materials like GaP, as shown in Fig. 4.143, but are uncommon and have been little studied.

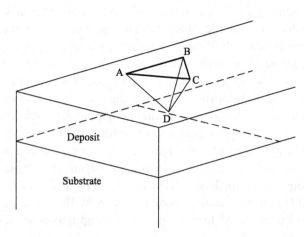

Figure 4.141   Schematic illustration of a tetrahedral stacking fault initiated at a point D on the substrate-deposit interface during epitaxial growth of Si on a (111) face of Si. The three faces DAC, DAB and DBC are all stacking faults which grew upward with the growing epitaxial layer. The dark triangle ABC of stacking fault edges can be revealed by etching as in Fig. 4.142. (After Pashley 1965. Reprinted with permission from *Advances in Physics*, **14**, pp. 327–416; Taylor & Francis Ltd., the Journal's web site: http://www.tandf.co.uk/journals.)

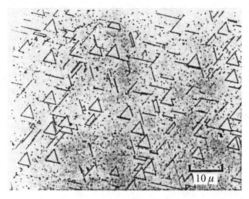

Figure 4.142   Light micrograph of an etched surface of a 4 μm thick layer of Si grown by evaporation in vacuum onto a (111) Si substrate. The lines and triangles mark the intersections of stacking faults with the surface. (After Unvala and Booker 1964. Reprinted with permission from *Philosophical Magazine*, **9** (1964), pp. 691–701; Taylor & Francis Ltd., the Journal's web site: http://www.tandf.co.uk/journals.)

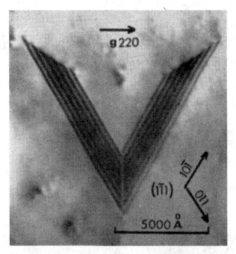

Figure 4.143   Transmission electron micrograph of a tetrahedral stacking fault in GaP taken with only the 220 reflection excited. One of the faults is invisible while the other two appear in fringe contrast. (After Chase and Holt 1972.)

Tripyramids

In the early growth of epitaxial silicon, in addition to stacking faults, microtwins and closely related tripyramids were found. Microtwins are small lamellae of first order twins on inclined {111} planes, ranging in thickness from a few atoms to perhaps a micron. They are equivalent to small regular stacks of stacking faults. Tripyramids are relatively large defects, visible by light microscopy as shown in Fig. 4.144a.

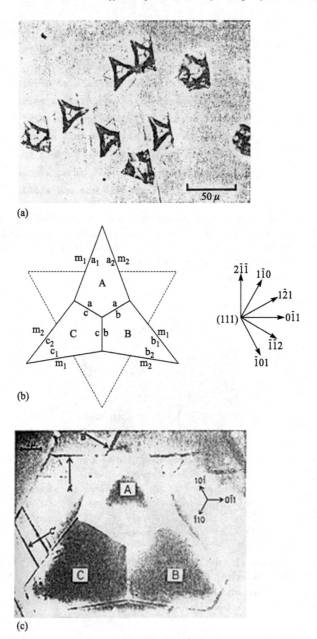

Figure 4.144 (a) Light micrograph of an etched epilayer of silicon showing tripyramids. (b) The ideal form of tripyramid in thick (111) Si deposits. The continuous lines outline the tripyramid and the broken lines indicate the associated defects. (c) Transmission electron micrograph of a tripyramid (A,B,C) and associated defects (A′, B′, C′) in a (111) Si epitaxial deposit. (After Booker 1965. Reprinted with permission from *Philosophical Magazine*, **11** (1965), pp. 1007–20; Taylor & Francis Ltd., the Journal's web site: http://www.tandf.co.uk/journals.)

Early studies (Theurer 1961, Miller *et al.* 1963) led to the suggestion that tripyramids consist of an array of doubly twinned material. Booker (1965) used transmission electron microscopy to show that tripyramids have the structure shown in Fig. 4.144b and 4.144c. Each of the regions A, B and C is doubly twinned with respect to the substrate orientation. The first twinning plane is (111), that of the substrate surface, and the second twinning occurs on a different one of the three inclined {111} planes of the Thomson tetrahedron, namely ($\bar{1}$11), (1$\bar{1}$1) and (11$\bar{1}$) in each of the three regions. Careful examination showed that the three twinned regions do not meet at a point, but a small triangular region, usually a few hundred nm across, is present at the triple junction. This Booker showed to consist of a singly twinned region on the (111) surface. The associated defects like A′, B′ and C′ in Fig. 4.144c are mainly singly twinned microtwin lamellae lying on the inclined {111} planes.

There is evidence that carbon contamination is also responsible for the nucleation of tripyramids. Unvala and Booker (1964) found that coal gas contamination of vacuum evaporated Si layers resulted in the growth of tripyramids. Booker and Joyce (1966) showed that carbon contamination resulted in the formation of surface SiC after annealing Si substrates at 1200° C and that subsequent growth resulted in multiple twinning.

Oval defects in epitaxial GaAs

Oval defects, sometimes referred to as 'coffee beans', from about 1 to 10 μm long, are characteristic of MBE-grown GaAs and other Ga-containing epitaxial layers (e.g. Mahajan 1997). There are a number of causes of these defects. Any particles contaminating the substrate surface or deposited during growth, especially Ga droplets produced by Ga effusion cells, can nucleate dislocations and/or stacking faults and twins or just lead to faster or irregular growth. The oval shape of these defects is a consequence of the symmetry of the (100) substrate orientation that is now universally used. There are two ⟨110⟩ directions at right angles in the (100) plane and they are of different polar character. One is a fast growth direction for inclined {111} planes and the other is a slow one. Consequently the defects grow out from a centre further in the one direction than the other resulting, generally, in a growth hillock with an oval outline on the surface. Often, growth from the original particle results in a visible central dot and a spine along the long axis of the oval so the defect resembles a coffee bean.

A particular, detailed model for oval defect development from a droplet is presented in Fig. 4.145. According to Mahajan (1997), the droplet dissolves the underlying substrate and produces a pit bounded by {111}$_{III}$ and {$\bar{1}\bar{1}\bar{1}$}$_V$ planes (see Fig. 4.145b). The pit is asymmetric since {$\bar{1}\bar{1}\bar{1}$}$_V$ planes are removed faster than {111}$_{III}$ planes (see Section 4.1.3). Concurrently, the droplet is exposed to arriving group III and V species. The group V species diffuse through the droplet to form an epitaxial III-V layer on the pit facets (see Fig. 4.145c). Another possibility is that

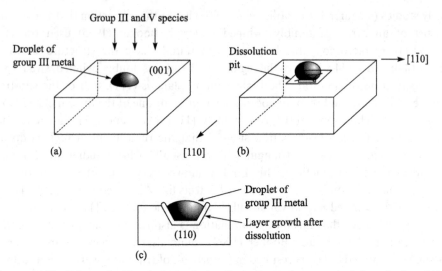

Figure 4.145 Schematic illustration of the formation of an oval defect during MBE growth of a III-V homoepitaxial layer: (a) a group-III metal droplet on a (001) III-V substrate, (b) a pit forms due to substrate dissolution, and (c) epitaxial growth on $\{111\}_{III}$ and $\{\bar{1}\,\bar{1}\,\bar{1}\}_V$ facets and the (001) plane in the presence of group-V species results in an oval defect. (After Mahajan 1997. Reprinted from *Progress in Materials Science*, **42**, Defects in epitaxial layers of compound semiconductors grown by OMVPE and MBE techniques, pages 341–55. Copyright 1997, with permission from Elsevier.)

polycrystalline semiconducting regions form on the droplet surface and group V species diffuse through it. If growth is terminated prior to complete transformation of the droplet into the semiconductor, during cooling polycrystalline and twinned regions form resulting in an oval defect (Mahajan 1997).

Oval defects usually cause surface and interface roughness, resulting in carrier scattering and reduced carrier mobility in HEMTs.

Grappes in InP

Another type of defect is rare except in LEC (liquid-encapsulated Czochralski) InP. First observed as clusters of etch pits, this was named the 'grappe' (French for bunch of grapes). Grappes were found in slices of LEC InP with several dopant impurities and any orientation (Stirland *et al.* 1983, Franzosi *et al.* 1984) but especially when the encapsulating boric oxide contained moisture (Brown *et al.* 1984). Grappes were studied by etching and stress birefringence infrared microscopy (see Section 3.3.1, Fig. 3.2 and Elliott and Regnault 1981, Elliott *et al.* 1982 and Stirland *et al.* 1983), x-ray topography (Stirland *et al.* 1983, Franzosi *et al.* 1984), SEM CL (Franzosi *et al.* 1984) and TEM (Augustus *et al.* 1983).

Grappes arise by 'prismatic punching' (Franzosi *et al.* 1984). When an indenter is driven into a single crystal slice of a ductile material, a plug of material is driven down through the slice. Such slip is called prismatic punching and occurs by dislocation loops slipping along prismatic surfaces down through the crystal. In LEC InP numerous centres of outward stress occur. These are precipitates or gas bubbles, some spherical and some triangular (Augustus *et al.* 1983, Brown *et al.* 1984 and Franzosi *et al.* 1984). As the crystal cools it contracts more rapidly than the precipitate, which causes prismatic punching outwards on all the slip systems. This sends out sets of dislocation loops, like smoke rings blown outward. The clusters of pits originally observed occur where the surface of the slice cuts the dislocation loops.

Growth hillocks in InP and GaN

Other growth defects that have received some attention include growth hillocks in InP (Gleichmann *et al.* 1990) and GaN (Herrera Zaldivar *et al.* 2001).

Electron microscopy studies found hillocks in epilayers formed mainly at stacking faults and dislocations (Gleichmann *et al.* 1990). Herrera Zaldivar *et al.* (2001) observed REBIC dark-bright contrast at the perimeter of growth hillocks in Si-doped GaN films (Fig. 4.146). Such dark-bright contrast was often seen at electrically charged grain boundaries in semiconductors (see Section 5.12.4). From their observations, Herrera Zaldivar *et al.* (2001) concluded that the hillock edges in this GaN material similarly had associated space-charge regions with corresponding energy band bending acting like Schottky barriers back-to back. They explained this on the basis of non-uniform distributions of charged defects and impurities at the hillocks (Herrera Zaldivar *et al.* 2001). This was corroborated by earlier CL measurements on the same films.

Micropipes and nanopipes in SiC and GaN

Silicon carbide (SiC) and gallium nitride (GaN) are promising wide-energy-gap materials for electronic and optoelectronic device applications. A major problem for these applications is the presence of micropipes in SiC (e.g. Heindl *et al.* 1997) and nanopipes in GaN (e.g. Qian *et al.* 1995, Cherns 2000).

Silicon carbide (SiC) is being developed for high-temperature, high-power, and high-frequency electronic devices, as well as high-brightness blue and green light-emitting devices. This is largely due to SiC's combination of a wide energy gap, high thermal conductivity, high saturation electron drift velocity, high breakdown electric field, and superior physical and chemical stability. SiC is also one of the best choices as a substrate for GaN-based light-emitting and power devices. A major problem is the formation of SiC micropipes (i.e. hollow tubes from about 0.1 to 10 μm diameter) penetrating the crystals along the growth direction. The presence of micropipes typically results in premature electrical failures in large-area (greater than 1 mm × 1 mm) SiC devices. For very-high-power device applications requiring much larger areas, the presence of even a single micropipe may be fatal. In SiC substrates, these defects act as sources to generate defects in epitaxial layers.

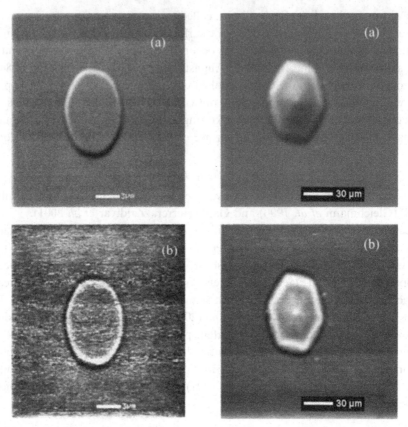

Figure 4.146   (a) Secondary electron image and (b) REBIC image at room temperature of a rounded hillock (left) and a pyramidal hillock (right) in Si-doped GaN films. (After Herrera Zaldivar *et al.* 2001. Reprinted with permission from *Journal of Applied Physics*, **90**, pp. 1058–60. Copyright 2001, American Institute of Physics.)

Gallium nitride (GaN) is another most promising semiconductor for optoelectronic devices. GaN with other III-V nitrides, like InN and AlN, all with the wurtzite structure, forms continuous alloy systems with direct energy gaps ranging from about 1.9 eV (for InN) through 3.4 eV (for GaN) to 6.3 eV (for AlN). In other words, such an alloy system would have a continuous range of energy-gap values throughout the visible and ultraviolet ranges. A major issue in the thin-film growth of these compounds, besides the possible presence of various defects, is the choice of a suitable substrate.

The mechanism of the formation of micropipes is controversial (for a review, with references on micropipes in monocrystalline semiconductors, see, Heindl *et al.* 1997).

The formation of micropipes was first considered by Frank (1951). According to this theory, dislocations with large Burgers vectors generate hollow cores, replacing the regions of high strain around dislocation lines. There is an additional (tube-shaped) free surface but the total energy is minimized. Thus, according to Frank's theory, the micropipe equilibrium radius is related to the magnitude of the Burgers vector:

$$r = \mu b^2 / 8\pi^2 \gamma \tag{4.47}$$

where r is the micropipe radius, μ the shear modulus, b the Burgers vector of the dislocation responsible for the micropipe, and γ the surface energy.

However, later studies found differences between Frank's predicted micropipe radii and experimental values (Qian *et al.* 1995, Cherns 2000). Several other micropipe and nanopipe formation mechanisms were proposed. These relate formation to the presence of inhomogeneities such as impurities (Augustine *et al.* 1997, Heindl *et al.* 1997, Lilienthal-Weber *et al.* 1997), inclusions (Dudley *et al.* 1999, Hofmann *et al.* 1999), liquid silicon droplets containing some carbon or graphite particles (Mahajan 2002), low-angle grain boundaries (Augustine *et al.* 1997), or surface steps (Dudley *et al.* 1999, Ohtani *et al.* 2001, 2002). Theoretical studies on the interaction between the micropipes (and nanopipes) and dislocations have been also published (e.g. Pirouz 1998, Sheinerman and Gutkin 2003).

Bundling and twisting of micropipes, as well as their coalescence resulting in the enlargement or annihilation of initial micropipes, were reported in SiC crystals (Gutkin *et al.* 2003).

It was demonstrated that defects such as micropipes, tilt boundaries, and inclusions in SiC substrates generate structural defects in GaN layers deposited on them (Poust *et al.* 2003) so it is essential to reduce the density of these defects.

Conditions and mechanisms for micropipe dissociation in SiC epitaxial layers grown by chemical vapour deposition have been established (Kamata *et al.* 2003), and the possibility of using a micropipe stop layer for the fabrication of high-power SiC devices was proposed (Tsuchida *et al.* 2003). Micropipes can dissociate into closed-core screw dislocations and there is a high probability of such dissociation in epitaxial layers grown with a relatively low C/Si ratio of source gases (Kamata *et al.* 2003, Tsuchida *et al.* 2003). The micropipe density can also be substantially reduced by optimizing the SiC bulk crystal growth conditions (Ohtani *et al.* 2002).

Micropipe-free substrate wafer surfaces for use in device applications can be obtained by first filling the micropipes (Khlebnikov *et al.* 1998). This is done using a thick film epitaxial growth technique of rapid crystallization during physical vapour deposition to facilitate crystal growth inside the micropipe. This micropipe filling allows the fabrication of large area Schottky devices (Rendakova *et al.* 1998a, Dmitriev *et al.* 1999).

4.8.2 Processing-induced defects

During fabrication of devices and integrated circuits in slices of semiconductors many things can go wrong to result in damaging the material, introducing defects or impurities. This generally causes deterioration in device performance which may result in rejection, i.e. low yields. Even more harmful is the possibility that the initial deterioration, although too small to cause rejection, may lead to premature failure. This will cause larger costs 'downstream' to the device manufacturers' customers. Much effort was expended each time such phenomena were encountered as semiconductor fabrication technology advanced. Some of the more important cases were dealt with in the literature and are discussed here. Others were dealt with 'in house' and kept as commercially sensitive information, not shared with competitors, and are difficult to obtain. Processing-induced defects may be introduced during semiconductor device fabrication, by oxidation, doping (by diffusion or ion implantation), wet and dry etching, metallization, and packaging. This field has been reviewed repeatedly, recording the defects induced by changing processing technologies (Schwuttke 1970, Ravi 1981, Kimerling and Patel 1985, Kolbesen and Strunk 1985, Kolbesen *et al.* 1989, 1991, 2001, Claeys and Vanhellemont 1993, Cerva *et al.* 2001, Ohashi *et al.* 2003, and Vanhellemont *et al.* 2004).

For the introduction of process-induced defects, Schwuttke (1970) classified stresses into three types, i.e. internal stress, external stress and stress-jumping. Internal stress can be generated by diffusion processes; it results in the formation of dislocation patterns causing non-uniform junctions. Thermal gradients across the wafer (during device processing) can induce stresses that generate dislocations. Stress-jumping occurs during such processing steps as oxidation, diffusion and metallization, which may produce in the substrate a reversal (from tensile to compressive or vice versa) and interaction of stresses. Such stress-jumping may raise the stress well above the yield strength, to cause dislocation generation along the mask window edges. (For more detail see Schwuttke 1970 and references therein.)

Stress-related problems and dislocation generation in silicon integrated circuit (IC) fabrication were outlined by Hu (1991). Localized stresses in ICs can arise due to thermal processing, film edges, embedded structural elements, oxidation of non-planar silicon surfaces, doped lattices, and heteroepitaxy (Hu 1991, Fahey *et al.* 1992 and Claeys and Vanhellemont 1993). The deposition of different films and their subsequent patterned etching during IC fabrication produces film edges. Differences between the elastic and thermal properties of these films and the substrate give rise to interface stresses during thermal processing. These stresses are concentrated at thin-film edges, and may be high enough for dislocation nucleation. To predict dislocation type, equilibrium shape and distribution at those edges, Vanhellemont *et al.* (2004) developed a semi-quantitative model. This related those characteristics to the substrate and film pattern geometry (i.e. orientation and pitch). Semiconductor devices may also include structures embedded in the substrate like highly doped regions, buried oxides, and isolation trenches. In these cases as well, the stresses are concentrated at sharp edges. The strain distributions around these structures can be obtained by employing finite element analysis methods. Computer simulations can provide useful insight into dislocation dynamics at film edges and corners (e.g. Schwarz and Chidambarrao 1999), as well as into dislocation accumulation in ULSI cells with shallow trench isolation (STI) structures (e.g. Ohashi *et al.* 2003). Schwarz and Chidambarrao (1999) used the finite element method to calculate the stresses due to a silicon-nitride film pad on a silicon substrate. Their numerical simulations of the dislocation motion on a variety of slip systems in different nitride pad locations determined the character of the dislocation activity. Ohashi *et al.* (2003) studied periodic structures of shallow trench isolation (STI) type ULSI cells. They analysed dislocation accumulation due to formation of oxide film in the STI structure. This demonstrated that high stress concentrations during oxidation, around the shoulder of the device and the bottom corners of the trench, result in dislocation generation and accumulation in those regions. With device miniaturization, a greater dislocation density accumulated at the bottom corners of the trench (Ohashi *et al.* 2003).

Cerva *et al.* (2001) addressed the problem of non-detrimental process-induced defects leading eventually to the formation of electrically harmful defects.

They also discussed the importance of analysing the local mechanical stresses that lead to the nucleation of dislocations. Such primary defects may be formed during plasma etching with chlorine- and bromine-based chemistry (Cerva *et al.* 2001). Increasing bias voltages, which result in greater kinetic energies of the impinging species on the surface, lead to more degradation of the substrate. The irregularities formed are possible sites leading to the formation of secondary defects, including dislocations and stacking faults. Cerva (1997) also described primary vacancy-type defects, forming due to high-dose amorphizing implants in the silicon substrate below implantation mask edges. These defects in integrated silicon devices act as nucleation sites for generating dislocations, which may glide due to stress fields into the substrate and thus degrade device performance (Cerva 1997, Cerva *et al.* 2001).

Kolbesen *et al.* (1991, 2001) described the formation mechanisms of some defects during wafer processing. These included dislocations at trenches, implantation-induced stacking faults, and tungsten wormholes.

Ion-implantation of dopant atoms, to form *p-n* junctions and source and drain regions in a semiconductor, also contributes to the nucleation of defects, like dislocations and stacking faults (e.g. Mahajan 1989, Kolbesen *et al.* 1991). The damage characteristics depend on the total dose and its rate, the energy and mass of the incident ions, and the semiconductor temperature. Ion implantation induces amorphized and highly-defective regions along the ion-implanation path. Dislocation loops typically form at the end of the ion range. Depending on the ion-implantation parameters, the amorphous region may either be buried within the material or extend to the material surface. Annealing, typically rapid thermal annealing (RTA), is used to activate the implanted dopants (by allowing dopant atoms to diffuse into substitutional lattice sites) and to repair the damage. Recrystallization annealing transforms the amorphous region into a volume containing defects such as stacking faults and twins (for more detail, see e.g. Mahajan 1989 and references therein). The reordering of the amorphous region proceeds through a solid-phase epitaxy process that occurs below the melt temperature.

Defects in materials other than Si and defects induced in not only production processing but during device operation are other major fields of study. There is, for example, a large literature on defects in lasers and their role in laser failure modes (Ueda 1996, 1999). Some of these topics will be covered in Section 5.11.

Mechanical damage

The simplest and most general early form of process-induced defect was mechanical damage, which can be introduced during slicing, lapping, grinding, and mechanical polishing of wafers (Stickler and Booker 1963, and Section 2.5.7). Such polishing involves dragging abrasive particles across the wafer surface and that introduces rows of dislocation loops and microcracks (see Fig. 2.31). In addition, mechanical damage may occur during device fabrication and assembly. For instance, mounting in a subassembly may involve die-attach or wire-bonding and defects can be introduced if excessive pressure is applied (e.g. Brantley and Harrison 1973). Scratches and cracks can occur during handling of the chips or at the edges during the sawing or cleaving of wafers. In practice, this can be resolved by replacing manual handling with automated process.

To achieve large-scale planarization (flatness) of silicon wafers, chemical-mechanical polishing (CMP), also referred to as chemomechanical polishing, is commonly employed (see Section 4.2.3). This employs both mechanical abrasion and surface polish-etching, which is intended to preferentially remove any damage as fast as it occurs, to attain smooth surfaces. However, if relatively large abrasive particles are used, significant damage is introduced. This can be remedied, in principle, by using smaller abrasive particles but this reduces the removal rate. A promising alternative is tribochemical polishing (TCP). This is a non-abrasive polishing method based on the removal of material by a friction-activated chemical dissolution process (Hah and Fischer 1998, Muratov and Fischer 2000). In contrast to CMP, TCP does not involve abrasive slurries but employs a hard polishing disk. This eliminates scratching and results in simpler post-polish cleaning.

As noted in Section 2.5.7, some defects can also be introduced by excessive pressure applied at too high a temperature in thermocompression bonding of wire leads to chips, or by excessively tight application of tweezers in handling slices or bare devices.

Dislocations can be introduced during the in-diffusion of dopant impurities (see Section 2.5.7). This can occur as the result of a steep change of the lattice constant at the diffusion front in the cases of the impurity atoms differing substantially in size from those of the host material and for high dopant concentration. The resulting large stress can exceed the elastic limit and lead to plastic deformation.

The backside of silicon wafers is sometimes intentionally mechanically damaged to introduce stacking faults or dislocations for the gettering of impurities. This is termed backside damage (BSD).

Another type of damage is that resulting from ion implantation and annealing and this too can introduce defects (see e.g. Cerva 1997).

Oxidation-induced stacking faults in silicon

The oxidation of silicon at high temperatures frequently results in the generation of stacking faults in the surface layers of the crystal as first reported by Thomas (1963). This is important because thermally grown oxide is routinely used as a diffusion mask in the production of planar integrated circuits and for passivating the device surface. Consequently oxidation-induced stacking faults (OISFs) were extensively studied by etching and light microscopy (Queisser and van Loon 1964, Fisher and Amick 1966, and Drum and van Gelder 1972), x-ray methods (Schwuttke *et al.* 1971, Matsui and Kawamura 1972) and transmission electron microscopy (Booker and Stickler 1965, Booker and Tunstall 1966, Jaccodine and Drum 1966). OISFs were also soon shown to be associated with leakage currents in silicon field effect transistors (Schwuttke *et al.* 1971).

Early TEM studies established the fact that the faults are extrinsic, lie in {111} planes inclined to the surface and are bounded by a Frank partial dislocation loop with a Burgers vector of $1/3\langle 111 \rangle$ (Booker and Tunstall 1966). Surface mechanical damage was found to be necessary in the early work, to nucleate these faults when a single oxidizing treatment was applied. Atmospheres varying in oxidizing strength from dry air through wet oxygen to pure steam will all produce the faults. The faults are nucleated at the start of the oxidation of mechanically damaged specimens and grow at the same rates so they are all the same size as can be seen in Fig. 4.147.

Booker and Stickler (1965) suggested a model for the production of the stacking faults which postulates (Sanders and Dobson 1969) that prismatic dislocations already present in the surface layers due to mechanical damage, dissociate to form a Shockley and a Frank partial dislocation bounding an area of extrinsic stacking fault according to the reaction

$$a/2[1\bar{1}0] \rightarrow a/3[1\bar{1}1] + a/6[1\bar{1}\bar{2}] \tag{4.48}$$

It will be remembered that a Shockley partial is one with its Burgers vector in the plane of the fault, $b = a/6[1\bar{1}\bar{2}]$ in this case, and consequently it can glide in that plane. A Frank partial, however, has its Burgers vector out of the fault plane. In the present case the Frank partial has $b = a/3[1\bar{1}1]$, which is perpendicular to the fault plane. This dislocation is therefore sessile, i.e. it is not mobile and cannot glide. Now providing the Frank dislocation is on the inward side of the fault, away from the surface, the fault can grow by the Frank partial advancing by climb into the crystal. The driving force for this climb was shown by Sanders and Dobson (1969) to

Figure 4.147 TEM image of oxidation-induced stacking faults in silicon. The four faults marked X lie in the four {111} planes inclined and normal to the (110) plane of the specimen. (After Booker and Tunstall 1966. Reprinted with permission from *Philosophical Magazine*, **13** (1966), p. 71; Taylor & Francis Ltd., the Journal's web site: http://www.tandf.co.uk/journals.)

be chemical in nature. It is the result of the undersaturation of vacancies in the crystal due to oxidation. Under conditions of vacancy undersaturation the extrinsic faults will enlarge by absorbing atoms, i.e. by emitting vacancies. Thus when the chemical driving force due to vacancy undersaturation is greater than the forces due to the line tension of the bounding partial dislocation and the stacking fault energy, which acts as a surface tension, the fault grows. When the driving force for growth is less than the combined shrinkage forces, the fault shrinks and may be annihilated altogether in time. Sanders and Dobson demonstrated by annealing experiments on thin TEM specimens that the faults grow on annealing in air at 1100° C and shrink on annealing in vacuum. The ratio of the actual to the equilibrium vacancy concentration at the oxide-silicon interface at 1100° C is $C/C_o = 0.8$. It is related to the oxidation-induced mass transfer and is a function of the oxidation conditions such as the temperature and the oxidizing ambient (dry or wet oxygen or steam).

Later work showed that OISFs could also be nucleated at the grown-in point-defect clusters whose concentration variations constitute 'swirl markings' (see Section 6.5.5) (Matsui and Kawamura 1972 and Ravi and Varker 1974).

Ravi (1974) carried out a transmission electron microscope study of the annihilation of OISFs in Si and found two mechanisms were involved. Faults could be removed by shrinkage due to climb of the bounding partial Frank dislocation loop. Alternatively an unfaulting reaction involving the nucleation and motion of Shockley partial dislocations was possible. For details of the various possible dislocation reactions that may occur see Ravi (1974). Hashimoto *et al.* (1976) found that OISFs could be annihilated by impurity diffusion during high-temperature heat treatment in a nitrogen atmosphere.

A study of the orientation dependence of the stacking fault nucleation in oxidized Si (Ravi 1975) led to the suggestion that this results from the dilatational stresses due to the oxidation being concentrated locally by defects in the damaged layer. In the absence of surface damage a different form of orientation dependence is observed.

Ravi (1972) showed that OISFs were decorated by precipitates in boron-diffused Si and the precipitates tended to be larger and more numerous at the bounding partial dislocations.

In present day silicon OISFs arise in a quite different way due to the lack of dislocations. Huff (2002) in a historical account of the development of microelectronics, pointed out that the elimination of dislocations from a semiconductor does not automatically result in improved semiconductor devices. This is the 'point-defect dilemma'. The absence of dislocations removes efficient internal sinks for gettering any excess point defects. Thermodynamics tells us that the equilibrium density of point defects falls with temperature. Therefore, crystal cooling after growth, in the absence of dislocation sinks, leads to an increasing supersaturation of vacancies or interstitials. These cluster to form microdefects (e.g. Sinno *et al.* 2000). Excess vacancies condense to form voids, whereas excess interstitials generate dislocation loops. Thus, it is now well established that these intrinsic point defects, as well as oxygen impurities, greatly influence the formation of growth microdefects throughout the bulk of the silicon. These microdefects include oxidation induced stacking faults OISFs (Voronkov 1982, and the review by Dornberger *et al.* 2001). These OISFs can, as before, lead to an increase in the leakage current of silicon devices.

These faults are thought to be generated in Czochralski (Cz) Si wafers as a result of the introduction of a supersaturation of Si interstitials during thermal oxidation (Hu 1974). Following diffusion and coalescence, these interstitials form areas of extrinsic stacking faults (see Section 2.3.7). The uniform size of bulk OISFs was related to the fact that the OISF nuclei are formed within less than 1 sec of the start of the oxidation process, and some cases of OISF size variations were related to a continuous generation of oxygen precipitates (Hu 1974).

According to the Voronkov model, which type of point defect is generated depends on the ratio v/G, where v is the pulling (or growth) rate and G is the axial thermal gradient near the growth interface (Voronkov 1982). If v/G is greater than a critical value, C_{crit}, vacancies predominate, but if $v/G < C_{crit}$, interstitials are dominant (Voronkov 1982). The critical value was determined to be $C_{crit} = 1.3 \times 10^{-3} \, \text{cm}^2 \, \text{min}^{-1} \, \text{K}^{-1}$ (Von Ammon *et al.* 1995) or $C_{crit} = 2.2 \times 10^{-3} \, \text{cm}^2 \, \text{min}^{-1} \, C^{-1}$ (Hourai *et al.* 1995). However, this model is not universally accepted.

Thermocouples were used to measure temperatures at several points in the melt during the Czochralski (CZ) growth of Si crystals (Abe 2000, and references therein). It was found that larger growth rates (v) produced smaller thermal gradients (G) in both FZ (floating zone) and CZ growth of Si crystals. Abe found that the type of point defect predominantly created was determined by the value of G near the interface during growth in both FZ and CZ growth. When G was less than about $20°\,C\,cm^{-1}$, vacancy rich CZ crystals were produced. When G exceeded $25°\,C\,cm^{-1}$, the species of point defects changed dramatically from vacancies to interstitials. Thus, unlike Voronkov (1982), Abe found that the growth velocity, v, played only a secondary role, in so far as it affects the value of G, the thermal gradient near the growth interface.

A commonly observed feature is a ring-like distribution of OISFs (Queisser and Van Loon 1964). The radius and position of such OISF rings, which emerge following thermal oxidation and etching, depend on growth conditions as shown in Fig. 4.148 (Von Ammon *et al.* 1995, and Sinno *et al.* 2000 and references therein).

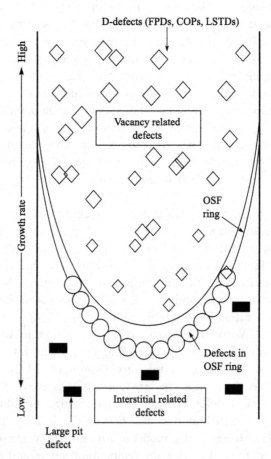

Figure 4.148 Illustration of the types of defect in Czochralski grown silicon crystals as a function of the growth (pulling) rate. The abbreviations are: FPD (flow pattern defect), COP (crystal originated pit), LSTD (light scattering tomography defect). (After Dornberger *et al.* 2001. Reprinted from *Journal of Crystal Growth*, **229**, Silicon crystals for future requirements of 300 mm wafers, pages 11–16. Copyright 2001, with permission from Elsevier.)

The dominant point defect species outside the OISF ring are silicon interstitials and interstitial-related defects, whereas inside the ring they are vacancies and vacancy-related microdefects (Von Ammon *et al.* 1995).

The distributions of the defects formed are illustrated in Fig. 4.148. For high growth (pulling) rates at the top of Fig. 4.148, vacancies predominate so, during cool-down, voids form. At low growth rates interstitial-related defects like stacking faults and dislocation loops are formed at the edges (Dornberger *et al.* 2001, and the review by Hurle and Rudolph 2004). An inner region that is vacancy rich is separated from an outer, interstitial-rich region by a ring that contains no excess of either point defect. Following oxidation at elevated temperatures (around 1000° C), high densities of OISFs appear at the boundary between these regions.

One can in principle 'engineer' intrinsic point defects in silicon by controlling Si growth to near equilibrium conditions. This avoids both vacancy agglomerates, i.e. voids, and self interstitials, i.e. dislocation loops (Falster and Voronkov 2000).

Mechanisms proposed for the nucleation of OISFs include mechanical damage, as discussed above (Fisher and Amick 1966, Ravi and Varker 1974), Si-O clusters (Mahajan *et al.* 1977), metallic impurities (Shimura *et al.* 1980), and nucleation at small vacancy clusters (Monson and Van Vechten 1999). Monson and Van Vechten (1999) pointed out that the annihilation of vacancies by dislocations during crystal growth affects the size and distribution, in the boule, of vacancy clusters, which nucleate OISFs.

Mahajan (1989) pointed out the similarities between OISFs and stacking faults due to the oxygen precipitation in Cz Si. Owing to their greater volume, oxygen precipitates produce Si interstitials that can agglomerate and cause extended defects.

Giri *et al.* (2001) employed PL spectroscopy and TEM to study recombination radiation from extended defects in Si and concluded that OISFs are nonradiative in Si.

References

Abe, T. (2000). The formation mechanism of grown-in defects in CZ silicon crystals based on thermal gradients measured by thermocouples near growth interfaces. *Materials Science and Engineering*, **B73**, 16–29.

Abrahams, M. S. and Buiocchi, C. J. (1967). Twins and stacking faults in vapor grown GaAs. *Journal of Physics and Chemistry of Solids*, **28**, 927.

Abrahams, M. S. and Ekstrom, L. (1960). Dislocations and brittle fracture in elemental and compound semiconductors. *Acta Metallurgica*, **8**, 654–62.

Abrahams, M. S. and Dreeben, A. (1965). Formation of dislocations around precipitates in single crystals of (Zn, Cd)S:Er. *Journal of Applied Physics*, **36**, 1688–92.

Abrahams, M. S. and Buiocchi, C. J. (1970). Mechanism of thermal annihilation of stacking faults in GaAs. *Journal of Applied Physics*, **41**, 2358–65.

Abrahams, M. S., Blanc, J. and Buiocchi, C. J. (1972). Like-sign asymmetric dislocations in zinc-blende structure. *Applied Physics Letters*, **21**, 185–6.

Ahearn, J. S., Mills, J. J. and Westwood, A. R. (1978a). Effect of electrolyte pH and bias voltage on the hardness of the (0001) ZnO Surface. *Journal of Applied Physics*, **49**, 96–102.

Ahearn, J. S., Mills, J. J. and Westwood, A. R. (1978b). Effect of bias voltage on the hardness of {1010} ZnO surfaces immersed in an electrolyte. *Journal of Applied Physics*, **49**, 614–17.

Ahearn, J. S., Mills, J. J. and Westwood, A. R. (1979). Chemomechanical effects in ZnO. *Journal de Physique*, **C6**, 173–6.

Ahlquist, C. N., Carrel, M. J. and Stroempl, P. (1972). The photoplastic effect in wurtzite and sphalerite structure II-VI compounds. *Journal of Physics and Chemistry of Solids*, **33**, 337–42.

Alexander, E., Kalman, Z. H., Mardix, S. and Steinberger, I. T. (1970). The mechanism of poytype formation in vapor-phase grown ZnS crystals. *Philosophical Magazine*, **21**, 1237–46.

Alexander, H. (1976). On the dislocation core structure in silicon and III-V compounds. In *Electron Microscopy 1976*, Vol. **I**, Proc. Sixth European Electron Microscopy Conference, Jerusalem (Israel: Tal Int. Publ.), pp. 208–11.

Alexander, H. (1979). Models of the dislocation structure. *Journal de Physique*, **C6**, 1–6.

Alexander, H. (1986). Dislocations in covalent crystals. In *Dislocations in Solids*, Vol. **7**, ed. Nabarro (Amsterdam: North-Holland), pp. 113–234.

Alexander, H. (1991). Dislocations in semiconductors. In *Poycrystalline Semiconductors* II Proc. In *Phys.*, **54**, eds. J. H. Werner and H. P. Strucnk, pp. 2–12.

Alexander, H. and Haasen, P. (1968). Dislocations and plastic flow in the diamond structure. *Solid State Physics*, **22**, 27–158.

Alexander, H. and Haasen, P. (1972). Dislocations in nonmetals. *Annual Review of Materials Science*, **2**, 291–312.

Alexander, H. and Teichler, H. (1991). Dislocations. In *Materials Science and Technology*, **4**, ed. W. Schroter (Weinheim: VCH), pp. 249–319.

Allen, J. W. (1957). On the mechanical properties of indium antimonide. *Philosophical Magazine*, **2**, 1475–81.

Amelinckx, S. (1979). Dislocations in particular structures. In *Dislocations in Solids*, **2**, ed. F. R. Nabarro (Amsterdam: North-Holland), pp. 67–460.

Amelinckx, S. and Dekeyser, W. (1959). The structure and properties of grain boundaries. *Solid State Physics*, **8**, 325–499.

Amelinckx, S., Strumane, G. and Webb, W. W. (1960). Dislocations in silicon carbide. *Journal of Applied Physics*, **31**, 1359–70.

Arlt, G. and Quadriflieg, P. (1968). Piezoelectricity in III-V compounds with a phenomenological analysis of the piezoelectric effect. *Physica Status Solidi*, **25**, 323–30.

Augustine, G., Hobgood, McD., Balakrishna, V., Dunne, G. and Hopkins, R. H. (1997). Physical vapor transport growth and properties of SiC monocrystals of 4H polytype. *Physica Status Solidi*, **B202**, pp. 137–48.

Augustus, P. D., Stirland, D. J. and Yates, M. (1983). Microstructure of grappe defects in InP. *Journal of Crystal Growth*, **64**, 121–8.

Austerman, S. B. and Gehman, W. G. (1966). The inversion twin: Prototype in BeO. *Journal of Materials Science*, **1**, 249–60.

Aven, M. and Prener, J. S. (1967). *Physics and Chemistry of II-VI Compounds*. Amsterdam: North-Holland.

Balluffi, R. W., Brokman, A. and King, A. H. (1982). CSL/DSC lattice model for general crystal boundaries and their line defects. *Acta Metallurgica*, **30**, 1453–70.

Barber, H. D. and Heasell, E. L. (1965a). Polarity effects in InSb alloyed p-n junctions. *Journal of Applied Physics*, **36**, 176–80.

Barber, H. D. and Heasell, E. L. (1965b). A technique for making alloy p-n junctions in InSb. *Solid State Electronics*, **8**, 113–17.

Bartels, W. J. and Nijman, W. (1977). Asymmetry of misfit dislocations in heteroepitaxial layers on (001) GaAs substrates. *Journal of Crystal Growth*, **37**, 204–14.

Beanland, R. (1995). Dislocation multiplication mechanisms in low-misfit strained epitaxial layers. *Journal of Applied Physics*, **77**, 6217–22.

Beanland, R., Dunstan, D. J. and Goodhew, P. J. (1996). Plastic relaxation and relaxed buffer layers for semiconductor epitaxy. *Advances in Physics*, **45**, 87–146.

Beaumont, B., Vennéguès, Ph. and Gibart, P. (2001). Epitaxial lateral overgrowth of GaN. *Physica Status Solidi*, **B227**, 1–43.

Bell, R. L. and Willoughby, A. R. (1966). Etch-pit studies of dislocations in InSb. *Journal of Materials Science*, **1**, 219–28.

Bell, R. L. and Willoughby, A. R. (1970). The effect of plastic bending on the electrical properties of indium antimonide 2. Four-point bending of n-type material. *Journal of Materials Science*, **5**, 198–217.

Berghezan, A., Fourdeux, A. and Amelinckx, S. (1961). Transmission electron microscopy studies of dislocations and stacking faults in a hexagonal metal-zinc. *Acta Metallurgica*, **9**, 464–90.

Berlincourt, D., Jaffe, H. and Shiozawa, L. R. (1963). Electroelastic properties of the sulfides, selenides and tellurides of zinc and cadmium. *Physical Review*, **129**, 1009–17.

Berner, K. and Alexander, H. (1967). Versetzungsdichte und lokale abgleitung in germanium-einkristallen. *Acta Metallurgica*, **15**, 933–41.

Bigger, J. R., McInnes, D. A., Sutton, A. P. *et al.* (1992). Atomic and electronic-structures of the 90-degree partial dislocation in silicon. *Physical Review Letters*, **69**, 2224–7.

Binsma, J. J., Giling, L. J. and Bloem, J. (1980). Phase relations in the system $Cu_2S-In_2S_3$. *Journal of Crystal Growth*, **50**, 429–36.

Binsma, J. J., Giling, L. J. and Bloem, J. (1981). Order-disorder behavior and tetragonal distortion of chalcopyrite compounds. *Physica Status Solidi*, **A63**, 595–603.

Black, J. F. and Jungbluth, E. D. (1967). Decorated dislocations and sub-surface defects induced in GaAs by the In-diffusion of zinc. *Journal of the Electrochemical Society*, **114**, 188–92.

Blanc, J. (1975). Thermodynamics of 'glide' and 'shuffle' dislocations in the diamond lattice. *Philosophical Magazine*, **32**, 1023–32.

Blank, H., Delavignette, P. and Amelinckx, S. (1962). Dislocations and wide stacking faults in wurtzite type crystals: zinc sulfide and aluminium nitride. *Physica Status Solidi*, **2**, 1660–9.

Blank, H., Delavignette, P., Gevers, R. and Amelinckx, S. (1964). Fault structures in wurzite. *Physica Status Solidi*, **7**, 747–64.

Blech, I. A., Meieran, E. S. and Sello, H. (1965). X-ray surface topography of diffusion-generated dislocations in silicon. *Applied Physics Letters*, **7**, 176–8.

Blistanov, A. A. and Geras'kin, V. V. (1970). Dislocations in single crystals with the wurtzite structure. *Soviet Physics Crystallography*, **14**, 550–3.

Bobb, L. C., Holloway, H., Maxwell, K. H. and Zimmerman, E. (1966). Oriented growth of semiconductors. III. Growth of gallium arsenide on germanium. *Journal of Applied Physics*, **37**, 4687–93.

Bollmann, W. (1967a). On the geometry of grain and phase boundaries. I. General theory. *Philosophical Magazine*, **16**, 363–81.

Bollmann, W. (1967b). On the geometry of grain and phase boundaries. II. Applications of general theory. *Philosophical Magazine*, **16**, 383–99.

Bollmann, W. (1970). *Crystal Defects and Crystalline Interfaces*. Berlin: Springer-Verlag.

Bollmann, W. (1972). The basic concepts of the O-lattice theory. *Surface Science*, **31**, 1–11.

Bollmann, W. (1974). Classification of crystalline interfaces by means of the O-lattice method. *Journal of Microscopy*, **102**, 233–9.

Booker, G. R. (1962). Growth structure difference on opposite {111} faces of a GaAs dendrite. *Journal of Applied Physics*, **33**, 75.

Booker, G. R. (1964). Crystallographic imperfections in Si. *Discussions of the Faraday Society*, **38**, 298–304.

Booker, G. R. (1965). Tripyramids and associated defects in epitaxial layers. *Philosophical Magazine*, **11**, 1007–20.

Booker, G. R. (1966). Stacking fault defects in epitaxial silicon layers. *Journal of Applied Physics*, **37**, 441–2.

Booker, G. R. and Howie, A. (1963). Intrinsic-extrinsic stacking-fault pairs in eitaxially grown silicon layers. *Applied Physics Letters*, **3**, 156–7.

Booker, G. R. and Joyce, B. A. (1966). A study of nucleation in chemically grown epitaxial silicon films using molecular beam techniques. II. Initial growth behaviour on clean carbon-contaminated silicon substrates. *Philosophical Magazine*, **14**, 301–15.

Booker, G. R. and Stickler, R. (1962). Crystallographic imperfections in epitaxially grown silicon. *Journal of Applied Physics*, **33**, 3281–90.

Booker, G. R. and Stickler, R. (1965). 2-dimensional defects in silicon after annealing in wet oxygen. *Philosophical Magazine*, **12**, 1303.

Booker, G. R. and Tunstall, W. J. (1966). Diffraction contrast analysis of 2-dimensional defects present in silicon after annealing. *Philosophical Magazine*, **13**, 71.

Booyens, H., Vermaak, J. S. and Proto, G. R. (1978). The asymmetric deformation of GaAs single crystals. *Journal of Applied Physics*, **49**, 5435–40.

Bourret, A., Thibault-Desseaux, J. and Lancon, F. (1983). The core structure of dislocations in Cz silicium studied by electron microscopy. *Journal de Physique*, **44**, C4–15 to C4–24.

Brack, K. (1965). X-ray method for the determination of polarity of SiC crystals. *Journal of Applied Physics*, **36**, 3560–2.

Braga, N., Buczkowski, A., Kirk, H. R. and Rozgonyi, G. A. (1994). Formation of cylindrical *n/p* junction diodes by arsenic enhanced diffusion along interfacial misfit dislocations in *p*-type epitaxial Si/Si(Ge). *Applied Physics Letters*, **64**, 1410–12.

Brafman, O., Alexander, E., Fraenkel, B. S., Kalman, Z. H. and Steinberger, I. T. (1964). Polar properties of ZnS crystals and the anomalous photovoltaic effect. *Journal of Applied Physics*, **35**, 1855–9.

Brandon, D. G., Ralph, B., Ranganathan, S. and Wald, M. S. (1964). A field ion microscope study of atomic configuration at grain boundaries. *Acta Metallurgica*, **12**, 813–21.

Brantley, W. A. and Harrison, D. A. (1973). Localized plastic-deformation of GaP and GaAs generated by thermocompression bonding. *Journal of the Electrochemical Society*, **120**, 1281–4.

Braun, C. and Helberg, H. W. (1986). Surface damage of CdTe produced during preparation and determination of dislocation types near microhardness indentations. *Philosophical Magazine*, **A53**, 277–84.

Bredikhin, S. I. and Shmurak, S. Z. (1974). Deformation-stimulated emission of ZnS crystals. *JETP Letters*, **19**, 367–8.

Bredikhin, S. I. and Shmurak, S. Z. (1975a). Effect of electric field on deformation induced light emission of ZnS crystals. *JETP Letters*, **21**, 156–7.

Bredikhin, S. I. and Shmurak, S. Z. (1975b). Deformation-stimulated electric-signal peaks produced by ZnS crystals. *Soviet Physics Solid State*, **17**, 1628–9.

Bredikhin, S. I., Osipiyan, Yu. A. and Shmurak, S. Z. (1975c). Effect of light on strain-stimulated light emission in ZnS crystals. *Soviet Physics JETP*, **41**, 373–5.

Bredikhin, S. I. and Shmurak, S. Z. (1977). The luminescence and electrical characteristics of ZnS crystals undergoing plastic deformation. *Soviet Physics JETP*, **46**, 768–73.

Bredikhin, S. I. and Shmurak, S. Z. (1979). Interaction of charged dislocations with luminescence centers in ZnS crystals. *Soviet Physics JETP*, **49**, 520–5.

Bredikhin, S. I., Osipiyan, Yu. A. and Shmurak, S. Z. (1975). Effect of light on strain-stimulated light emission in ZnS crystals. *Soviet Physics JETP*, **41**, 373–5.

Brongersma, H. H. and Mull, P. M. (1973). Absolute configuration assignment of molecules and crystals in discussion. *Chemical Physics Letters*, **19**, 217–20.

Brown, G. T., Cockayne, B., Elliott, C. R. *et al.* (1984). A detailed microscopic examination of dislocation clusters in LEC InP. *Journal of Crystal Growth*, **67**, 495–506.

Bulatov, V. V. (2001). Bottomless complexity of core structure and kink mechanisms of dislocation motion in silicon. *Scripta Materialia*, **45**, 1247–52.

Bulatov, V. V., Justo, J. F., Cai, W. and Yip, S. (1997). Kink asymmetry and multiplicity in dislocation cores. *Physical Review Letters*, **79**, 5042–5.

Bullough, R. and Newman, R. C. (1963). The interaction of impurities with dislocations in silicon and germanium. In *Progress in Semiconductors*, eds. A. F. Gibson and R. E. Burgess, **7** (London: Heywood), pp. 100–34.

Bullough, R. and Newman, R. C. (1970). The kinetics of migration of point defects to dislocations. *Reports on Progress in Physics*, **33**, 101–48.

Bullough and Tewari (1979). Lattice theories of dislocations. In *Dislocations in Solids*, **2**, ed. F. R. Nabarro (Amsterdam: North-Holland), pp. 1–65.

Burr, K. F. and Woods, J. (1971). The anomalous dispersion of x-rays by single crystal ZnSe. *Journal of Materials Science*, **6**, 1007–11.

Cai, W., Bulatov, V. V., Chang, J., Li, J. and Yip, S. (2004). Dislocation core effects on mobility. In *Dislocations in Solids*, **12**, ed. F. R. Nabarro (Amsterdam: North-Holland), pp. 2–117.

Carlsson, L. (1971). Orientation and temperature dependence of the photoplastic effect in ZnO. *Journal of Applied Physics*, **42**, 676–80.

Carlsson, L. and Ahlquist, C. N. (1972). Photoplastic behaviour of CdTe. *Journal of Applied Physics*, **43**, 2529–36.

Carlsson, L. and Svensson, C. (1969). Increase of flow stress in ZnO under illumination. *Solid State Communications*, **7**, 177–9.

Carlsson, L. and Svensson, C. (1970). Photoplastic effect in ZnO. *Journal of Applied Physics*, **41**, 1652–6.

Caveney, R. J. (1968). Hg diffusion-induced defects in CdS crystals. *Philosophical Magazine*, **18**, 939–44.

Cerva, H. (1997). Defects below mask edges in silicon induced by amorphizing implantations. *Defect Diffusion Forum*, **148**, 103–21.

Cerva, H., Engelhardt, M., Hierlemann, M., Pölzl, M. and Thenikl, T. (2001). Misfortune, challenge, and success: defects in processed semiconductor devices. *Physica B: Condensed Matter*, **308–310**, 13–17.

Chadderton, L. T., Fitzgerald, A. G. and Yoffe, A. D. (1963). Stacking faults in zinc sulphide. *Philosophical Magazine*, **8**, 167–73.

Chadderton, L. T., Fitzgerald, A. G. and Yoffe, A. D. (1964). Disordering of defects in single crystals of zinc sulfide. *Journal of Applied Physics*, **35**, 1582–6.

Charig, R. M., Joyce, B. A., Stirland, D. J. and Bicknell, R. W. (1962). Growth mechanism and defect structures in epitaxial silicon. *Philosophical Magazine*, **7**, 1847–60.

Chase, B. D. and Holt, D. B. (1972). Transmission electron microscope observations on GaP electroluminescent diode materials. *Journal of Materials Science*, **7**, 265–78.

Chaudhuri, A. R., Patel, J. R. and Rubin, L. G. (1962). Velocities and densities of dislocations in germanium and other semiconductor crystals. *Journal of Applied Physics*, **33**, 2736–46.

Chaudhuri, A. R., Patel, J. R. and Rubin, L. G. (1963). Correction. *Journal of Applied Physics*, **34**, 240.

Chen, J., Sekiguchi, T., Yang, D. *et al.* (2004). Electron-beam-induced current study of grain boundaries in multicrystalline silicon. *Journal of Applied Physics*, **96**, 5490–5.

Cherns, D. (2000). The structure and optoelectronic properties of dislocations in GaN. *Journal of Physics: Condensed Matter*, **12**, 10205–12.

Chikawa, J. (1964). Faults in wurtzite type CdS crystals. *Japanese Journal of Applied Physics*, **3**, 229–30.

Chikawa, J. and Nakayama, T. (1964). Dislocation structure and growth mechanism of cadmium sulfide crystals. *Journal of Applied Physics*, **35**, 2493–501.

Chikawa, J. I. and Austerman, S. B. (1968). X-ray diffraction contrast of inversion twin boundaries in BeO crystals. *Journal of Applied Crystallography*, **1**, 165–71.

Chisholm, M. F., Maiti, A., Pennycook, S. J. and Pantelides, S. T. (1998). Atomic configurations and energetics of arsenic impurities in a silicon grain boundary. *Physical Review Letters*, **81**, 132–5.

Chisholm, M. F., Buczko, R., Mostoller, M. *et al.* (1999). Atomic structure and properties of extended defects in silicon. *Solid State Phenomina*, **67–8**, 3–13.

Choi, S. K. and Mihara, M. (1972). Impurity effects on the dislocation velocity in gallium arsenide. *Journal of the Physical Society of Japan*, **32**, 1154.

Choi, S. K., Mihara, M. and Ninomiya, T. (1977). Dislocation velocities in GaAs. *Japanese Journal of Applied Physics*, **16**, 737–45.

Choi, S. K., Mihara, M. and Ninomiya, T. (1978). Dislocation velocities in InAs and GaSb. *Japanese Journal of Applied Physics*, **17**, 329–33.

Churchill, J. N. and Watt, L. A. (1969). Polarity effects in the heat treatment of InSb. *Journal of Applied Physics*, **40**, 3872–3.

Churchman, A. T., Geach, G. A. and Winton, J. (1956). Deformation twinning in materials of the A4 (diamond) structure. *Proceedings of the Royal Society*, **A238**, 194–203.

Claesson, A. (1979). Effect of disorder and long-range strain field on the electron-states. *Journal de physique*, **40**, Suppl. 6, 39–41.

Claeys, C. and Vanhellemont, J. (1993). Recent progress in the understanding of crystallographic defects in silicon. *Journal of Crystal Growth*, **126**, 41–62.

Cline, C. F. and Kahn, J. S. (1963). Microhardness of single crystals of BeO and other wurzite compounds. *Journal of the Electrochemical Society*, **110**, 773–5.

Cockayne, D. J. and Hons, A. (1979). Dislocations in semiconductors as studied by weak-beam electron microscopy. *Journal de Physique*, **C6**, 11–18.

Cockayne, D. J., Hons, A. and Spence, J. C. (1980). Gliding dissociated dislocations in hexagonal CdS. *Philosophical Magazine*, **A42**, 773–81.

Coster, D., Knol, K. S. and Prins, J. A. (1930). Unterschide in der intensitaet der rontgenstrahlen reflexion an den beiden 111-flaechen der zinblende. *Zeitschrift fur Physik*, **63**, 345–69.

Cottrell, A. H. and Bilby, B. A. (1949). Dislocation theory of yielding and strain ageing of iron. *Proceedings of the Physical Society*, **A62**, 49–62.

Csányi, G., Ismail-Beigi, S. and Arias, T. A. (1998). Paramagnetic structure of the soliton of the 30 degree partial dislocation in silicon. *Physical Review Letters*, **80**, 3984–7.

Cullis, A. G. (1973). Transmission electron-microscope observations of extended and unextended dislocation nodes in Si and Ge/Si layers using weak-beam technique. *Journal of Microscopy*, **98**, 191–5.

Cullis, A. G., Robbins, D. J., Barnett, S. J. and Pidduck, A. J. (1994). Growth ripples upon strained SiGe epitaxial layers on Si and misfit dislocation interactions. *Journal of Vacuum Science and Technology*, **12**, 1924–31.

Cullis, A. G., Pidduck, A. J. and Emeny, M. T. (1995). Misfit dislocation sources at surface ripple troughs in continuous heteroepitaxial layers. *Physical Review Letters*, **75**, 2368–71.

Czaja, W. (1966a). Conditions for the generation of slip by diffusion of phosphorus into silicon. *Journal of Applied Physics*, **37**, 3441–6.

Czaja, W. (1966b). Response of Si and GaP p-n junctions to a 5- to 40-keV electron beam. *Journal of Applied Physics*, **37**, 4236–48.

Czaja, W. and Wheatley, G. H. (1964). Simultaneous observation of diffusion-induced dislocation slip patterns in Si with electron beam scanning and optical means. *Journal of Applied Physics*, **35**, 2782–3.

Czernuska, J. T. (1989). Effect of surface charge on the dislocation mobility in semiconductors. In *International Symposium on Structural Properties of Dislocations in Semiconductors*, Oxford Conference Series No. 104 (Bristol: Institute of Physics), pp. 315–19.

Dangor, A. E. and Holt, D. B. (1962). Direct observations of the geometry of defects in germanium. *Philosophical Magazine*, **8**, 1921–35.

D'Aragona, F. S., Delavignette, P. and Amelinckx, S. (1966). Direct evidence for the mechanism of the phase transition wurtzite-sphalerite. *Physica Status Solidi*, **14**, K115–K118.

Darling, D. F. and Field, B. O. (1973). Molecular orientation by surface forces. 2. On prediction of orientation of crystal overgrowths. *Surface Science*, **36**, 630–40.

Das, B. N. and Weinstein, M. (1967). Crystal imperfections in vapor-grown CdS. In *II-VI Semiconducting Compounds*, 1967. Proceedings of International Conference, ed. D. G. Thomas (New York: Benjamin), pp. 147–66.

Dash, W. C. (1957). The observation of dislocations in silicon. In *Dislocations and Mechanical Properties of Crystals*, eds. J. C. Fisher, W. G. Johnston, R. Thomson and T. Vreeland, Jr (New York: Wiley), pp. 57–68.

Dash, W. C. (1958). The growth of silicon crystals free from dislocations. In *Growth and Perfection of Crystals*, eds. R. H. Doremus, B. W. Roberts and D. Turnbull (New York: Wiley), pp. 361–85.

Datta, S., Yacobi, B. G. and Holt, D. B. (1977). Scanning electron-microscope studies of local variations in cathodoluminescence in striated ZnS platelets. *Journal of Materials Science*, **12**, 2411–20.

Delavignette, P., Kirkpatrick, H. B. and Amelinckx, S. (1961). Dislocations and stacking faults in aluminium nitride. *Journal of Applied Physics*, **32**, 1098–100.

Delin, A., Ravindran, P., Eriksson, O. and Wills, J. M. (1998). Full-potential optical calculations of lead chalcogenides. *International Journal of Quantum Chemistry*, **69**, 349–58.

Dewald, J. F. (1957). The kinetics and mechanism of formation of anode films on single-crystal InSb. *Journal of the Electrochemical Society*, **104**, 244–51.

Dewald, J. F. (1960). The charge and potential distributions at the zinc oxide electrode. *Bell System Technical Journal*, **39**, 615–39.

Dillon, J. A. (1960). The interaction of oxygen with silicon carbide surfaces. In *Silicon Carbide*, eds. J. R. O'Connor and J. Smiltens (Oxford: Pergamon), pp. 235–40.

Dillon, J. A. (1962). Assymetries in the surface properties of the (111) and ($\bar{1}\bar{1}\bar{1}$) faces of ZnTe. *Journal of Applied Physics*, **33**, 669–72.

Dingley, D. J., Randle, V. and Baba-Kishi, K. Z. (1993). *Atlas of Backscattering Patterns*. Bristol: Adam Hilger.

Dmitriev, V., Rendakova, S., Kuznetsov, N. *et al.* (1999). Large area silicon carbide devices fabricated on SiC wafers with reduced micropipe density. *Materials Science and Engineering*, **B61–2**, 446–9.

Dodson, B. W. and Tsao, J. Y. (1987). Relaxation of strained layer semiconductor structures by plastic flow. *Applied Physics Letters*, **51**, 1325–7.

Dornberger, E., Virbulis, J., Hanna, B. *et al.* (2001). Silicon crystals for future requirements of 300 mm wafers. *Journal of Crystal Growth*, **229**, 11–16.

Drum, C. M. (1965). Intersecting faults on basal and prismatic planes in aluminium nitride. *Philosophical Magazine*, **11**, 313–34.

Drum, C. M. and van Gelder, W. (1972). Stacking-faults in (100) epitaxial silicon caused by HF and thermal oxidation and effects on p-n-junctions. *Journal of Applied Physics*, **43**, 4465.

Dudley, M., Huang, X. R., Huang, W. *et al.* (1999). The mechanism of micropipe nucleation at inclusions in silicon carbide. *Applied Physics Letters*, **75**, pp. 784–6.

Duesbery, M. S. (1989). The dislocation core and plasticity. In *Dislocations in Solids*, **8**, ed. F. R. Nabarro (Amsterdam: North-Holland), pp. 67–173.

Duesbery, M. S. and Richardson, G. Y. (1991). The dislocation core in crystalline materials. *CRC Critical Reviews in Solid State and Materials Science*, **17**, 1–46.

Dunstan, D. J. (1997). Strain and strain relaxation in semiconductors. *Journal of Materials Science: Materials Electronics*, **8**, 337–75.

Durose, K. and Russell, G. J. (1987). Lateral twins in the sphalerite structure. In *Microscopy of Semiconducting Materials 1987*. Conference Series No. 87 (Bristol: Institute of Physics), pp. 327–32.

Durose, K. and Russell, G. J. (1990). Twinning in CdTe. *Journal of Crystal Growth*, **101**, 246–50.

Eaglesham, D. J., Kvam, E. P., Maher, D. M., Humphreys, C. J. and Bean, J. C. (1989). Dislocation nucleation near the critical thickness in GeSi/Si strained layers. *Philosophical Magazine*, **A59**, 1059–73.

Ebina, A. and Takahashi, T. (1967). Crystal structure and stacking disorder of ZnS single crystals grown from the melt. *Journal of Applied Physics*, **38**, 3079–86.

Ehrenreich, H. and Hirth, J. P. (1985). Mechanism for dislocation density reduction in GaAs crystals by indium addition. *Applied Physics Letters*, **46**, 668–70.

Elliott, C. R. and Regnault, J. C. (1981). Birefringence studies of defects in III-V semiconductors. In *Microscopy of Semiconducting Materials 1981*. Conference Series No. 60 (Bristol: Institute of Physics), pp. 365–70.

Elliott, C. R., Regnault, J. C. and Wakefield, B. (1982). Applications of polarised infrared microscopy in the evaluation of InP and related compounds. In *Proceedings of the International Symposium on GaAs and Related Compounds*, Conference Series No. 65 60 (Bristol: Institute of Physics), pp. 553–60.

Ellis, S. G. (1959). On the growth of gallium arsenide crystals from the melt. *Journal of Applied Physics*, **30**, 947–8.

Ellis, W. C. and Treuting, R. G. (1951). Atomic relationships in the cubic twinned state. *Transactions of the AIME*, **189**, 53–5.

Eremenko, V. G. and Fedorov, A. V. (1995). New effect of interaction between moving dislocation and point defects in silicon. *Materials Science Forum*, **196/201**, 1219–23.

Eremenko, V., Abrosimov, N. and Fedorov, A. (1999). The origin and properties of new extended defects revealed by etching in plastically deformed Si and SiGe. *Physica Status Solidi*, **A171**, 383–8.

Erofeeva, S. A. and Osipiyan, Yu. A. (1973). Mobility of dislocations with the sphalerite lattice. *Soviet Physics Solid State*, **15**, 538–40.

Erofeev, V. N. and Nikitenko, V. I. (1971). Velocity of single dislocations in germanium single crystals. *Soviet Physics Solid State*, **13**, 241.

Erofeev, V. N., Nikitenko, V. I. and Osvenskii, V. B. (1969). Effect of impurities on the individual dislocation mobility in silicon. *Physica Status Solidi*, **35**, 79–88.

Ewing, R. E. and Greene, P. E. (1964). Influence of vapour composition on the growth rate and morphology of GaAs epitaxial films. *Journal of the Electrochemical Society*, **111**, 1266–9.

Fahey, P. M., Mader, S. R., Stiffler, S. R. *et al.* (1992). Stress-induced dislocations in silicon integrated-circuits. *IBM Journal of Research and Development*, **36**, 158–82.

Falster, R. and Voronkov, V. V. (2000). The engineering of intrinsic point defects in silicon wafers and crystals. *Materials Science and Engineering*, **B73**, 87–94.

Farvacque, J. L., Bougrioua, Z. and Moerman, I. (2001). Free-carrier mobility in GaN in the presence of dislocation walls. *Physical Review*, **B63**, 115202 (8 pages).

Faust, J. W. and Sagar, A. (1960). Effect of the polarity of the III-V intermetallic compounds on etching. *Journal of Applied Physics*, **31**, 331–3.

Faust, J. W. and John, H. F. (1962). Growth facets on III-V intermetallic compounds. *Journal of Physics and Chemistry of Solids*, **23**, 1119–22.

Faust, J. W. and John, H. F. (1964). The growth of semiconductor crystals from solution using the twin-plane re-entrant-edge mechanism. *Journal of Physics and Chemistry of Solids*, **25**, 1407–15.

Feklisova, O. V., Yakimov, E. B. and Yarykin, N. (2003). Contribution of the disturbed dislocation planes to the electrical properties of plastically deformed silicon. *Physica*, **B340–342**, 1005–8.

Feklisova, O. V., Pichaud, B. and Yakimov, E. B. (2005). Annealing effect on the electrical activity of extended defects in plastically deformed p-Si with low dislocation density. *Physica Status Solidi*, **A202**, 896–900.

Figielski, T. (1965). Dislocations as traps for holes in germanium. *Physica Status Solidi*, **9**, 555.

Finch, R. H., Queisser, H. J., Thomas, G. and Washburn, J. (1963). Structure and origin of stacking faults in epitaxial silicon. *Journal of Applied Physics*, **34**, 406–15.

Fisher, A. W. and Amick, J. A. (1966). Defect structure on silicon surfaces after thermal oxidation. *Journal of the Electrochemical Society*, **113**, 1054–60.

Fitzgerald, E. A. (1991). Dislocations in strained-layer epitaxy – theory, experiment, and applications. *Materials Science Reports*, **7**, 91–142.

Fitzgerald, E. A., Watson, G. P., Proano, R. E. *et al.* (1989). Nucleation mechanisms and the elimination of misfit dislocations at mismatched interfaces by reduction in growth area. *Journal of Applied Physics*, **65**, 2220–37.

Fitzgerald, A. G. and Mannami, M. (1966). The analysis of defects by observation of Moiré fringes from overlapping crystals – Application to zinc sulphide crystals. *Proceedings of the Royal Society*, **A293**, 469–78.

Fitzgerald, A. G., Mannami, M., Pogson, E. N. and Yoffe, A. D. (1966). Structure of zinc selenide crystals and defects introduced during growth. *Philosophical Magazine*, **14**, 197–200.

Fitzgerald, A. G., Mannami, M., Pogson, E. N. and Yoffe, A. D. (1967). Crystal growth and defect structure of zinc sulfide and zinc selenide platelets. *Journal of Applied Physics*, **38**, 3303–10.

Fletcher, N. H. (1967). Structure and energy of crystal interfaces. 2. A simple explicit calculation. *Philosophical Magazine*, **16**, 159–64.

Fletcher, N. H. (1969). General principles governing the structure and energy of interfaces between crystals. In *Interfaces*, Proceedings of the International Conference, ed. R. C. Gifkins (Melbourne: Butterworths), pp. 1–18.

Fletcher, N. H. and Adamson, P. L. (1966). Structure and energy of crystal interfaces. I. Formal development. *Philosophical Magazine*, **14**, 99–110.

Fletcher, N. H. and Lodge, K. W. (1975). Energies of Interfaces between crystals: an *Ab Initio* approach. In *Epitaxial Growth Part B*, ed. J. W. Matthews (New York: Academic Press), pp. 529–57.

Frank, F. C. (1949). Answer by Frank in discussion of a paper by N. F. Mott that introduced what became known as Frank's rule. *Physica*, **15**, 131–3.

Frank, F. C. (1951). Capillary equilibria of dislocated crystals. *Acta Crystallographica*, **4**, 497–501.

Frank, F. C. and Nicholas, J. F. (1953). Stable dislocations in the common crystal lattices. *Philosophical Magazine*, **44**, 1213–35.

Frank, F. C. and Read, W. T. (1950). Multiplication processes for slow moving dislocations. *Physical Review*, **79**, 722–3.

Frank, F. C. and van der Merwe, J. H. (1949a). One-dimensional dislocations. 1. Static theory. *Proceedings of the Royal Society London Series*, **A198**, pp. 205–16.

Frank, F. C. and van der Merwe, J. H. (1949b). One-dimensional dislocations. 2. Misfitting monolayers and oriented overgrowth. *Proceedings of the Royal Society London Series*, **A198**, 216–25.

Frank, F. C. and Turnbull, D. (1956). Mechanism of diffusion of copper in germanium. *Physical Review*, **104**, 617–18.

Franzosi, P., Salviati, G., Cocito, M., Taiariol, F. and Ghezzi, C. (1984). Inclusion-like defects in Czochralski grown InP single-crystals. *Journal of Crystal Growth*, **69**, 388–98.

Freund, L. B. (1990). The driving force for glide of a threading dislocation in a strained epitaxial layer on a substrate. *Journal of the Mechanics and Physics of Solids*, **38**, 657–79.

Friedel, G. (1926). *Lecons de cristallographie* (Paris: Berger Levrault), pp. 250–2.

Frisch, H. J. and Patel, J. R. (1967). Chemical influence of holes and electrons on dislocation velocity in semiconductors. *Physical Review Letters*, **18**, 784–7.

Gai, P. L. and Howie, A. (1974). Dissociation of dislocations in GaP. *Philosophical Magazine*, **30**, 939–43.

Gatos, H. C. and Lavine, M. C. (1960a). Characteristics of the {111} surfaces of the III-V intermetallic compounds. *Journal of the Electrochemical Society*, **107**, 327–433.

Gatos, H. C. and Lavine, M. C. (1960b). Etching and inhibition of the {111} surfaces of the III-V intermetallic compounds: InSb. *Journal of Physics and Chemistry of Solids*, **14**, 169–74.

Gatos, H. C. and Lavine, M. C. (1965). Chemical behaviour of semiconductors: Etching characteristics. In *Progress in Semiconductors*, **9**, eds. A. F. Gibson and R. E. Burgess (New York: Heywood), pp. 1–45.

Gatos, H. C., Moody, P. L. and Lavine, M. C. (1960). Growth of InSb crystals in the ⟨111⟩ polar direction. *Journal of Applied Physics*, **31**, 212–13.

Gatos, H. C., Lavine, M. C. and Warekois, E. P. (1961). Characteristics of the {111} surfaces of the III-V intermetallic compounds. II. Surface damage. *Journal of the Electrochemical Society*, **108**, 645–49.

George, A. (1997). Plastic deformation of semiconductors: some recent advances and persistent challenges. *Materials Science and Engineering*, **A233**, 88–102.

George, A. and Yip, S. (2001). Preface to the viewpoint set on: dislocation mobility in silicon. *Scripta Materialia*, **45**, 1233–8.

George, A. and Rabier, J. (1987). Dislocations and plasticity in semiconductors. I – Dislocation structures and dynamics. *Revue de Physique Appliquee*, **22**, 941–66.

George, A., Schröter, W., Escaravage, C. and Champier, G. (1972). Velocities of screw and 60 degrees dislocations in silicon. *Physica Status Solidi*, **B53**, 483.

Gezci, S. and Woods, J. (1972). Dislocation etch pits in zinc selenide. *Journal of Materials Science*, **7**, 603–8.

Ghezzi, C. and Servidori, M. (1974). X-ray study of heterogeneous nucleation of dislocations in P-diffused silicon. *Journal of Materials Science*, **9**, 1797–802.

Gilman, J. J. (1959). Plastic anisotropy of LiF and other rocksalt-type crystals. *Acta Metallurgica*, **7**, 608–13.

Gilman, J. J. (1961). Nature of dislocations. In *Mechanical Behaviour of Materials at Elevated Temperatures*, ed. J. E. Dorn (New York: McGraw-Hill), pp. 17–44.

Gilman, J. J. (1969). *Micromechanics of Flow in Solids*. New York: McGraw-Hill.

Giri, P. K., Coffa, S., Raineri, V., *et al.* (2001). Photoluminescence and structural studies on extended defect evolution during high-temperature processing of ion-implanted epitaxial silicon. *Journal of Applied Physics*, **89**, 4310 –17.

Gleichmann, R., Frigeri, C. and Pelosi, C. (1990). Hillock formation in InP epitaxial layers: a mechanism based on dislocation/stacking fault interactions. *Philosophical Magazine*, **A62**, 103–14.

Gomez, A. M. and Hirsch, P. B. (1977). On the mobility of dislocations in germanium and silicon. *Philosophical Magazine*, **36**, 169–79.

Gomez, A. M. and Hirsch, P. B. (1978). Dissociation of dislocations in GaAs. *Philosophical Magazine*, **A38**, 733–7.

Gomez, A., Cockayne, D. J., Hirsch, P. B. and Vitek, V. (1975). Dissociation of near-screw dislocations in germanium and silicon. *Philosophical Magazine*, **31**, 105–13.

Gosling, T. J., Jain, S. C. and Harker, A. H. (1994). The kinetics of strain relaxation in lattice-mismatched semiconductor layers. *Physica Status Solidi*, **A146**, 713–34.

Gottschalk, H., Patzer, G. and Alexander, H. (1978). Stacking fault energy and ionicity of cubic III-V compounds. *Physica Status Solidi*, **A45**, 207–17.

Grimmer, H., Bollmann, W. and Warrington, D. H. (1974). Coincidence-site lattices and complete pattern-shift lattices in cubic crystals. *Acta Crystallographica*, **A30**, 197 –207.

Grovenor, C. R. (1985). Grain boundaries in semiconductors. *Journal of Physics*, **C18**, 4079–119.

Guruswamy, S., Rai, R. S., Faber, K. T., *et al.* (1989). Influence of solute doping on the high-temperature deformation behavior of GaAs. *Journal of Applied Physics*, **65**, 2508–12.

Gutkin, M. Yu., Sheinerman, A. G., Argunova, T. S., *et al.* (2003). Micropipe evolution in silicon carbide. *Applied Physics Letters*, **83**, 2157–9.

Gutmanas, E. Y. and Haasen, P. (1979a). Photoplastic effect in CdTe. *Journal de Physique*, **C6**, 169–72.

Gutmanas, E. Y., Travitzky, N. and Haasen, P. (1979b). Negative and positive photoplastic effect in CdTe. *Physica Status Solidi*, **A51**, 435–44.

Haasen, P. (1957a). Twinning in indium antimonide. *Transactions of the AIME*, **209**, 30–3.

Haasen, P. (1957). On the plasticity of germanium and indium antimonide. *Acta Metallurgica*, **5**, 598.

Haasen, P. (1975). Kinkenbildung in geladenen versetzungen. *Physica Status Solidi*, **A28**, 145–55.

Haasen, P. (1979). Kink formation and migration as dependent on the Fermi level. *Journal de Physique*, **C6**, 111–16.

Haasen, P. and Seeger, A. (1958). In *Halbleiterprobleme IV*, ed. W. Schottky (Braunschweig: Vieweg), pp. 68–118.

Haasen, P. and Alexander, H. (1968). Dislocations and plastic flow in the diamond structure. *Solid State Physics*, **22**, 27–158.

Haasen, P. and Alexander, H. (1972). Dislocations in nonmetals. *Annual Review of Materials Science*, **2**, 291–312.

Hah, S. R. and Fischer, T. E. (1998). Tribochemical polishing of silicon nitride. *Journal of the Electrochemical Society*, **145**, 1708–14.

Hall, R. N. (1968). Photomechanical and electomechanical effects in semiconductors. In *Proceedings of Ninth International Conference on Physics of Semiconductors*, **I** (Leningrad: Nauka).

Haneman, D. (1960). Behaviour of InSb surfaces during heat treatment. *Journal of Applied Physics*, **31**, 217–18.

Haneman, D. (1962a). Free bonds in semiconductors. In *Reports on Conference on Physics of Semiconductors*, Exeter (London: Institute of Physics), pp. 842–7.

Haneman, D. (1962b). Behaviour of InSb surfaces during heat treatment. In *Semiconducting Compounds*, **I**, *Preparation of III-V Compounds*, eds. R. K. Willardson and H. L. Goering (New York: Reinhold), pp. 432–5.

Haneman, D., Russel, G. J. and Ip, H. K. (1964). Bonding and decomposition in III-V compounds. In *Proceedings of International Conference on Semiconductors*, Paris (Paris: Dunod), pp. 1141–5.

Hanneman, R. E. and Jorgenson, P. J. (1967). On the existence of electromechanical and photomechanical effects in semiconductors. *Journal of Applied Physics*, **38**, 4099–100.

Hanneman, R. E. and Westbrook, J. H. (1968). Effects of adsorption on the indentation deformation of non-metallic solids. *Philosophical Magazine*, **18**, 73–88.

Hanneman, R. E., Ginn, M. C. and Gatos, H. C. (1962). Elastic strain energy associated with the 'A' surfaces of the III-V compounds. *Journal of Physics and Chemistry of Solids*, **23**, 1554–6.

Hartmann, H., Mach, R. and Selle, B. (1982). Wide gap II–VI compounds as electronic materials. In *Current Topics in Materials Science*, **9**, ed. E. Kaldis. Amsterdam: North Holland.

Hashimoto, H., Shibayama, H., Masaki, H. and Ishikawa, H. (1976). Annihilation of stacking-faults in silicon by impurity diffusion. *Journal of the Electrochemical Society*, **123**, 1899–902.

Häussermann, F. and Schaumburg, H. (1973). Extended dislocations in germanium. *Philosophical Magazine*, **27**, 745–51.

Heggie, M. I., Jones, R., Lister, G. M. and Umerski, A. (1989). Interaction of impurities with dislocation cores in silicon. *Institute of Physics Conference Series*, **104**, 43–6.

Heiland, G. and Kunstmann, P. (1969). Polar surfaces of ZnO crystals. *Surface Science*, **13**, 72–84.

Heindl, J., Strunk, H. P., Heydemann, V. D. and Pensl, G. (1997). Micropipes: hollow tubes in silicon carbide. *Physica Status Solidi*, **A162**, 251–62.

Heindl, J., Dorsch, W., Strunk, H. P., *et al.* (1998). Dislocation content of micropipes in SiC. *Physical Review Letters*, **80**, pp. 740–1.

Henneke, H. L. (1965). Comment on 'Polarity Effects in InSb-Alloyed p-n Junctions'. *Journal of Applied Physics*, **36**, 2967–8.

Herrera Zaldivar, M., Fernández, P. and Piqueras, J. (2001). Study of growth hillocks in GaN:Si films by electron beam induced current imaging. *Journal of Applied Physics*, **90**, 1058–60.

Hinkley, E. D., Rediker, R. H. and Lavine, M. C. (1964). Inversion of {111} surfaces in single crystal regrowth during interface-alloying of intermetallic compounds. *Applied Physics Letters*, **5**, 110–12.

Hirsch, P. B. (1980). Structure and electrical properties of dislocations in semiconductors. *Journal of Microscopy*, **118**, 3–12.

Hirsch, P. B. (1981). Electronic and mechanical properties of dislocations in semiconductors. In *Defects in Semiconductors*, Proceedings of the Materials Research Society 1980 (New York: North-Holland), pp. 257–71.

Hirsch, P. B. (1985). Dislocations in semiconductors. *Materials Science and Technology*, **1**, 666–77.

Hirsch, P. B., Pirouz, P., Roberts, S. G. and Warren, P. D. (1985). Indentation hardness and polarity of hardness on {111} faces of GaAs. *Philosophical Magazine*, **B52**, 759–84.

Hirth, J. P. and Lothe, J. (1968). *Theory of Dislocations*. New York: McGraw-Hill.

Hofmann, D., Bickermann, M., Eckstein, R., *et al.* (1999). Sublimation growth of silicon carbide bulk crystals: experimental and theoretical studies on defect formation and growth rate augmentation. *Journal of Crystal Growth*, **198–199**, 1005–10.

Holt, D. B. (1960). Filled and empty dangling bonds in III-V compounds. *Journal of Applied Physics*, **31**, 2231–2.

Holt, D. B. (1962). Defects in the sphalerite structure. *Journal of Physics and Chemistry of Solids*, **23**, 1353–62.

Holt, D. B. (1964). Grain boundaries in the sphalerite structure. *Journal of Physics and Chemistry of Solids*, **25**, 1385–95.

Holt, D. B. (1966). Misfit dislocations in semiconductors. *Journal of Physics and Chemistry of Solids*, **27**, 280–95.

Holt, D. B. (1969). Antiphase boundaries in semiconducting compounds. *Journal of Physics and Chemistry of Solids*, **27**, 1053–67.

Holt, D. B. (1974). The growth and structure of epitaxial films and heterojunctions of II-VI compounds. *Thin Solid Films*, **24**, 1–53.

Holt, D. B. (1996). The role of defects in semiconductor materials and devices. *Scanning Microscopy*, **10**, 1047–78.

Holt, D. B. and Dangor, A. E. (1963). Direct observations of defects in germanium. *Philosophical Magazine*, **8**, 1921–36.

Holt, D. B. and Culpan, M. (1970). Scanning electron microscope studies of striations in ZnS. *Journal of Materials Science*, **5**, 546–56.

Holt, D. B. and Brada, Y. (1997). EBIC studies of the electrical barriers in striated ZnS platelets exhibiting the anomalous photovoltaic effect. In *Microscopy of Semiconducting Materials 1997*, Conference Series No. 157 (Bristol: Institute of Physics), pp. 629–34.

Holt, D. B. and Salviati, G. (1990). Twinning and impurity segregation in Cr-doped and Fe-doped LEC InP. *Journal of Crystal Growth*, **100**, 497–507.

Holt, D. B. and Salviati, G. (1993). Twins in SI InP. *Microscopy of Semiconducting Materials* (1993), Institute of Physics Conference Series (134), pp. 739–42.

Holt, D. B. and Wilcox, D. M. (1971). Crystallographic defects in epitaxial layers of cadmium sulphide. *Journal of Crystal Growth*, **9**, 193.

Holt, D. B. and Wilcox, D. M. (1972). The effect of substrate orientation on the structure of epitaxial films of II-VI compounds. *Thin Solid Films*, **10**, 141–7.

Holt, D. B., Abdalla, M. I., Gejji, F. H. and Wilcox, D. M. (1976). Crystallography of the phase transfomed structures in epitaxial (110) films of CdS, CdSe and CdTe and ordering in II-VI compounds. I. The domain-form structure. *Thin Solid Films*, **37**, 91–107.

Holt, D. B., Hardingham, C., Lazzarini, L. *et al.* (1996). Properties and structure of antiphase boundaries in GaAs/Ge solar cells. *Materials Science and Engineering*, **B42**, 204–7.

Hornstra, J. (1958). Dislocations in the diamond lattice. *Journal of Physics and Chemistry of Solids*, **5**, 129–41.

Hornstra, J. (1959). Models of grain boundaries in the diamond lattice. I. Tilt about ⟨110⟩. *Physica*, **25**, 409–22.

Hornstra, J. (1960). Models of grain boundaries in the diamond lattice. II. Tilt about ⟨001⟩ and theory. *Physica*, **26**, 198–208.

Houghton, D. C. (1991). Strain relaxation kinetics in $Si_{1-x}Ge_x/Si$ heterostructures. *Journal of Applied Physics*, **70**, 2136–51.

Hourai, M., Kajita, E., Nagashima, T. *et al.* (1995). Growth parameters determining the type of grown-in defects in Czochralski silicon crystals. *Materials Science Forum*, **196–201**, 1713–18.

Hu, S. M. (1974). Formation of stacking faults and enhanced diffusion in the oxidation of silicon. *Journal of Applied Physics*, **45**, 1567–73.

Hu, S. M. (1991). Stress-related problems in silicon technology. *Journal of Applied Physics*, **70**, R53–R80.

Huff, H. R. (2002). An electronics division retrospective (1952–2002) and future opportunities in the twenty-first century. *Journal of the Electrochemical Society*, **149**, S35–S58.

Hull, R. (1999). Misfit strain and accommodation in SiGe heterostructures. In *Germanium Silicon: Physics and Materials*, eds. R. Hull and J.C. Bean. This is volume **56** in the series *Semiconductors and Semimetals*, eds. R. K. Willardson and A. C. Beer (San Diego: Academic Press), pp. 101–67.

Hull, R., Bean, J. C. and Buescher, C. (1989). A phenomenological description of strain relaxation in $Ge_xSi_{1-x}/Si(100)$ heterostructures. *Journal of Applied Physics*, **66**, 5837–43.

Hull, R., Gray, J., Wu, C. C., Atha, S. and Floro, J. A. (2002). Interaction between surface morphology and misfit dislocations as strain relaxation modes in lattice-mismatched heteroepitaxy. *Journal of Physics: Condensed Matter*, **14**, 12829–41.

Hulme, K. F. and Mullin, J. B. (1962). InSb: A review of its preparation, properties and device applications. *Solid State Electronics*, **5**, 211–47.

Humphreys, C. J., Maher, D. M., Eaglesham, D. J., Kvam, E. P. and Salisbury, I. G. (1991). The origin of dislocations in multilayers. *Journal de Physique (III)*, **1**, 1119–30.

Hurle, D. T. and Rudolph, P. (2004). A brief history of defect formation, segregation, faceting and twinning in melt-grown semiconductors. *Journal of Crystal Growth*, **264**, 550–64.

Igarashi, O. (1971). Crystallographic orientations and interfacial mismatches of single-crystal CdS films deposited on various faces of zinc-blende-type crystals. *Journal of Applied Physics*, **42**, 4035–42.

Imai, M. and Sumino, K. (1983). In situ x-ray topographic study of the dislocation mobility in high-purity and impurity-doped silicon crystals. *Philosophical Magazine*, **A47**, 599–621.

Iunin, Yu. L. and Nikitenko, V. I. (2001). Modes of kink motion on dislocations in semiconductors. *Scripta Materialia*, **45**, 1239–46.

Jaccodine, R. J. and Drum, C. M. (1966). Extrinsic stacking faults in silicon after heating in wet oxygen. *Applied Physics Letters*, **8**, 29.

Jagodzinski, H. (1949). *Acta Crystallographica*, **2**, 201–7.

Jain, S. C., Willander, M. and Maes, H. (1996). Stresses and strains in epilayers, stripes and quantum structures of III-V compound semiconductors. *Semiconductor Science and Technology*, **11**, 641–71.

Jain, S. C., Willander, M., Narayan, J. and Van Overstraeten, R. (2000). III–nitrides: Growth, characterization, and properties. *Journal of Applied Physics*, **87**, 965–1006.

James, R. W. (1948). *The Optical Principles of the Diffraction of X-Rays*. London: G. Bell and Sons.

Jepps, N. W. and Page, T. F. (1983). Polytypic transformations in silicon carbide. *Progress in Crystal Growth and Characterization*, **7**, 259–307.

Jesser, W. A. and van der Merwe, J. H. (1994). An assessment of the roles of climb and glide in misfit strain relief. *Journal of Applied Physics*, **75**, 872–8.

Jesson, D. E., Chen, K. M., Pennycook, S. J., Thundat, T. and Warmack, R. J. (1995). Crack-like sources of dislocation nucleation and multiplication in thin films. *Science*, **268**, 1161–3.

Jones, R. (1979). Theoretical calculations of electron states associated with dislocations. *Journal de Physique*, **C6**, 33–8.

Jones, R. (2000). Do we really understand dislocations in semiconductors? *Materials Science and Engineering*, **B71**, 24–9.

Jones, R. and Blumenau, A. T. (2001). Interaction of dislocations in Si with intrinsic defects. *Scripta Materialia*, **45**, 1253–8.

Jones, R., Umerski, A., Stitch, P., Heggie, M. I. and Oberg, S. (1993). Density-functional calculations of the structure and properties of impurities and dislocations in semiconductors. *Physica Status Solidi*, **A138**, 369–81.

Joshi, M. L. (1965). Effect of fast cooling on diffusion-induced imperfections in silicon. *Journal of the Electrochemical Society*, **112**, 912–16.

Joshi, M. L. and Wilhelm, F. (1965). Diffusion induced imperfections in silicon. *Journal of the Electrochemical Society*, **112**, 185–8.

Jowett, C. E. (1979). Failure mechanisms and analysis procedures for semiconductor devices. *Microelectronics Journal*, **9**, 5–13.

Jungbluth, E. D. and Wang, P. (1965). Process-introduced structural defects and junction characteristics in npn Si epitaxial planar transistors. *Journal of Applied Physics*, **36**, 1967–73.

Justo, J. F., de Koning, M., Cai, W. and Bulatov, V. V. (2000). Vacancy interaction with dislocations in silicon: The shuffle-glide competition. *Physical Review Letters*, **84**, 2172–5.

Kabler, M. N. (1963). Dislocation mobility in germanium. *Physical Review*, **131**, 54.

Kamata, I., Tsuchida, H., Jikimoto, T., Miyanagi, T. and Izumi, K. (2003). Conditions for micropipe dissociation by 4H-SiC CVD growth. *Materials Science Forum*, **433–436**, 261–4.

Kamiya, T., Durrani, Z. A. and Ahmed, H. (2002). Control of grain-boundary tunneling barriers in polycrystalline silicon. *Applied Physics Letters*, **81**, 2388–90.

Kannan, V. C. and Washburn, J. (1970). Direct dislocation velocity measurement in silicon by x-ray topography. *Journal of Applied Physics*, **41**, 3589.

Karstensen, F. (1957). Preferential diffusion of Sb along small angle boundaries in Ge and the dependence of this effect on the direction of the dislocation lines in the boundary. *Journal of Electronics and Control*, **3**, 305–7.

Kazmerski, L. L. (1982). Chemical, compositional, and electrical properties of semiconductor grain boundaries. *Journal of Vacuum Science and Technology*, **20**, 423–9.

Kazmerski, L. L. (1993). Microcharacterization to nanocharacterization of semiconductor grain boundaries. *Surface Science Reports*, **19**, 169–89.

Kazmerski, L. L. (1994). Auger electron spectroscopy. In *Microanalysis of Solids*, eds. B. G. Yacobi, D. B. Holt and L. L. Kazmerski (New York: Plenum Press), pp. 99–146.

Kazmerski, L. L. and Dick, J. R. (1984). Determination of grain boundary impurity effects in polycrystalline silicon. *Journal of Vacuum Science and Technology*, **A2**, 1120–2.

Kazmerski, L. L., Ireland, P. J. and Ciszek, T. F. (1980). Evidence for the segregation of impurities to grain boundaries in multigained silicon using AES and SIMS. *Applied Physics Letters*, **36**, 323–5.

Khlebnikov, I., Madangarli, V. P., Khan, M. A. and Sudarshan, T. S. (1998). Thick film SiC epitaxy for 'filling up' micropipes. *Materials Science Forum*, **264–2**, 167–70.

Kimerling, L. C. (1978). Recombination enhanced defect reactions. *Solid State Electronics*, **21**, 1391–1401.

Kimerling, L. C. and Patel, J. R. (1985). Silicon defects: Structures, chemistry, and electrical properties. In *VLSI Electronics Microstructure Science*, **12**, eds. N. G. Einspruch and H. Huff (New York: Academic Press), pp. 223–67.

Kirichenko, L. G., Petrenko, V. F. and Uimin, G. V. (1978). Nature of the dislocation charge in ZnSe. *Soviet Physics JETP*, **47**, 389–94.

Kishino, S., Isomae, S., Tamura, M. and Maki, M. (1978). Suppression of oxidation-stacking-fault generation by preannealing in N_2 atmosphere. *Applied Physics Letters*, **32**, 1–3.

Kittler, M., Seifert, W., Stemmer, M. and Palm, J. (1995). Interaction of iron with a grain boundary in boron-doped multicrystalline silicon. *Journal of Applied Physics*, **77**, 3725–8.

Knippenberg, W. F. (1963). Growth phenomena in silicon carbide. *Philips Research Reports*, **18**, 161–274.

Kohn, J. A. (1958). Twinning in diamond-type structures – a proposed boundary-structure model. *American Mineralogist*, **43**, 263–84.

Kolar, H. R., Spence, J. C. and Alexander, H. (1996). Observation of moving dislocation kinks and unpinning. *Physical Review Letters*, **77**, 4031–4.

Kolbesen, B. O. and Strunk, H. P. (1985). Analysis, electrical effects, and prevention of process-induced defects in silicon integrated circuits. In *VLSI Electronics Microstructure Science*, **12**, eds. N. G. Einspruch and H. Huff (New York: Academic Press), pp. 143–222.

Kolbesen, B. O., Bergholz, W., Cerva, H. *et al.* (1989). Effects of extended lattice defects on silicon semiconductor devices. In *International Symposium on Structural Properties of Dislocations in Semiconductors*, Oxford, Conference Series No. 104 (Bristol: Institute of Physics), pp. 421–30.

Kolbesen, B. O., Bergholz, W., Cerva, H., *et al.* (1991). Process-induced defects in VLSI. *Nuclear Instrumentation and Methods in Physics Research*, **B55**, 124–31.

Kolbesen, B., Cerva, H. and Zoth, G. (2001). Defects and contamination in microelectronic device production: State-of-the-art and prospects. *Solid State Phenomena*, **76–77**, 1–6.

Kressel, H. (1975). The application of heterojunction structures to optical devices. *Journal of Electronic Materials*, **4**, 1081–141.

Krivanek, O. L., Isoda, S. and Kobayashi, K. (1977). Lattice imaging of a grain boundary in crystalline germanium. *Philosophical Magazine*, **36**, 931–40.

Kroemer, H. (1987). Sublattice allocation and antiphase domain suppression in polar-on-nonpolar nucleation. *Journal of Vacuum Science and Technology*, **B5**, 1150–4.

Krüger, O., Seifert, W., Kittler, M. and Vyvenko, O. F. (2000). Extension of hydrogen passivation of intragrain defects and grain boundaries in cast multicrystalline silicon. *Physica Status Solidi*, **B222**, 367–78.

Ku, S. M. (1963). The preparation and properties of vapour-grown GaAs-GaP alloys. *Journal of the Electrochemical Society*, **110**, 992–5.

Kubo, I. and Tomiyama, N. (1971). Polarity of ZnO crystal. III Inversion twin boundaries on {101̄0}. *Japanese Journal of Applied Physics*, **10**, 952.

Küsters, K. H. and Alexander, H. (1983). Photoplastic effect in silicon. *Physica*, **116B**, 594–9.

Laister, D. and Jenkins, G. M. (1968). Stacking-faults in tellurium-doped gallium arsenide. *Journal of Materials Science*, **3**, 584–9.

Laister, D. and Jenkins, G. M. (1969). Image contrast of triple loops in tellurium-doped gallium arsenide. *Philosophical Magazine*, **20**, 361.

Laister, D. and Jenkins, G. M. (1971). Electrical and electron microscope studies of annealing of tellurium-doped gallium arsenide. *Philosophical Magazine*, **23**, 1077–1100; 4 layer stacking faults in gallium arsenide. *Philosophical Magazine*, **24**, 705.

Laister, D. and Jenkins, G. M. (1973). Deformation of single crystals of gallium arsenide. *Journal of Materials Science*, **8**, 1218–32.

Latanision, R. M. and Fourie, J. T. (eds.) (1977). Surface effects in crystal plasticity (NATO Advanced Study Institutes Series, Series E: Applied Science No. 17) (Leyden: Nordhoff).

Lavine, M. C., Rosenberg, A. J. and Gatos, H. C. (1958). Influence of crystal orientation of the surface behaviour of InSb. *Journal of Applied Physics*, **29**, 1131–2.

Leamy, H. J., Frye, R. C., Ng, K. K. *et al.* (1982). Direct observation of grain boundary effects in polycrystalline silicon thin-film transistors. *Applied Physics Letters*, **40**, 598–600.

Lee, D. B. (1974). The push-out effect in silicon n-p-n diffused transistors. *Philips Research Reports Supplement*, **5**, 1–131.

Leipner, H. S., Hubner, C. G., Staab, T. E. M. *et al.* (2000). Vacancy clusters in plastically deformed semiconductors. *Journal of Physics: Condensed Matter*, **12**, 10071–8.

Leipner, H. S., Wang, Z., Gu, H. *et al.* (2004). Defects in silicon plastically deformed at room temperature. *Physica Status Solidi*, **A201**, 2021–8.

Levine, E. and Tauber, R. N. (1968). The preparation and examination of PbTe by transmission electron microscopy. *Journal of the Electrochemical Society*, **115**, 107–8.

Levine, E., Washburn, J. and Thomas, G. (1967a). Diffusion-induced defects in silicon. I. *Journal of Applied Physics*, **38**, 81–7.

Levine, E., Washburn, J. and Thomas, G. (1967b). Diffusion-induced defects in silicon. II. *Journal of Applied Physics*, **38**, 87–95.

Likhtman, V. I., Rebinder, P. A. and Karpenko, G. V. (1958). *Effect of a surface active medium on the deformation of metals.* (London: H. M. O.).

Lilienthal-Weber, Z., Chen, Y., Ruvimov, S., Swider, W. and Washburn, J. (1997). Nano-tubes in GaN. In *Materials Research Society Symposium Proceedings*, **449**, 417–22.

Liliental-Weber, Z., Chen, Y., Ruvimov, S. and Washburn, J. (1997). Nanotubes and pinholes in GaN and their formation mechanism. *Materials Science Forum*, **258–2**, 1659–64.

Liu, X. W., Hopgood, A. A., Usher, B. F., Wang, H. and Braithwaite, N. St. J. (1999). Formation of misfit dislocations during growth of In$_x$Ga$_{1-x}$As/GaAs strained-layer heterostructures. *Semiconductor Science and Technology*, **14**, 1154–60.

Liu, X. W., Hopgood, A. A., Usher, B. F., Wang, H. and Braithwaite, N. St. J. (2003). Formation of misfit dislocations in strained-layer GaAs/In$_x$Ga$_{1-x}$As/GaAs heterostructures during postfabrication thermal processing. *Journal of Applied Physics*, **94**, 7496–501.

Lo, Y. H. (1991). New approach to grow pseudomorphic structures over the critical thickness. *Applied Physics Letters*, **59**, 2311–13.

Look, D. C. and Sizelove, J. R. (1999). Dislocation scattering in GaN. *Physical Review Letters*, **82**, 1237–40.

Look, D. C., Stutz, C. E., Molnar, R. J., Saarinen, K. and Liliental-Weber, Z. (2001). Dislocation-independent mobility in lattice-mismatched epitaxy: application to GaN. *Solid State Communications*, **117**, 571–75.

Lopatin, S., Pennycook, S. J., Narayan, J. and Duscher, G. (2002). Z-contrast imaging of dislocation cores at the GaAs/Si interface. *Applied Physics Letters*, **81**, 2728–30.

Louchet, F. and Thibault-Desseaux, J. (1987). Dislocation cores in semiconductors. From the 'shuffle or glide' dispute to the 'glide and shuffle' partnership. *Review de Physique Appliquee*, **22**, 207–19.

Louchet, F. and Thibault, J. (1989). On the shuffle-glide controversy. In *International Symposium on Structural Properties of Dislocations in Semiconductors*, Oxford. Conference Series No. 104 (Bristol: Institute of Physics), pp. 47–8.

Lu, G. and Cockayne, D. J. (1983). Dislocation-structures and motion in II-VI semiconductors. *Physica*, **116 B&C**, 646–49.

Lu, J., Wagener, M., Rozgonyi, G., Rand, J. and Jonczyk, R. (2003). Effects of grain boundary on impurity gettering and oxygen precipitation in polycrystalline sheet silicon. *Journal of Applied Physics*, **94**, 140–4.

Madelung, O., Schulz, M. and Weiss, H. (eds.) (1985). *Landolt Börnstein New Series* (Berlin: Springer) **17h**, pp. 27 and 68.

Maeda, K. and Sakamoto, K. (1977). Reversible hardening induced by electron-beam irradiation in CdS single-crystals. *Journal of the Physical Society of Japan*, **42**, 1914–17.

Maeda, K., Ueda, O., Murayama, Y. and Sakamoto, K. (1977). Mechanical properties and photomechanical effect in GaP single crystals. *Journal of Physics and Chemistry of Solids*, **38**, 1173–9.

Maeda, K., Nakagawa, K., Takeuchi, S. and Sakamoto, K. (1981). Cathodoluminescence studies of dislocation-motion in IIB-VIB compounds deformed in SEM. *Journal of Materials Science*, **16**, 927–34.

Maeda, K., Suzuki, K., Yamashita, Y. and Mera, Y. (2000). Dislocation motion in semiconducting crystals under the influence of electronic perturbations. *Journal of Physics: Condensed Matter*, **12**, 10079–91.

Mahajan, S. (1989). Growth and processing-induced defects in semiconductors. *Progress in Materials Science*, **33**, 1–84.

Mahajan, S. (1997). Defects in epitaxial layers of compound semiconductors grown by OMVPE and MBE techniques. *Progress in Materials Science*, **42**, 341–55.

Mahajan, S. (2000). Defects in semiconductors and their effects on devices. *Acta Materialia*, **48**, 137–49.

Mahajan, S. (2002). Origins of micropipes in SiC crystals. *Applied Physics Letters*, **80**, 4321–3.

Mahajan, S., Rozgonyi, G. A. and Brasen, D. (1977). Model for formation of stacking-faults in silicon. *Applied Physics Letters*, **30**, 73–5.

Mardix, S. (1984). The formation of macroscopic polytypic regions in ZnS crystals. *Journal of Applied Crystallography*, **17**, 328–30.

Mardix, S. (1986). Polytypism: A controlled thermodynamic phenomenon. *Physical Review*, **B33**, 8677–84.

Mardix, S. (1991). Symmetry and martensitic transformations in ZnS crystals. *Acta Crystallographica*, **A47**, 177–80.

Mardix, S. and Steinberger, I. T. (1970). Tilt and structure transformation in ZnS. *Journal of Applied Physics*, **41**, 5339–41.

Mardix, S., Kalman, Z. H. and Steinberger, I. T. (1968). Periodic slip processes and the formation of polytypes in zinc sulphide crystals. *Acta Crystallographica*, **A24**, 464–9.

Mariano, A. N. and Hanneman, R. E. (1963). Crystallographic polarity of ZnO crystals. *Journal of Applied Physics*, **34**, 384–8.

Marklund, S. (1979). Electron states associated with partial dislocations in silicon. *Physica Status Solidi*, **B92**, 83–9.

Mathis, S. K., Wu, X. H., Romanov, A. E. and Speck, J. S. (1999). Threading dislocation reduction mechanisms in low-temperature-grown GaAs. *Journal of Applied Physics*, **86**, 4836–42.

Mathis, S. K., Romanov, A. E., Chen, L. F. *et al.* (2000). Modeling of threading dislocation reduction in growing GaN layers. *Physica Status Solidi*, **A179**, 125–45.

Matsui, J. and Kawamura, T. (1972). Spotty defects in oxidized floating-zoned dislocation-free silicon crystals. *Japanese Journal of Applied Physics*, **11**, 197.

Matthews, J. W. (1966). Accommodation of misfit across interface between single-crystal films of various face-centred cubic metals. *Philosophical Magazine*, **13**, 1207–12.

Matthews, J. W. (1975). Defects associated with the accommodation of misfit between crystals. *Journal of Vacuum Science and Technology*, **12**, 126–33.

Matthews, J. W. (1979). Misfit dislocations. In *Dislocations in Solids*, **2**, ed. F. R. Nabarro (Amsterdam: North-Holland), pp. 461–545.

Matthews, J. W. and Isebeck, K. (1963). Dislocations in evaporated lead sulphide films. *Philosophical Magazine*, **8**, 469–85.

Matthews, J. W. and Blakeslee, A. E. (1974). Defects in epitaxial multilayers. 1. Misfit dislocations. *Journal of Crystal Growth*, **27**, 118–25.

Matthews, J. W., Mader, S. and Light, T. B. (1970). Accommodation of misfit across interface between crystals of semiconducting elements or compounds. *Journal of Applied Physics*, **41**, 3800–4.

Maurice, J.-L. and Colliex, C. (1989). Fast diffusers Cu and Ni as the origin of electrical activity in a silicon grain boundary. *Applied Physics Letters*, **55**, 241–3.

McHugo, S. A. and Sawyer, W. D. (1993). Impurity decoration of defects in float zone and polycrystalline silicon via chemomechanical polishing. *Applied Physics Letters*, **62**, 2519–21.

Meieran, E. S. (1965). Transmission electron microscope study of gallium arsenide. *Journal of Applied Physics*, **36**, 2544.

Meingast, R. and Alexander, H. (1973). Dissociated dislocations in germanium. *Physica Status Solidi*, **A17**, 229–36.

Mendelson, S. (1964a). Stacking fault nucleation in epitaxial silicon on variously oriented silicon substrates. *Journal of Applied Physics*, **35**, 1507–81.

Mendelson, S. (1964b). Growth and imperfections in epitaxially grown silicon on variously oriented silicon substrates. In *Single Crystal Films*, eds. M. H. Francombe and H. Sato (Oxford: Pergamon), pp. 251–81.

Mendelson, S. (1967). Defect formation in epitaxial films on native and foreign substrates. *Surface Science*, **6**, 233–45.

Merz, W. J. (1958). Photovoltaic effect in striated ZnS single crystals. *Helvetica Physica Acta*, **31**, 625–35.

Mihara, M. and Ninomiya, T. (1968). Dislocation velocity in indium antimonide. *Journal of the Physical Society of Japan*, **25**, 1198.

Mihara, M. and Ninomiya, T. (1975). Dislocation velocities in indium antimonide. *Physica Status Solidi*, **A32**, 43–52.

Millea, M. F. and Kyser, D. F. (1965). Thermal decomposition of GaAs. *Journal of Applied Physics*, **36**, 308–13.

Miller, D. P., Harper, J. G. and Perry, T. R. (1961). High temperature oxidation and vacuum dissociation studies on the A{111} and B{1̄1̄1̄} surfaces of GaAs. *Journal of the Electrochemical Society*, **108**, 1123–6.

Miller, D. P., Watelski, S. B. and Moore, C. R. (1963). Structure defects in pyrolytic silicon epitaxial fillms. *Journal of Applied Physics*, **34**, 2813–21.

Minamoto, M. T. (1962). Significance of crystallographic polarity in the fabrication of junctions in InSb. *Journal of Applied Physics*, **33**, 1826–29.

Möller, H. J. (1993). *Semiconductors for Solar Cells*. Boston: Artech House.

Monson, T. K. and Van Vechten, J. A. (1999). On the origins of oxidation-induced stacking faults in silicon. *Journal of the Electrochemical Society*, **146**, 741–3.

Moody, P. L., Gatos, H. C. and Lavine, M. C. (1960). Growth of GaAs crystals in the ⟨111⟩ polar direction. *Journal of Applied Physics*, **31**, 1696–7.

Mooney, P. M. (1996). Strain relaxation and dislocations in SiGe/Si structures. *Materials Science and Engineering*, **R17**, 105–46.

Mooney, P. M. and Chu, J. O. (2000). SiGe technology: Heteroepitaxy and high-speed microelectronics. *Annual Review of Materials Science*, **30**, 335–62.

Morizane, K. (1977). Antiphase domain-structures in GaP and GaAs epitaxial layers grown on Si and Ge. *Journal of Crystal Growth*, **38**, 249–54.

Morizumi, T. and Takahashi, K. (1970). Epitaxial vapour growth of ZnTe on InAs. *Japanese Journal of Applied Physics*, **9**, 849–50.

Moustakas, T. D., Iliopoulos, E., Sampath, A. V. *et al.* (2001). Growth and device applications of III-nitrides by MBE. *Journal of Crystal Growth*, **227–228**, 13–20.

Mueller, R. K. and Jacobson, R. L. (1961). Growth twins in indium antimonide. *Journal of Applied Physics*, **32**, 550–1.

Mugge, O. (1914). *Neues JB*, **1**, 48.

Murarka, S. P., Levinstein, H. J., Marcus, R. B. and Wagner, R. S. (1977). Oxidation of silicon without formation of stacking faults. *Journal of Applied Physics*, **48**, 4001–3.

Muratov, V. A. and Fischer, T. E. (2000). Tribochemical polishing. *Annual Review of Matetials Science*, **30**, 27–51.

Nabarro, F. R., Basinski, Z. S. and Holt, D. B. (1964). The plasticity of pure single crystals. *Advances in Physics*, **13**, 192–323.

Nadeau, J. S. (1964). The photomechanical effects in alkali halide crystals. *Journal of Applied Physics*, **35**, 669–77.

Nagai, H. (1972). Anisotropic bending during epitaxial growth of mixed crystals on GaAs substrates. *Journal of Applied Physics*, **43**, 4254–6.

Narayan, J. and Oktyabrsky, S. (2002). Formation of misfit dislocations in thin film heterostructures. *Journal of Applied Physics*, **92**, 7122–7.

Neave, J. H., Larsen, P. K., Joyce, B. A., Gowers, J. P. and van der Veen, J. F. (1983). Some observations on Ge:GaAs(001) and GaAs:Ge(001) interfaces and films. *Journal of Vacuum Science and Technology*, **BI**, 668–75.

Negrii, V. D. and Osipyan, Y. A. (1978). Influence of dislocations on radiative recombination processes in cadmium-sulfide. *Soviet Physics Solid State*, **20**, 432–6.

Negrii, V. D. and Osipiyan, Yu. A. (1979). Dislocation emission in CdS. *Physica Status Solidi*, **A55**, 583–8.

Negrii, V. D. and Osipiyan, Yu. A. (1982). Distinctive features of the luminescence of cadmium sulfide deformed at low temperatures. *Soviet Physics Solid State*, **24**, 197–9.

Nickel, N. H. (1999). Hydrogen in semiconductors II. In *Semiconductors and Semimetals*, Volume **61**, eds. R. K. Willardson, A. C. Beer and E. R. Weber (San Diego: Academic Press).

Nicolaeva, A. Z., Tonoyan, A. A., Semeneva, L. A. and Dolomanov, L. A. (1975). Differences in the depth of the damage layer on polar faces of GaAs single crystals. (As abstracted in Diffusion and Defect Data **12**, 208.)

Nikitenko, V. I., Farber, B. Ya. and Iunin, Yu. L. (1987). Formation kinetics and behaviour for nonlinear excitations limiting dislocation mobility in semiconductor single crystals. *Bulletin of the Academy of Sciences of the USSR Division of Physical Science*, **51**, 81–6.

Ninomiya, T. (1979). Velocities and internal-friction of dislocations in III-V compounds. *Journal de Physique*, **40**, Suppl. 6, 143–5.

Ohashi, T., Sato, M., Maruizumi, T. and Kitagawa, I. (2003). Simulation of dislocation accumulation in ULSI cells with STI structure. *Applied Surface Science*, **216**, 340–6.

Ohtani, N., Fujimoto, T., Katsuno, M., Aigo, T. and Yashiro, H. (2001). Surface step model for micropipe formation in SiC. *Journal of Crystal Growth*, **226**, 254–60.

Ohtani, N., Fujimoto, T., Katsuno, M., Aigo, T. and Yashiro, H. (2002). Growth of large high-quality SiC single crystals. *Journal of Crystal Growth*, **237–239**, 1180–6.

Oldham, W. G. and Milnes, A. G. (1964). Interface states in abrupt semiconductor heterojunctions. *Solid State Electronics*, **7**, 153–65.

Olsen, A. and Spence, J. C. (1981). Distinguishing dissociated glide and shuffle set dislocations by high-resolution electron-microscopy. *Philosophical Magazine*, **A43**, 945–65.

Olsen, G. H., Abrahams, M. S. and Zamerowski, T. J. (1974). Assymetric cracking in III-V compounds. *Journal of the Electrochemical Society*, **121**, 1650–6.

Osipiyan, Yu. A. (1989). Electrical and optical phenomena of II-VI semiconductors associated with dislocations. In *International Symposium on Structural Properties of Dislocations in Semiconductors*, Oxford 1989 Cof. Series No. 104 (Institute of Physics: Bristol), pp. 109–18.

Osipiyan, Yu. A. and Petrenko, V. F. (1973a). Nature of the photoplastic effect. *Soviet Physics JETP*, **36**, 916–20.

Osipiyan, Yu. A. and Petrenko, V. F. (1973b). Experimental observation of the influence of an electric field on the plastic deformation of ZnSe crystals. *JETP Letters*, **17**, 399–400.

Osipiyan, Yu. A. and Petrenko, V. F. (1976a). The effect of illumination on the deformation currents in ZnS. *Soviet Physics Doklady*, **21**, 87–8.

Osipiyan, Yu. A. and Petrenko, V. F. (1976b). Short-circuit effect in plastic deformation of ZnS and motion of charged dislocations. *Soviet Physics JETP*, **42**, 695–9.

Osipiyan, Yu. A. and Savchenko, I. B. (1968). Experimental observation of the influence of light on plastic deformation of cadmium sulphide. *JETP Letters*, **7**, 100–2.

Osipiyan, Yu. A. and Savchenko, I. B. (1973). Kinetics of the photoplastic effect and its dependence on orientation. *Soviet Physics Solid State*, **14**, 1723–5.

Osipiyan, Yu. A. and Shikhsaidov, M. Sh. (1973). Negative photoplastic effect in cadmium sulfide. *Soviet Physics Solid State*, **15**, 2475–6.

Osipiyan, Yu. A. and Smirnova, I. S. (1968). Perfect dislocations in the wurtzite lattice. *Physica Status Solidi*, **30**, 19–29.

Osipiyan, Yu. A. and Smirnova, I. S. (1971). Partial dislocations in the wurtzite lattice. *Journal of Physics and Chemistry of Solids*, **32**, 1521–30.

Osipiyan, Yu. A., Petrenko, V. F. and Savchenko, I. B. (1968). Infrared quenching of the photoplastic effect in cadmium sulphide. *JETP Letters*, **13**, 442–4.

Osipiyan, Yu. A., Petrenko, V. F. and Strukova, G. K. (1973). Study of the photoplastic effect at α and β dislocations in CdS. *Soviet Physics Solid State*, **15**, 1172–4.

Osipiyan, Yu. A., Petrenko, V. F. and Shikhsaidov, M. Sh. (1974). Impurity photoplastic effect in ZnS:Al single crystals. *JETP Letters*, **20**, 163–4.

Osipiyan, Yu. A., Petrenko, V. F., Zaretskii, A. V. and Whitworth, R. W. (1986). Properties of II–VI semiconductors associated with moving dislocations. *Advances in Physics*, **35**, 115–88.

Osvenskii, V. B. and Kholodnyi, L. P. (1972). Mobility of single dislocations in GaAs. *Fizika Tverdogo Tela*, **14**, 3330.

Osvenskii, V. B., Kholodnyi, L. P. and Milvidskii, M. G. (1969). Effect of doping impurities on the anisotropy of plastic deformation in GaAs single crystals. *Soviet Physics Doklady*, **14**, 144–6.

Osvenskii, V. B., Kholodnyi, L. P. and Milvidskii, M. G. (1973). Influence of dopants on the velocity of dislocations in GaAs single crystals. *Soviet Physics Solid State*, **15**, 661–2.

Ourmazd, A., Hull, R. and Tung, R. T. (1991). Interfaces. In *Materials Science and Technology*, Vol. **4** Electronic Structure and Properties of Semiconductors, ed. W. Schroter (Weinheim: VCH), pp. 379–448.

Packeiser, G. and Haasen, P. (1977). Constrictions in the stacking faults of dislocations in germanium. *Philosophical Magazine*, **35**, 821–7.

Pandey, D. and Krishna, P. (1983). The origin of polytype structures. *Progress in Crystal Growth and Characterization*, **7**, 213–58.

Pankove, J. I. (1999). GaN: from fundamentals to applications. *Materials Science and Engineering*, **B61–62**, 305–9.

Pankove, J. I. and Johnson, N. M. (1991). Hydrogen in semiconductors. In *Semiconductors and Semimetals*, **34**, eds. R. K. Willardson and A. C. Beer (San Diego: Academic Press).

Pashley, D. W. (1956). The study of epitaxy in thin surface films. *Advances in Physics*, **5**, 173.

Pashley, D. W. (1965). The nucleation, growth, structure and epitaxy of thin surface films. *Advances in Physics*, **14**, 327–416.

Patel, J. R. (1970). Burgers vector of dislocations generated for dislocation velocity measurements in semiconductors. *Journal of Applied Physics*, **41**, 2814.

Patel, J. R. and Chaudhuri, A. R. (1966). Charged impurity effects on the deformation of dislocation-free germanium. *Physical Review*, **143**, 601–8.

Patel, J. R. and Freeland, P. E. (1967). Change of dislocation velocity with Fermi level in silicon. *Physical Review Letters*, **18**, 833–5.

Patel, J. R. and Freeland, P. E. (1970). Burgers vector of dislocations generated for dislocation velocity measurements in semiconductors. *Journal of Applied Physics*, **41**, 2814.

Patel, J. R. and Testardi, L. R. (1977a). Electronic effects on dislocation velocities in heavily doped germanium. *Applied Physics Letters*, **30**, 3–5.

Patel, J. R. and Testardi, L. R. (1977b). Reply to comments on 'Electronic effects on dislocation velocities in heavily doped silicon'. *Physical Review*, **B15**, 4124–5.

Patel, J. R., Testardi, L. R. and Freeland, P. E. (1976). Electronic effects on dislocation velocities in heavily doped silicon. *Physical Review*, **B13**, 2548–3557.

Partridge, P. G. (1967). The crystallography and deformation modes of hexagonal close-packed metals. *International Metal Reviews*, No. **118**, pp. 169–94.

Pashinkin, A. S., Tishchenko, I. V., Korneeva and Ryzhenko, B. N. (1960). Concerning the polymorphism of some chalcogenides of zinc and cadmium. *Soviet Physics Crystallography*, **5**, 243–8.

Pearton, S. J., Corbett, J. W. and Stavola, M. (1992). *Hydrogen in Crystalline Semiconductors*. Berlin: Springer.

Peierls, R. E. (1940). The size of a dislocation. *Proceedings of the Physical Society London*, **52**, 34–7.

Peissker, E., Haasen, P. and Alexander, H. (1961). Anisotropic plastic deformation of indium antimonide. *Philosophical Magazine*, **7**, 1279–303.

Pennycook, S. J. and Jesson, D. E. (1991). High-resolution Z-contrast imaging of crystals. *Ultramicroscopy*, **37**, 14–38.

Pennycook, S. J., Chisholm, M. F., Yan, Y., Duscher, G. and Pantelides, S. T. (1999). A combined experimental and theoretical approach to grain boundary structure and segregation. *Physica*, **B273–274**, 453–7.

Perovic, D. D. and Houghton, D. C. (1993). Spontaneous nucleation of misfit dislocations in strained epitaxial layers. *Physica Status Solidi*, **A138**, 425–30.

Perovic, D. D. and Houghton, D. C. (1995). The introduction of dislocations in low misfit epitaxial systems. In *Microscopy of Semiconducting Materials 1995*, Inst. Phys. Conf. Ser. 146 (Bristol: Institute of Physics), pp. 117–34.

Perovic, D. D., Weatherly, G. C., Baribeau, J. M. and D. C. Houghton, D. C. (1989). Heterogeneous nucleation sources in molecular beam epitaxy-grown Ge_xSi_{1-x}/Si strained layer superlattices. *Thin Solid Films*, **183**, 141–56.

Petrenko, V. F. and Whitworth, R. W. (1980). Charged dislocations and the plastic deformation of II-VI compounds. *Philosophical Magazine*, **41**, 681–99.

Petrenko, V. F., Kiriichenko, L. G. and Strukova, G. K. (1982). Unpublished, quoted by Osipiyan *et al.* (1986).

Phillips, J. C. (1973). *Bonds and Bands in Semiconductors*. New York: Academic Press.

Pichaud, B., Putero, M. and Burle, N. (1999). Elemental dislocations mechanisms involved in the relaxation of heteroepitaxial semiconducting systems. *Physica Status Solidi*, **A171**, 251–65.

Pirouz, P. (1987). Deformation mode in silicon, slip or twinning? *Scripta Metallurgica*, **21**, 1463–8.

Pirouz, P. (1989a). Dislocation Mechanisms for Twinning and Polytypic Transformations in Semiconductors. In *Structure and Properties of Dislocations in Semiconductors 1989*. Conf. Series No. 104 (Bristol: Institute of Physics), pp. 49–56.

Pirouz, P. (1989b). On twinning and polymorphic transformations in compound semiconductors. *Scripta Metallurgica*, **25**, 401–6.

Pirouz, P. (1998). On micropipes and nanopipes in SiC and GaN. *Philosophical Magazine*, **A78**, 727–36.

Pizzini, S. (1999). Chemistry and physics of segregation of impurities at extended defects in silicon. *Physica Status Solidi*, **171**, 123–32.

Pizzini, S. (2002). Chemistry and physics of defect interaction in semiconductors. *Solid State Phenomena*, **85–86**, 1–66.

Pizzini, S., Borsani, F., Sandrinelli, A. and Narducci, D. (1989). Effect of impurity segregation on the electrical properties of grain boundaries in polycrystalline silicon. In *Point and Extended Defects in Semiconductors*, eds. G. Benedek, A. Cavallini and W. Schroter, Nato ASI Series Series B Physics, **202** (New York: Plenum), pp. 105–21.

Ponce, F. A. (1997). Defects and interfaces in GaN epitaxy. *MRS Bulletin*, **22**, 51–7.

Pond, R. C. (1985). The geometrical character of extended interfacial defects in semiconducting materials. In *Polycrystalline Semiconductors*, ed. G. Harbecke (Berlin: Springer-Verlag), pp. 27–45.

Pond, R. C. (1989a). Line defects in interfaces. In *Dislocations in Solids*, Vol. **8**, ed. F. R. Nabarro (Amsterdam: North-Holland), pp. 1–66.

Pond, R. C. (1989b). Symmetry and crystallography: Implications for structure. In *International Symposium on Structural Properties of Dislocations in Semiconductors*, Oxford Conf. Series No. 104, pp. 25–36.

Pond, R. C. and Bollmann, W. (1979). The symmetry and interfacial structure of bicrystals. *Proceedings of the Royal Society*, **A292**, 449–72.

Pond, R. C. and Vlachavas, D. S. (1983). Bicrystallography. *Proceedings of the Royal Society*, **A386**, 5–143.

Pond, R. C. and Bastaweesy, A. (1984). The theory of crystallographic defects in periodic interfaces. *Journal de Physique*, **C4**, 225–30.

Pond, R. C., Gowers, J. P., Holt, D. B. *et al.* (1984). A general treatment of antiphase domain formation and identification at polar-nonpolar semiconductor interfaces. In *Materials Research Society Symposium Proceedings*, **25**, pp. 273–8.

Poust, B. D., Koga, T. S., Sandhu, R. *et al.* (2003). SiC substrate defects and III-N heteroepitaxy. *Journal of Physics*, **D36**, A102–A106.

Powell, A. R., Iyer, S. S. and LeGoues, F. K. (1994). New approach to the growth of low dislocation relaxed SiGe material. *Applied Physics Letters*, **64**, 1856–8.

Prussin, S. (1961). Generation and distribution of dislocations by solute diffusion. *Journal of Applied Physics*, **32**, 1876–81.

Pugh, E. N., Westwood, A. R. and Hitch, T. T. (1966). Effects of liquid metals on the fracture strength of germanium. *Physica Status Solidi*, **15**, 291–7.

Qian, W., Rohrer, G. S., Skowronski, M. *et al.* (1995). Open-core screw dislocations in GaN epilayers observed by scanning force microscopy and high-resolution transmission electron microscopy. *Applied Physics Letters*, **67**, 2284–6.

Queisser, H. J. (1969). Observations and properties of lattice defects in silicon. In *Semiconductor Silicon*, ed. R. R. Haberecht and E. L. Kern (New York: Electrochemical Society), pp. 585–95.

Queisser, H. J., Hubner, K. and Shockley, W. (1961). Diffusion along small-angle grain boundaries in silicon. *Physical Review*, **123**, 1245–54.

Queisser, H. J., Finch, R. H. and Washburn, J. (1962). Stacking faults in epitaxial silicon. *Journal of Applied Physics*, **33**, 1536–7.

Queisser, H. J. and Van Loon, P. G. G. (1964). Growth of lattice defects in silicon during oxidation. *Journal of Applied Physics*, **35**, 3066–7.

Queisser, H. J. and Haller, E. E. (1998). Defects in semiconductors: Some fatal, some vital. *Science*, **281**, 945–50.

Randle, V. (1993). *The Measurement of Grain Boundary Geometry*. Bristol: Adam-Hilger.

Randle, V. (2001). The coincidence site lattice and the 'Sigma Enigma'. *Materials Characterization*, **47**, 411–16.

Randle, V. and Engler, O. (2000). *Introduction to Texture Analysis: Macrotexture, Microtexture and Orientation Mapping*. Amsterdam: Gordon and Breach.

Ranganathan, S. (1966). On geometry of coincidence-site lattices. *Acta Crystallographica*, **21**, 197–9.

Ravi, K. V. (1972). Generation of dislocations and stacking-faults at surface heterogeneities in silicon. *Journal of Applied Physics*, **43**, 1785.

Ravi, K. V. (1974). Annihilation of oxidation induced stacking-faults in silicon. *Philosophical Magazine*, **30**, 1081–90.

Ravi, K. V. (1975). Orientation dependance of stacking-fault nucleation in silicon. *Philosophical Magazine*, **31**, 405–10.

Ravi, K. V. (1981). *Imperfections and Impurities in Semiconductor Silicon*. New York: Wiley.

Ravi, K. V. and Varker, C. J. (1974). Oxidation-induced stacking-faults in silicon.1. Nucleation phenomenon. *Journal of Applied Physics*, **45**, 263–71.

Ray, I. L. F. and Cockayne, D. J. H. (1970). The observation of dissociated dislocations in silicon. *Philosophical Magazine*, **22**, 853–6.

Ray, I. L. F. and Cockayne, D. J. H. (1971). The dissociation of dislocations in silicon. *Proceedings of the Royal Society*, **A325**, 543–54.

Ray, I. L. F. and Cockayne, D. J. H. (1973). Investigation of dislocation geometries in the diamond cubic structure. *Journal of Microscopy*, **98**, 170–3.

Raza, B. (1994). Ph.D. Thesis, University of London.

Raza, B. and Holt, D. B. (1991). EBIC contrast of grain boundaries in polycrystalline solar cells. In *Polycrystalline Semiconductors II*. Springer Proceedings in *Physics*, **54**, eds. J. H. Werner and H. P. Strunk (Berlin: Springer-Verlag), pp. 72–6.

Raza, B. and Holt, D. B. (1995). EBIC studies of grain boundaries. In *Microscopy of Semiconducting Materials 1995*. Conference Series No. 146 (Bristol: Institute of Physics), pp. 107–10.

Read, W. T. (1954). Theory of dislocations in germanium. *Philosophical Magazine*, **45**, 775–96.

Rebinder, P. A. (1928). In Proceedings of Sixth Physics Conference (Moscow: State Press), p. 29 as quoted by Westwood *et al.* (1981).

Rebinder, P. A., Schreiner, L. A. and Zhigach, K. F. (1944). Hardness reducers in drilling (Acad. Sci. USSR: Moscow).

Rendakova, S., Ivantsov, V. and Dmitirev, V. (1998a). High quality 6H- and 4H-SiC pn structures with stable electric breakdown grown by liquid phase epitaxy. *Materials Science Forum*, **264–268**, 163–6.

Rendakova, S. V., Nikitina, I. P., Tregubova, A. S. and Dmitriev, V. A. (1998b). Micropipe and dislocation density reduction in 6H-SiC and 4H-SiC structures grown by liquid phase epitaxy. *Journal of Electronic Materials*, **27**, 292–5.

Reppich, B., Haasen, P. and Ilschner, B. (1964). Kreichen von silizium-einkristallen. *Acta Metallurgica*, **12**, 1283–8.

Romanov, A. E., Pompe, W., Beltz, G. E. and Speck, J. S. (1996). Modeling of threading dislocation density reduction in heteroepitaxial layers. 1. Geometry and crystallography. *Physica Status Solidi*, **B198**, 599–613.

Romanov, A. E., Pompe, W., Beltz, G. E. and Speck, J. S. (1997). Modeling of threading dislocation density reduction in heteroepitaxial layers. 2. Effective dislocation kinetics. *Phys. Status Solidi*, **B199**, 33–49.

Romanov, A. E., Pompe, W., Mathis, S., Beltz, G. E. and Speck, J. S. (1999). Threading dislocation reduction in strained layers. *Journal of Applied Physics*, **85**, 182–92.

Rosner, S. J., Carr, E. C., Ludowise, M. J., Girolami, G. and Erikson, H. (1997). Correlation of cathodoluminescence inhomogeneity with microstructural defects in epitaxial GaN grown by metalorganic chemical-vapor deposition. *Applied Physics Letters*, **70**, 420–2.

Rozgonyi, G. A. and Foster, N. F. (1967). Orientation inversions in polycrystalline CdS bulk crystals and thin films. *Journal of Applied Physics*, **38**, 5172–6.

Sagar, A., Lehman, W. and Faust, J. W. (1968). Etchants for ZnSe. *Journal of Applied Physics*, **39**, 5336–8.

Sakai, A., Sunakawa, H. and Usui, A. (1997). Defect structure in selectively grown GaN films with low threading dislocation density. *Applied Physics Letters*, **71**, 2259–61.

Sanders, I. R. and Dobson, P. S. (1969). Oxidation, defects and vacancy diffusion in silicon. *Philosophical Magazine*, **20**, 881.

Sato, M. and Sumino, K. (1977). In situ tensile tests of silicon crystals at elevated temperatures in a high voltage electron microscope. In *Proceedings Fifth International Conference on High Voltage Electron Microscopy*, Kyoto, pp. 459–62.

Sazhin, N. P., Milvidskii, M. G., Osvenskii, V. B. and Stolyarov, O. G. (1966). Influence of doping on the plastic deformation of GaAs single crystals. *Soviet Physics Solid State*, **8**, 1223–7.

Schäfer, S. (1967). Messung von versetzungsgeschwindigkeiten in germanium. *Physica Status Solidi*, **19**, 297.

Schaumburg, H. (1970). Velocities measurements on screw-dislocations and 60 degrees-dislocations in germanium. *Physica Status Solidi*, K1.

Schaumburg, H. (1972). Velocities of screw-dislocations and 60 degrees dislocations in germanium. *Philosophical Magazine*, **25**, 1429.

Schlossberger, F. (1955). Controlled preparation and x-ray investigation of cadmium sulfide. *Journal of the Electrochemical Society*, **102**, 22–6.

Schmidt, W., Pilgermann, B., Kuehn, G. and Fischer, P. (1973). Kristallographische polaritaet von $A^{III}B^{V}$-kristallen. *Kristall und Technik*, **8**, 913–21.

Schreiber, J., Höring, L., Uniewski, H., Hildebrandt, S. and Leipner, H. S. (1999). Recognition and distribution of A(g) and B(g) disloctions in indentation deformation zones on {111} and {110} surfaces of CdTe. *Physica Status Solidi*, **A171**, 89–97.

Schröter, W. (1980). Electric and dynamic properties of dislocations in the elemental and compound semiconductors. In *Electronic Structure of Crystal Defects and Disordered Systems*, Summer School, Aussois, 1980 (Les Ulis Cedex: Les Editions de Physique), pp. 129–74.

Schröter, W. and Haasen, P. (1977). The chemomechanical effect in semiconductors. In *NATO Advanced Study Institutes Series*, Series E: Applied Science No. 17. Surface Effects in Crystal Plasticity, eds. R. M. Latanision and J. F. Fourie, pp. 681–8.

Schröter, W. and Cerva, H. (2002). Interaction of point defects with dislocations in silicon and germanium: Electrical and optical effects. *Solid State Phenomena*, **85–86**, 67–143.

Schröter, W., Labusch, R. and Haasen, P. (1977). Comment on 'Electronic effects on dislocation velocities in heavily doped silicon' by J. R. Patel, L. R. Testardi and P. E. Freeland. *Physical Review*, **B15**, 4121–3.

Schwarz, K. W. and Chidambarrao, D. (1999). Dislocation dynamics near film edges and corners in silicon. *Journal of Applied Physics*, **85**, 7198–208.

Schwuttke, G. H. (1970). Silicon material problems in semiconductor device technology. *Microelectronics and Reliability*, **9**, 397–412.

Schwuttke, G. H. and Queisser, H. J. (1962). X-ray observations of diffusion-induced dislocations in silicon. *Journal of Applied Physics*, **33**, 1540–2.

Schwuttke, G. H. and Rupprecht, H. (1966). X-Ray analysis of diffusion-induced defects in gallium arsenide. *Journal of Applied Physics*, **37**, 167.

Schwuttke, G. H., Brack, K. and Hearn, E. W. (1971). Influence of stacking faults on leakage currents of FET devices. *Microelectronics and Reliability*, **10**, 467.

Seager, C. H. (1985). Grain boundaries in polycrystalline silicon. *Annual Review of Materials Science*, **15**, 271–302.

Seager, C. H. and Ginley, D. S. (1979). Passivation of grain boundaries in polycrystalline silicon. *Applied Physics Letters*, **34**, 337–40.

Seager, C. H., Ginley, D. S. and Zook, J. D. (1980). Improvement of polycrystalline silicon solar cells with grain-boundary hydrogenation techniques. *Applied Physics Letters*, **36**, 831–3.

Seager, C. H. and Ginley, D. S. (1981). Studies of the hydrogen passivation of silicon grain boundaries. *Journal of Applied Physics*, **52**, 1050–5.

Secco D'Aragona, F. and Delavignette, P. (1966). Fautes de croissance dans la wurtzite. *Journal de Physique*, **C3**, 121–7.

Secco D'Aragona, F., Delavignette, P. and Amelinckx, S. (1966). Direct evidence for the mechanism of the phase transition wurtzite-sphalerite. *Physica Status Solidi*, **14**, K115–K118.

Seidensticker, R. G. and Hamilton, D. R. (1963). The dendrite growth of InSb. *Journal of Physics and Chemistry of Solids*, **24**, 1585–91.

Seifert, W., Morgenstern, G. and Kittler, M. (1993). Influence of dislocation density on recombination at grain boundaries in multicrystalline silicon. *Semiconductor Science and Technology*, **8**, 1687–91.

Seki, Y., Matsui, J. and Watanabe, H. (1976). Impurity effect on the growth of dislocation-free InP single crystals. *Journal of Applied Physics*, **47**, 3374–6.

Seki, Y., Watanabe, H. and Matsui, J. (1978). Impurity effect on grown-in dislocation density of InP and GaAs crystals. *Journal of Applied Physics*, **49**, 822–8.

Seltzer, M. S. (1966). Influence of charged defects on mechanical properties of lead sulphide. *Journal of Applied Physics*, **37**, 4780–4.

Shachar, G. and Brada, Y. (1968). Negative differential photovoltages in ZnS crystals. *Journal of Applied Physics*, **39**, 1701–4.

Shay, J. L. and Wernick, J. H. (1975). *Ternary Chalocpyrite Semiconductors: Growth, Electronic Properties and Applications*. Oxford: Pergamon.

Sheftal, N. N. and Magumedov, Kh. A. (1967). Morphological aspects of epitaxial growth of GaAs crystals in the polar direction. In *Crystal Growth*. Proceedings International Conference on Crystal Growth, Boston, ed. H. S. Peiser (Oxford: Pergamon), pp. 533–6.

Sheinerman, A. G. and Gutkin, M. Y. (2003). Elastic fields of a screw superdislocation with a hollow core (pipe) perpendicular to the free crystal surface. *Physics of the Solid State*, **45**, 1694–700.

Sheldon, P., Jones, K. M., Al-Jassim, M. M. and Yacobi, B. G. (1988). Dislocation density reduction through annihilation in lattice-mismatched semiconductors grown by MBE. *Journal of Applied Physics*, **63**, 5609–11.

Shimizu, H. and Sumino, K. (1970). Anisotropy in hardness on (111) and ($\bar{1}\bar{1}\bar{1}$) surfaces in InSb. *Journal of the Physical Society of Japan*, **29**, 1096.

Shimura, F., Tsuya, H. and Kawamura, T. (1980). Surface-micro-defect and inner-micro-defect in annealed silicon-wafer containing oxygen. *Journal of Applied Physics*, **51**, 269–73.

Shiraki, H. (1974). Silicon wafer annealing effect in loop defect generation. *Japanese Journal of Applied Physics*, **13**, 1514–23.

Shockley, W. (1953). Dislocations and edge states in the diamond crystal structure. *Physical Review*, **91**, 228.

Shockley, W. and Read, W. T. (1949). Quantitative predictions from dislocation models of crystal grain boundaries. *Physical Review*, **75**, 692.

Shockley, W. and Read, W. T. (1950). Dislocation models of crystal grain boundaries. *Physical Review*, **78**, 275–89.

Shoeck (1980). Thermodynamics and thermal activation of dislocations. In *Dislocations in Solids*, **3**, ed. F. R. N. Nabarro (Amsterdam: North-Holland), pp. 63–163.

Siethoff, H. (1970). The effect of charged impurities on the yield point of silicon. *Physica Status Solidi*, **40**, 153–61.

Siethoff, H. and Brion, H. G. (2003). The interaction of boron and phosphorus with dislocations in silicon. *Materials Science and Engineering*, **A355**, 311–14.

Sinno, T., Dornberger, E., von Ammon, W., Brown, R. A. and Dupret, F. (2000). Defect engineering of Czochralski single-crystal silicon. *Materials Science and Engineering*, **R28**, 149–98.

Sirtl, E. and Adler, A. (1961). Chromsaeure-fluszsaeure als spezifisches system zur aetzgrubenentwicklung auf silizium. *Zeitschrift fur Metallkunde*, **52**, 529–31.

Smith, D. A. and Pond, R. C. (1976). Bollmann's O-lattice theory: A geometrical approach to interface structure. *International Metals Reviews*, **21**, 61–74.

Speake, C. C., Smith, P. J., Lomer, T. R. and Whitworth, R. W. (1978). The glide plane of dislocations in zinc sulphide. *Philosophical Magazine*, **A38**, 603–6.

Speck, J. S., Brewer, M. A., Beltz, G., Romanov, A. E. and Pompe, W. (1996). Scaling laws for the reduction of threading dislocation densities in homogeneous buffer layers. *Journal of Applied Physics*, **80**, 3808–16.

Speck, J. S. and Rosner, S. J. (1999). The role of threading dislocations in the physical properties of GaN and its alloys. *Physica*, **B273–274**, 24–32.

Spence, J. and Koch, C. (2001). Experimental evidence for dislocation core structures in silicon. *Scripta Materialia*, **45**, 1273–8.

Steinberger, I. T. (1983). Polytypism in zinc sulphide. *Progress in Crystal Growth and Characterization*, **7**, 7–53.

Steinberger, I. T. and Mardix, S. (1967). Polytypism in ZnS crystals. In *II-VI Semiconducting Compounds*, ed. D. G. Thomas (New York: Benjamin), pp. 167–78.

Steinberger, I. T., Kiflawi, I., Kalman, Z. H. and Mardix, S. (1973). The stacking faults and partial dislocations involved in structure transformations of ZnS crystals. *Philosophical Magazine*, **27**, 159–75.

Steinhard, H. and Shäfer, S. (1971). Dislocation velocities in indium antimonide. *Acta Metallurgica*, **19**, 65–70.

Steinmann, A. and Zimmerli, U. (1963). Growth peculiarities of GaAs single crystals. *Solid State Electronics*, **6**, 597–604.

Stevens, R. (1972a). Defects in silicon carbide. *Journal of Materials Science*, **7**, 517–21.

Stevens, R. (1972b). Neutron irradiation damage in SiC whiskers. *Philosophical Magazine*, **25**, 523–8.

Stevens, R. (1972c). Twin morphology in silicon carbide. *Journal of Materials Science*, **7**, 723–6.

Stickler, R. and Booker, G. R. (1963). Surface damage on abraded silicon specimens. *Philosophical Magazine*, **8**, 859–76.

Stirland, D. J., Hart, D. G., Clark, S., Regnault, J. C. and Elliott, C. R. (1983). Characterization of defects in InP substrates. *Journal of Crystal Growth*, **61**, 645–57.

Stirpe, M. B., Perovic, D. D., Lafontaine, H. L. and Goldberg, R. D. (1997). Controlling misfit dislocation generation in strained layer epitaxy by point defect injection. In *Microscopy of Semiconducting Materials 1997*, Conf. Ser. No. 157 (Bristol: Institute of Physics), pp. 127–30.

Stolwijk, N. A., Poisson, Ch. and Bernardini, J. (1996). Segregation-controlled kinetics of fast impurity diffusion in polycrystalline solids. *Journal of Physics: Condensed Matter*, **8**, 5843–56.

Stowell, M. J. (1975). Defects in epitaxial deposits. In *Epitaxial Growth* Part B, ed. J. W. Matthews (New York: Academic Press), pp. 437–92.

Strukova, G. K. (1977). Unpublished as quoted in Osipiyan *et al.* (1986).

Sturner, H. W. and Bleil, C. E. (1964). Optical studies of defect structures in cadmium sulfide and cadmium selenide. *Applied Optics*, **3**, 1015–21.

Sugahara, T., Sato, H., Hao, M. *et al.* (1998). Direct evidence that dislocations are non-radiative recombination centers in GaN. *Japanese Journal of Applied Physics* (Part 2) **37**, L398–L400.

Sugiura, L. (1997). Dislocation motion in GaN light-emitting devices and its effect on device lifetime. *Journal of Applied Physics*, **81**, 1633–8.

Sumino, K. (1987). Dislocations in GaAs crystals. In *Defects and Properties of Semiconductors: Defect Engineering*, eds. J. Chikawa, K. Sumino and K. Wada (Tokyo: KTC Publishers), pp. 3–24.

Sumino, K. (1997). Kinetic properties of dislocations in semiconductors revealed by x-ray topography. *Il Nuovo Cimento*, **19D**, 137–46.

Sumino, K. (1999). Impurity reaction with dislocations in semiconductors. *Physica Status Solidi*, **A171**, 111–22.

Sumino, K. (2003). Basic aspects of impurity gettering. *Microelectronics Engineering*, **66**, 268–80.

Sumino, K. and Harada, H. (1981). In situ x-ray topographic studies of the generation and the multiplication processes of dislocations in silicon crystals at elevated temperatures. *Philosophical Magazine*, **A44**, 1319–34.

Sumino, K. and Shimizu, H. (1975a). Polarity in bending deformation of InSb crystals. I Experiments. *Philosophical Magazine*, **32**, 123–42.

Sumino, K. and Shimizu, H. (1975b). Polarity in bending deformation of InSb crystals. II Theory and supplementary experiments. *Philosophical Magazine*, **32**, 143–57.

Sumino, K. and Yonenaga, I. (2002). Interactions of impurities with dislocations: mechanical effects. *Diffusion and Defect Data Part B (Solid State Phenomena)*, **85–86**, 145–76.

Sumino, K., Kodaka, S. and Kojima, K. (1974). Dynamical state of dislocations in germanium crystals during deformation. *Materials Science and Engineering*, **13**, 263–8.

Sutton, A. P. and Balluffi, R. W. (1995). *Interfaces in Crystalline Materials*. Oxford: Oxford University Press.

Suzuki, H. (1952). Chemical interaction of solute atoms with dislocations. *Science Reports of Research Institute, Tohoku University*, **A4**, 455–63.

Suzuki, T. (2000). Relation between the suppression of the generation of stacking faults and the mechanism of silicon oxidation during annealing under argon containing oxygen. *Journal of Applied Physics*, **88**, 1141–8.

Swaminathan, V. (1982). Defects in GaAs. *Bulletin of Materials Science (India)*, **4**, 403–43.

Swaminathan, B., Saraswat, K. C., Dutton, R. W. and Kamins, T. I. (1982). Diffusion of arsenic in polycrystalline silicon. *Applied Physics Letters*, **40**, 795–8.

Tachikawa, M. and Yamaguchi, M. (1990). Film thickness dependence of dislocation density reduction in GaAs-on-Si substrates. *Applied Physics Letters*, **56**, 484–6.

Takeuchi, S. and Suzuki, K. (1999). Stacking fault energies of tetrahedrally coordinated crystals. *Physica Status Solidi*, **A171**, 99–103.

Tersoff, J. and LeGoues, F. K. (1994). Competing relaxation mechanisms in strained layers. *Physical Review Letters*, **72**, 3570–3.

Theurer, H. C. (1961). Epitaxial silicon films by the hydrogen reduction of $SiCl_4$. *Journal of the Electrochemical Society*, **108**, 649–53.

Thibault, J., Rouviere, J. L. and Bourret, A. (1991). Grain boundaries in semiconductors. In *Materials Science & Technology*, **4**, Electronic Structure and Properties of Semiconductors, ed. W. Schroter (Weinheim: VCH).

Thomas, D. J. D. (1963). Surface damage and Cu precipitation in Si. *Physica Status Solidi*, **3**, 2261–73.

Trigunayat, G. C. (1991). A survey of the phenomenon of polytypism in crystals. *Solid State Ionics*, **48**, 3–70.

Tsuchida, H., Kamata, I., Jikimoto, T., Miyanagi, T. and Izumi, K. (2003). 4H-SiC epitaxial growth for high-power devices. *Materials Science Forum*, **433–436**, 131–6.

Tuppen, C. G., Gibbings, C. J. and Hockly, M. (1989). The effects of misfit dislocation nucleation and propagation on Si/Si$_{1-x}$Ge$_x$ critical thickness values. *Journal of Crystal Growth*, **94**, 392–404.

Tweet, A. G. and Gallagher, C. J. (1956). Structure sensitivity of Cu diffusion in Ge. *Physical Review*, **103**, 828.

Ueda, O. (1996). *Reliability and Degradation of III-V Optical Devices*. Boston: Artech House.

Ueda, O. (1999). Reliability issues in III-V compound semiconductor devices: optical devices and GaAs-based HBTs. *Microelectronics Reliability*, **39**, 1839–55.

Unvala, B. A. and Booker, G. R. (1964). Growth of epitaxial silicon layers by vacuum evaporation. I Experimental procedure and initial assessment. *Philosophical Magazine*, **9**, 691–701.

van der Merwe, J. H. and Ball, C. A. B. (1975). Energies of interfaces between crystals. In *Epitaxial Growth*, Part B, ed. J. W. Matthews (New York: Academic Press), pp. 493–528.

van der Merwe, J. H. (2001). Interfacial energy: bicrystals of semi-infinite crystals. *Progress in Surface Science*, **67**, 365–81.

van der Merwe, J. H. (2002). Misfit dislocations in epitaxy. *Metallurgical and Materials Transactions*, **A33**, 2475–83.

Vanhellemont, J., De Gryse, O. and Clauws, P. (2004). Extended defects in silicon: an old and new story. *Solid State Phenomena*, **95–96**, 263–72.

van Landuyt, J. and Amelinckx, S. (1971). Stacking faults in silicon carbide (6H) as observed by means of transmission electron microscopy. *Material Research Bulletin*, **6**, 613–20.

van der Walt, C. M. and Sole, M. J. (1967). The plastic behaviour of crystals with the NaCl-structure. *Acta Metallurgica*, **15**, 459–62.

Vdovin, V. I. (1999). Misfit dislocations in epitaxial heterostructures: Mechanisms of generation and multiplication. *Physica Status Solidi*, **A171**, 239–50.

Venables, J. D. and Broudy, R. M. (1959). Photo-anodization of InSb. *Journal of Applied Physics*, **30**, 1110–11.

Venables, J. D. and Broudy, R. M. (1960). Anodization of InSb. *Journal of the Electrochemical Society*, **107**, 296–8.

Verma, A. J. and Krishna, P. (1966). *Polymorphism and Polytypism in Crystals*. New York: Wiley.

Von Ammon, W., Dornberger, E., Oelkrug, H. and Weidner, H. (1995). The dependence of bulk defects on the axial temperature gradient of silicon crystals during Czochralski growth. *Journal of Crystal Growth*, **151**, 273–7.

Voronkov, V. V. (1982). The mechanism of swirl defect formation in silicon. *Journal of Crystal Growth*, **59**, 625–43.

Vyvenko, O. F., Krüger, O. and Kittler, M. (2000). Cross-sectional electron-beam-induced current analysis of the passivation of extended defects in cast multicrystalline silicon by remote hydrogen plasma treatment. *Applied Physics Letters*, **76**, 697–9.

Walter, H. U. (1977). Generation and propagation of defects in indium antimonide. *Journal of the Electrochemical Society*, **124**, 250–8.

Walton, A. J. (1977). Triboluminescence. *Advances in Physics*, **26**, 887–948.

Warburton, W. K. and Turnbull, D. (1975). Fast diffusion in metals. In *Diffusion in Solids: Recent Developments*, eds. A. S. Nowick and J. J. Burton. New York: Academic Press, Chap. 4, pp. 171–229.

Warekois, E. P. and Metzger, P. H. (1959). X-ray method for the differentiation of {111} surfaces in III-V semiconducting compounds. *Journal of Applied Physics*, **30**, 960–2.

Warekois, E. P., Lavine, M. C. and Gatos, H. C. (1960). Damaged layers and crystalline perfection in the {111} surfaces of III-V intermetallic compounds. *Journal of Applied Physics*, **31**, 1302–3.

Warekois, E. P., Lavine, M. C., Mariano, A. N. and Gatos, H. C. (1962). Crystallographic polarity in the II-VI compounds. *Journal of Applied Physics*, **33**, 690–6.

Warren, P. D., Pirouz, P. and Roberts, S. G. (1984). Simultaneous observation of α- and β-dislocation movement and their effect on the fracture behaviour of GaAs. *Philosophical Magazine*, **A50**, L23 to L28.

Warren, P. D., Roberts, S. G. and Hirsch, P. B. (1987). Microhardness anisotropy and polarity in elemental semiconductors and in $A^{III}B^{V}$ semiconductor compounds. *Bulletin of the Academy of Sciences of the USSR Division of Physical Science*, **51**, 168–72.

Washburn, J., Thomas, G. and Queisser, H. J. (1964). Diffusion-induced dislocations in silicon. *Journal of Applied Physics*, **35**, 1909–14.

Weber, J. (1994). Correlation of structural and electronic properties from dislocations in semiconductors. *Solid State Phenomena*, **37–38**, 13–24.

Weertman, J. and Weertman, J. R. (1980). Moving Dislocations. In *Dislocations in Solids*, **3**, ed. F. R. N. Nabarro (Amsterdam: North-Holland), pp. 1–59.

Weinstein, M. and Wolff, G. A. (1967). Mechanisms of epitaxial growth of compound semiconductors. In *Crystal Growth*. Proceedings of International Conference on Crystal Growth, Boston, ed. H. S. Peiser (Oxford: Pergamon), pp. 537–41.

Weinstein, M., Wolff, G. A. and Das, B. N. (1965). Growth of wurtzite CdTe and sphalerite type CdS single-crystal films. *Applied Physics Letters*, **6**, 73.

Weimann, N. G., Eastman, L. F., Doppalapudi, D., Ng, H. M. and Moustakas, T. D. (1998). Scattering of electrons at threading dislocations in GaN. *Journal of Applied Physics*, **83**, 3656–9.

Weiss, B. L. and Hartnagel, H. L. (1977). The influence of dopants on the hardening of GaAs. *Journal of Applied Physics*, **48**, 3614–15.

Wessel, K. and Alexander, H. (1977). Mobility of partial dislocations in silicon. *Philosophical Magazine*, **35**, 1523–36.

Westbrook, J. H. (1968). Surface effects on the mechanical properties of non-metals. In *Surfaces and Interfaces*, **II** (Syracuse, N. Y.: Syracuse University Press), pp. 3–138.

Westwood, A. R. C. (1974). Tewksbury lecture: Control and application of environment-sensitive fracture processes. *Journal of Materials Science*, **9**, 1871–95.

Westwood, A. R. C. and Latanision, R. M. (1976). Surface and environmental effects in deformation. *Materials Science and Engineering*, **25**, 225–31.

Westwood, A. R. C., Ahearn, J. S. and Mills, J. J. (1981). Developments in the theory and application of chemomechanical effects. *Colloids and Surfaces*, **2**, 1–35.

White, J. G. and Roth, W. C. (1959). Polarity of gallium arsenide crystals. *Journal of Applied Physics*, **30**, 946–7.

Whitworth, R. W. (1975). Charged dislocations in ionic crystals. *Advances in Physics*, **24**, 203–304.

Wilkes, P. (1969). Defects in epitaxial layers of ZnS on Si. *Journal of Materials Science*, **4**, 91–3.

Willis, J. R., Jain, S. C. and Bullough, R. (1990). The energy of an array of dislocations – implications for stain relaxation in semiconductor heterostructures. *Philosophical Magazine*, **A62**, 115–29.

Willoughby, A. F. W. (1968). Anomalous diffusion effects in silicon (A review). *Journal of Materials Science*, **3**, 89–98.

Wilson, R. B. (1966). Variation of electromechanical coupling in hexagonal CdS. *Journal of Applied Physics*, **37**, 1932–3.

Wolfe, C. M., Nuese, C. J. and Holonyak, N. (1965). Growth and dislocation structure of single-crystal $Ga(As_{1-x}P_x)$. *Journal of Applied Physics*, **36**, 3790–801.

Wolff, G. A., Frawley, J. and Hietanen, J. (1964). On the etching of II-VI and III-V compounds. *Journal of the Electrochemical Society*, **111**, 22–7.

Wong, C. Y., Grovenor, C. R. M., Batson, P. E. and Smith, D. A. (1985). Effect of arsenic segregation on the electrical properties of grain boundaries in polycrystalline silicon. *Journal of Applied Physics*, **57**, 438–42.

Woods, J. (1960). Etch pits and dislocations in cadmium sulphide crystals. *British Journal of Applied Physics*, **11**, 296–302.

Yacobi, B. G. and Holt, D. B. (1990). *Cathodoluminescence Microscopy of Inorganic Solids*. (New York: Plenum Press), pp. 214–19.

Yagi, K., Takayanagi, K., Kobayashi, K. and Honjo, G. (1971). In situ observation of formation of misfit dislocations in pseudomorphic monolayer overgrowth of metals on non-metals. *Journal of Crystal Growth*, **9**, 84–97.

Yan, Y., Albin, D. and Al-Jassim, M. M. (2001). Do grain boundaries assist S diffusion in polycrystalline CdS/CdTe heterojunctions? *Applied Physics Letters*, **78**, 171–3.

Yarykin, N. and Steinman, E. (2003). Comparative study of the plastic deformation- and implantation-induced centres in silicon. *Physica*, **B340–342**, 756–9.

Yonenaga, I. (1998). Dynamic behavior of dislocations in InAs: in comparison with III-V compounds and other semiconductors. *Journal of Applied Physics*, **84**, 4209–13.

Yonenaga, I. (2001). Dislocation behavior in heavily impurity doped Si. *Scripta Materialia*, **45**, 1267–72.

Yonenaga, I. (2003). Dislocation–impurity interaction in Si. *Materials Science in Semiconductor Processing*, **6**, 355–8.

Yoenaga, I. and Sumino, K. (1978). Dislocation dynamics in the plastic-deformation of silicon-crystals. 1. Experiments. *Physica Status Solidi*, **A50**, 685–93.

Yonenaga, I. and Sumino, K. (1992). Impurity effects on the mechanical-behavior of GaAs crystals. *Journal of Applied Physics*, **71**, 4249–57.

Yonenaga, I. and Sumino, K. (1993). Effects of dopants on dynamic behavior of dislocations and mechanical strength in InP. *Journal of Applied Physics*, **74**, 917–24.

Yonenaga, I. and Sumino, K. (1996). Influence of oxygen precipitation along dislocations on the strength of silicon crystals. *Journal of Applied Physics*, **80**, 734–8.

Yonenaga, I., Sumino, K. and Hoshi, K. (1984). Mechanical strength of silicon crystals as a function of the oxygen concentration. *Journal of Applied Physics*, **56**, 2346–50.

Zare, R., Cook, W. R. and Shiozawa, L. R. (1961). X-ray correlation of the A-B layer order of CdSe with the sign of the polar axis. *Nature*, **189**, 217–19.

Zheleva, T. S., Nam, O. H., Bremser, M. D. and Davis, R. F. (1997). Dislocation density reduction via lateral epitaxy in selectively grown GaN structures. *Applied Physics Letters*, **71**, 2472–4.

Zimin, D., Alchalabi, K. and Zogg, H. (2002). Heteroepitaxial PbTe-on-Si pn-junction IR-sensors: correlations between material and device properties. *Physica E: Low-Dimensional Systems and Nanostructures*, **13**, 1220–3.

Zunger, A. (1987). Order-disorder transformation in ternary tetrahedral semiconductors. *Applied Physics Letters*, **50**, 164–6.

Symbols

Latin symbols	
b	is the Burgers vector
B_o	is a constant in the expression for the dislocation velocity [Equation (4.7)]
B_1	is a constant in the expression for the dislocation velocity [Equation (4.8)]
C	is the constant, in Equation (4.21), $= \mu b/\{2(1 - v)\} \approx \mu b/3$.
E	is the electric field in the specimen
E_d	is the activation energy for dislocation movement
K	is a rate constant
l	is the direction of the dislocation line
m	is the exponent of the shear stress in the expression for the dislocation velocity [Equation (4.8)] and is close to one and varies from one semiconductor to the other.
k	is Boltzmann's constant
q	is the charge per unit length of dislocation
T	is the temperature in degrees Kelvin
v	is the dislocation velocity
v_{av}	is the average velocity of the moving dislocations
v_0	is a material constant in the expression for dislocation velocity, Equation (4.8)
Greek symbols	
ε	is the shear strain
ρ	is the dislocation density (length per unit volume)
ρ_m	is the density of moving dislocations
σ_{ij}	is the elastic stress tensor

Greek symbols

τ	is the shear stress
τ_{appl}	is the stress due to the externally applied load
τ_{eff}	is the effective stress experienced by the moving dislocations
τ_{int}	is the internal elastic back-stress
τ_{ly}	is the lower yield stress
τ_0	is a stress constant in the expression for dislocation velocity [see Equation (4.8)], $= 1\,\text{MPa}$
τ_{uy}	is the upper yield stress

5

The electrical, optical and device effects of dislocations and grain boundaries

5.1 Introduction to the electrical effects of dislocations and other defects in semiconductors

In this chapter the historically important influence of high densities of dislocations on the electrical properties of semiconductors and on device performance is first outlined. (In the early days of semiconductor studies, high densities were generally either grown-in or introduced into the material by plastic deformation.) The mechanisms giving rise to the electronic and luminescence properties of dislocations and other defects are next treated. The role of defects in devices is discussed. The electrical properties of grain boundaries in polycrystalline semiconductors are also treated.

5.1.1 Introduction

The first short paper reporting that plastic deformation of Ge and Si was possible at raised temperatures, also reported that this increased the resistivity of Ge and the lifetime of photo-injected carriers was greatly reduced (Gallagher 1952). Further early studies revealed that <1% plastic strain would eliminate all the electrons in lightly doped *n*-type Ge and turn it *p*-type. (Ge was the important semiconductor at that time.) The effects of dislocations were so important that much basic dislocation theory is due to workers in the pioneering group at Bell Laboratories, especially Shockley (who introduced 'dangling bonds' and Shockley partial dislocations) and Read (of the Frank-Read source and the first theory of the electrical effects of dislocations). However, the industrial laboratories lost interest once it was found possible to grow low or zero dislocation density Si and effectively avoid or passivate process-induced dislocations.

The electrical effects of the large numbers of varied types of dislocations and point defects introduced by plastic deformation of Ge and Si continued to be studied, however, by academic groups. Much was learned but some things remained obscure despite a great deal of subsequent work (for reviews of the early work see Labusch and Schröter 1978 and Figielski 2002).

Later work concentrated on the study of dislocations introduced by slow deformation at lower temperatures to minimize impurity contamination. However, it has since been found that much more stringent precautions are required to avoid contamination (see Section 5.5). Recently much progress has been made through the use of microscopical methods especially SEM (scanning electron microscope) EBIC (electron beam induced current) and CL (cathodoluminescence) to study the electronic properties of single dislocations. Still unanswered questions relate to the responsiblity for the semiconductor electronic properties of the dislocation core and of the other defects generated in introducing them.

Such unsolved aspects of the electrical and optical properties of dislocations were outlined by Weber (1994), who presented experimental evidence of the correlation of the dislocation structure with electrical and optical properties. As will be shown in the following sections (and mentioned repeatedly throughout this book), impurities and native point defects generally play an important role in determining the electrical and optical effects of dislocations and grain boundaries. This complication also explains some of the experimental irreproducibility in the literature.

The properties of grain boundaries in semiconductors also have always been of theoretical and practical interest. At first this was because a number of semiconductors were only available in polycrystalline form. Later it was because of their role in the polycrystalline materials used in e.g. TFT (thin film transistor) displays and photovoltaic cells for terrestrial power generation.

5.1.2 Device effects

Important device parameters and device yields improved with falling defect densities (and only with them) in the early Ge-based industry. For example, the germanium room-temperature minority carrier lifetime increased with decreasing dislocation density as shown in Fig. 5.1. Dislocations were found to have the same effect in many other semiconductors (Table 5.1). Thus dislocations act as traps or recombination centres. The so-called $1/f$ electrical noise was orders of magnitude greater in germanium after plastic deformation (Brophy 1956, 1959) showing that dislocations act as generation-recombination centres (Morrison 1956). For reviews of the early work see Bardsley (1960), Figielski (2002) and Mil'shtein (2002). Hence, dislocation density was used as the basic indicator of the quality of semiconductor material. These defect correlations motivated much of the classic research on the crystallography, origin and properties of dislocations at the great American industrial electronics laboratories from the late 1940s to the 1960s.

Dislocation density correlations are still important in evaluating new materials and device technologies. For any given device-technology and material combination, there is a critical dislocation density 'threshold', ρ_{crit}. Provided that the dislocation density, $\rho < \rho_{crit}$, the yield of working devices is satisfactory. Etch pit densities (EPDs), i.e. dislocation densities, are still routinely determined to monitor

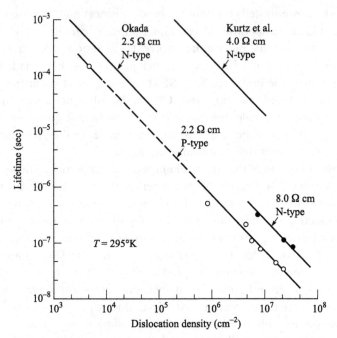

Figure 5.1 Early Ge data showing that the minority carrier lifetime falls exponentially with the dislocation density above about $10^4\,\text{cm}^{-2}$ in both n- and p-type material. [After Wertheim and Pearson 1957. Reprinted with permission from *Physical Review*, **107**, pp. 694–8, 1957 (http://publish.aps.org/linkfaq.html). Copyright 1957 by the American Physical Society.]

Table 5.1. *Some early studies reporting that increasing densities of dislocations reduce the minority carrier lifetimes*

Material	References
Silicon	Kurtz *et al.* (1956), Hunter *et al.* (1973)
Germanium	Okada (1955), Kurtz *et al.* (1956)
GaAs	Ettenberg (1974)
GaP	Stringfellow *et al.* (1974),
	Werkhoven *et al.* (1977, 1978/79)

the quality of Si wafers. For most purposes, EPDs below $10^4\,\text{cm}^{-2}$ are adequate and more expensive zero EPD wafers are not needed. However, some devices operate satisfactorily despite relatively large densities of defects, e.g. GaN blue LEDs. In the mid 1990s, high-efficiency, bright, long-lived blue light emitting diodes (LEDs) were announced by a Japanese firm (Nichia Chemical Company) using GaN grown on sapphire. These LEDs were then about 100 times brighter than previous commercially available SiC blue LEDs (Cook and Schetzina 1995). Lester *et al.* (1995) pointed out that LED efficiency varied with etch pit density (ρ) in the III-V materials previously used to make red and yellow and green LEDs, as shown in Fig. 5.2. Astonishment was, therefore, caused by their report that the bright, long-lived GaN LEDs contained 10^{10} dislocations per cm^2. That is, the new LEDs achieved greatly improved performance despite a dislocation density from 4 to 6 orders of magnitude higher than the long-established GaP and GaAs based materials could tolerate!

Unfortunately, from the research point of view, but fortunately from the practical standpoint, many defect-related problems disappear due to 'good housekeeping'. That is, the adoption of the best practice based on experience, especially in the mature Si industry, solves many defect-related problems without these ever being studied or understood. Moreover, if the problem does not 'disappear' quickly, the process or device is often abandoned for some alternative rather than fund the time-and-money-consuming research needed to solve it. Therefore, the dominant Si industry lost much of its initial interest in dislocations and relatively few studies of the role of defects in devices have been made.

5.1.3 The electrical properties of dislocations

Dislocations affect electrical and optoelectronic properties because they introduce localized levels or bands into the forbidden energy gaps of semiconductors.

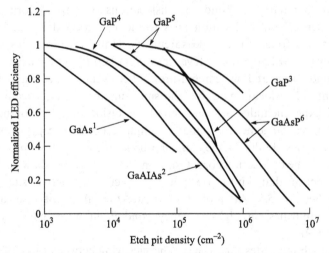

Figure 5.2 The efficiencies of III-V visible-light LEDs fall with increases in the dislocation density. (After Lester *et al.* 1995, where references to the original sources of the data may be found.) Reprinted with permission from *Applied Physics Letters*, **66**, pp. 1249–51. Copyright 1995, American Institute of Physics.)

They must do so because they 'break the symmetry' of the crystal structure, as we discussed in Section 2.4, and that determines the form of the electron energy bands.

Shockley's 'dangling bonds' (e.g. Fig. 2.23a) in the defect cores result in deep (near mid-gap) recombination levels. Due to deformation potential effects, the elastic strain field of dislocations pulls shallow trap states out of the conduction band.

Moreover impurity atoms in the defect core can give rise to modified energy levels in the band gap. We now outline the mechanisms for the electronic effects of dislocations.

5.1.4 Dangling bonds and mid-gap states: the evidence from ESR and DLTS

Shockley (1953) introduced the idea of 'dangling bonds', i.e. covalent bonding orbitals that do not reach and bond to a second atom, occupied by only one electron in an intrinsic crystal at low temperatures. He suggested that these would occur in the core of dislocations, like the 60° shuffle set unit dislocation of Fig. 2.23a. The wave functions of adjacent bonds would overlap, he suggested, to form a one-dimensional, half-filled band.

There is much evidence, to be discussed later, that indicates that dislocation lines are often charged negatively. This is evidence for the presence of deep trap states presumably associated with broken bonds. Such states can also act as donors as well as acceptors and so affect the carrier density. Broken bonds should be chemically reactive as well.

If neighbouring dangling bonds on dislocations overlap to form a half-filled dislocation energy band dislocation lines would be expected to conduct. No evidence of this was found in d.c. measurements. However, Grazhulis et al. (1977a and b), Osipiyan (1981, 1983) and Kveder et al. (1985) found evidence of ultra-high frequency conduction at low temperatures. This occurred back and forth, along short segments of dislocations only. The small number of electrically active deep levels found experimentally per unit length of dislocation led to the suggestion that Hornstra rearrangement eliminates dangling bonds except at 'special sites' such as kinks, jogs, faults in the dangling-bond reconstruction process or point defect centres in the dislocation core (Osipiyan 1981, 1983). Wilshaw and co-workers provided evidence that kinks are not the electrically active sites as will be discussed in Section 5.5. Attempts to separate the effects of dislocations, point defects and impurities on high-frequency conduction were made by Brohl and Alexander (1989).

The dangling bond states should have energies lying between those of the bonding states of the valence band and the antibonding states of the conduction band. Thus dangling bond levels or mini-bands should occur around the middle of the forbidden energy gap. Such deep dislocation states were originally thought to cause the electrical effects of dislocations. Dangling bonds are sp^3 hybrid orbitals containing a single electron and can be detected by electron spin resonance (ESR) due to the unpaired spins of these electrons.

To observe ESR, a constant magnetic field is applied to the semiconductor crystal in a certain direction together with variable-frequency microwave radiation incident at right angles to it. Unpaired electrons were originally pictured as charged particles spinning about an axis, to explain the fact that they have a magnetic moment. This aligns with the applied field and precesses around it. When the microwave frequency equals that of a characteristic resonance, energy is absorbed to produce transitions between spin states. In paramagnetic materials, resonance absorption thus appears as peaks in the microwave absorption spectrum at particular values of the incident microwave frequency and applied magnetic field. The spectrum of unpaired electrons, like those in dangling bonds, depends on the symmetry of their surroundings. Therefore ESR spectra can yield information on the local environments of the defects containing unpaired electrons.

A broad ESR absorption line was found in Si plastically deformed at 750–840° C (Alexander *et al.* 1965). Material deformed at a lower temperature (650–700° C) was then found to contain a set of overlapping lines (Grazhulis and Osipyan 1970, Grazhulis 1979). The complex sets of lines eventually resolved were found to be distinguishable by symmetry, photo-ESR, dependence on doping and line width to correspond to centres K1, K2 and Y (Suezawa and Sumino 1989) associated with dislocations and K3–K5 due to point defect clusters (Alexander 1986). The details of all the dangling bond centres involved are still not entirely clear (see the discussions of the differing interpretations of these by Osipyan 1988 and Kisielowski-Kemmerich *et al.* 1985). More recently Kisielowski-Kemmerich (1989) carried out LCAO modelling which led him to suggest that the dislocation centres were attributable to vacancies and vacancy-impurity complexes formed during dislocations motion and impurity gettering during deformation. In particular he identified Ki with a particular form of vacancy (V_{3c}) trapped at a reconstruction (bond rearrangement) defect and K2 with a multi-vacancy complex

DLTS (deep level transient spectroscopy) can detect electron and hole traps and determine their density, capture cross-section and energy level(s) in the forbidden gap. Comparisons of results obtained by ESR and other techniques are helpful. Thus, comparing ESR and DLTS, it was found that none of the near band edge centres found by DLTS are present in Si deformed below 550° C (about 0.48 of the melting point) nor are the K3–K5 centres which are point-defect clusters (Alexander 1986). Yarykin and Steinman (2003) studied silicon samples, plastically deformed at 630° C and others that were Si-ion implanted, using DLTS and PL (photoluminescence). They found evidence that most of the DLTS lines observed in deformed material were due to small interstitial clusters whereas PL D1–D4 emission bands were only found in the presence of dislocations. Thus dislocation and interstitial-cluster related states could be distinguished. Feklisova *et al.* (2003, 2005) compared DLTS and SEM EBIC on Si plastically deformed to introduce a low density of dislocations. From EBIC measurements, they observed that the concentration of recombination centres in the dislocation trails were greater than near the dislocations. The DLTS measurements revealed that the total concentration of the deep-level centres was related to the concentration of defects in the dislocation trails (Feklisova *et al.* 2003, 2005). (For a summary and further discussion of the formation and importance of point defect debris trails in plastically deformed semiconductors see Section 4.7 and Section 5.9.1.) As discussed in Section 5.5.7, ESR, DLTS and SDLTS studies revealed that the electrical effects of plastic deformation of semiconductors are mostly related to native point defect debris and not to dislocations.

Kisielowski *et al.* (1991) published a quantitative analysis for three distinguishable effects contributing to inhomogeneities of the band gap of doped and plastically deformed silicon. These were charge carrier capture by deep traps associated with (i) point-like defects and with (ii) dislocations as well as (iii) capture by shallow states due to the stress fields of dislocations. In material deformed at a low temperature (700° C) the first effect is the most important and the contribution from (ii) can only be distinguished by detailed analysis of the ESR spectra of the samples and it does not result in any accumulation of charge, delocalized along the dislocation. A contribution from (iii), i.e. charge carrier capture by shallow states, in specimens specially deformed so as to introduce straight dislocation segments, could be detected by the appearance of electric dipole spin resonances.

Thus there is agreement that two important conclusions can be drawn from ESR studies of deformed Si. Firstly, there are fewer than 5% of the dislocation dangling bond like centres than models like that of Fig. 2.23a predict. This is believed to be due to energy-reducing bond rearrangements, that reconnect dangling bonds as suggested by Hornstra (1958) (see Section 4.1.1), along mainly dissociated dislocations. Secondly, equal or greater numbers of unpaired electrons in dangling bond like states are due to point defects like vacancies and clusters (Alexander 1986) and 'trail' defects left as debris behind dislocations when they move in silicon.

The electrical effects of the recently discovered trail debris left on slip planes, behind moving dislocations, in plastically deformed Si were studied by Eremenko and his co-workers.

Eremenko and Yakimov (2004) used SEM EBIC and etching to study slip planes in Si plastically deformed by four-point bending at 800–950° C and having a Schottky contact for charge collection. Under the Schottky barrier the slip planes appeared in the usual EBIC dark contrast but there was bright EBIC contrast extending for several mm outside it, as shown in Fig. 5.3. As they pointed out, this means that the slip planes collected minority carriers and transferred them along to the contact to register as EBIC signal. This current was found to fall exponentially with distance from edge of the contact as shown in Fig. 5.4. Just the same behaviour is exhibited by grain boundaries running from under a Schottky contact on Si

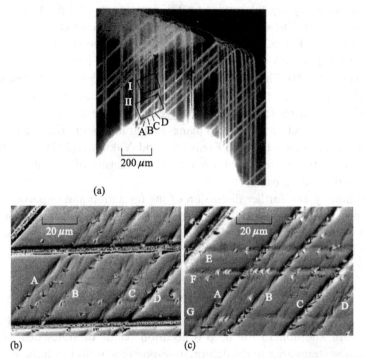

(a)

(b)   (c)

Figure 5.3   (a) EBIC image of intersecting bright slip bands near a Schottky barrier. Light microscope interference-contrast images of the etched surface in the areas marked I (b) and II (c) in (a) and showing the distribution of dislocations (etch pits) and the extended defects (long ridges) in the slip bands labelled A to D on one set of planes and E to G on the other. (After Eremenko and Yakimov 2004.)

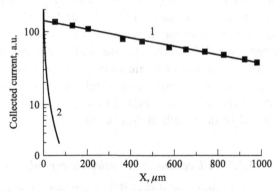

Figure 5.4   The decay of EBIC current with distance from the edge of the Schottky contact along a bright slip plane (curve 1) and far from any such plane (curve 2). (After Eremenko and Yakimov 2004.)

into the surrounding crystal producing the same form of bright defect contrast (see Section 5.10.4). These slip bands were characterized by etching and found to appear as shown in Fig. 5.3. They contain a high linear density of long ridges, which are the extended defects or trails found by Eremenko and Fedorov (1995) and Eremenko *et al.* (1999). The close similarity of the bright contrast of these slip bands and grain boundaries (Section 5.10.4) as well as the enhanced conductivity often seen along grain boundaries (Section 5.10) suggests that the transmission along and, perhaps, charge collection to, the slip planes and grain boundaries is due, at least in part, to the dislocations present. Eremenko and Yakimov (2004) reported a correlation between the brightness of a band and the density of the ridges along it, so trail defects also play a part.

For a summary and further discussion of the formation and importance of point defect debris trails in plastically deformed semiconductors see Section 4.7 and Section 5.9.1. Additional ESR, DLTS and SDLTS studies on dislocations are presented in Section 5.5.7.

5.1.5 The deformation potential and shallow states

Landauer (1954) first treated the interaction of electrons with the strain field of the dislocation. In addition to the deep dislocation bands of states due to dangling bonds, shallow states arise in the deformation-potential wells produced by the strain fields of dislocations (e.g. Claesson 1979, Jones 1979, Jones 2000). It is thought that these levels allow carrier generation and recombination and are important for electrical noise and breakdown initiation.

5.1.6 Dislocation excitons

Bonch-Bruevich and Glasko (1961), on the basis of Russian studies of the luminescence of deformed semiconductors (like Ge and CdS), published a theory of excitons bound by the deformation potential of dislocations. Emtage (1967) independently treated such dislocation excitons and calculated that these should have a sufficiently high recombination rate to give readily observable luminescent emission. For a more recent treatment and references to additional Russian theoretical work see Bozhokin *et al.* (1982). Dislocation cathodoluminescence and electroluminescence will be dealt with in Section 5.6.

5.1.7 Dislocations in piezoelectric crystals

Piezoelectricity is a third source of dislocation electronic effects. Semiconducting elements with the centrosymmetric diamond structure (Si, Ge and diamond) cannot be piezoelectric but sphalerite and wurtzite structure compounds are non-centrosymmetric and piezoelectric. Therefore, in the compounds, the elastic strain fields of edge and 60° dislocations produce electric fields (a piezoelectrically induced polarization vector) (Merten 1964a, 1964b, Booyens *et al.* 1977, 1978a, 1978b).

Evidence of dislocation piezoelectric effects was found in studies of V_t, the threshold voltage of GaAs MESFETs (metal semiconductor field-effect transistors). V_t varies across GaAs slices and this is attributed to piezoelectric charge densities in the substrate due to residual stresses in processing overlayers. For references to this and related work see McNally *et al.* (1995). Support for this was found in a study of the effects of externally applied elastic stresses on V_t in chips of GaAs containing MESFETs in several crystallographic orientations (McNally *et al.* 1988). They showed that observed shifts of tens of mV could be accounted for by a piezoelectric model. Large additional shifts, ΔV_t, also occur, however, and these correlate with proximity to crystal defects (e.g. Miyazawa and Hyuga 1986). McNally *et al.* (1995) modelled the piezoelectric effect of edge (α and β) and 60° (α and β) dislocations on V_t of GaAs MESFETs. They obtained good agreement with experimental results. Shifts, ΔV_t, of opposite signs occurred and were attributed to the presence of dislocations with edge components of opposite polarities, α and β. Screw dislocations have no piezoelectric effect.

5.1.8 Dislocations and impurities

Dislocations interact strongly with impurity atoms, elastically and electrostatically at long range and by chemical bonding in the core (Section 4.7.1). Therefore, if the material spends any time at a high temperature, electrically active impurity atoms migrating to the dislocation lines must profoundly alter their electronic states. The early work on the effects of small amounts of plastic strain studied specimens deformed at high temperatures (>2/3 of the melting point) where semiconductors are ductile at the strain rates used in normal testing. The dislocation densities introduced were large so the effects were large. The results in such cases, however, were strongly influenced by the numerous point defects that were also produced.

Later fundamental work concerned smaller densities of dislocations introduced by very slow deformation at low temperatures (1/2 to 2/3 the melting point). For Si, plastic deformation at normal testing rates can only take place above about 800° C but slow 'easy glide' (strain on one carefully oriented slip system only, to avoid work hardening) is possible down to about 650° C or even 450° C. Wessel and Alexander (1977) introduced a high and low T technique. This applies a very small strain at high temperature (~950° C) to initiate slip followed by deformation at low temperature (~650° C). Later workers used lower values for both their high and low temperatures. It was thought that relatively little impurity diffusion should occur during the low-T deformation so the dislocations produced should be comparatively clean.

More recently it was found, however, that the electrical properties of dislocations so introduced are still dominated by traces of Cu or Ni. This is so unless impurity-free Si ingots are specially selected, using a heat treatment and PL spectral test for traces of transition metals (Higgs *et al.* 1990a). Moreover, the deformation then has to be carried out, by the high and low T technique, in special quartz deformation rigs to avoid contamination during deformation from hot machine parts. Dislocations produced in this way, free of transition metal contamination, produce very little electron-hole recombination, i.e. no detectable SEM EBIC dark contrast nor SEM CL D (dislocation related) line emission (Higgs *et al.* 1990a,b). A fuller account of this important work will be given in Section 5.6.

The cumulative evidence from DLTS (deep level transient spectroscopy), ESR (electron spin resonance) and SEM EBIC is that charged deep states occur only at distances of many atoms along dislocation lines. It is clear, however, that these electronic states exist and produce marked electrical and optical effects under many circumstances.

5.1.9 The polarity of dislocations in semiconducting compounds

We saw in Section 4.1.3 that, to each type of dislocation (except screws) in the diamond cubic structure, there correspond two opposite polarity types of dislocation, α and β, in the sphalerite structure and their core structures may be glide or shuffle set in form or a mixture of the two (Section 4.1.2). These α and β dislocations differ in polarity, i.e. in the element of the atoms occurring in their cores. Therefore dislocations of opposite polarity are expected to differ in their electronic properties (see e.g. the piezoelectric effects in Section 5.1.7). The observed electrical effects of deformation, producing both types, in sphalerite-structure compound semiconductors, therefore, are more complex. Moreover, the semiconducting compound material available is of poorer quality than the best silicon. It will have some degree of non-stoichiometry (see Section 6.3.2) and contain residual unwanted impurities. For these reasons much less is known about the effects of polar dislocations than about those of dislocations in the elemental materials such as Si.

It is to be expected that bonds dangling from A and B atoms in the cores of dislocations in AB semiconducting compounds (e.g. III-V or II-VI materials) would result in different electrical and luminescent properties.

An important feature is the possibility of two discernible defects present due to the polarity of semiconducting compounds. In GaAs, for example, the extra half-plane of an edge dislocation can end in a row of either Ga or As atoms. The dangling bonds and their effects are expected to be different in these two cases. Dislocations with dangling bonds from A atoms (e.g. Ga) are referred to as α-edges and those from B atoms (e.g. As) are called β-edges. Fig. 5.5 presents the relation between crystallographic polarity, the α or β form of edge dislocations and their 'glide' or 'shuffle' set type. By using polar bending (see Section 4.1.3), numbers of similar,

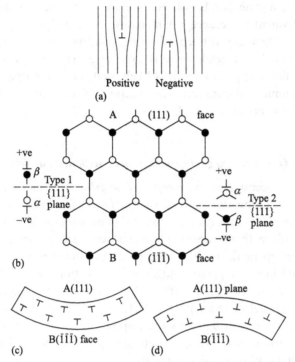

Figure 5.5 The relation between crystallographic polarity, the α or β form of edge dislocations and their 'glide' or 'shuffle' type. (a) Positive and negative edge dislocations. (b) The atoms with dangling bonds in the cores of positive and negative shuffle and glide set dislocations in sphalerite-structure materials. (c) and (d) The effect of plastic bending of a {111} slab of an AB sphalerite-structure material is to introduce an excess of edge dislocations of one sign. For example, for the sign of bending in (c) the dislocations are negative and will be of α form if they are of shuffle type but will be of β form if they are of glide type (compare b). In (d) the positive edges will either be β shuffle or α glide types.

parallel edge dislocations can be introduced. The difference between +ve and −ve edge dislocations is that between α and β character. In the α dislocations, the dangling bonds are from III-valent atoms, whereas in the β dislocations, the bonds dangle from V-valent atoms. The bonds from the III-valent atoms can be expected to contain 3/4 electron in the neutral dislocation, whereas the bonds from the V-valent atoms will contain 5/4 electrons in the neutral dislocation. While the +ve dislocations are β-type (V-valent atom dangling bonds) and the −ve dislocations are α-type (III-valent atom dangling bonds), this reverses if the cores are glide rather than shuffle set as originally assumed. In other words, if the dislocations are glide set, the +ve dislocations will be α-type and the −ve dislocations have β character. Therefore, it has been recommended that both the polarity and the core type assumed in determining it is specified (e.g. α_g, i.e. α polarity assuming the core to be glide set) when dislocation types are discussed.

The V-valent dangling bonds will be expected to be more donor in character, whereas the III-valent more acceptor in nature, so the sign or magnitude of the electrical effects of bending will depend on the polarity of the bending.

It should be noted that, while there is some evidence that dislocations of polar character have different properties, it is likely that, as in the case of Si, these are dominated by impurity decoration. The reactions of impurity atoms with α- and β-edge dislocations should also be different.

5.1.10 Quantized magneto-conductance oscillations at dislocations

Dislocations in semiconductors were reported to exhibit a quantum effect of a type typically attributed to low-dimensional systems (Figielski *et al.* 1998, Figielski *et al.* 2000, Wosinski *et al.* 2002). This was observed in measurements of low-temperature magneto-conductance in GaAsSb/GaAs heterostructures. During these measurements a strong magnetic field was applied parallel to the axes of misfit dislocations, in the heterointerface. Magneto-conductance oscillations were observed with a period dependent on the angle between the magnetic field direction and the dislocation axes (Figielski *et al.* 1998, Figielski *et al.* 2000, Wosinski *et al.* 2002). These oscillations were attributed to the trapping of charge carriers on localized quasi-stationary orbits surrounding the charged dislocations (Figielski *et al.* 1998, Figielski *et al.* 2000, Wosinski *et al.* 2002).

5.1.11 The electrical properties of grain boundaries

Small angle grain boundaries (GBs) consist of walls of dislocations (see Section 2.3.10). Similarly (Section 4.5.2) large angle, general GBs consist of regions of good and bad misfit (the O lattice, etc.) so some form of local atomic misfit occurs. Thus all GBs involve symmetry breaking and will introduce energy levels in the forbidden gap. Grain boundaries, like dislocations, may contain dangling bonds and have strain fields that produce localized states via the deformation potential (see Section 5.10 and Section 5.12) and these will be affected by any impurity contamination. Dangling bonds in GBs can be detected by electron spin resonance (ESR, see Section 5.1.4) but as discussed by Grovenor (1985), the interpretation of ESR results on GBs is not straightforward.

The electrical properties of grain boundaries were studied from the beginning of semiconductor science [Matare (1955) and Matare and Wegener (1957) (Ge); Mackintosh (1956), Mueller (1959a,b), Mueller and Jacobsen (1959, 1962), Mueller and Maffitt (1964) (InSb); Holt *et al.* (1958), Alfrey and Wiggins (1960) (GaP); De Nobel (1959), Lang *et al.* (1963) (CdTe) and Thornton (1963) (GaAs)]. Both impurity segregation and carrier trapping at GBs influence the transport properties of polycrystalline semiconductors (see Section 5.10). In particular, the presence of trapped charge in GBs results in band bending and energy barriers to electrical transport. This topic was extensively reviewed by Grovenor (1985), Seager (1985), Greuter and Blatter (1990), Möller (1993), Sutton and Baluffi (1995), and Kamins (1998) and will be discussed in more detail in Section 5.10 and Section 5.12.

5.2 The electrical effects of the deformation of semiconductors: the Read theory

Early experimental work showed that dislocations introduced by processing the earliest device material had relatively large effects on the conductivity of *n*-type Ge but relatively small effects in *p*-type Ge. Read (1954a, 1954b) therefore assumed that dangling bond energy states deep in the forbidden gap act as acceptors, able to take up a second electron. No evidence of d.c. conduction along dislocation lines could be found so Read, contrary to Shockley's original idea, assumed the dangling bonds introduced discrete acceptor levels, rather than a band (arising from the overlap of the dangling bond wave functions), in *n*-type Ge.

In Read's model, therefore, dislocations in *n*-type Ge have a negative charge, Q, per unit length which is screened by an equal positive space charge in a surrounding volume of radius r_d (Fig. 5.6). The spacing of dangling bonds along a 60° dislocation of the form shown in Fig. 5.7 is the spacing of neighbouring atoms in a $\langle 110 \rangle$ direction = $\mathbf{b}$, the Burgers vector. The line charge is the number of dangling bonds per unit length times the occupation function, f (fraction occupied by a second electron). Hence, for unit length of 60° dislocation we can write:

$$Q = \pi r_d^2 q N_D = q \, f/b \tag{5.1}$$

where N_D is the density of donor atoms (positively charged ions at room temperature).

In Ge, $b = 0.397$ nm so there are $1/b = 2.5 \times 10^7$ dangling bonds per cm of such a dislocation. For a dislocation density of $10^9 \, \text{cm}^{-2}$, i.e. $10^9 \, \text{cm}$ of dislocation line per cm^3, there are, therefore, 2.5×10^{16} dangling bonds per cm^3. At room temperature,

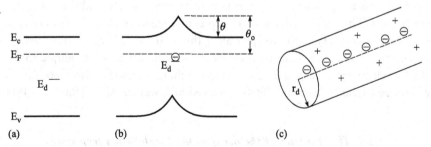

Figure 5.6   The Read model for the electrical effects of dislocations assumes a negative line charge. Energy band diagrams of (a) uncharged and (b) charged dislocations and (c) the screening space charge cylinder around the charged dislocation in *n*-type material.

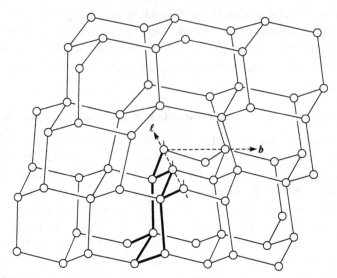

Figure 5.7   In the original Shockley model of a 60° dislocation, adopted by Read, there was a broken bond dangling from each atom along the core. ℓ is the dislocation line vector and b is its Burgers vector. This is the undissociated shuffle set model of the dislocation core. (After Hornstra 1958. Reprinted from *Journal of Physics and Chemistry of Solids*, **5**, Dislocations in the diamond lattice, pp. 129–41, Copyright 1958, with permission from Elsevier.)

Read's various treatments of the statistical problem of the occupation of dislocation states indicated that about one bond in ten would be occupied by a second electron. Thus a density of 10^9 60°-dislocations per cm^2 would accept 2.5×10^{15} electrons per cm^3. This is greater than the total number of charge carriers in 1.7 ohm-cm n-type Ge. Experimentally, plastic deformation of a fraction of a percent, introducing such dislocation densities as this, did turn such lightly doped n-type Ge intrinsic or even p-type.

Substituting the value of N_d for 1.7 ohm-cm Ge leads to a value of $r_d = 1$ μm. This means that all the material becomes intrinsic due to incorporation into the space charge cylinders for a dislocation density of about $10^8 \, cm^{-2}$. To see this suppose all the dislocations are parallel, form a square array seen end on, and that the cylinders are in contact. Then each dislocation occupies an area normal to the lines of $(2 \, μm)^2$. The dislocation density is the reciprocal of this $= 0.25 \times 10^8 \, cm^{-2}$. This value is in order of magnitude agreement with the density found above to be ample to absorb all the free electrons. The large size of the space charge cylinders also accounts for the large charge carrier scattering effects observed at lower densities (Bardsley 1960).

5.2.1 *The statistics of the occupation of dislocation trap states*

Read (1954a, 1954b) used several approximations to find f, the fraction of dislocation states occupied. The simplest neglected configurational entropy and assumed the single, minimum energy configuration of electrons evenly spaced along the line. The free energy of the dislocation plus crystal system falls linearly with the number of accepted electrons, i.e. with f as each electron drops in energy from the conduction band to the dislocation level (Fig. 5.8). As the line becomes negatively charged the energy bands bend up at the dislocation and the electrostatic energy increases roughly as f^2. The equilibrium value of f is found by minimizing the free energy. The energy decrease when an electron drops into the dislocation is $E_F - E_d$ since E_F, the Fermi energy measured from the valence band edge at a distance from the disloca- tion, is the free energy (electrochemical potential) of the carriers. The total increase in free energy of the system due to the presence of the dislocation divided by the number of electrons on the dislocation, i.e. the dislocation energy per electron, is:

$$U_e = E_d - E_F - \varepsilon_s(f) \tag{5.2}$$

where $\varepsilon_s(f)$ is the electrostatic energy per electron. (Remember that in diagrams like those in Figs. 5.8 (a and b), the electrostatic potential (energy) increases downwards.) The free energy per site (most of which are empty) is:

$$f U_e = f \left\{ E_d - E_F - \varepsilon_s(f) \right\} \tag{5.3}$$

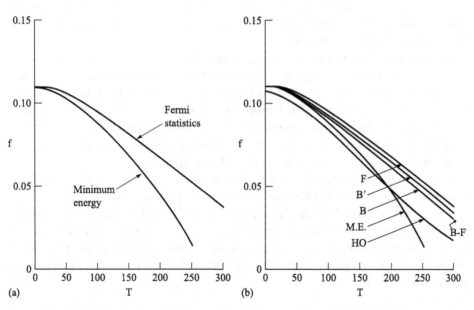

Figure 5.8   (a) The variation with temperature of the fraction of filled dislocation acceptors, f, on the Fermi and minimum energy approximations and (b) on all the approximations developed by Read (1954b). (The letters B = Boltzmann and HO = Harmonic Oscillator are labels assigned by Read to two of his approximations.) (After Read 1954b. Reprinted with permission from *Philosophical Magazine*, **45** (1954), pp. 1119–28; Taylor & Francis Ltd., the Journal's web site: http://www.tandf.co.uk/journals.)

To minimize the free energy we set the first derivative of $f U_e$ with respect to f, equal zero to obtain:

$$E_F - E_d = \frac{d}{df} f \varepsilon_s (f) = \varepsilon^* \tag{5.4}$$

Read derived $\varepsilon_s(f)$ for a row of electrons evenly spaced along the axis of a cylinder of positive space charge from electrostatics. This utilized the expression for the spacing of dangling bonds along dislocations of the atomic form shown in Fig. 5.7 (now known as undissociated shuffle set dislocations) as a function of the angle, α, between the line direction l and b:

$$c = b \sin 60° / \sin \alpha = 0.866\, b \operatorname{cosec} \alpha \tag{5.5}$$

This gives the variation of dangling bond spacing with dislocation type (α) with the result that e.g. $c = b$ for $\alpha = 60°$ dislocations and c is infinite for $\alpha = 0°$, i.e. there are no dangling bonds in screw dislocation cores on this oversimplified (wrong) model. (Screws are now known to have electrical effects not very different from edges.) From these equations

$$E_F - E_d = \varepsilon_o f \{3 \ln(f/f_c) - 0.232\} \tag{5.6}$$

where $\varepsilon_o = q^2 / \kappa c$, $f_c = c (\pi N_D)^{1/3}$, q is the charge on the electron, κ is the dielectric constant and the final numerical factor in Equation (5.6) arises from the particular expression for c (Equation 5.5).

According to Read's minimum energy approximation (Equation 5.6), f varies with temperature as shown in Fig. 5.8a. The occupation fraction is lower than the value given by Fermi statistics, at all temperatures, due to the electrostatic repulsion of the electrons for each other. The temperature dependence is due to the Fermi level, in material outside the space charge cylinders, moving down toward the intrinsic (roughly mid-gap) level. The other approximations gave values for f falling generally between the Fermi and minimum energy values (Fig. 5.8b).

For a particular specimen N_D can be determined before the dislocations are introduced and an average value for the dangling bond spacing along the dislocations, c, can be used. The Fermi level can be calculated from the Hall coefficient in the starting material so Equation (5.6) can be solved for f as a function of T (E_F varies with T in the material unaffected by dislocations) for a range of assumed positions of the dislocation trap level E_d. By fitting the experimental curves to the computed ones E_d can be determined.

5.2.2 Experimental tests of the Read theory

To relate the theory to the electrical transport properties of deformed Ge, Read (1955) treated the scattering of electrons by an array of parallel dislocations such as might be introduced by plastic bending. He assumed the electrons to make elastic collisions with impenetrable space charge cylinders round the dislocations.

The component of carrier velocity parallel to the dislocation lines would be unchanged by such collisions. The conductivity parallel to the dislocations, $\sigma_{\parallel}^*$, is affected only by the reduction in the concentration of conduction electrons, n, due to trapping by the acceptor states. Let $\langle n \rangle$ be the average electron density in the deformed material. Then $\langle n \rangle = n(1 - \varepsilon)$ where $\varepsilon = \rho \pi r_d^2$ is the fraction of the volume occupied by the space charge cylinders. Since:

$$\sigma_{\parallel}^* / \sigma_{\parallel} = \langle n \rangle / n \tag{5.7}$$

measurements of the conductivity parallel to the dislocations in deformed ($\sigma_{\parallel}^*$) and undeformed ($\sigma_{\parallel}$) control samples (to obtain n) in principle give $\langle n \rangle$ and $n - \langle n \rangle =$ the number of dislocation-trapped electrons. Hence if ρ and c (the spacing of the dangling bonds) are known, f can be found since the total number of trapped electrons is also given by:

$$n\varepsilon = n - \langle n \rangle = f\rho/c \tag{5.8}$$

For current flow normal to the dislocations other factors reduce the conductivity in addition to the reduction in carrier density. Suppose first that the mean free path of the carriers is much less than the dislocation spacing. The dislocations will then have a negligible effect on the carrier mean free path and mobility. They still produce an additional reduction in conductivity because the electrons cannot drift in straight lines, i.e. the dislocation cylinders distort the current flow lines. Read (1955) expressed the reduced conductivity in the perpendicular flow case, $\sigma_{\perp}^*$, as:

$$\sigma_{\perp}^* / \sigma_{\perp} = g\langle n \rangle / n \tag{5.9}$$

where g is a distortion factor given by the ratio of the average electric field in the remaining normal n-type material to the average field in the whole specimen. In addition, dislocation scattering may reduce the mean free path and time and so also reduce the free carrier mobility μ. Explicit expressions were obtained by Read for all these effects.

5.2.3 The effect of charged dislocations on the Hall coefficient

Read considered the influence of dislocations on the Hall coefficient using his model. He obtained the expression:

$$R_{\perp} = \mu_H / \mu \langle n \rangle q \tag{5.10}$$

so the reduced average carrier concentration $\langle n \rangle$ can be obtained directly from R, the Hall coefficient for current flow and magnetic field perpendicular to the dislocations.

5.2.4 Experimental verification of the Read theory

The geometry of plastic bending of crystals is shown in Fig. 5.9. The bending is accommodated by introducing an excess of edge dislocations of one sign over those of the opposite sign, as shown in Fig. 5.9b, to a density:

$$\rho = 1/(r\, b \cos \omega) \qquad (5.11)$$

where r is the radius of curvature to which the crystal was plastically bent and ω is the angle between the slip plane and the neutral plane (the middle plane in the crystal which does not change its width).

In principle, after thus introducing a known density of parallel dislocations, ρ_{maj}, into a specimen of initial carrier density $n = N_D\, \mathrm{cm}^{-3}$, the average carrier concentration $\langle n \rangle$ can be obtained from the Hall coefficient using Equation (5.10). The fraction of occupied sites, f, can then be obtained from Equation (5.8) by assuming a reasonable value for the average spacing of dangling bond sites, c. By fitting this experimental value of f and its temperature dependence to the theoretical curves given by Read's approximations to the occupation statistics (Fig. 5.8), the dislocation energy level E_d can be obtained.

The first experimental work to attempt this was that of Logan *et al.* (1959). The Read treatment contains a number of unmeasurable parameters so they recast it slightly as follows. They confirmed Equation (5.10) for $R_\perp$ and the similar expression for the Hall coefficient for current flow parallel to the dislocation array.

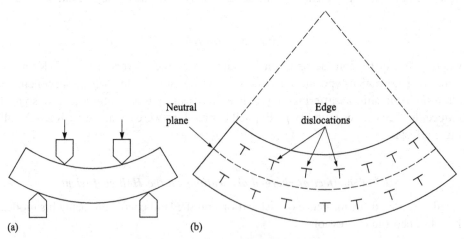

Figure 5.9   (a) Four-point plastic bending, as shown here, produces a more constant radius of curvature and therefore a more uniform dislocation density than does three-point bending (using one descending knife edge). (b) Plastic bending results in the introduction of a calculable (excess) density of edge dislocations all of the same sign.

Previous work had shown that $\mu_{\mathrm{H}}/\mu \cong 1$ so they wrote $R_{\perp} = R_{\parallel} = 1/\langle n \rangle q$. Similarly the Hall coefficient in undeformed material is $R_{\mathrm{o}} = 1/nq$, so:

$$R_{\perp}/R_{\mathrm{o}} = R_{\parallel}/R_{\mathrm{o}} = n/\langle n \rangle = 1/(1-\varepsilon) \tag{5.12}$$

(This relation has been derived theoretically in various ways.)

Logan *et al.* (1959) measured the Hall coefficients of control specimens of n-type Ge and others deformed by four-point bending (Fig. 5.10) at temperatures from $4\,\mathrm{K}$ to $300\,\mathrm{K}$ to obtain values of ε. In Equation (5.8) neither ρ nor c could be accurately determined so it was rewritten in the form:

$$\varepsilon(T) = \lambda f(T) \tag{5.13}$$

where:

$$\lambda = \rho/cN_{\mathrm{D}} \tag{5.14}$$

was treated as an adjustable parameter. They obtained a value of λ by fitting measured curves of ε to theoretical curves of f for various values of E_{d}. Their results gave values of $E_{\mathrm{c}} - E_{\mathrm{d}} = 0.179$ to $0.225\,\mathrm{eV}$. From the value of λ obtained by curve fitting, values of ρ were found from Equation (5.14). These were about eight times greater than the value obtained from the radius of curvature via Equation (5.11), which was in good agreement with the value obtained from etch pit counts.

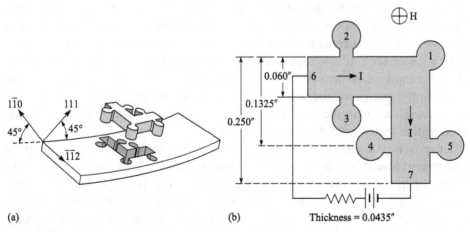

(a)   (b)   Thickness = 0.0435"

Figure 5.10   (a) Samples of n-type Ge with the orientation shown here were bent about the [112] axis to introduce arrays of dislocations aligned in this direction. (b) Specimens for electrical measurements were then cut out in the form shown. The dislocation alignment is indicated by the fine lines and measurement circuit used is given schematically. (After Logan *et al.* 1959; Reprinted with permission from *Journal of Applied Physics*, **30**, pp. 885–95. Copyright 1959, American Institute of Physics.)

5.2.5 The Read theory: conclusions

The result of the work of Logan *et al.* proved to be characteristic. A large body of careful work on Ge (Broudy 1963) and InSb (Bell *et al.* 1966, Bell and Willoughby 1966, 1970) also resulted in dislocation densities calculated from electrical measurements that were nearly an order of magnitude larger than those found by etch pit counting. The starting point of the Read model, the Shockley model of the core of an undissociated shuffle set 60° dislocation core, first suggested the existence of dangling bond deep trap states. However, these states were found by ESR to number only a few percent of those in the undissociated Shockley shuffle set model (Section 5.1.4). It is believed that fundamental bond rearrangements occur in the core to minimize the number of dangling bonds per unit length (Hornstra 1958). It is now generally agreed that dangling bonds only occur at occasional 'special sites' along dislocation lines (Hirsch 1981, 1985).

A second drawback of the theory is that the value of the dislocation acceptor level energy is very sensitive to the occupation statistics chosen. Subsequently different workers interpreting their experimental results in terms of Read's model were able to deduce positions for the dislocation levels in Ge and in Si distributed from near the bottom to near the top of the forbidden gap in each material. Due to the repulsion of trapped electrons for each other, only a few percent of the sites in the Read model would have been expected to be occupied (see Section 5.2.1). Hence the lower number of sites actually available for occupation makes little difference to the charge on the electron or the size of the space charge cylinder. The relatively good agreement between theory and experiment on the numbers of dislocations required to remove all the electrons and turn lightly doped n-type Ge intrinsic or p-type can thus be understood. It now appears that the difference between the experimental and Read-theory-deduced densities of dislocations in deformed semiconductors were due to impurities and/or the additional contribution made by large densities of deep trap states associated with the point defect debris produced by the dislocations during deformation (Section 5.1.4 and Section 5.9.1).

Subsequently a great deal of work was done by academic research groups in Germany (Labusch and Schröter 1978, Schröter *et al.* 1980, Alexander 1986, 1991). These led to two modifications to the Read theory. Firstly, in typical semiconductors at room temperature all donors and acceptors are permanently ionized. Consequently the screening of line charges in Ge, InSb and Si involves mobile charge carriers as well. Secondly, as originally suggested by Shockley, the dangling bond wave functions are now believed to produce narrow energy bands rather than sharp levels in the forbidden energy gap. The charge on the dislocation then depends on the position of the Fermi level in relation to the dislocation band or bands. Zero dislocation line charge corresponds to one electron in each bond. The empty levels in the dislocation band act as acceptors when the Fermi level lies higher, and the filled levels act as donors when it lies lower. Both positive and negative line charges are possible and the effects actually found in both *p*- and *n*-type material can be accounted for. This later, modified, band model is more acceptable theoretically and it yields better agreement with experiment over a wider range of materials including *p*-type Ge and over a wider range of phenomena than the earlier theory (Labusch and Schröter 1978).

However, Read's theory is the simplest first approximation treatment. It accounts for the type of effect produced by deformation and its temperature dependence and it (just) gives order of magnitude agreement with experiment. Moreover Read's basic model of a charged dislocation line and large (~1 μm) screening space charge cylinder is the basis for all subsequent developments in the theory of the electrical properties of dislocations and other defects as we discuss in Section 5.2.6.

The large electrical effects of plastic deformation were characteristic for early material when care was taken to minimize segregation of impurities to the defects. In material subjected to normal schedules of device processing the dislocations are often largely electrically neutralized (their energy states are compensated presumably by point defect decoration of the dislocation core). It is now clear that dislocations in the purest Si are electrically inactive and only produce enhanced local carrier recombination when they are deliberately contaminated with Cu, Fe and other 'heavy' metals (see Section 5.5.8 and Section 5.6.4). In Si, especially, they can be deliberately passivated (Section 5.5.8), i.e. made rendered electrically inactive by H or Cl decoration. This is a major reason that the industry has little interest in dislocations today, although a low dislocation density is used as a figure of merit to show that starting material is of good quality (densities in commercial slices are from 0 to 10^4cm^{-2}).

5.2.6 Later developments

The ESR evidence is that unpaired spins, corresponding to single electrons in dangling bonds occur in deformed Si but in numbers that are about two orders of magnitude less than expected on the undissociated shuffle set model (Fig. 5.7) for the density of dislocations measured to be present (see Section 5.1.4, Section 5.5.7 and Section 5.9). Hence dislocation dissociation and, as expected, bond rearrangements in the core as suggested by Hornstra (1958) minimize the number of dangling bonds per unit length. Thus, it is generally agreed that dangling bonds only occur at occasional 'special sites' along dislocation lines (Hirsch 1981, 1985, Schröter and Cerva 2002, Cai *et al.* 2004). One hypothesis is that these correspond to kinks in partial dislocations. (Jogs are short lengths of dislocation line up or down out of the slip plane, to a neighbouring slip plane. Kinks are short lengths in the slip plane roughly normal to the main dislocation line direction. Kinks are essential for movement from one Peierls potential trough to another (Section 2.5.2) as indicated in Fig. 2.24.)

It was originally puzzling that d.c. conduction along dislocation lines was not found. Dislocation energy bands exist but d.c. effects are not observed because the lengths along which there are continuous bands are short. High frequency a.c. conduction, back and forth along line segments, is observed at low temperatures (Grazhulis *et al.* 1977a,b, Osipiyan 1981, 1983, Kveder *et al.* 1985). The evidence is that there is conduction along dislocation segments of lengths corresponding to a few to about 20 dangling bonds in communication. Changes of line direction, kinks, jogs, decorating impurity atoms and nodes joining dislocations into networks could be the causes of the breaks between segments (Brohl and Alexander 1989).

It is well known that dislocations act as effective scatterers, thus affecting carrier mobilities in semiconductors and degrading their transport characteristics. This scattering is treated by using the Read model of charged dislocations and space-charge cylinders. The topic was reviewed by Jaszek (2001), who offered an updated critical analysis of carrier scattering by dislocations in semiconductors, as well as by misfit dislocations in heterostructures (see also Section 5.8.2). He outlined the different models for carrier scattering in semiconductors and discussed many of the issues. These included (i) the incomplete understanding of the dislocation core structure, (ii) the nature of dislocation energy levels, (iii) screening and strain fields, (iv) the validity of the relaxation-time approximation for the anisotropic scattering potentials related to dislocations, and (v) the equation (Pödör 1966) relating the mobility perpendicular to the dislocation lines, $\mu_\perp$, to the dislocation density N and line charge $Q = qf/b$.

The Read model of a charged dislocation surrounded by a compensating cylindrical space charge region giving rise to band bending was also successfully used to account for dislocation photoconductivity (Section 5.3.1) and in the charge-controlled theory of dislocation EBIC dark contrast (Section 5.5.2).

5.2.7 The electrical effects of polar dislocations

Polar dislocations can occur in semiconducting compounds with the wurtzite and sphalerite structures (Section 4.1.3). This raises the possibility of observing differences in electrical properties between dislocations of differing polarity and using this information to throw light on the core structures of these defects and on the mechanisms determining their line charges and recombination behaviour. Such studies were carried out on the II-VI compounds which are more ionically bonded than are the III-Vs.

The charge on dislocations in II-VI compounds

The large line charge, q per unit length, of moving dislocations constitutes a significant current during plastic flow as was first shown by Osipiyan and Petrenko (1975) (for reviews of the later studies see Osipiyan *et al.* 1986, Osipiyan 1989). In II-VI compounds, one type of dislocation, e.g. Se(g) in ZnSe, will be much more mobile than the dislocations of the opposite polarity, e.g. Zn(g), so the current results from the motion of the mobile type only (see Fig. 4.71). It is found that the dislocation current is directly proportional to the plastic strain rate, $\dot{\varepsilon}_p$.

It was seen in Section 4.3 that the dislocation current I_d, reaching an electrometer in a measurement like that shown in Fig. 4.73, could be written in terms of the line charge q as (Equation 4.12):

$$I_d = \frac{dQ}{dt} = q\frac{d}{b_z}\frac{dh_P}{dt} \tag{5.15}$$

where Q is the charge reaching the electrometer contact, enabling q to be determined.

If there are significant densities of screening charge carriers, these can move with the dislocations and reduce the effective value of q. This is the case in lower resistivity samples of II-VI compounds as well as, generally, the III-V and elemental semiconductors. The possibility of applying this method to the II-VI compounds is also partly a consequence of the fact that, uniquely, they deform plastically at room temperature and below, unlike the III-V semiconductors and Si and Ge, which are plastic only at relatively high temperatures. Thus this method is applicable only to relatively high resistivity, e.g. semi-insulating, II-VI compounds.

Petrenko and Whitworth (1980, see Section 4.3) studied the seven types of II-VI compound crystals listed in Table 5.2.

Petrenko and Whitworth (1980) used the method of Fig. 4.73 to find both the sign of dislocation charge in each of the cases listed in Table 5.3. In the *n*-type materials, ZnO, ZnS, ZnSe, CdS, CdSe and CdTe all the dislocations were negatively charged, regardless of polarity, while in the one *p*-type material available (ZnTe), they were positively charged. Hence, they concluded, the charge on dislocations, even in the more ionic II-VIs, is not determined by the charges of the ions in the core but by electron trapping. This is further supported by experimental and theoretical studies of the anomalously large line charges on moving dislocations in II-VI semiconductors. The accepted model to account for these large charges is based on the idea that

Table 5.2. *The II-VI crystals studied by Petrenko and Whitworth*

Material	Structure	Conductivity type (undoped, as-grown)	Slip plane	Equally favoured slip directions
ZnS	Sphalerite	*n*-type	(111)	[10$\bar{1}$] and [01$\bar{1}$]
ZnSe	"	"	"	"
ZnTe	"	*p*-type	"	"
ZnO	Wurtzite	*n*-type	(0001)	[21$\bar{1}$0] and [11$\bar{2}$0]
CdS	"	"	"	"
CdSe	"	"	"	"
CdTe	"	"	"	"

Table 5.3. *Typical charges on mobile dislocations in II-VI compounds for a strain rate of $10^{-5} s^{-1}$, 300° K. Here q is the charge per unit length on a dislocation line, b is the Burgers vector and e is the charge on the electron. From Osipiyan* et al. *(1986) (data from Petrenko and Whitworth 1980, Zaretskii* et al. *1977 and Petrenko 1982)*

Conductivity type and material	Structure	Dislocation type	Mean charge in the dark q_o $10^{-10} cm^{-1}$	Mean charge in the dark (qb/e)	Maximum charge in light q_1 $10^{-10} cm^{-1}$	(qb/e)
n-ZnSe	sphalerite	60° unit Se(g)	−2.0	0.50	−3.7	0.92
p-ZnTe	sphalerite	60° unit Te(g)	+1.6	−0.42	+2.9	−0.78
n-CdTe	sphalerite	60° unit Te(g)	−0.42	0.12	-	-
n-ZnO	wurtzite	60° unit Zn(g)	−2.7	0.54	-	-
n-CdSe	wurtzite	60° unit Cd(g)	−0.63	0.17	−1.3	0.35
n-CdS	wurtzite	60° unit Cd(g)	−1.4	−0.36	−2.1	0.54
		60° unit S(g)	−2.7	0.70	-	-

broken bonds in the dislocation cores are filled with electrons from point centres swept through during plastic deformation.

In ZnS the Osipiyan group concluded that the dislocations involved in slip were mainly partials that removed the lamellae of wurtzite structure material as they moved. We have therefore not included data for this material here. For a discussion of this complication in interpreting the data for ZnS, see Petrenko and Whitworth (1980).

The dislocation charge, under illumination with light of the frequency that produced the greatest change in that charge (0.9 of that corresponding to the band gap energy, i.e. $hv = E_g$) was also determined with the results listed in Table 5.3.

The charge on the dislocations increased linearly with the band gap as shown in Fig. 5.11.

The application of an external electric field to a II-VI crystal exerts a force on the charge dislocations producing the electroplastic effects, discussed in Section 4.3.4.

5.3 Recombination at dislocations

As noted in Section 5.1, dislocations can introduce electronic states in the energy gap of a material due to dangling bonds, to their strain field via the deformation potential, or to point defects like interstitials, vacancies and impurities in their cores. These states affect electronic processes in semiconductors including recombination which is important in light-emitting devices. Here dislocations are typically detrimental as they usually act as non-radiative recombination centres competing

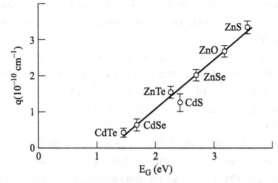

Figure 5.11 Charge on dislocations in II-VI compounds moving in the dark plotted against the forbidden energy gap. The charge on moving dislocations in *p*-type ZnTe is positive, for all the other II-VI compounds, which are *n*-type it is negative. [After Petrenko and Whitworth 1980. Reprinted with permission from *Philosophical Magazine*, **41** (1980), pp. 681–99; Taylor & Francis Ltd., the Journal's web site: http://www.tandf.co.uk/journals.]

with the desired radiative recombination channels. (There are proposals for employing dislocation-related luminescence in Si light-emitting diodes; this is discussed in Section 5.12.2). In addition to defects intrinsic to dislocations, like dangling bonds and kinks, dislocations are also likely to getter impurities and native point defects (attract and remove them from solution in the crystal). This often renders the impurities electrically ineffective (see Section 5.12.1). Studying recombination at dislocations elucidates their electronic properties.

Typically, dangling bonds at dislocations collect majority carriers so dislocations become electrically charged. Consequently, a cylindrical space charge (the Read cylinder) of opposite sign is formed (see Section 5.2). Charged dislocations attract minority carriers, and can significantly reduce the minority carrier lifetime.

The role of the impurities that often decorate dislocations is important and difficult to determine. However Section 5.5.8 and Section 5.6.4 presented conclusive evidence of the importance of transition metal atoms at dislocations in silicon. This started with material, proved free of even trace contamination, and added minute amounts of a number of metallic impurities. The dislocations only exhibited SEM EBIC and CL contrast due to enhanced recombination when decorated with transition metal atoms. Other semiconductors are not available so pure as Si, however. For them, it is not so easy to determine to what extent recombination is intrinsic (involving dislocation states alone), or is extrinsic (due to impurity decoration). However, it is well established that most electronic and optical properties of dislocations depend on the interaction of point defect and dislocation states (e.g. Schröter and Cerva 2002). Kveder *et al.* (2001) presented a quantitative model of minority carrier recombination at decorated dislocations. They assumed that strain-field-induced shallow levels, and deep levels induced by impurity atoms or by dislocation-core defects, could exchange carriers. Thus, impurity atoms in the dislocation core can significantly increase carrier recombination at the dislocation.

A study of the effect of copper contamination on misfit dislocation recombination in SiGe/Si heterostructures led to the conclusion that light contamination (about 1 part per billion) results only in shallow trap levels at the dislocation core, whereas greater contamination (about 15 parts per billion) produces deep states (Kittler *et al.* 1995).

The Read theory (see Section 5.2) employed an equilibrium dislocation line charge. The dislocation energy levels act as generation/recombination centres. Both electrons and holes are captured and emitted giving a dynamic equilibrium charge. Dislocations therefore contribute to generation-recombination noise in devices (see Section 5.1.2). In early studies, Gippius and Vavilov (1963, 1965a and b, Gippius *et al.* 1965) observed the infrared electroluminescence (recombination radiation) spectrum of dislocations in Ge by comparing the emission spectra of deformed and undeformed specimens into which carriers were injected. This led to a good deal of Russian theoretical work on dislocation exciton recombination, which is believed to be responsible for the observed emission (see Section 5.6.3).

Enhanced recombination at dislocations also produces dark contrast in the scanning electron microscope (SEM) electron beam induced current (EBIC) and cathodoluminescence (CL) images. This is an active research topic at present.

Here we will discuss the effects of dislocations on photoconductivity and on luminescence in semiconductors which are important for both characterization and applications. Recombination-enhanced dislocation motion is also outlined in this section.

5.3.1 *Dislocations and photoconductivity*

The conductivity of semiconductors increases when illuminated by light of photon energies $hv > E_g$. This is called photoconductivity. Such photons, when absorbed, promote electrons into the conduction band, generating hole-electron pairs. This increases the charge carrier density and so the conductivity. When the light is cut off the carriers gradually disappear by recombination. The rate of recombination is increased by dislocations and this was an early dislocation research topic. Extensive experimental work was carried out by a Polish group and the phenomenon was treated by an extension of the Read model (for a review see Figielski 1978).

Both generation and recombination occur at dislocations. In *n*-type material, for example, there will be a flux of electrons into the dislocation deep acceptor states and another of electrons thermally activated back out to the conduction band as well as a flux of holes captured to recombine with electrons via the dislocation centre (Fig. 5.12).

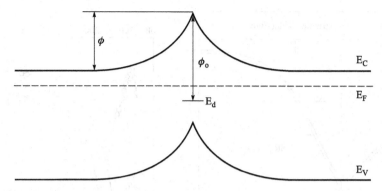

Figure 5.12 Energy band structure and potential barriers of a negatively charged dislocation in *n*-type material (After Wilshaw and Booker 1987).

The net rate of electron capture per unit volume and time can then be written (Figielski 1978) as the sum of the fluxes into and out of the dislocation states, i.e.

$$R_e = C_e \, \rho \, [n \exp(-q\phi/kT) - n_o \exp(-q\phi_o/kT)] \tag{5.16}$$

where C_e is the probability of an electron transition between the dislocation energy level and the conduction band, ρ is the dislocation density (total length of dislocation lines in unit volume), n and n_o are the carrier densities respectively far from and near the dislocation and ϕ and ϕ_o are the potential barriers against electrons respectively entering and leaving the dislocation states from the conduction band (Fig. 5.12). A similar expression can be written for the net rate of capture of holes by the dislocations. Hence expressions were obtained for the excess majority carrier density n in terms of the free (Δp) and dislocation-captured (Δm) densities of generated holes:

$$\Delta n = \Delta p + \Delta m \tag{5.17}$$

By applying this approach to extrinsic (dislocation) photoconductivity Figielski showed that the temperature dependence of the excess carrier density n should be of the form of Fig. 5.13a, following, in different temperature ranges, either the density of excess free minority carriers (holes, i.e. Δp) or the volume density of holes captured by the dislocations (Δm) whichever is the larger. The relative excess (photo)conductivity $\Delta\sigma/\sigma$ depends on Δn and so should have the same form of temperature dependence. This was found experimentally to be so (Fig. 5.13b). Similarly, applying this model to the decay of photoconductivity predicted a universal decay curve of Δn versus time (Fig. 5.14). This also was borne out by experiments.

Unlike the results of measurements of the effects of dislocations on conductivity and Hall mobility, measurements of extrinsic photoconductivity in different laboratories and on samples prepared in different ways are in good agreement.

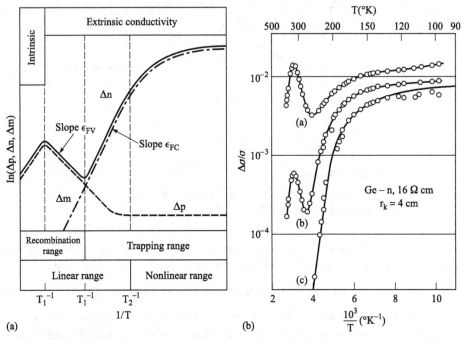

Figure 5.13   (a) Predicted form of dependence of excess charge carrier concentration on reciprocal temperature for the recombination potential barrier model of Fig. 5.12. Δn and Δp are the excess (photon generated) free electron and hole densities and Δm is the number of holes per unit volume captured by the dislocations. (b) The measured temperature dependence of the photoconductivity in plastically bent Ge at different rates of generation of electron-hole pairs (different intensities of illumination): $G = $ (a) 10^{17} (b) 5×10^{15} and (c) $10^{14}\,cm^{-3}\,s^{-1}$. For the lowest injection rate, the photoconductivity, i.e. Δn, follows the curve for Δm at all temperatures. (After Figielski 1978. Reprinted from *Solid State Electronics*, **21**, Recombination at dislocations, pp. 1403–12, Copyright 1978, with permission from Elsevier.)

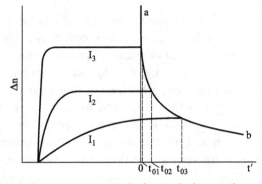

Figure 5.14   Universal decay curve a to b for majority carriers after cutting off the illumination. The photoconductivity time curves are plotted for increasing intensities of illumination ($I_1 < I_2 < I_3$). (After Figielski 1978. Reprinted from *Solid State Electronics*, **21**, Recombination at dislocations, pp. 1403–12, Copyright 1978, with permission from Elsevier.)

Photoconductivity measurements along dislocations and from the dislocation core to the bulk were reviewed by Labusch (1997). The core of a straight dislocation was considered to behave as a one-dimensional quantum system with quasi-metallic properties for the case of one-dimensional bands partially filled with electrons or holes (Labusch 1997). Conduction along dislocation cores was found for 60°-dislocations in Ge and screw dislocations in CdS (see Labusch 1997 and references cited therein).

5.3.2 Dislocations and luminescence

Dislocations typically reduce luminescence as they usually act as non-radiative recombination channels. Recently, however, efforts were made to employ radiative recombination due to dislocations, which is observed e.g. in Si and Ge, in light-emitting devices (see Section 5.12.2). Dislocation luminescence in Si is of special interest, since it produces room-temperature electroluminescence in the important optical fibre communications wavelength range from 1.1 to 1.6 microns. Prominent luminescence bands associated with dislocations in Si were labelled D1, D2, D3 and D4 (Drozdov *et al.* 1976). Presently, these D-bands are ascribed (somewhat debatably) in the literature as follows: D1 (0.81 eV) and D2 (0.87 eV) are associated with jogs, kinks and stacking faults in dislocations, self-interstitial and metal contamination, as well as with oxygen precipitates at the dislocation; D4 (1.0 eV) is related to transitions involving the shallow one-dimensional dislocation bands (due to the dislocation strain field) and D3 (0.94 eV) is attributed to its phonon replica (for a review, see e.g. Higgs *et al.* 2000). The intensities of these bands depend on the concentration of impurities present in the material, and researchers report differing results. This is not surprising because of the difficulty of identifying the optically active centres and understanding the influence of impurities and point defects. Jones *et al.* (2000) suggested that the formation of multi-vacancy and multi-interstitial defects in dislocation cores could cause the D-band photoluminescence associated with dislocations in Si and SiGe.

Despite dislocation densities of the order of $10^{10}\,\mathrm{cm}^{-2}$, practical GaN-based optoelectronic devices were successfully developed (see Section 5.1). Nevertheless, it is important to determine the role of defects (dislocations, intrinsic point defects, and impurities) in the luminescence of GaN. The exceptional behaviour of GaN-based devices containing high dislocation densities was related to the fact that dislocations in GaN cluster in the walls of cellular structures so relatively large regions of the material are dislocation free. The diffusion length of injected carriers is relatively small, so most carriers recombine radiatively before reaching the dislocations (see Section 4.6.7). Another major controversy in the luminescence of both bulk and epitaxial GaN concerns the defects responsible for yellow luminescence (YL). This is seen in PL spectra as a broad luminescence band around 2.3 eV (see, e.g. the III-nitrides review by Jain *et al.* 2000). This has been attributed to intrinsic and extrinsic point defects such as Ga and N vacancies, Ga interstitials, N antisites and carbon atoms, as well as to extended defects inside grains and at low-angle grain boundaries. The YL band is also controversially attributed to Ga vacancy complexes. Elsner *et al.* (1998) showed that in GaN these defects are especially stable at the core of the dislocation and that their formation is more likely at threading-edge dislocations.

Cathodoluminescence (CL) microscopy, especially in a STEM or TEM, provides a powerful means for investigating radiative and non-radiative recombination at individual dislocations (e.g. Petroff *et al.* 1977, 1980a and Pennycook *et al.* 1980 employing STEM; and Booker *et al.* 1979 and Myhajlenko *et al.* 1984 using TEM). From these, and numerous later investigations, it appears that both dislocation core and the Cottrell atmospheres of impurities and/or native point defects around dislocations are needed to explain all cases of dislocation recombination.

Another useful microscopic technique for the study of recombination mechanisms is time-resolved CL. In this, the decay time is measured by using an electron-beam blanking system and a fast detector. Generally, the luminescence intensity decays exponentially as:

$$L(t) = L(0)\exp\left(-t/\tau\right) \tag{5.18}$$

where $L(t)$ is the light intensity at a time t after the electron beam was cut off and $L(0)$ is the intensity under electron bombardment. The measured values of τ are effective lifetimes $1/\tau_{eff} = 1/\tau_{bulk} + 1/\tau_{surf}$ where τ_{bulk} and τ_{surf} are the bulk and surface recombination times, respectively. The minority carrier lifetime measured near a dislocation depends, in addition, on the recombination time near the dislocation, τ_{disl}, as:

$$1/\tau_{eff} = 1/\tau_{bulk} + 1/\tau_{surf} + 1/\tau_{disl} \tag{5.19}$$

Fig. 5.15 presents the minority carrier lifetime at single dislocations in GaP measured by Davidson and Rasul (1977) and Rasul and Davidson (1977). It shows CL decays obtained at the dislocation (using electron beam voltage, $V_b = 25\,\text{kV}$) and away from it (using V_b of 15 and 25 kV).

5.3.3 Recombination-enhanced dislocation motion

Radiation-enhanced dislocation glide (REDG), or recombination-enhanced dislocation motion (REDM) is dislocation glide in a semiconductor enhanced by carrier injection. Carriers can be injected due to laser light excitation, electron-beam irradiation, or current injection (for reviews, see Maeda and Takeuchi 1996, and Vanderschaeve *et al.* 2001). This process can play an important role in the degradation of semiconductor devices.

The main degradation modes of optoelectronic devices are rapid- or gradual-degradation, and catastrophic failure. Recombination-enhanced dislocation climb and glide are responsible for rapid degradation (e.g. Ueda 1999).

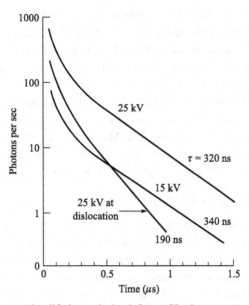

Figure 5.15 Minority carrier lifetimes derived from CL decay rates in GaP away from a dislocation (for electron beam voltages of 15 and 25 kV) and at a dislocation (for a beam voltage of 25 kV). (After Rasul and Davidson 1977.)

The influence of irradiation on the properties of defects is generally referred to as the photoplastic effect for light illumination and the cathodoplastic effect for electron-beam irradiation (for reviews, see e.g. Osipyan *et al.* 1986, Maeda and Takeuchi 1996). These effects in wide energy gap semiconductors are related to dislocation-induced electronic states involved in electronic processes in a material.

Dislocation glide in semiconductors (at low to medium temperatures) is limited by lattice friction through the Peierls mechanism. In this, dislocation glide occurs through kink pair formation on the dislocation line and transport of the kinks along the line leading to a continuous motion (see Section 2.5.2). Such thermally activated dislocation motion is observed in deformation experiments in TEMs, which allow the determination of the activation parameters (see Table 5.4 below).

Recombination-enhanced dislocation motion has been attributed to lattice friction reduction under excitation. The non-radiative recombination of electron-hole pairs at dislocations reduces the activation energy for kink formation.

The effects of parameters like temperature, dislocation character and electron beam intensity on the dislocation mobility were examined during TEM *in situ* deformation to elucidate the role of electronic excitation on dislocation motion (e.g. Vanderschaeve *et al.* 2001). Other methods, such as cathodoluminescence scanning electron microscopy (CL-SEM) in an instrument equipped with a bending apparatus can also be used (Maeda *et al.* 1983). The usefulness of CL-SEM was also demonstrated by Schreiber and Vasnyov (2004) and Vasnyov *et al.* (2004).

They employed CL video recording, computer-aided acquisition and fast processing of image data to investigate dislocation dynamics and recombination in several semiconductors. Their data indicated that the dynamic and recombination properties depend on the dislocation structure and the specific material studied (Schreiber and Vasnyov 2004, Vasnyov *et al.* 2004).

Maeda *et al.* (1983) studied dislocations in *n*-GaAs single crystals. They found that, above a critical temperature, T_c, electron irradiation did not affect dislocation velocity, whereas below T_c it enhanced it. Measurements of the dislocation velocity as a function of irradiation intensity I and stress τ, led Maeda *et al.* (1983) and Maeda and Takeuchi (1996) to express the dislocation velocity v as:

$$v = v_d^0(\tau)\exp\left[-(E_d/kT)\right] + v_i^*(\tau)\,I\exp[-(E_i/kT)] \qquad (5.20)$$

where E_d is the activation energy without excitation (i.e. beam intensity $I=0$), $E_i = E_d - \Delta E$ is the activation energy under excitation, τ is the applied shear stress and $v_d^0(\tau)$ and $v_i^*(\tau)$ are pre-exponential factors which vary approximately linearly with τ. Thus, the velocity in the presence of excitation is expressed as the sum of a thermal activation term and an enhancement term with a reduced activation energy and a pre-exponential factor proportional to the excitation intensity I (Maeda *et al.* 1983, Maeda and Takeuchi 1996). These observations are illustrated in Fig. 5.16. Above a critical temperature T_c, the first term dominates and no effect of excitation is observed. At low temperature the second term describes the dislocation motion. This is recombination enhanced defect motion (REDM) (Maeda *et al.* 1983).

To summarize, numerous observations of radiation-enhanced dislocation glide in several semiconductors reveal that (e.g. Maeda *et al.* 2000, Vanderschaeve *et al.* 2001): (i) velocity enhancement is observed below a critical temperature T_c, (ii) the effect of excitation is reversible, (iii) the activation energy under excitation is independent of the irradiation intensity, (iv) the dislocation velocity obeys an Arrhenius-type temperature dependence with the pre-exponential factor and the activation energy being reduced under excitation, and (v) the pre-exponential factor

Table 5.4. *Parameters from Equation (5.20) derived from in situ deformation measurements, describing electron-irradiation-enhanced dislocation glide in four semiconductors (see Vanderschaeve* et al. *2001, and references cited therein)*

Material (dislocation type)	E_d (eV)	E_i (eV)	ΔE (eV)	E_g (eV)	Reference
Ge (60°)	1.7	0.7	1.0	0.67	Yonenaga *et al.* 1999
Si (60°)	2.2	1.6	0.6	1.12	Werner *et al.* 1995
GaAs (α)	1.0	0.3	0.7	1.42	Maeda *et al.* 1983, 1996
GaAs (β, screw)	1.7	0.6	1.1	1.42	Maeda *et al.* 1983, 1996
ZnS (α, β, screw)	1.2	0.3	0.9	3.68	Levade *et al.* 1994

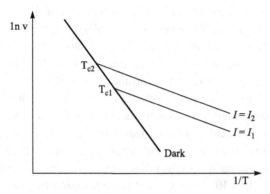

Figure 5.16 Illustration of radiation enhanced dislocation glide according to Equation (5.20). The irradiation reduces the activation energy by a magnitude that is independent of the irradiation intensity. The pre-exponential factor is proportional to the excitation intensity I. The critical temperatures for two levels of irradiation intensity are also indicated. (Adapted from Maeda and Takeuchi 1996).

increases approximately linearly with the excitation power at low intensities but saturates at high intensities. Saturation at high beam intensities can be expected due to the saturation of the recombination centres at high rates of electron-hole pair generation (e.g. Vanderschaeve *et al.* 2001).

Mechanisms that may be involved in this phenomenon were outlined by Maeda and Takeuchi (1996). The most plausible, the phonon-kick mechanism of defect migration, is related to the emission of phonons (due to non-radiative electron-hole recombination at dislocations sites), which contribute to the activation of dislocation motion through a modification of the Peierls mechanism (Maeda and Takeuchi 1996). Other possible mechanisms include (for details and literature references see Maeda and Takeuchi 1996): (i) charge state mechanisms (the defect migration energy is dependent on the Fermi level), which appear to be incompatible with radiation-enhanced glide, (ii) the excited state mechanism (a defect, having an electron excited to upper orbital, feels a reduced migration barrier, as compared to one in the ground state), (iii) a phonon softening mechanism, proposed by Van Vechten *et al.* (high densities of electron-hole pairs induce lattice softening that leads to a non-thermal atomic rearrangement; see Maeda and Takeuchi 1996 and references cited therein).

From *in situ* TEM studies of the effect of electron beam irradiation on dislocation motion in several semiconductors exhibiting enhanced dislocation glide, the reduction in activation energies can be derived (see Table 5.4).

ΔE is in all cases (but Ge) lower than the energy gap E_g (Table 5.4). Yonenaga *et al.* (1999) found recombination enhanced defect motion (REDM) during *in-situ* TEM experiments on dislocation motion in Ge. They attributed the reduction in activation energy for dislocation motion to recombination-assisted kink formation. However, the reduction in activation energy of about 1 eV is greater than the energy band gap, which cannot be explained by recombination of thermalized carriers only. As a plausible explanation, Yonenaga *et al.* (1999) suggested that the high-energy electrons of the incident beam generate hot carriers (with energies greater than thermal). These may interact with dislocations prior to thermalization and release energies above E_g, thus leading to the formation of kinks.

The relatively long operating life of GaN-based devices containing high dislocation densities is mainly related to their low dislocation mobility. This is due to the much greater critical resolved shear stress required for dislocation motion (caused by internal misfit strain, thermal strain, and external mechanical strain) and the small radiation enhancement effect (Sugiura 1997). Consequently, the dislocation glide velocity is less than 10^{-23} cm/s, and the degradation by dislocation glide is very slow. This contrasts with GaAlAs/GaAs light-emitting devices. These exhibit a strong recombination enhancement effect and relatively fast dislocation glide (of the order of 10^{-4} cm/s at room temperature), so they degrade rapidly by dislocation glide (Sugiura 1997).

5.4 The effect of dislocations on optical absorption

The generation of carriers by dislocations in material under illumination involves the absorption of photons of energy sufficient to raise electrons from a filled to an empty level. Consequently it has been useful to study the effects of dislocations on the optical absorption in semiconductors at photon energies below the absorption edge (for valence to conduction band electron transitions).

The earliest observations were made by Lipson *et al.* (1955) who observed a shift of the fundamental absorption edge and a long wavelength tail in compressed Ge. Meyer *et al.* (1967) found that the apparent extended tail in deformed Ge and Si was markedly dependent on the direction of light propagation relative to the active slip planes. They reported that the apparent extended absorption tail was due to strong, anisotropic scattering from the active slip planes.

Subsequently Barth and Güth (1970) found in deformed Ge an infrared absorption line at a photon energy of 0.1 eV. Barth and Elsässer (1971) found the absorption was markedly dichroic. That is, the absorption of plane polarized light was a maximum when the polarization vector was parallel to the Burgers vector of the main glide system.

Güth's (1972) energy band calculations showed that the shallow states split off the energy bands by the stress near dislocation gave a polarization vector for optical transitions that agreed with the observed orientation. Barth *et al.* (1976) observed the optical absorption of 60° dislocations in Ge. They interpreted this in terms of two shallow bands along the lines of Güth's (1972) theory. Deep dangling bond states were not observed in optical absorption, it was suggested, because of the weak overlapping of dangling bond states with energy band states. Kamieniecki (1979) interpreted the dichroic absorption in terms of two shallow bands, one 0.1 eV below the conduction band and one 0.1 eV above the valence band.

Spectrally distinguishable photoconductivity arises due to carriers activated from dislocation energy states by absorption of photons of less than band gap energy (Kamieniecki (1979)). Yakimov *et al.* (1976) reported infrared quenching of dislocation photoconductivity in Si, which occurred only at temperatures below 180° K.

The effect of threading dislocations on optical absorption in III-V compounds was studied by Iber *et al.* (1996) and Peiner *et al.* (2002), who investigated mismatched heteroepitaxial GaAs and InP grown on Si substrates. They considered the enhancement of optical absorption (at energies below the band gap) caused by internal electric fields due to charged dislocations to derive an expression for the absorption coefficient (Iber *et al.* 1996, Peiner *et al.* 2002).

Their derivation was based on the Dow and Redfield (1972) theory of the optical absorption due to internal electric microfields. According to that theory, the absorption coefficient, in the presence of a uniform field F, shows an exponential spectral dependence (i.e. the Urbach rule) and it can be expressed as (Dow and Redfield 1972):

$$\alpha = \exp[C(\hbar\omega - E_g)/F] \tag{5.21}$$

where C is a constant. For non-uniform fields, the absorption coefficient is derived by multiplying the uniform-field absorption coefficient by the probability distribution $P(F)$ of these fields and integrating over F (Dow and Redfield 1972), thus:

$$\alpha(\hbar\omega) = \int_0^\infty P(F)\,\alpha(\hbar\omega, F)\,dF \tag{5.22}$$

Iber *et al.* (1996) expressed the electric field due to dislocations as:

$$F(r) = Q_l/(2\pi\varepsilon_0\varepsilon_r r) \tag{5.23}$$

where Q_l is the line charge density. Assuming that only the dislocations nearest to a given point contribute to the field there, Iber *et al.* (1996) derived the probability distribution $P(F)$ and so the absorption coefficient α_d:

$$\alpha_d = \alpha_0\left\{1 - \frac{C}{F_0}\frac{\sqrt{\pi}}{2}(E_g - \hbar\omega)\exp\left[\left(\frac{E_g - \hbar\omega}{2}\frac{C}{F_0}\right)^2\right]\mathrm{erfc}\left(\frac{E_g - \hbar\omega}{2}\frac{C}{F_0}\right)\right\} \tag{5.24}$$

where α_0 and C/F_0 are fitting parameters that can be obtained from experiments (Iber *et al.* 1996).

The absorption coefficient of heteroepitaxial layers of InP and GaAs grown on Si substrates (Fig. 5.17), containing dislocations, were measured by spectroscopic ellipsometry. The results were found to be in close agreement with the above model (Iber *et al.* 1996 and Peiner *et al.* 2002). From spectroscopic ellipsometry

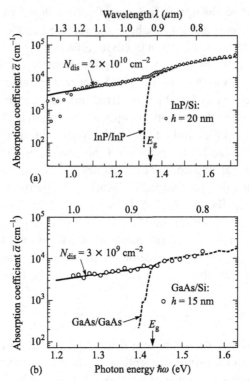

Figure 5.17   Optical absorption coefficients of InP and GaAs layers grown on Si substrates. (After Peiner *et al.* 2002, *Journal of Physics: Condensed Matter*, **14**, pp. 13195–201, 2002.)

measurements (combined with anodic stripping), the dislocation-density profile in the InP/Si layers near the heterointerface was also determined (Iber *et al.* 1996, Peiner *et al.* 2002).

5.5 SEM EBIC microscopy of individual dislocations

One difficulty in studying the electrical effects of plastic deformation was that these were produced by the large numbers of dislocations of various types and point defect debris invariably introduced by plastic deformation. For fundamental understanding it was better to study the electrical effects of single well-characterized dislocations, i.e. 60° or edge, dissociated or constricted and either impurity decorated or not. The techniques that have the necessary spatial resolution for this are electron beam induced current (EBIC) and cathodoluminescence (CL) scanning electron microscopy and more recently developed scanning probe techniques. The results relate to the atomic core structure, which can now be seen using atomic resolution transmission electron microscopy, and to electronic states for the core calculated using quantum mechanics.

A phenomenological theory of EBIC contrast was developed by Donolato (1978/79), see e.g. the reviews by Donolato (1985) and Holt (1989) and a model of recombination at dislocations was successfully developed and applied by Wilshaw *et al.* (1989) and Wilshaw and Fell (1989) as will be discussed in Section 5.5.1 and Section 5.5.2 and an alternative Shockley-Read-Hall (SRH) recombination model was proposed (Section 5.5.5).

Fig. 5.18 represents the use of a surface Schottky barrier on *n*-type material to observe dislocation EBIC contrast. The contrast is the relative reduction in collected current in the presence of the defect, so the charge collection current (i.e. EBIC) in perfect material must first be calculated. The electron beam dissipates energy in a generation volume, producing a density $g(r)$ of hole-electron pairs at each point. This distribution can be represented by the Everhart-Hoff depth-dose distribution, by the uniform generation density sphere or by the results obtained using a Monte Carlo simulation programme (see e.g. Joy 1988, 1995). The excess minority carriers (i.e. holes in *n*-type material) are subject to the continuity (diffusion) equation:

$$D\nabla^2 p(r) - p(r)/\tau = -g(r) \tag{5.25}$$

This represents the effects of generation, g, diffusion (D is the minority carrier diffusion coefficient) and recombination at a rate reciprocal to the minority carrier lifetime, τ. The density of minority carriers, p, falls exponentially with distance, r, from a point source due to recombination during diffusion. The EBIC current is obtained by integrating the carrier density over the charge collecting surface subject to appropriate boundary conditions. The recombination velocity v_s is

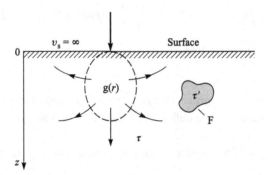

Figure 5.18   The electron beam in an SEM generates electron-hole pairs at a rate $g(r)$ per unit volume. The carriers diffuse away, recombining as they go. An electrical barrier, such as a Schottky contact on the surface, will attract carriers of one sign to 'collect' an EBIC current. This is reduced by extra recombination via defects like that at F when the beam is incident nearby. [After Donolato 1985. In *Polycrystalline Semiconductors. Physical Properties and Applications*, ed. G. Harbeke (Berlin: Springer-Verlag), pp. 138–54 (1985). With kind permission of Springer Science and Business Media.]

assumed infinite at $z = 0$ in Fig. 5.18 to represent the effect of charge collection by the contact. This gives the charge collection probability for a point source of carriers as:

$$\Phi = \exp(-z_0/L) \tag{5.26}$$

Multiplying the source strength g (number of carriers generated per second) at each point by the collection probability (fraction of carriers collected) and integrating over the generation volume gives the EBIC current. By symmetry this is essentially a one-dimensional problem. Thus:

$$I_o = \int_0^\infty \exp(-z/L)g(z)\,dz \tag{5.27}$$

5.5.1 Phenomenological theory of dislocation EBIC contrast

The phenomenological theory of EBIC dislocation contrast in semiconductors is due to Donolato (1978/79, 1983), whose account is followed here. EBIC micrographs show dislocations as dark lines and this is attributed to enhanced recombination. Defects are therefore modelled as reducing the minority carrier lifetime from the bulk value τ_o to τ' inside a volume F. The Donolato first order theory assumes that the carrier density, p_o, is not significantly reduced in volume F. Then the net carrier generation rate can be written:

$$G(r) = g(r)(V) - \frac{1}{\tau'}p_o(r)(F) \tag{5.28}$$

where V is the total volume. Hence the EBIC current is:

$$I = \int G(r)\Phi(z)\,dV = I_o - \frac{1}{\tau'}\int p_o(r)\Phi(z)\,dV = I_o - I^* \tag{5.29}$$

Contrast (see Fig. 5.19) is defined as:

$$i^* = I^*/I_0 \tag{5.30}$$

An advantage of this first order theory is that it is applicable to point, line and surface defects, simply by modelling the defect as a volume of the appropriate geometry.

For dislocations, F is a cylinder of radius r_d, so $dV = \pi r_d^2\,dl$. Hence:

$$I^* = \gamma \int p_o(r)dl, \tag{5.31}$$

where $\gamma = \pi r_d^2/D\tau'$. Hence maximum contrast $C \propto \{1/(D\tau')\}$.

In this treatment, γ, with dimensions $[cm^2\,s^{-1}]$, is the 'strength' or line recombination velocity of the dislocation, and determines the magnitude of contrast produced by a particular dislocation. Detailed treatments show that the contrast

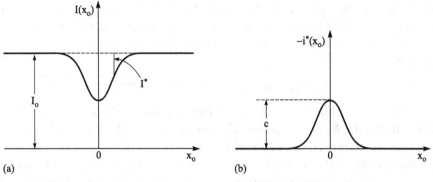

Figure 5.19   (a) Typical EBIC profile across a defect, and (b) the derived contrast profile. [After Donolato 1985. In *Polycrystalline Semiconductors. Physical Properties and Applications*, ed. G. Harbeke (Berlin: Springer-Verlag), pp. 138–54 (1985). With kind permission of Springer Science and Business Media.]

profile (Fig. 5.19) for a line scan across a dislocation at a constant depth can be written in the form:

$$i^*(x) = \gamma f\left(x, E_0, \text{Collection surface geometry}\right) \tag{5.32}$$

where E_0 is the energy of the incident beam electron. Given an explicit theoretical expression for the geometrical and SEM parameter function, f, therefore, values of the strength, γ, can be obtained from the measured EBIC contrast profile.

The theory was used to compute simulations of EBIC images for series of increasing beam voltages (Donolato and Klann 1980). The simulated images agreed well with experiment for dislocation half loops and tetrahedral stacking faults. Predictions of the theory concerning resolution (dislocation image widths) were quantitatively verified by Toth (1981).

Pasemann (1981) applied a perturbative approach to include higher order effects in the theory. Experimental evidence of predicted second order effects was found by Pasemann *et al.* (1982) and agreement between theoretical and experimental dislocation EBIC contrast profiles was good. They also measured the EBIC strengths of individual dislocations in Si and found they all fell within a few percent of one of the values: 0.29 (ascribed to 60° unit dislocations), 0.68 (dissociated 60° dislocations) and 0.02 (screw dislocation).

5.5.2 Physical (charge controlled recombination) theory of dislocation EBIC contrast

Beyond the phenomenological theory of dislocation EBIC contrast, the physics of recombination at dislocations must be treated. Wilshaw and Booker (1985) and Wilshaw *et al.* (1989) deduce EBIC contrast using the model of energy band bending at a negatively charged dislocation, seen previously in Fig. 5.12. We are now considering the effect of a single dislocation rather than a density ρ of dislocations (as in Equation 5.16) so the current of captured electrons can be written as the difference between the flux of electrons in and that out of the dislocation states (Fig. 5.20):

$$J_e = J_i - R_e = C_e N_d[(1-f)n_o\exp(-q\phi/kT) - fN_c\exp(-q\phi_0/kT)] \tag{5.33}$$

Here C_e is the probability of an individual electron making the transition to or from a dislocation state; N_d is the number of states per unit length of the dislocation; f is the dislocation state occupancy factor (the fraction of these states that are occupied); N_c is the density of states in the conduction band, and ϕ and ϕ_o are the potential barriers against electrons entering and leaving the dislocation states from the conduction band, respectively (Fig. 5.12).

Donolato modelled the dislocation as a cylinder of radius r_d in which the minority carrier lifetime is reduced from the bulk value τ_o to τ'. Then the rate of hole capture per unit length of dislocation:

$$J_h = \pi r_d^2 \Delta p / \tau' \tag{5.34}$$

where Δp is the density of holes. Now, Donolato's dislocation strength, γ, from Equation (5.31) can be written as:

$$\gamma = \{(\pi r_d^2)/(D_h\tau')\} = \{J_h/(\Delta p D_h)\} \tag{5.35}$$

For equilibrium we must have:

$$J_h = \gamma \Delta p D_h = J_e = J = \text{the dislocation recombination current} \tag{5.36}$$

Hence from Equation (5.33):

$$\gamma \Delta p D_h = C_e N_d[(1-f)n_o\exp(-q\phi/kT) - fN_c\exp(-q\phi_0/kT)] \tag{5.37}$$

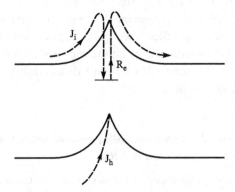

Figure 5.20 Carrier fluxes into and out of a dislocation on the charge-controlled recombination theory. J_i and R_e are the fluxes of electrons (numbers per unit time) into and out of the dislocation core respectively. J_h is the rate of capture of holes by the dislocation.

Hence:

$$-(q\phi/kT) = \ln\{[\gamma\Delta pD_h + C_eN_dfN_c\exp(-q\phi_0/kT)]/[C_eN_d(1-f)n_o]\} \quad (5.38)$$

Substituting Equation (5.36) into (5.38) and using the fact that $\ln(1/x) = -\ln(x)$ we obtain

$$\phi = (kT/q)\ln\{C_eN_d[(1-f)n_o]/[J+R_e]\} \quad (5.39)$$

where

$$R_e = C_eN_dfN_c\exp(-q\phi_0/kT) \quad (5.40)$$

is the rate of re-emission of electrons from the dislocation per unit length (see Equation 5.33).

We recall that the dislocation line charge: $Q = \pi r_d^2 qN_D = q\,f/b$ (see Equation 5.1) where N_D, the doping density, $= n_o$, is the equilibrium carrier density. Substituting πr_d^2 from this equation in Equation (5.35) we obtain:

$$\gamma = [Q/(n_oqD_h\tau')] \quad (5.41)$$

We assume initially for simplicity that ϕ is proportional to the dislocation line charge per unit length, i.e. we write $\phi = AQ$, following Wilshaw *et al.* (1989). Substituting for Q in Equation (5.41) from Equation (5.39) gives:

$$\gamma = [kT/(An_oq^2D_h\tau')]\ln\{[C_eN_d[(1-f)n_o]/[J+R_e]\} \quad (5.42)$$

$$\text{i.e. } \gamma \propto \left(\frac{T}{n_o}\right)\ln\left(\frac{(1-f)n_oC_eN_d}{J+R_e}\right) \quad (5.43)$$

Donolato had shown that EBIC contrast, C, is proportional to defect recombination strength, γ, so:

$$C = \left(\frac{\beta T}{n_o}\right)\ln\left(\frac{(1-f)n_oC_eN_d}{J+R_e}\right) \quad (5.44)$$

where β is the constant of proportionality which depends both on the specimen geometry, e.g. the defect depth and distance from the charge-collecting barrier and on the SEM operating parameters such as the beam voltage.

EBIC contrast, according to Wilshaw's charge-controlled recombination theory, is given by Equation (5.44). It depends, through the definitions of J and R_e (Equations 5.33, 5.36 and 5.40), on the fundamental dislocation parameters, $\phi_0 = E_c - E_d$, N_d the line density of recombination states, and C_e the probability of transitions from E_c to E_d and vice versa, plus fundamental materials parameters like N_c the conduction band density of states. We shall see how they may be experimentally evaluated in Section 5.5.4. Equations (5.41) and (5.44) mean that γ, and so C, is approximately proportional to Q, the dislocation line charge, hence the name of the theory.

5.5.3 Temperature dependence of dislocation EBIC contrast

The behaviour of the EBIC contrast (Equation 5.44) depends on the relative magnitudes of J and R. For large beam currents and hence large excess minority carrier densities Δp, the flow of holes and electrons into the dislocation, J will be large. If there is a small number of dislocation states in the gap, N_d, or the dislocation levels are deep within the gap (i.e. large ϕ_o), R, the rate of escape of electrons from the defect will be small. Then $J \gg R$ and Equation (5.44) reduces to:

$$C = \frac{\beta T}{n_o} \ln \left\{ \frac{(1-f)n_o C_e N_d}{\Delta p \gamma D_h} \right\} \tag{5.45}$$

Under these conditions the charge on the dislocation is increased above its equilibrium value (since J, the electron capture rate $\gg R$ the re-emission rate). The contrast is dependent on the beam current ($\Delta p \propto I_b$, so $C \propto -\ln I_b$). Fig. 5.21 presents experimental data supporting this prediction in the right-hand half of the plots. From Equation (5.45), for constant beam current, C is approximately proportional to T (regime 1 in Fig. 5.22d), since Wilshaw and Fell (1989) argued convincingly that D_h is nearly independent of temperature. This prediction was also found to agree with experiment (Fig. 5.23) for relatively clean dislocations (produced by high–low T deformation but not in transition-metal-free material) in Si at low temperatures and high beam currents by Wilshaw and Fell (1989).

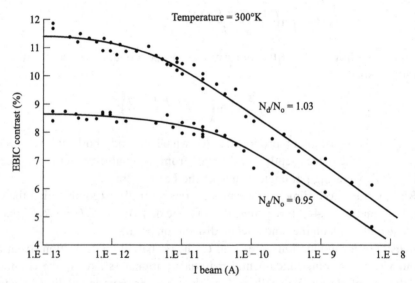

Figure 5.21 EBIC contrast versus $\ln(I_b)$ for a screw dislocation in silicon. The solid curves are theoretical. [(After Wilshaw *et al.* 1989. In *Structure and Properties of Dislocations in Semiconductors*, 1989. Conf. Series No.104 (Bristol: Institute of Physics), pp. 85–96.]

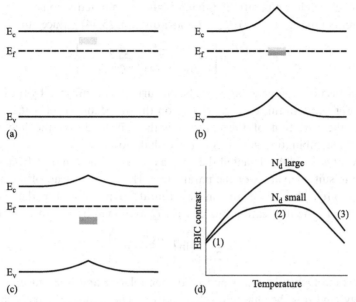

(a)

(b)

(c)

(d)

Figure 5.22   The narrow band of dislocation trap states (the stippled box) can lie (a) above, (b) be pinned to or (c) fall below the Fermi level. (After Wilshaw and Fell 1991.) In case (a) the dislocation would be uncharged and there would be no charge-controlled EBIC contrast. (d) The Wilshaw theory predicted the three regimes of variation of contrast with temperature shown here. Regime (1), for low T and large beam currents and so large J is governed by Equation 5.45 as discussed in the text. In regime (2), for small N_d, the situation is as shown in (c) and C is independent of T. In regime (3), the movement of E_F down toward the mid-gap position causes the line charge and hence C, to fall. This holds for both large N_d, with the dislocation in the situation shown in (b), and eventually, even for small N_d. (After Wilshaw *et al.* 1989 and Holt 1996.)

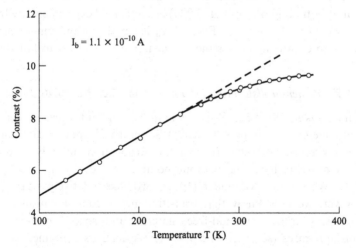

Figure 5.23   EBIC contrast versus temperature for a screw dislocation in silicon. The curve is theoretical. (After Wilshaw *et al.* 1989.)

At relatively high temperatures (giving high thermal activation for R) and small beam currents (giving small J) $R \gg J$ and Equation (5.44) reduces to:

$$C = \frac{\beta T}{n_o} \ln \left\{ \frac{(1-f)n_o C_e N_d}{C_e N_d f N_c \exp(-q\phi_0/kT)} \right\} \tag{5.46}$$

The contrast is now independent of the beam current as confirmed by the low current region of Fig. 5.21. In this case the charge on the dislocation is near the equilibrium value and the variation of the contrast with T depends on the line density of dislocation recombination states, N_d, as we shall now see.

Wilshaw *et al.* (1989) calculated the contrast C as a function of T from the theory above, using suitable values of the parameters. However, instead of the simplifying relation $\phi \approx AQ$, they used the more accurate expression for the electrostatic potential due to the dislocation line charge, Q, obtained by Masut *et al.* (1982):

$$\phi = \frac{Q}{2\pi\varepsilon\varepsilon_o} \ln \left\{ \frac{\lambda_D Q}{q} - 0.5 \right\} \tag{5.47}$$

where λ_D is the Debye length, ε and ε_o are the relative and free space permitivities, respectively, and q is the charge on the electron. The simulated curves showed that variation of contrast with temperature could take two different forms depending on the relative density of traps, N_d, along the dislocation line (Fig. 5.22d).

Firstly, if N_d is large the defect states will be pinned at the Fermi level (Fig. 5.22b). As T rises, the Fermi level moves toward the mid-gap position so the occupation of the states and hence the charge and contrast of the dislocation falls, giving the behaviour in regime 3 in Fig. 5.22d. Secondly, if N_d is small, all the states may be occupied without the excess line charge and consequent band bending being sufficient to raise the states to the Fermi level (Fig. 5.22c). Then for small I_b and high T, the dislocation line charge and hence contrast (since $C \propto \gamma \propto Q$) are independent of both (regime 2 in Fig. 5.22d) as confirmed experimentally by the high temperature portion of Fig. 5.23. Even if N_d is small, the movement of E_F down towards mid-gap eventually causes the line charge and contrast to fall in regime 3.

5.5.4 Fundamental dislocation parameters obtained from EBIC contrast

Wilshaw and Booker (1985) and Wilshaw *et al.* (1989) made measurements on high purity, float-zone-grown *n*-type ($10^{15}\,\text{cm}^{-3}$) Si deformed under clean conditions in two stage compression at 850°C (to generate dislocations initially) and 420°C to multiply and move dislocations without point defect diffusion. This procedure, developed by Wessel and Alexander (1977), produces relatively clean dislocation cores. However, we now know that transition metal contamination can only be avoided by using transition-metal-free material and special non-contaminating deformation apparatus (see Section 5.5.8). The specimens contained large crystallographically aligned dislocation loops.

By measuring EBIC contrast, C, as a function of T under experimental conditions such that one of the three regimes of Fig. 5.22 applies or by measuring C as a function of I_b with T held constant as in Fig. 5.21, experimental data can be obtained that can be fitted to theoretical curves. These were calculated as described above from Equation (5.46) modified by the use of (5.47). The best fit was found to families of curves simulated for different values of the parameters involved. Hence, best-fit values were obtained for some of the fundamental parameters. Not all the parameters can be obtained at once. More detailed accounts of the procedures and difficulties in obtaining the fundamental properties of dislocations in this way can be found in the papers of Wilshaw and co-workers. The main results of all this work are summarized below.

Wilshaw and Booker (1985) fitted the data they obtained under the experimental conditions quoted above. They found the number of dislocation states per unit length, to be $N_d \approx 2 \times 10^6$ cm^{-1} for the screw dislocations and about 10% more for the 60° dislocations studied. Their accuracy they estimated to be not better than 20 to 30%. (These low line densities are in agreement with the values found by deep level transient spectroscopy by several groups.) This value for N_d is low compared with the density of dangling bonds in the original Shockley model ($= 2.5 \times 10^7$ cm^{-1} in Ge, see Fig. 5.7). Wilshaw and Booker (1987) noted that the dislocation EBIC contrast was too weakly dependent on ϕ_0, the depth of the dislocation states below the conduction band edge, for this to be evaluated. However, the dislocation charge per unit length from the data was $Q = 2.6 \times 10^{-13}$ C cm^{-1}. Hence the band bending can be calculated from Equation (5.47). Since the dislocation has a negative charge, the defect states must lie below the Fermi level, which can be calculated for 10^{15} cm^{-3} material at 350° C. This leads to the conclusion that the defect level for both screw and 60° dislocations is 0.55 eV below the conduction band edge, i.e. at the mid-gap position. This is in agreement with the DLTS results of various groups on bulk-deformed material. However, the line density of states is much lower than even recent theories suggest.

Wilshaw and Fell (1989, 1991) compared the 13% EBIC contrast of dislocations introduced, as above, by deformation at 420° C, with the 2% contrast of others produced by deformation at 650° C. The 650° C dislocations had an even lower line concentration of states $N_d \approx 5 \times 10^5$ cm^{-1} as against $\sim 2 \times 10^6$ cm^{-1} for those produced at 420° C. The depth of these states could only be determined to be more than 0.3 eV below the conduction band edge. These results also were in agreement with those obtained by DLTS and electron paramagnetic resonance (EPR) on similar specimens (Weber and Alexander 1983, Omling *et al.* 1985, Wilshaw and Fell 1989). Measurements on the curved corners of the dislocation loops (Fig. 2.25) and on the straight side segments gave the same value for N_d. This is evidence that dislocation kinks do not introduce deep recombination centres. This follows since there must be high line densities of kinks along the curved regions to take the dislocation line out of the crystallographically aligned Peierls troughs of the straight segments. Therefore, kinks cannot be the 'special sites' where the dangling-bond like centres is thought to be located.

The effects of impurities on EBIC contrast were studied by Wilshaw and Fell (1991), Fell and Wilshaw (1991), Fell *et al.* (1993) and Wilshaw *et al.* (1991, 1997). Wilshaw and Fell (1991) and Wilshaw *et al.* (1997) showed that deliberate Cu contamination of dislocations, introduced into Si at 650°C, increased the EBIC contrast to 7%. This was higher than for similar uncontaminated dislocations (2%) but less than the 13% magnitude for uncontaminated 420°C dislocations.

Wilshaw and Fell (1991) and Wilshaw *et al.* (1997) also showed that dislocations could be produced with sufficiently low values of N_d (states per unit length of dislocation) that they could fill without pinning the Fermi level so it could lie well above the dislocation energy level, as in Fig. 5.22c. The regime 2 behaviour of Fig. 5.22d, i.e. contrast independent of temperature, is exhibited by such dislocations. Cu-contamination can add sufficient further states per unit length to give Fermi level pinning, i.e. the situation of Fig. 5.22b. Dislocations in this state exhibit the large N_d regime 3 behaviour of Fig. 5.22d. They also studied oxidation-induced stacking faults (OISFs, see Section 4.8.2) which, when not deliberately contaminated gave no EBIC contrast. OISFs gave strong contrast at the bounding partial only, when Fe-contaminated. Weaker OISF contrast arose from the fault also when Cu-contaminated. They ascribed this difference to differences in the precipitation of the two impurities.

The results that had been obtained by Wilshaw in his Ph.D. work on samples deformed at 420°C in the Cologne laboratory of Alexander, and those that were obtained on specimens similarly deformed in Oxford many years later, were reported to be 'identical' by Fell and Wilshaw (1991), i.e. they demonstrated sample-independent reproduceability. They continued the analysis of the contamination results. Their conclusion was that Cu contamination altered neither N_d nor E_d, the energy level of the dislocation recombination state of 420°C dislocations. This they suggested showed that the high stresses needed for deformation of silicon at 420°C produces an intrinsic deep (> 0.53 eV) level in the dislocation core. For further details see Wilshaw *et al.* (1991).

Dislocation loops were introduced into *n*-type (5×10^{15} cm^{-3}) GaAs by microhardness indentor by Wilshaw *et al.* (1989). The EBIC contrast, they found, increased with temperature. The contrast fell with I_b but not in proportion to $\ln(I_b)$ (as in Fig. 5.21 for the case of Si dislocations in regime 1 of Fig. 5.22d). α and β dislocations were introduced by microindentation and bending of specimens of GaAs. These were of LEC (liquid encapsulated Czochralski), MOCVD (metal-organic chemical vapour deposition) and LPE (liquid phase epitaxy) materials. The recombination strength of dislocations in the LEC and LPE materials increased with decreasing temperature. The results for the purer MOCVD material were different, e.g. the recombination strength decreased with decreasing temperature. In the MOCVD material contrast increased with doping concentration. Only small differences were found between α, β and screw dislocations.

5.5.5 *The Shockley-Read-Hall recombination model for dislocation EBIC contrast*

Kittler and Seifert (1993a and b) reported two distinct types of dislocation dark EBIC contrast in silicon. These they called class 1 and 2, characterized by different forms of variation with beam current and temperature, I_b and T, as shown in Fig. 5.24. Wilshaw *et al.* had analysed Class 1 contrast, i.e. this is the charge controlled recombination (CCR) like behaviour discussed above.

A particularly convincing observation by Kittler *et al.* (1994) is Fig. 5.25. Opposite signs of variation of EBIC contrast, C, with temperature is exhibited by adjacent, parallel misfit dislocations in Cu-contaminated *p*-type SiGe/Si. The contrast, C, of dislocation 1 increased as T increased while dislocation 3 decreased in contrast and dislocation 2 fell to invisibility. C versus T is plotted in Fig. 5.26a for these dislocations.

An alternative theory for both the forms of behaviour of Fig. 5.24 was proposed by Kittler and Seifert (1994). Donolato's expression for the EBIC contrast of a dislocation:

$$C \propto 1/(D\tau') \qquad (5.48)$$

follows from Equations (5.30) and (5.31) and Fig. 5.19b. Kittler and Seifert started with Equation (5.48) and assumed recombination in the defect region to be due to a single type of recombination centres obeying Shockley-Read-Hall (SRH) theory. That is, recombination takes place via independent centres, not at interacting

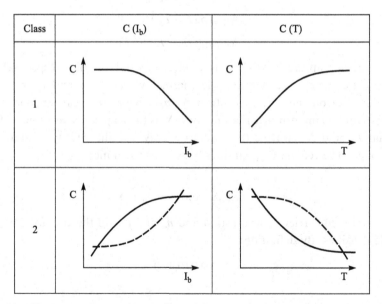

Figure 5.24 Classes 1 and 2 are the two forms of variation of dislocation dark EBIC contrast with beam current and temperature that were observed by Kittler and Seifert (1993a).

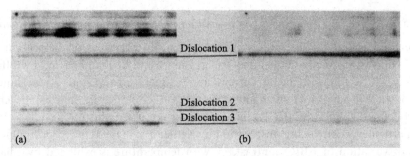

Figure 5.25   Misfit dislocations in the interface of p-type, Cu-contaminated Si$_{0.98}$Ge$_{0.02}$/Si seen in EBIC micrographs recorded at (a) 80° and (b) 300° K. The contrast variation of dislocation 1 is of class 1 and dislocation 2 is of class 2. Dislocation 3 exhibits intermediate behaviour. (After Kittler *et al.* 1994. Reprinted from *Materials Science and Engineering*, **B24**, Recombination activity of 'clean' and contaminated misfit dislocations in Si(Ge) structures, pp. 52–5, Copyright 1994, with permission from Elsevier.)

charge-controlled centres as in the Wilshaw theory. The SRH recombination theory (see, e.g. Sze, 1985, pp. 44 ff) gives the net rate of carrier recombination R as:

$$R = \left\{ (np - n_i^2)/[(n + n_1)\tau_{po} + (p + p_1)\tau_{no}] \right\} \tag{5.49}$$

$$\text{where } \tau_{po} = \frac{1}{N_t \sigma_p \langle v_{th} \rangle_p} \tag{5.50}$$

$$\tau_{no} = \frac{1}{N_t \sigma_n \langle v_{th} \rangle_n} \tag{5.51}$$

$$n_1 = N_C \exp[(E_t - E_C)/(kT)] \tag{5.52}$$

$$p_1 = N_V \exp[(E_V - E_t)/(kT)] \tag{5.53}$$

σ_n and σ_p are the capture cross-sections for electrons and for holes, respectively, $\langle v_{th} \rangle_n$ and $\langle v_{th} \rangle_p$ are their respective average thermal velocities, n and p are the non-equilibrium electron and hole concentrations, n_i is the intrinsic carrier concentration, τ_{no} and τ_{po} are the minority-carrier lifetimes, N_t is the trap concentration, E_t the trap level, and E_V and N_V are the (maximum) energy and the density of states in the valence band. The remaining quantities have their usual meanings.

Now:

$$n = n_o + \Delta n \text{ and } p = p_o + \Delta n \tag{5.54}$$

Δn is the injected-carrier concentration and n_o and p_o are the equilibrium electron and hole densities. By definition:

$$\tau' \equiv (\Delta n)/(R) \tag{5.55}$$

Hence:

$$\tau' = [(n_o + n_1 + \Delta n)\tau_{po} + (p_o + p_1 + \Delta n)\tau_{no}]/[n_o + p_o + \Delta n] \tag{5.56}$$

For the p-type material of Fig. 5.25 it can be assumed that $\tau_{po} = \tau_{no} = \tau_o$ and that n_o is negligible so Equation (5.56) becomes:

$$\tau' = \tau_o[1 + 2\beta + (n_1/p_o) + (p_1/p_o)]/(1+\beta) \qquad (5.57)$$

where $\beta = \Delta n/p_o$ is the injection level.

The variation of dislocation contrast with T and the injection level, i.e. with I_b was simulated using Equation (5.57) substituted into Equation (5.48). Kittler and Seifert (1994) assumed in their calculations that D was independent of T. Since $D \propto \mu T$, this implies that $\mu \propto T^{-1}$. On this model, they found that the form of variation of C with T depends essentially on the depth of the independent recombination centres (see Fig. 5.26). Due to the thermal velocity variation with T, $C \propto T^{1/2}$ for 0.55 eV (mid-gap) and similar deep centres. Shallow centres, like those at a depth of 0.075 eV in Fig. 5.26b, give T dependence due to changes in the occupation of the centres.

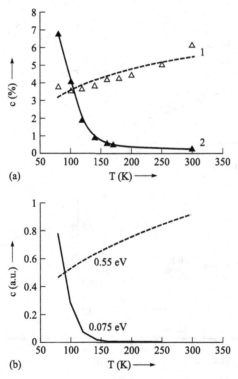

Figure 5.26   (a) Experimental plots of EBIC contrast versus temperature for dislocations 1 and 2 in Fig. 5.25, and (b) $C(T)$ curves calculated for dislocations with independent (Shockley-Read-Hall) recombination centres at energy levels depths of 0.075 and 0.55 eV. (After Kittler and Seifert 1994. Reprinted from *Materials Science and Engineering*, **B24**, Two types of electron-beam-induced current behaviour of misfit dislocations in Si(Ge): Experimental observations and modelling, pp. 78–81. Copyright 1994, with permission from Elsevier.)

This occupation change is caused by the Fermi level movement, which does not affect the occupation of the deep centres.

The contrast, C, varied with I_b, i.e. with injection level, β, for the two dislocations as shown in Figs. 5.27a and c. Assuming the same shallow and deep levels as before, this dependence followed from the model, as simulations showed (Figs. 5.27b and d). For both shallow and deep centres at low temperatures, the form of variation of C with I_b (β) is the same. C increases slightly with I_b (β) at room temperature in the shallow centre case.

Agreement between the experimental and simulated curves in Figs. 5.26 and 5.27 is good. The ability of the SRH model to account for both forms of behaviour is valuable. It will be important to test the model on a greater number of experimental observations, however. This should include curve fitting to obtain values for the parameters of the model.

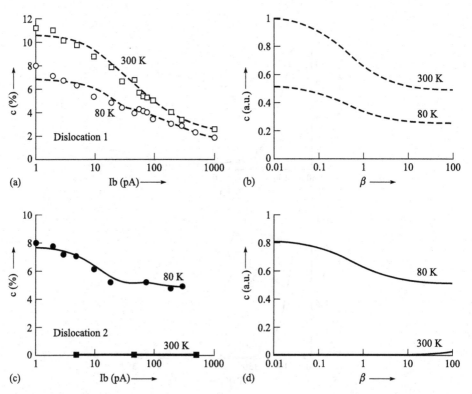

Figure 5.27   (a) and (c) experimental variation of EBIC contrast with I_b for dislocations 1 and 2 respectively. The calculated variation of C with injection level β for dislocations with independent recombination centres at energy level depths of (b) 0.55 eV and of (d) 0.075 eV at the two temperatures marked. (After Kittler and Seifert 1994. Reprinted from *Materials Science and Engineering*, **B24**, Two types of electron-beam-induced current behaviour of misfit dislocations in Si(Ge): Experimental observations and modelling, pp. 78–81. Copyright 1994, with permission from Elsevier.)

Charge controlled recombination or independent recombination centres at dislocations?

The contrast of similar and neighbouring dislocations can exhibit either a class 1 or 2 dependence on temperature (see Fig. 5.24), like those labelled 1 and 2 in Fig. 5.25. The form of variation of class 1 dislocations in Fig. 5.24 can be accounted for either by the charge-controlled or the deep-independent-centre models of recombination. Only the shallow-independent-centre model, however, can account for class 2 dislocation EBIC contrast behaviour.

Do dislocation recombination centres, in fact, act independently in accordance with Shockley-Read-Hall theory or do they interact as in the CCR model?

There are several reasons for thinking that dislocation traps do interact. At the accepted trap spacings the trap-state wave functions overlap so the energy levels broaden into a narrow band as Shockley (1961) and Labusch and Schröter (1978) assumed. Conduction along dislocation lines due to this delocalization was found experimentally by Osipiyan (1981, 1983) at low temperatures and microwave frequencies. The idea of charge-controlled recombination is also supported by the success of charged dislocation models in accounting for electrical noise (Morrison 1956), the influence of dislocations on conductivity (Labusch and Schröter 1978) and on photoconductivity (Figielski 1978).

Consider the magnitude of the space charge cylinder surrounding any line-charged dislocation. The fraction of dislocation traps occupied by electrons was calculated by Read (1954b). He used the Shockley (1953) model with one trap per atomic distance along the dislocations and developed several detailed treatments leading to the similar results (Fig. 5.8). The electrons drop from the conduction band into dislocation traps to lower their energy. Eventually trapping became energetically unfavourable as the result of the rising electrostatic energy (band bending) due to the dislocation line charge. Read concluded that about one in ten traps would be filled at room temperature. It is now known that trap states only occur at intervals of about 100 atomic spacings along partial dislocations, so trap states occur about every 50 atomic distances along extended unit dislocations (see the end of Section 5.2.5 and Section 5.2.6). Hence, complete trap filling is possible without producing the same electrostatic potential rise. As the Fermi level rises through the trap energy as the temperature is lowered, the traps can go from empty to full (provided the density N_d is small in Fig. 5.22d, i.e. the trap states all capture electrons on the CCR model.) The space charge cylinder radius is then about $0.15\,\mu m$ (Wilshaw, private communication) for lightly doped ($10^{15}\,cm^{-3}$) material, rather than the $1\,\mu m$ given by the Read model with 100 times higher trap density. Even 100 interatomic distances between charged traps is only $38.4\,nm$ ($100\ a_o/\sqrt{2}$ where a_o is the lattice constant of silicon) so the space charge cylinder radius is still much larger. Thus, the charged line and surrounding space charge cylinder model of the CCR theory is self-consistent and involves reasonable magnitudes for the parameters of the model.

A large amount of experimental work on class 1 EBIC contrast dislocations has been reported (Section 5.5.2 to 5.5.4). Those results were accounted for by the CCR theory. Moreover, curve fitting lead to values for dislocation parameters like N_d, which are in good agreement with values obtained by independent methods (ESR in the case of N_d). This strongly supports the CCR theory.

The SRH recombination theory of dislocation EBIC contrast has not been so fully worked out or tested as the CCR theory. Moreover, implausible magnitudes for some parameters appear to be required. Using SRH theory (Kittler and Seifert 1981), the rate of recombination for a unit length of dislocation, γ, can be written as:

$$\gamma = N_d \sigma v_{th} \qquad (5.58)$$

where N_d is the number of traps per unit length of dislocation, σ is the capture cross-section of the independent traps (assumed to be atom-sized so $\sigma \approx 10^{-14} \, cm^{-2}$), and v_{th} is the thermal velocity of the minority carriers. Using this expression, from dislocation recombination strengths γ they found values for N_d. To account for even the minimum EBIC contrast observable (which they took to be $C = 0.5\%$), required a minimum $N_d = 170 \, \mu^{-1}$. Dislocations are frequently observed with EBIC contrast an order of magnitude greater. For such dislocations the SRH theory would require $N_d = 1700 \, \mu m^{-1}$ making the trap spacing only one or two atoms as against the presently accepted value of about 100, i.e. $N_d = 26 \, \mu m^{-1}$. Even if we take the capture cross-section, σ, to be an order of magnitude larger than assumed above, the SRH model still requires large trap densities compared to those found by ESR. Furthermore, at such high trap densities, trap wave functions should overlap and broaden into a band so the traps would not be independent.

The height of the potential barrier ϕ (Fig. 5.12), i.e. the amount of band bending produced by charged traps at dislocation lines can be estimated. Let E_d be small (shallow traps), however, N_d is large on the SRH model. Hence, as the temperature falls, the number of trapped electron charges per unit length will become significant, so band-bending results in E_F pinned to E_d. From room to liquid nitrogen temperatures the movement of E_F is much greater than kT, so that of ϕ will also be. Majority carriers will require thermal activation over such high potential barriers to be captured as is assumed by the CCR theory (Equation 5.33).

It is unclear what determines whether a dislocation exhibits class 1 or 2 EBIC contrast. The Wessel and Alexander (1977) two-temperature deformation method, that Wilshaw and co-workers used to produce their dislocations, was thought to ensure that they were not impurity contaminated. The work of Higgs *et al.* (Section 5.5.8 and Section 5.6.4) showed that they are. Kusanagi *et al.* (1992) found 60° and screw dislocations produced by the two-temperature method had different forms of temperature dependence of EBIC contrast, as did the apparently similar dislocations 1, 2 and 3 in Fig. 5.25. The existence of dislocations, whose

EBIC contrast exhibits class 2 behaviour, not the class 1 (CCR-like) behaviour, requires explanation. At present only SRH recombination can provide an explanation. Further theoretical and experimental work is needed, however, to provide a firm foundation for SRH theory.

For an additional review of this field see Alexander (1994). It is hoped that the results of further experimental work, including that on III-V materials, and its analysis, should resolve the matter.

5.5.6 Bright EBIC defect contrast

As will be discussed in Section 5.6.1 and Section 5.6.3, in SEM CL, cases of both dark and bright dislocation contrast occur and the CL bright contrast is particularly 'enlightening'. Similarly, in SEM EBIC, cases of both dark and bright dislocation contrast occur. However, EBIC bright contrast does not occur through a single mechanism. Nevertheless, it has been studied and a number of the mechanisms have been identified and provide information concerning electrical conditions near the dislocation line.

Blumtritt *et al.* (1989) had done considerable work on this and summarized their observations and interpretations of four different sources of bright EBIC defect contrast as follows.

Lifetime contrast of getter zones

Around decorated defects zones of bright contrast appear (see Fig. 5.28). This is due to the increased minority carrier diffusion length arising from the zone being denuded of impurity by the getter action of the defect. Under optimum conditions

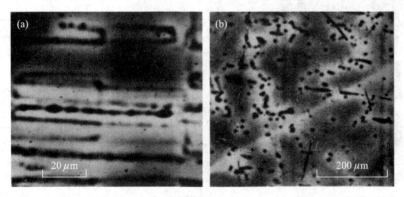

Figure 5.28 (a) SEM EBIC micrograph taken at a beam accelerating voltage of 10 kV showing bright getter zones around heavily decorated, dissociated 60° dislocations in a processing-contaminated p⁺-n diode and (b) a 20 kV SEM EBIC image of gold-doped, deformed and annealed n-type Fz (floating zone) Si. [After Blumtritt *et al.* 1989. In *International Symposium on Structural Properties of Dislocations in Semiconductors*, Institute of Physics Conference Series 104, 1989 (Bristol: Institute of Physics), pp. 233–8.]

increases as small as a few percent can be detected in this way. Comparisons of the effectiveness of gettering by different kinds of defects can also be made through the strength of this form of contrast.

Charge collection by space charge regions in depletion layers

According to Blumtritt *et al.* a major reason for bright contrast in EBIC images is the creation of space charge regions in the denuded zones adjacent to decorated and charged defects such as grain boundaries, as shown in the diagram on the right of Fig. 5.29.

Contrast due to a repulsive defect potential

Negatively charged defects will repel electrons and positively charged ones will repel holes. Depending on the geometry this can result in increased charge collection and bright contrast. Any such effect in dislocations is apparently too small to observe in EBIC images but it can be seen in extended planar defects and precipitates (Gleichmann *et al.* 1983 and Jakubowicz and Habermeier 1985).

In Fig. 5.30a bright contrast arises because the negative charge on the shallow triangular CuSi precipitate repels electrons from the hole-electron pair generation volume (shown as a grey circle in Fig. 5.30c) down toward the deep charge-collecting junction, to add to the signal. When the surface Schottky barrier is used, these

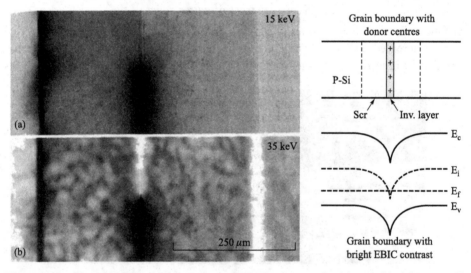

Figure 5.29   The vertical line in the centre of EBIC micrographs (a) and (b) show a small-angle grain boundary in p-type hot-pressed Si. The upper portion appears in bright contrast due to additional charge collection in the adjacent depletion barriers, as shown in the diagram. [After Blumtritt *et al.* 1989. In *International Symposium on Structural Properties of Dislocations in Semiconductors*, Institute of Physics Conference Series 104, 1989 (Bristol: Institute of Physics), pp. 233–8.]

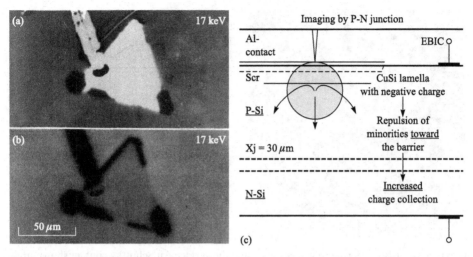

Figure 5.30   SEM EBIC micrographs of a negatively charged CuSi planar defect in p-type Si (a) imaged by the 30 μ deep p-n junction and (b) by the Al Schottky barrier on the surface as shown in the (c) diagram of the cross-sectional structure of the specimen. [After Blumtritt *et al.* 1989. In *International Symposium on Structural Properties of Dislocations in Semiconductors*, Institute of Physics Conference Series 104, 1989 (Bristol: Institute of Physics), pp. 233–8.]

downward-repelled carriers are lost from the signal however. The locally reduced signal appears as a dark area.

Contrast due to doping inhomogeneities

The charge collection property of an electrical barrier depends e.g. on the width of the space charge region which depends on the doping density locally. Hence doping inhomogeneities can affect the percentage of the minority carriers collected in EBIC images. Depending on the conditions either an increased or a decreased local doping density can give rise to bright contrast, i.e. greater signal strength.

Fig. 5.31 shows EBIC micrographs of oxygen-induced defects in Czochralski Si that was annealed at 1100° C for 20 h. After thermal donor formation at 450° C the sample became *n*-type and bright contrast (halos) were seen round the dark defects for beam electron energy greater than 15 kV as shown in Fig. 5.31a. After destruction of the thermal donors the bright contrast disappeared (Fig. 5.31b). Moreover, the width of the bright zones corresponded to the oxygen diffusion length. Hence Blumtritt *et al.* (1989) concluded that this bright contrast was due to a 50% decrease in the net doping leading to an increase in the width of the space charge region around the defects, as shown in the diagram in Fig. 5.31. The thermal donor distribution also reflects the reduction in interstitial oxygen round the defects (Seifert and Kittler 1987).

A second type of bright doping contrast can be seen at low beam energies and high injection levels when the pair generation volume under the beam (shown as a tear

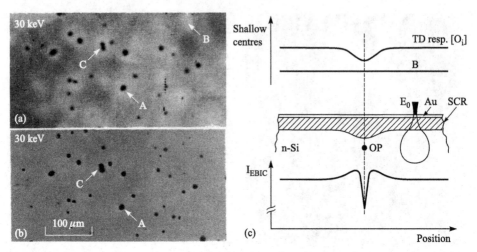

Figure 5.31   EBIC micrographs recorded using the surface Au Schottky contact shown in the diagram. (a) Image showing the contrast at oxygen induced [O$_i$] defects after thermal donor formation, (b) image recorded after thermal donor destruction and (c) diagram of a cross-section through the specimen. [After Blumtritt *et al.* 1989. In *International Symposium on Structural Properties of Dislocations in Semiconductors*, Institute of Physics Conference Series 104, 1989 (Bristol: Institute of Physics), pp. 233–8.]

drop shape in the diagram in Fig. 5.32) is inside the space charge region and the electron hole plasma is relatively stable. In such cases space charge regions of stronger electrical field, i.e. of higher doping concentrations, will collect larger currents giving local bright regions (Leamy 1982).

This form of contrast is shown in the case of a grain boundary (GB) in a thick phosphorus-doped silicon-on-insulator (SOI) layer, in Fig. 5.32. There is a striking change in contrast of the top, horizontal GB and the right-most vertical GB between Figs. 5.32a and b. In (a) the beam conditions do not exist for this form of contrast. However, in (b) the beam energy is low and the injection (beam current density) is high so higher field regions in the GB space charge layers appear bright. This is evidence that the bright regions in Fig. 5.32b are enriched in phosphorus with the results shown in the cross-sectional diagram. This interpretation was supported by the fact that the positions of these two GBs were where two solidification fronts met during the recrystallization process. Consequently, segregation effects should be enhanced at these GBs.

As discussed in Section 5.1.4, Eremenko and Yakimov (2004) observed bright EBIC contrast along slipped planes in plastically bent Si outside the area of round charge collecting Schottky barriers on the surface. These are the bright areas at the bottom right in Fig. 5.33a and at the bottom of Fig. 5.3a (see Section 5.1.4). The bright contrast lines at the top of Fig. 5.33a continue in the usual dark contrast where they run under the Schottky barrier which is clear evidence that both dark and

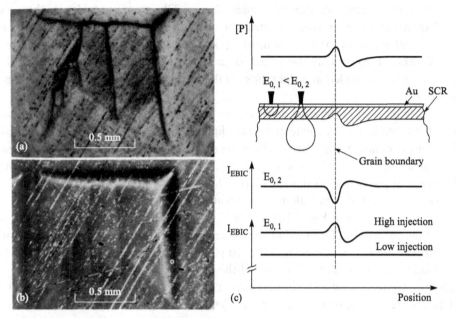

Figure 5.32 Bright contrast due to enhanced doping at a grain boundary in a SOI layer. EBIC micrographs recorded with (a) a 30 kV beam under low injection conditions, and (b) a 5 kV beam under high injection conditions, and (c) explanatory diagram. The parallel, diagonal lines across the micrographs are due to surface scratches. [After Blumtritt *et al.* 1989. In *International Symposium on Structural Properties of Dislocations in Semiconductors*, Institute of Physics Conference Series 104, 1989 (Bristol: Institute of Physics), pp. 233–8.]

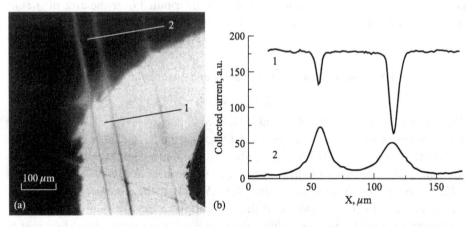

Figure 5.33 (a) EBIC micrograph of a bent Si sample, and (b) line scan profiles recorded along the lines marked 1 (dark contrast) and 2 (bright contrast). The half-widths for the dark lines are smaller than for those for bright contrast. Moreover, the relative amplitudes of the two dark and two bright lines are reversed. (After Eremenko and Yakimov 2004.)

bright contrast arise at slipped planes. Fig. 5.33b presents evidence that the mechanisms of the two forms of contrast, however, are different. It can be seen there, that the half widths of the dark-contrast, appearing as dips in the upper line scan, are considerably narrower than the bright-contrast peaks in the lower one. Moreover the relative depths of the left and right dips are the reverse of the relative heights of these two peaks.

The distances from the edge of the charge collecting Schottky barrier, over which detectable bright contrast extends along slipped planes, are large, i.e. carriers can flow long distances to the barrier along the slipped lines. For example, Fig. 5.4 (see Section 5.1.4) is a plot of the decreasing strength of the collected current with distance along such a slipped plane (curve 1). It can be seen that the decay length, for the current to fall to a half along this exponential curve, was 750 μm.

Evidence is shown in Fig. 5.70 (Section 5.6.4) that CL emission corresponding to two of the dislocation 'D' emission lines arises from the dislocation lines while that of the other two 'D' lines comes from slipped planes between the dislocations. This was assumed to be due to debris left behind the dislocations as they moved. Debris has generally been assumed to be in the form of vacancies or interstitials and/or dipoles of interstitial or vacancy character (see Section 2.3.12). Eremenko and Yakimov (2004), however, quote much new evidence that (i) 'quasi-two-dimensional defects' are formed behind moving dislocations and alter the crystal properties in the swept area of the glide plane (Bondarenko *et al.* 1981), (ii) they can be revealed by selective chemical etching (Bondarenko *et al.* 1981 and Eremenko *et al.* 1997), (iii) they affect the photoluminescence (Negrii and Osipyan 1982a), (iv) they alter the dislocation mobility anisotropically (Nikitenko *et al.* 1981), and (v) the slipped plane shows strong EBIC contrast (Bondarenko *et al.* 1986 and Alexander *et al.* 1990). Finally, a new type of 'needle like' defect was found to be distributed over the area of the slip plane across which dislocations moved (Eremenko *et al.* 1999).

5.5.7 *The electronic energy levels and bands of dislocations*

Early studies

The earliest studies of the electrical effects and energy levels of dislocations in semiconductors are discussed in Section 5.1 and Section 5.2 (see also Figielski 2002).

Grazhulis *et al.* (1977a) studied conductivity and the Hall effect in Si crystals containing high densities of dislocations and deduced that energy levels lie about 0.42 eV above the valence band and 0.43 eV below the conduction band and attributed them to states localized at the dislocation cores.

Diagrams for the trap states of dislocations in Si were put forward by Shröter *et al.* (1980) using the Read model to interpret the electrical effects of plastic deformation (Section 5.2) and by Osipiyan (1981, 1983) based on ESR and microwave low temperature conductivity studies. Kveder *et al.* (1982) combined ESR and DLTS results to deduce energy band diagrams. These are shown in Figs. 5.34(a and b) for dislocations due to deformation below $0.6\,T_m$ (T_m is the melting point) and in Fig. 5.34c for dislocations annealed at or above $0.6\,T_m$. Dislocation energy levels at several depths in the forbidden gap were also reported by Ono and Sumino (1985) and by Omling *et al.* (1985). The roles of both dislocation and point defect states (the latter can be eliminated by annealing without reducing the dislocation density) and the methods of deriving the diagrams are discussed in these papers.

ESR, DLTS and SDLTS studies

Extensive ESR studies were carried out by Alexander and co-workers (Alexander 1991, Weber 1994) on silicon plastically deformed 5% in compression at 650° C. The results led to the conclusion that most (65% to 80%) of the paramagnetic centres (unpaired electrons, i.e. dangling bonds) produced were due to point defects or point defect clusters (debris). The dislocations were related to only a small number of deep states, about $5 \times 10^{14}\,\mathrm{cm}^{-3}$ for 1% strain, in the middle and upper half of the forbidden energy band gap as well as shallow (200 meV) states. This agrees with the result in Fig. 5.34c for Si after the point defects were annealed out. Thus the electrical effects of plastic deformation are mainly due to native point defect debris and not to dislocations (see Section 5.1.4 and Section 5.9.1).

Schröter *et al.* (1989) carried out computer simulations of capacitance transient spectroscopy of dislocations in semiconductors. They solved the rate equation (representing the capture and emission of both types of carriers at the dislocation)

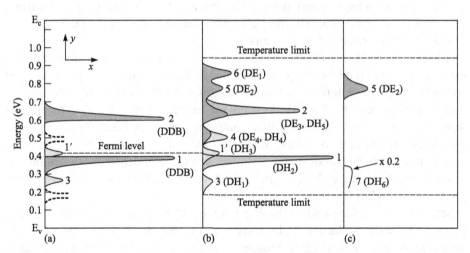

Figure 5.34 Energy plotted against density of states for dislocations in Si deformed at temperatures below $0.6\,T_m$. (a) The results deduced from ESR and (b) from DLTS observations. The results in (c) were derived from DLTS for dislocations annealed at or above $0.6\,T_m$ (to eliminate point defect debris left behind the dislocations on the slip planes). (After Kveder *et al.*, 1982, in which the meaning of the labels of the bands is explained.)

and inserted it into the solution of Poisson's equation linking charge variations in the space charge region with the capacitance change, ΔC, of the Schottky contact (Schröter *et al.* 1989). The dislocation capture rate is modified by an electrostatic potential, resulting from the interaction between charge carriers at the dislocation and free carriers. Schröter *et al.* concluded that in transient capacitance spectra, dislocations (or point defect clouds around them) are characterized by broadened lines and features, in contrast with isolated point defects. The computer simulations were used to examine these differences in terms of (i) modifications of capture and emission rates or (ii) the spectral distribution of localized dislocation states.

Omling *et al.* (1985) employed capacitance transient spectroscopy for defect energy levels in plastically deformed *n*-type silicon. They concluded that two types of defect levels are generated: (i) those of point defects around (or inside) dislocations, and (ii) those directly related to dislocations (Omling *et al.* 1985). They reported that the characteristic levels in *n*-type Si are: the A line ($E_c - 0.19$ eV), B line ($E_c - 0.29$ eV), C1 line ($E_c - 0.52$ eV), C2 line ($E_c - 0.51$ eV) and D line ($E_c - 0.54$ eV). Again the dislocation-related lines were substantially broader than those of point defects. The position reported in the literature for the well-known C line, varies widely.

Cavalcoli *et al.* (1997), using DLTS, found the same A, B, C, and D defect states in plastically deformed *n*-type Si. From the dependence of these levels on the deformation characteristics and annealing procedures, they concluded that only the C line at ($E_c - 0.40$) eV could be ascribed to dislocations, since it scaled with the dislocation density and was thermally stable. The traps A, B, and D were attributed to deformation-induced point defects that may not be localized at dislocations (Cavalcoli *et al.* 1997). These authors also discussed problems in the DLTS analysis of plastically deformed semiconductors.

Schröter *et al.* (1995) demonstrated, using DLTS, that the electronic states of extended defects could be categorized as band-like or localized by the rate R_i at which the levels reach internal equilibrium, relative to the carrier emission rate, R_e, and the capture rate, R_c. Thus, defect states are bandlike if $R_i \gg R_e, R_c$ or localized if $R_i \ll R_e, R_c$. Schröter *et al.* (1995) fitted computer simulation results to experimental DLTS data for Si and concluded, similarly to Cavalcoli *et al.* (1997), that DLTS signals ascribed to 60° dislocations are in fact due to point-defect clouds around the dislocations. Thus, this technique can be employed in studies of interactions between point defects and dislocations.

Schröter *et al.* (2002a) studied the shape of the DLTS C line in n-type Si, on which previous researchers disagreed. In some experiments the C line is nearly symmetrically broadened, in others it broadens asymmetrical with low-temperature tails. Schröter *et al.* (2002a) suggested that symmetric broadening is due to point defects at the dislocation core, whereas the low-temperature tails are due to point defect clouds in the strain field of edge-type dislocations. Thus, the C line could be characteristic of point defects in both the core and the far-field region of dislocations.

Knobloch *et al.* (2003) used DLTS to study the effect of contamination on the C1 line, and concluded that it could only be observed with gold contamination (see Section 5.5.8).

Wosinski and Figielski (1989) used DLTS to study plastically deformed GaAs crystals. The observed level at $(E_c - 0.68)$ eV, which controls carrier recombination, was attributed to the dislocation cores. They also outlined the generation of As antisite defects during plastic deformation. The concentration of EL6 electron traps at $(E_c - 0.35)$ eV increased substantially with deformation but not that of EL2 (Wosinski and Figielski 1989). This was attributed to the quality of the crystal used, compared with those giving the different results of other researchers.

Yastrubchak *et al.* (2001) employed DLTS to investigate deep-level traps related to lattice-mismatch generated defects in GaAs/InGaAs heterostructures. They identified an (ED1) electron trap at $(E_c - 0.64)$ eV and ascribed it to electron states associated with threading dislocations in the ternary compound, and a hole trap (HD3) at $(E_v + 0.67)$ eV. The latter was attributed to misfit dislocations at the heterostructure interface (Yastrubchak *et al.* 2001). This ED1 trap was that observed by Wosinski and Figielski (1989). Yastrubchak *et al.* (2001), like Schröter *et al.* (1995), analysed the dependence of DLTS-line shape on the trap filling time. They concluded that the ED1 electron trap states are localized, whereas the HD3 hole trap states are bandlike.

Microscopic methods of study

SEM EBIC and CL provide means for studying the electrical and luminescence properties of individually resolved dislocations. Because of their importance, each is discussed in a section of its own (Section 5.5 and Section 5.6).

Fell and Wilshaw (1989) introduced an innovation in which the position of dislocation energy levels was derived by an analysis of the dark EBIC contrast of dislocations in the depletion region of a surface Schottky barrier as at (b) and (c) in Fig. 5.35 (see Section 5.5.2 and Section 5.5.4). Electron capture to the dislocation level occurs via thermal excitation over the repulsive barrier ϕ of the space charge region of the charged dislocation. Hence the electron capture rate and so the EBIC contrast is $J_e \propto \exp(-q\phi/kT)$ (Fell and Wilshaw 1989). In the case of recombination at dislocations in the depletion region of an electrical barrier like (b) and (c) in Fig. 5.35, to be captured electrons must be thermally excited over an additional repulsive potential, V, due to the depletion region field. Hence (Fell and Wilshaw 1989):

$$J_e \propto \exp[-q(\phi + V)/kT] \tag{5.59}$$

The energy band model for a dislocation in the depletion region (see Fig. 5.35) shows that the dislocation charge (and recombination efficiency) can be expected to be zero for $V = \phi_0$, where ϕ_0 is the energy difference between the conduction band edge and the Fermi level. The value of ϕ_0 can thus be determined from the measured

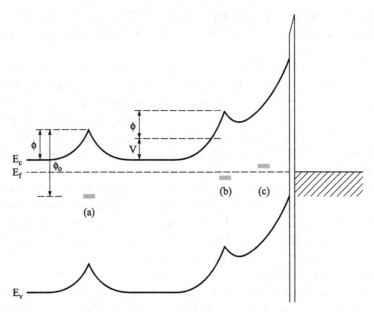

Figure 5.35   Band structure diagram of a Schottky barrier to an n-type semiconductor, including the energy levels of (a) a charged dislocation in the bulk, (b) a charged dislocation in the depletion region, and (c) an uncharged dislocation in the depletion region. [(After Fell and Wilshaw 1989). In *International Symposium on Structural Properties of Dislocations in Semiconductors*, Institute of Physics Conference Series **104**, 1989 (Bristol: Institute of Physics), pp. 227–32.]

dependence of V as a function of depth and the observation of the disappearance of the dislocation contrast. One can increase V by applying a reverse bias to the Schottky barrier. Hence, a dislocation at a given depth with a level below the Fermi level (so it is charged) may be made neutral by applying the appropriate reverse bias. Thus an alternative method for finding the value of ϕ_0 is to determine the value of V for which the dislocation contrast disappears, so the position of the dislocation energy levels in the band gap of a semiconductor can be found from EBIC observations (Fell and Wilshaw 1989).

A more direct method for determining dislocation energy levels was developed by Cavallini and co-workers. This they described as quenched (by) infrared beam induced current (QIRBIC) (Castaldini *et al.* 1987, Castaldini and Cavallini 1989, Castaldini *et al.* 1989 and Cavallini and Castaldini 1991). In this, dislocations are observed either by LBIC (light or laser beam induced current, using photons with energies greater than the bandgap) or by EBIC, scanning the top surface. Simultaneously, the Si sample is flood illuminated from below with monochromatic infrared light of variable, less than band gap, photon energy (Fig. 5.36). Absorption occurs whenever this photon energy becomes equal to one of the activation energies for transitions between a dislocation energy level and either the conduction or valence bands or between two dislocation levels. This affects the charge state of dislocations and so its LBIC or EBIC contrast.

Fig. 5.37 shows that dislocations do have a number of energy levels. The argument is that, on the Donolato theory (Section 5.5.1), contrast is proportional to dislocation strength. Moreover, on the Wilshaw theory (Section 5.5.2), dislocation strength is proportional to the charge per unit length of dislocation, Q. Fig. 5.37, therefore, on CCR theory, is a plot of the variation of Q with photon energy with peaks at certain activation energies. The defect energy spectrum is observed directly and the technique can spatially resolve individual dislocations. Measurements of

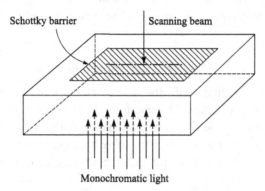

Figure 5.36   Arrangement for QIRBIC (quenched infrared beam induced current) imaging of dislocations in silicon. [After Castaldini and Cavallini 1989). In *Point and Extended Defects in Semiconductors*, eds. G. Benedek, A. Cavallini, W. Schroter (New York: Plenum Press), pp. 257−68 (1989). Copyright 1989 Plenum Press. With kind permission of Springer Science and Business Media.]

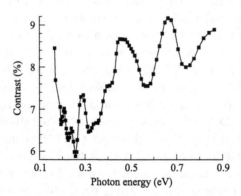

Figure 5.37   Dislocation contrast in QIRBIC versus the photon energy of the monochromatic infrared illumination, from below the specimen in Fig. 5.36. [After Castaldini and Cavallini, 1989). In *Point and Extended Defects in Semiconductors*, eds. G. Benedek, A. Cavallini, W. Schroter (New York: Plenum Press), pp. 257−268 (1989). Copyright 1989 Plenum Press. With kind permission of Springer Science and Business Media.]

contrast versus the intensity of the quenching light from below, gives information about the fractional occupation of the energy level.

SDLTS (scanning deep level transient spectroscopy) is a second scanning beam technique that can resolve individual dislocations and determine their energy levels directly (Petroff and Lang 1977, Petroff *et al.* 1978a,b,c, Breitenstein and Heydenreich 1985, Breitenstein 1989, Schröter *et al.* 1989). SDLTS combines the spectral resolution for trap energy levels of DLTS with the spatial resolution of the SEM but attaining the sensitivity required is not easy. However, some valuable results have been obtained by SDLTS.

DLTS has also been employed in conjunction with EBIC to elucidate the role of impurities in dislocation activity in semiconductors (e.g. Kittler *et al.* 2003, Knobloch *et al.* 2003). This is discussed in the following section.

5.5.8 The role of impurities

Impurity atoms or ions, that misfit the host crystal, are attracted to dislocation or grain boundary cores by elastic interactions. This gives rise to the well-known Cottrell atmospheres (see Section 4.7). Impurities at the defect core can create or alter energy levels (see Section 4.7.1). Thus there has always been a need to determine how much observed dislocation and grain boundary properties are affected by impurities.

The EBIC and XBIC contrast of dislocations in Si and the nearly universal role of traces of transition metals in Si

Extensive and conclusive work on the problem in Si was carried out by Lightowlers, Higgs and their co-workers using a combination of SEM EBIC and CL with other powerful methods of analysis. Higgs *et al.* (1990a) used a heat treatment and PL (photoluminescence) test that had been shown to be able to detect transition metal with high sensitivity, down to concentrations of $10^{10}\,\text{cm}^{-3}$ in the case of surfaces traces of copper (Canham *et al.* 1989). They were then able to find an ingot of FZ (floating zone) Si that contained no detectable amounts of transition metals. They used specimens cut from this pure ingot and a deformation rig made of quartz to avoid contamination. Deforming these specimens by the high-then-low temperature deformation method of Wessel and Alexander (1977) they introduced dislocations that were undetectable by SEM EBIC, i.e. exhibited no effective locally enhanced recombination. However, when such a specimen was brushed with a copper wire and then heated, the dislocations showed strong dark EBIC contrast, i.e. significantly increased minority carrier recombination was present at such minimally Cu-contaminated dislocations. The combination of CL and PL was essential to reinforce and extend the results obtainable with EBIC so further discussion of these important results will be left to Section 5.6.4.

Another powerful form of charge-collection microscopy is x-ray beam induced current (XBIC). It employs a high-intensity (about 10^{10} photons/s) synchrotron x-ray beam focused into an area of about $1-2\,\mu m^2$ (Vyvenko *et al.* 2002a, Vyvenko *et al.* 2004). The sample, mounted in an XRF (x-ray fluorescence) holder, is scanned in the X and Y directions with a 0.1 μm step size, under the stationary beam by computer-controlled stepping motors. The XBIC and XRF signals are stored in a computer as a function of the (X, Y) coordinates for later analysis. XBIC provides microscopic information on electrically active defects and the x-ray fluorescence is a microprobe (μ-XRF) of metal impurity concentrations and precipitates with parts-per-million (ppm) sensitivity and micron-scale spatial resolution. Thus, XBIC/μ-XRF directly correlates the recombination at defects with their chemical cleanliness or contamination (Vyvenko *et al.* 2002a). Thus, it can simultaneously map the distribution of metal impurities and the minority carrier diffusion length. The sensitivity of μ-XRF is several orders of magnitude higher than that of EDS (energy dispersive spectroscopy of X-rays) in an electron-beam instrument.

Fig. 5.38 shows the correspondence between the defects imaged in EBIC and in XBIC micrographs of an area of a multicrystalline Si solar cell (Vyvenko *et al.* 2002a). The differences in contrast of individual defects in these maps are attributed to their depth below the surface and the difference of penetration depth between the electron- (several μm) and x-ray beams (on the order of 100 μm). Thus, in the case of this multicrystalline Si solar cell, XBIC is better for identifying performance-limiting defects than EBIC (but the spatial resolution of EBIC is better). Further results obtained by this method, on recombination active defects in multicrystalline Si solar cells were reported by Buonassisi *et al.* (2003).

EBIC   XBIC

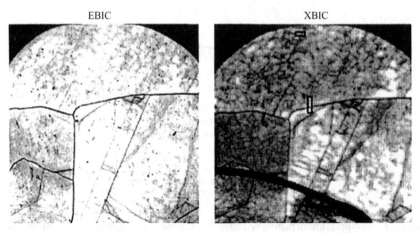

Figure 5.38   EBIC and XBIC images of the same area of a multicrystalline Si sample. Both scans show an area of approximately 3×3 mm. (After Vyvenko *et al.* 2002a. Reprinted with permission from *Journal of Applied Physics*, **91**, pp. 3614–17. Copyright 2002, American Institute of Physics.)

Kittler *et al.* (1995) measured the EBIC contrast of misfit dislocations in SiGe/Si epilayers, as a function of temperature between 80 and 300 K. They observed that copper contamination of the epilayer modified contrast as follows. In uncontaminated samples dislocations displayed weak contrast at low temperature only. With contamination of about 1 ppb, a noticeable increase in contrast was observed at low temperatures, but it was undetectable at room temperature. This they attributed to shallow trap levels at the dislocations. With the increase of a contamination level to 15 ppb, most dislocations showed contrast at all temperatures. This was attributed to near-midgap centres at the dislocation. Increasing contamination to the ppm range produced strong contrast enhancement over the entire temperature range (Kittler *et al.* 1995).

Experimental studies of the influence of Cu, Ni, Fe or Au contamination on the temperature dependence of EBIC dislocation contrast, $C_{dis}(T, I_b)$, in Si, established four categories (types) of contrast (see Fig. 5.39) (e.g. Kittler *et al.* 1995, Kittler *et al.* 2003, Kusanagi *et al.* 1995, and Kveder *et al.* 2001). These are, listed in order of increasing metal contamination, type L1 (clean), type L2, mixed type, type H1, and type H2. Although these types were related to the concentration of metal atoms decorating the dislocation, it did not matter which metal it was. In the case of type

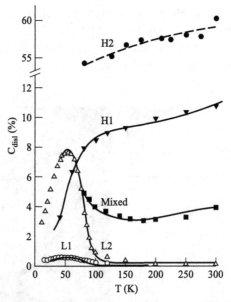

Figure 5.39   Temperature dependence of the EBIC contrast of dislocations in Si for different amounts of contamination. Contrast type change with increasing metal contamination is shown for type L1 (clean), type L2, mixed, type H1, and type H2. [After Kveder *et al.* 2001. Reprinted with permission from *Physical Review*, **B 63**, 115208–1 to 115208–11, 2001 (http://publish.aps.org/linkfaq.html). Copyright 2001 by the American Physical Society.]

H2 (the most heavily doped), the dislocations were decorated with metal silicide precipitates (Kittler *et al.* 1995).

It was also found that hydrogen passivation (Higgs and Kittler 1994) and phosphorus diffusion gettering (Seifert *et al.* 1997) of impurities may convert dislocation contrast from type H1 to L2. Measurement of the $C_{dis}(T)$ type can be used to determine impurity densities per unit dislocation length using the model of Kveder *et al.* (2001). On this model, the densities of impurities per unit dislocation length are as follows. For type L1 there are no impurities. For type L2 there are of the order of 10^4–10^5 impurities per cm. For type H1 10^6 impurities/cm or more are present. For mixed type contrast impurity densities are between those for types L2 and H1 (Kveder *et al.* 2001). This model considers the exchange of electrons and holes between shallow 1D bands, due to the dislocation strain field, and deep electronic levels, due to impurity atoms segregated at the dislocations. This is plausible for impurity atoms located within a few nm of the dislocation core. Their wave functions will overlap the 1D dislocation bands. Thus, the presence of low concentrations of impurity atoms at the dislocation would result in a significant increase of the recombination of carriers captured at the 1D dislocation bands (for details see Kveder *et al.* 2001). Calculated curves of the EBIC contrast support the experimental observations as shown in Fig. 5.39 (Kveder *et al.* 2001).

Kittler *et al.* (2003) used DLTS and EBIC to examine the limits of passivation and external gettering in contaminated Si. As in the earlier studies mentioned above, they determined by DLTS that greater amounts of impurities form clouds in the dislocation strain fields. There they are weakly bound and can be passivated by hydrogenation or removed by gettering. However, at lower concentrations the few impurity atoms are strongly bound to the dislocation core and thus are unaffected by those treatments (Kittler *et al.* 2003).

The power of the XBIC/XRF combination of techniques is again demonstrated in Fig. 5.40 (Vyvenko *et al.* 2002b). It shows that in this intentionally Cu-contaminated SiGe/Si misfit dislocation structure, the Cu forms precipitates along these dislocations (these appear as dark regions in Fig. 5.40b) and these are also dark in XBIC, i.e. they are points of reduced diode current due to increased recombination (Vyvenko *et al.* 2002b).

The XBIC/XRF combination can also produce a quantitative correlation between the recombination at the precipitates and the concentration of metals (Vyvenko *et al.* 2002b). This is demonstrated in Fig. 5.41 by the plot of XBIC contrast against the Cu concentration derived from the XRF maps (for details see Vyvenko *et al.* 2002b). A linear dependence of XBIC contrast on Cu concentration could be understood by assuming that all, or some constant fraction, of the Cu precipitates was electrically active (Vyvenko *et al.* 2004).

Vyvenko *et al.* (2004) also demonstrated that they could estimate the depth of metal precipitates by comparing several XBIC/XRF images recorded for different sample orientations (for details see Vyvenko *et al.* 2004).

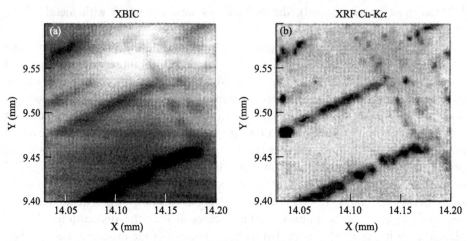

Figure 5.40   (a) An XBIC and (b) an XRF image of an area of Cu-doped $Si_{0.98}Ge_{0.02}/Si$ containing misfit dislocations. The XRF image (b) is a map of Cu-Kα intensity. [(After Vyvenko *et al.* 2002b). *Journal of Physics: Condensed Matter*, **14**, 13079–86, 2002.]

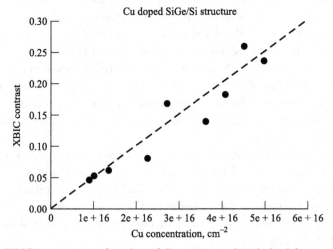

Figure 5.41   XBIC contrast as a function of Cu concentration derived from maps like those in Fig. 5.40b. [(After Vyvenko *et al.* 2002b). *Journal of Physics: Condensed Matter*, **14**, 13079–86, 2002.]

Knobloch *et al.* (2003) examined the effect of gold contamination on the dislocation-related C1 DLTS line in *n*-type silicon. They showed that, although undetectable for clean 60° dislocations, the C1 line emerged with gold contamination. They demonstrated that the C1 line consists of a distribution of energy levels (probably due to point defects or impurities). The width of that distribution increased with the contamination level (Knobloch *et al.* 2003). These results were attributed to the presence of impurities in both the dislocation strain field and in the dislocation core (Knobloch *et al.* 2003).

Schröter and Cerva (2002) presented an extensive review of the interaction of point defects with dislocations in silicon and germanium and their typically strong influence on the defect electronic properties. The range and strength of the interaction are distinguished between (i) the short-range (core) region with strong potentials and (ii) the far-field region with weaker potentials due to the long-range elastic and electrical fields related to dislocations (Schröter and Cerva 2002). Experimental results obtained by DLTS, ESR, Hall effect, EBIC, spectral photoconductivity and photoluminescence were reviewed. The electrical and optical properties of dislocations are mostly related to the point defects at the dislocation or to the interaction between the levels due to the point defects and dislocation states (Schröter and Cerva 2002).

Schröter and Cerva (2002) also pointed out that there is still a problem determining (either by measurements or calculations) the density of dislocation-induced electronic states $N_d(E)$ in the band gap. The difficulty is due to the many-electron nature of the dislocation and its interactions with point defects. Thus, typically point defects around a dislocation both add a specific distribution of levels to that of the dislocation and eliminate its spectral characteristics (Schröter and Cerva 2002 and references therein). The variations of the DLTS line shape with electron filling of the defect states allow us to distinguish the electronic states of extended defects as band-like or localized (Schröter *et al.* 1995). The existence of 1D dislocation core states in Ge has been experimentally established. However, it appears that this may not be the case for Si, where theoretical work revealed that 30°- and 90°-partial dislocations have reconstructed cores and thus no deep 1D states in the band gap (Schröter and Cerva 2002). (In Si, deep localized states related to dislocations are due to kinks, jogs, reconstruction defects, and point defects and impurities). Both theoretical and experimental research has confirmed the existence of the shallow 1D dislocation bands (due to the dislocation strain field). These are an empty band below the conduction band edge and a filled one above the valence band edge. The wave functions of these extended shallow bands overlap with deep levels associated with point defects at dislocations. Thus, the latter may have a substantial influence on the recombination processes (Schröter and Cerva 2002 and references therein). It was established that these shallow 1D bands and deep levels, due to impurity atoms segregated at the dislocations, could exchange electrons and holes. Thus, impurity atoms in the dislocation core can significantly increase carrier recombination there, which was revealed in experimental studies by the significant enhancement of the EBIC contrast with increasing concentration of metal atoms decorating the dislocation, as was shown above in Fig. 5.39 (Kveder *et al.* 2001).

Finally, Schröter and Cerva (2002) also outlined some examples of the influence of extended defects on the performance of Si electronic devices. These included the increase of the reverse leakage current. Also, during device processing, dislocations may facilitate the diffusion of various species (e.g. dopants and transition metals). This may result in enhanced reverse leakage current and development of shorts in the case of heavily decorated dislocations (Schröter and Cerva 2002 and references therein). (See also Section 5.11)

Passivation of dangling bonds in dislocations and grain boundaries

Extended defects in crystalline semiconductors contain dangling bonds that introduce deep levels in the energy band gap, which act as efficient recombination centres. An effective method of reducing the number of such recombination centres (in addition to intrinsic bond rearrangements) is the hydrogen passivation of dangling bonds, i.e. neutralizing the dangling bonds by atomic hydrogen attachment (for general overviews, see Pankove and Johnson 1991, Pearton *et al.* 1992, and Nickel 1999). The hydrogen passivation of extended defects (i.e. dislocations, grain boundaries, and surfaces or interfaces) in semiconductor devices improves their performance. For example, hydrogen passivation of grain boundaries in polycrystalline silicon photovoltaic (PV) cells improves their photovoltaic efficiency. The hydrogen also passivates dangling bonds at vacancies and dislocations, further improving the performance of such PV cells. This is similar to the passivation of dangling bonds in amorphous silicon.

Simple annealing in an H_2 atmosphere is not used because efficient diffusion and passivation require atomic hydrogen. This can be obtained by heating H_2 to about 2000° C, or by an H_2 glow or plasma discharge (see, for example, Pankove and Johnson 1991). Thus samples for hydrogenation can be placed in a glow discharge (plasma) chamber to provide the atomic hydrogen and heating them to about 100 to 350° C to promote the in-diffusion of the hydrogen. The temperature should not exceed the dissociation temperature of the Si-H bond formed. Alternatively, hydrogenation can be produced by ion implantation or electrochemical techniques. A drawback of the plasma and ion implantation methods is that they produce surface damage (especially in III-V compounds).

In early studies of passivation of polycrystalline silicon, Seager *et al.* (1979, 1980, 1981) showed that hydrogenation significantly reduces minority carrier recombination at grain boundaries and improves diode characteristics. Similarly, Hanoka *et al.* (1983) observed substantial improvements in solar cell efficiency in studies of hydrogen passivation of electrically active defects. Hydrogen passivation of the many grain boundaries and intragrain defects in multicrystalline silicon (for terrestrial solar cells) was further studied by Krüger *et al.* (2000).

Metastability is a problem in hydrogen-passivated semiconductors, like polycrystalline silicon. The effect is similar to that in hydrogenated amorphous silicon which manifests itself in reversible changes in spin density due to the dangling bonds. In hydrogenated semiconductors such as amorphous silicon, light may result in the generation of metastable defects. In hydrogenated-amorphous-silicon solar cells, this leads to degradation with illumination. Annealing irradiated samples in the range between about 150 and 200° C typically leads to a recovery of the physical properties. Defect metastability in hydrogen-passivated polycrystalline silicon is not as pronounced as in hydrogenated amorphous silicon (see, for example, Nickel 1999).

Hydrogen passivation of extended defects has been studied in polycrystalline Si and GaAs (for a review, see Zozime and Castaing 1996) and in the case of dislocations in GaAs/Si (Pearton *et al.* 1987), GaAs/InP (Chakrabarti *et al.* 1990), InGaAs/GaAs (Matragrano *et al.* 1993), and InP/GaAs heterostructures (Chatterjee *et al.* 1994, Chatterjee and Ringel 1995, Ringel 1997). Defects (and surfaces) can be passivated by the post-deposition annealing of hydrogen-containing silicon nitride layers, SiN$_x$:H (see, e.g. Lüdemann 1999, Aberle 2001, Duerinckx and Szlufcik 2002, and Jiang *et al.* 2003).

An example of hydrogen passivation of dislocations in InP/GaAs heterostructures is shown in Fig. 5.42, which presents DLTS spectra as a function of hydrogen plasma exposure time (Chatterjee and Ringel 1995, Ringel 1997). The concentrations of hole traps (HT1A, HT1B, and HT2), which were attributed to dislocations in heteroepitaxual p-InP, were greatly reduced by hydrogen plasma exposure, i.e. the dislocations were passivated (Chatterjee and Ringel 1995).

This reduction of the concentration of dislocation-related trap states by hydrogen passivation also results in a significant reduction of the diode leakage current, as shown in Fig. 5.43 (Chatterjee and Ringel 1995). The activation energies ΔE (shown in the figure) of the dominant generation-recombination centres were derived from a relationship for the reverse (leakage) current:

$$J_{\mathrm{rev}} \propto \exp(-\Delta E/kT) \tag{5.60}$$

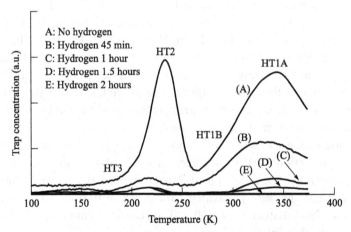

Figure 5.42   DLTS spectra of InP/GaAs heterostructures as a function of hydrogen plasma exposure time. (After Chatterjee and Ringel 1995. Reprinted with permission from *Journal of Applied Physics*, **77**, pp. 3885–98. Copyright 1995, American Institute of Physics.)

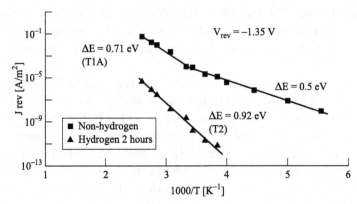

Figure 5.43   Reverse leakage current of non-hydrogenated and hydrogenated heteroepitaxial InP diodes as a function of temperature. (After Chatterjee and Ringel 1995. Reprinted with permission from *Journal of Applied Physics*, **77**, pp. 3885–98. Copyright 1995, American Institute of Physics.)

As can be seen from Fig. 5.43, the two-hour hydrogen passivation treatment resulted in the reduction of the leakage current density by several orders of magnitude (Chatterjee and Ringel 1995).

Scanning electron microscope EBIC and CL are also often employed to study passivation. EBIC can be used to characterize the electrical activity of extended defects and measure the minority carrier diffusion length, and CL provides spectroscopic information. Thus Higgs and Kittler (1994) concluded from photoluminescence and EBIC measurements on hydrogen-plasma treated misfit dislocations in SiGe/Si epilayers that although deep midgap dislocation levels were passivated, most shallow levels were not. Wilshaw *et al.* (1997) studied the EBIC contrast and recombination efficiency of extended defects and the differences in recombination activity in these cases were attributed to the position of the defect states in the energy gap. They found, like Higgs and Kittler (1994), that hydrogen passivation is effective at removing deep- but not shallow-levels in the energy gap. Hydrogen plasma treatments were shown to increase the diffusion length of polycrystalline silicon thin films (Ballutaud *et al.* 2002). Both hydrogenation and phosphorus diffusion were shown to strongly reduce the recombination activity of dislocations in silicon at room temperature (Seifert *et al.* 1997). This was ascribed to hydrogen passivation of deep levels and impurity gettering due to phosphorus diffusion. Cross-sectional EBIC (see Section 3.7.2) observations on samples given hydrogen plasma treatment for 1 h at 310° C showed that grain boundaries and intragrain dislocations were passivated to a depth of 100 μm (Krüger *et al.* 2000, Vyvenko *et al.* 2000, Kittler *et al.* 2001). The recombination activity of the defects in this case was substantially reduced.

A different passivation method, the phosphidization of dislocations in GaAs grown on Si substrate (GaAs/Si), was employed by Wang *et al.* (2001). In this case, exposure to PH_3 plasma resulted in the improvement of GaAs/Si solar cell efficiency. It was determined that the incorporation of P atoms in GaAs layers during PH_3 plasma exposure passivated the dislocation-related recombination centres in the GaAs/Si.

5.6 SEM CL microscopy of individual dislocations

Scanning electron microscope (SEM) panchromatic (all wavelengths) cathodoluminescence (CL) images are those in which the whole of the emitted spectrum is detected and displayed. This emission is generally dominated by the fundamental or edge emission for which the photon energy $hv \approx E_g$. This is (conduction) band to (valence) band edge recombination radiation. At sufficiently low temperatures that excitons are not thermally disrupted, much or all of this may be found to be due to exciton recombination and consequently to occur at a slightly lower photon energy than E_g. Recombination via shallow or deep levels associated with defects will occur at still lower photon energies, i.e. still longer wavelengths.

Numerous early studies on the important elemental (Ge and Si) and compound semiconductors showed that increasing densities of dislocations reduced minority carrier lifetimes (Fig. 5.1 and Table 5.1). That is, dislocations enhance recombination. If the increased recombination at a defect is non-radiative, it reduces the intensity of the panchromatic emission as it does in the case of the electroluminescence from LEDs (Fig. 5.2). Such a defect is darker than the surroundings in panchromatic CL SEM micrographs, and is described as appearing in dark contrast. Conversely, if the defect recombination is radiative, the dislocation will appear in bright contrast in monochromatic CL micrographs recorded using the wavelength emitted by the defect. Experimental studies showed that defects in some materials do appear in bright CL contrast, i.e. they act as radiative recombination centres (e.g. Brummer and Schreiber 1972, 1974, 1975, Batstone and Steeds 1985).

5.6.1 Localized recombination and dark and bright dislocation CL contrast

The Donolato phenomenological model for EBIC dark defect contrast, (see Section 5.5.1), treats the dislocation as a small cylindrical region in which the recombination rate is higher than elsewhere, i.e. in which the minority carrier lifetime is reduced from the defect-free bulk value τ_o to a smaller value τ'. Any enhanced total (mainly non-radiative) recombination, at and near dislocations that produces dark EBIC contrast, also produces dark contrast in SEM CL images. The CL contrast, C_{CL}, is defined, in the same way as the EBIC contrast, as the relative reduction of the CL signal when the beam is incident on a point above the dislocation, i.e.

$$C_{CL} = (L_{CL} - L_{CLD})/L_{CL} \qquad (5.61)$$

where L_{CL} and L_{CLD} are the CL signal levels (proportional to the emitted light intensity) at points far from any defect and at the defect, respectively.

Lohnert and Kubalek (1983, 1984) were the first to apply the Donalato model to CL dark contrast. Hastenrath *et al.* (1979) had measured both the CL yield (intensity) and the relaxation time along line scans across a CL-dark dislocation line seen end on in GaAs:Se and found the minimum in the CL emission at the dislocation line corresponded to a minimum in the relaxation time (Fig. 5.44). This is as expected since the dark contrast is due, as assumed by the theory, to a reduced non-radiative recombination time near the dislocation line. Steckenborn *et al.* (1981) however made similar measurements across the end of a dark appearing dislocation in the same type of material and reported that the yield and lifetime varied *inversely*. This discrepancy was accounted for by Balk *et al.* (1976) who showed that the contrast could be reversed by varying the beam current (Fig. 5.45). This result was for a dislocation around which their measurements had revealed a Se-denuded zone. This is believed to be a thermal effect. The final conclusion was that the defects appear in dark contrast in general because they act as centres of non-radiative recombination (Lohnert *et al.* 1979).

Lohnert *et al.* (1979) also studied dislocations appearing in dark CL contrast in green-emitting GaP:N. The CL spectrum of this diode is shown in Fig. 5.46.

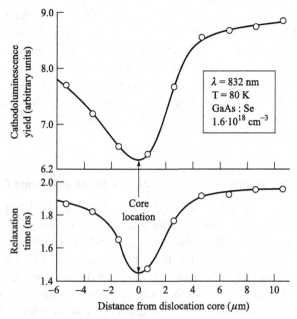

Figure 5.44 Variation of the CL intensity and of the relaxation time for a line scan across a dislocation perpendicular to the surface in GaAs doped with Se. (After Hastenrath *et al.* 1979.)

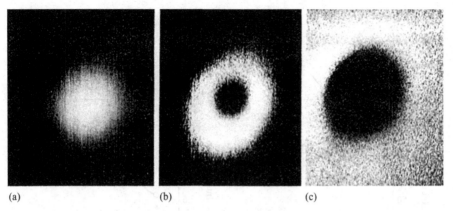

(a)   (b)   (c)

Figure 5.45 The contrast at the point of emergence of a single dislocation reversed from (a) bright for a relatively low electron beam current of 9.3×10^{-6} A, through (b) dark-and-bright for 1.01×10^{-5} A, to (c) dark for a larger current, 1.09×10^{-5} A. The material was GaAs doped with 4.5×10^{18} cm^{-3} Se. (After Balk *et al.* 1976.)

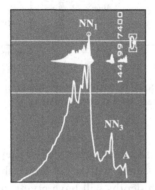

Figure 5.46 Typical CL spectrum of a green-emitting GaP LED at 90 K. The beam was set at 30 kV and about 50 nA. The wavelengths of the lines marked were $NN_1 = 568.5$ nm, $NN_3 = 548.5$ and $A = 536$ nm. (After Lohnert *et al.* 1979.)

The result of measurements of the intensities of the three lines NN_1, NN_3 and A along scans across the end of such a dislocation is shown in Fig. 5.47. This is strong direct evidence of an increase in nitrogen concentration toward the dislocation line. (The concentration of the A-centre is proportional to the nitrogen concentration and that of the NN-pair centres to the square of the N concentration.)

Hergert and Pasemann (1984), Pasemann and Hergert (1986) and Jakubowicz (1986) independently realized that although both EBIC and CL dark dislocation contrast can be treated by the Donolato model there was one significant difference. This is that C_{CL} is dependent on the depth of the dislocation (or other defect) while the EBIC contrast is not. This depth sensitivity arises from the self-absorption of CL light by the material. There will be some depth such that from this or greater depths,

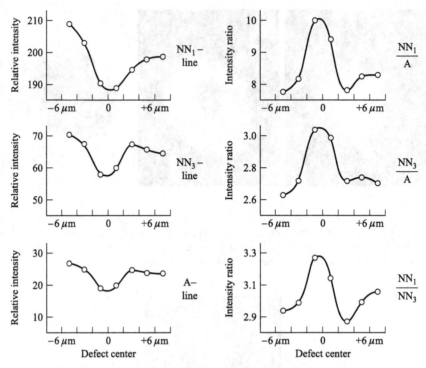

Figure 5.47   Line intensities and intensity ratios along a line through the centre of a dislocation (its position is marked zero) perpendicular to the surface of the LED of Fig. 5.46. (After Lohnert *et al.* 1979.)

all CL photons emitted upward are absorbed before they reach the free surface of the specimen. Hence, no CL signal will arise from this, or any greater, depth. Consequently defects at such depths will have no effect and will be invisible in CL micrographs although they can give dark contrast in EBIC (Fig. 5.48).

As the dislocation depth increases toward the disappearance level its contrast will fall because the dislocation is able to affect less and less of the CL that is emitted to contribute to the signal. Jakubowicz (1986), Pasemann and Hergert (1986) and Hildebrandt *et al.* (1991) used the Donolato model to treat simultaneous EBIC and CL dark contrast measurements to obtain information about the dislocation depth. Later Hergert *et al.* (1987) and Hildebrandt and Hergert (1990) treated CL, EBIC, PL and LBIC contrast together. Schreiber and Hildebrandt (1991) and Schreiber *et al.* (1991) discussed the determination of dislocation recombination strength from combined SEM-CL and −EBIC data, and Hildebrandt *et al.* (1998) outlined the theoretical fundamentals of the CL and EBIC contrast of defects in semiconductors.

This approach has seldom been used to study the properties of dislocations because, in most cases, dark CL contrast is due, mainly or entirely, to enhanced non-radiative recombination. The information obtainable is then essentially the same

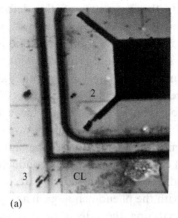 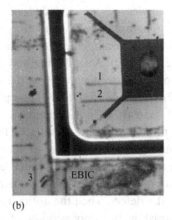

(a)   (b)

Figure 5.48   (a) CL and (b) EBIC micrographs of an area on an early GaAs LED. Dislocation 1 is visible in EBIC but not in CL and dislocation 3 is much less visible in CL. This is because 1 is deeper than the disappearance depth at which self-absorption eliminates all the CL before it can be emitted and dislocation 3 is near this depth. Compare the shallow dislocation, 2, which is visible in both EBIC and CL. [(After Schreiber and Hergert 1989.) In *International Symposium on the Structure and Properties of Dislocations in Semiconductors*, 1989. Conf. Series No. 104 (Bristol: Institute of Physics), pp. 97–107.]

(except for the dislocation depth, as we shall see) as in the phenomenological analysis of EBIC defect contrast, which has been more extensively employed (Section 5.5). However, there is a literature of CL contrast studies of dislocations [see in addition to the papers mentioned above, e.g. Hergert and Pasemann 1984, Hergert *et al.* 1987, Bode *et al.* 1987, Jakubowicz *et al.* 1987, and the review in Yacobi and Holt 1990 (pp. 121–46)].

The orientation, shape and depth of dislocations in semiconductors, can be found by simultaneous CL and EBIC studies (Jakubowicz 1986, Jakubowicz *et al.* 1987). The method is to compare the experimental results with theoretical curves derived for the ratio of the CL to the EBIC contrast expressed as (Jakubowicz 1986, Jakubowicz *et al.* 1987):

$$\frac{C_{\text{CL}}}{C_{\text{EBIC}}} = \frac{1 - e^{-H(\alpha - 1/L)}}{1 - e^{-h(\alpha - 1/L)}} \tag{5.62}$$

where H is the position of the defect, h is the penetration depth of the electron beam, α is the absorption coefficient, and L is the minority carrier diffusion length (see Fig. 5.49). This equation allows one to separate the contribution of the defect configuration to the contrast from that due to local changes in recombination properties. For $H > h$, $C_{\text{CL}} > C_{\text{EBIC}}$; for $H < h$, $C_{\text{CL}} < C_{\text{EBIC}}$, and for $H = h$ or $\alpha \gg 1/L$, $C_{\text{CL}} = C_{\text{EBIC}}$. Thus, the defect depth H can be found by comparing C_{CL} and C_{EBIC} at any distance from the defect, provided that h, α, and L are known. By obtaining the best fit for the experimental line data to simulated curves for different inclination angles of an extended defect, the geometry of the defect can be determined. The numerical calculations were carried out by Jakubowicz (1986) for an inclined dislocation intersecting the surface of a semi-infinite semiconductor at an angle β (see Fig. 5.50). A similar method by Pasemann and Hergert (1986) considered a dislocation positioned parallel to the surface at a particular depth. Jakubowicz *et al.* (1987) and Bode *et al.* (1987) compared experimental observations of both CL and EBIC contrast at non-radiative dislocations in GaAs with the calculations according to the Jakubowicz model (1986). These experiments showed that the method could be used effectively to determine the orientation, shape and depth (as well as the local recombination properties) of dislocations in semiconductors.

An additional difficulty is that both radiative and non-radiative recombination may occur at a defect. Then the difficulty with the phenomenological analysis of CL image contrast is the impossibility of separating the effects of the two. In the

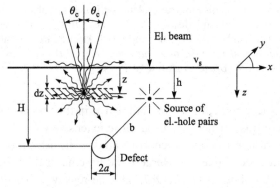

Figure 5.49   The geometrical configuration of the Jakubowicz model. (After Jakubowicz 1986. Reprinted with permission from *Journal of Applied Physics*, **59**, pp. 2205–9. Copyright 1986, American Institute of Physics.)

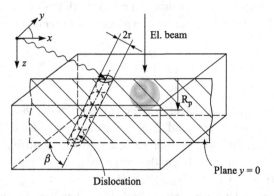

Figure 5.50   A dislocation at an angle β to the surface of a semi-infinite semiconductor. (After Jakubowicz 1986. Reprinted with permission from *Journal of Applied Physics*, **59**, pp. 2205–9. Copyright 1986, American Institute of Physics.)

phenomenological theory the contrast depends on the change of the minority carrier lifetime from the perfect material to the dislocation. This depends on the combined radiative and non-radiative recombination times. The observable lifetime, τ, is given by:

$$\frac{1}{\tau} = \frac{1}{\tau_{rr}} + \frac{1}{\tau_{nr}} \tag{5.63}$$

where τ_{rr} is the effective radiative recombination time due to all the radiative mechanisms operating and τ_{nr} is the non-radiative recombination time due to all the non-radiative recombination paths present. The radiative and non-radiative recombination times cannot be separated so information about the luminescence properties of the dislocation is not obtainable.

Fortunately, however, as noted at the end of the previous section, sometimes the radiative recombination is strong enough, i.e. τ_{rr} is short enough, that the dislocation (or other defect) is the source of detectable CL emission. Then bright CL contrast occurs in monochromatic CL images recorded at the emission wavelength. Such defect emission has a photon energy less than the band gap of the material. Spectral analysis of the emission provides information about the defect CL mechanism (Section 5.6.3).

Of course the quantitative evaluation of the equations derived from phenomenological theories for application to experimental situations requires the use of some analytical expression for the distribution of e-h pairs in the generation volume. This is discussed, e.g. by Hergert and Hildebrandt (1988), Hildebrandt and Hergert (1990, 1991). A popular alternative to the use of analytical approximations to the e-h depth-dose distribution is the use of Monte Carlo electron trajectory simulations (see Czyzewski and Joy 1990, Holt and Napchan 1994, Chim *et al.* 1992, Chan *et al.* 1993, Pey *et al.* 1993a,b, 1995a,b, and especially Joy 1988 and 1995).

5.6.2 *Other forms of dislocation contrast*

Dot-and-halo contrast, i.e. dislocations seen end on as dark dots surrounded by brighter regions, was observed by Kyser and Wittry (1964) and Casey (1967) in SEM CL images of GaAs heavily doped with Te (e.g. Fig. 5.51). Casey explained this as due to segregation of the dopant to the dislocations. It was later shown that a denuded zone was left around the dislocations (Shaw and Thornton 1968, Balk *et al.* 1976, Hastenrath *et al.* 1979 and Lohnert *et al.* 1979). The luminescence efficiency of GaAs goes through a maximum for a doping density of about 10^{18} cm^{-3}. In the commonest case, segregation results in a high concentration close to the dislocation line corresponding to a low CL emission in the black 'dot', and a surrounding denuded zone giving the bright 'halo'. Dark dot only, bright halo only and more complex appearances are also possible. This can also occur in GaAs doped with Se or Si although the smaller atomic misfits in these cases limit the tendency to

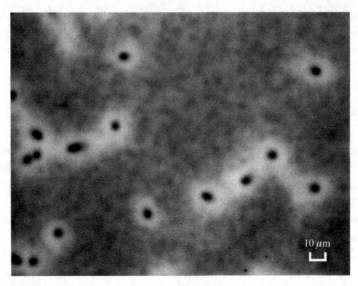

Figure 5.51 SEM CL micrograph showing dot-and-halo contrast at dislocations in GaAs doped with 10^{18} Te atoms cm^{-3}.

segregate to dislocations (Shaw and Thornton 1968) and in GaAs doped with S (Fornari *et al.* 1985). Thus, this well-established effect of dislocation-impurity interaction affecting bright CL defect contrast was complex, producing several different forms of contrast including the strongly marked dot-and-halo type. For a more detailed account, see e.g. Yacobi and Holt (1990, pp. 122–4). Dot and halo CL contrast was also seen in ZnTe after Li was diffused in (Bensahel and Dupuy 1979).

Similar dark and surrounding bright contrast occurred at small angle boundaries in CdS (Brümmer and Schreiber 1974), ZnTe (Bensahel and Dupuy 1979) and CdTe (Chamonal *et al.* 1983). This can also give rise to the appearance of dislocations as bright lines (halo only). It arises from variation of the panchromatic (overall, but mainly near-bandgap) CL radiation due to dopant concentration variations.

It was thought that this mechanism, however, was not responsible for the dot and halo contrast seen in some substrate-quality GaP. It was thought that the gettering effect of the dislocations reduced the concentration of non-radiative centres, thereby increasing the probability of radiative recombination (Davidson *et al.* 1975, Darby and Booker 1977).

A case in which there was good evidence to show that the contrast of dislocations was not due to impurities but due to the core structure of the defect itself was studied by Petroff *et al.* (1980a). They observed misfit dislocations in the interface between a top layer of $0.7\,\mu m$ of $Ga_{0.8}Al_{0.2}As_{0.75}P_{0.25}$ and a $0.2\,\mu m$ layer of $Ga_{0.92}Al_{0.08}As$. These layers were grown on $0.6\,\mu m$ of $Ga_{0.5}Al_{0.5}As$ on a GaAs substrate. For such studies, misfit dislocations were preferred to glide dislocations as the latter are always accompanied by point defects generated during movement. Also the epitaxial growth temperature would have been relatively low and the material purity high, making impurity decoration of the dislocations less likely. The samples were thinned to enable them to be examined in a STEM machine capable of carrying out TEM-like STEM analyses as well as EBIC and CL. It was found that all the numerous 60° dislocations, identified by STEM, appeared in dark contrast in both EBIC and CL images but the one sessile edge-type dislocation showed no contrast in EBIC or CL. The invisibility of this dislocation indicated that its core was not associated with any recombination levels, radiative (to give bright CL contrast) or non-radiative.

CL microscopy and spectroscopy are also useful for the microcharacterization of semiconductors. They can map defects to determine dislocation distributions and densities, as well as study impurity segregation. For example, semi-insulating (SI) GaAs exhibiting cellular dislocation arrays was studied by Wakefield *et al.* (1984). They observed in the CL spectra (see Fig. 5.52) peaks at 1.494 eV (due to residual carbon acceptors) and 1.514 eV (due to shallow donors or bound excitons). Monochromatic 1.494 eV CL micrographs (see Fig. 5.53) provide clear evidence of the segregation of carbon to the dislocation walls. Although dark spots along the dislocation walls were not detected in this study, they were later observed by Warwick *et al.* (1985). Such an inhomogeneous distribution of impurities could result in non-uniform electrical transport in the semiconductor and thus is highly undesirable for, e.g. GaAs integrated circuits. The uniformity in such cases could

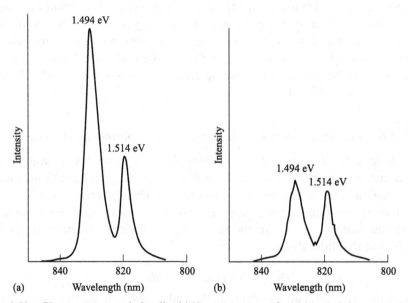

Figure 5.52 CL spectra, recorded at liquid He temperature, from (a) a bright area and (b) an adjacent dark area in semi-insulating GaAs. (After Wakefield *et al.* 1984. Reprinted with permission from *Applied Physics Letters*, **45**, pp. 66–8. Copyright 1984, American Institute of Physics.)

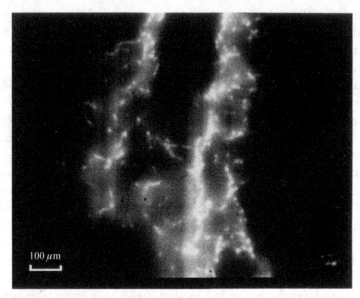

Figure 5.53 Monochromatic CL micrograph, recorded at liquid He temperature, of a cell boundary in semi-insulating GaAs using the 1.494-eV emission due to residual carbon. (After Wakefield *et al.* 1984. Reprinted with permission from *Applied Physics Letters*, **45**, pp. 66–8. Copyright 1984, American Institute of Physics.)

be improved, e.g. by indium doping to reduce the grown-in dislocation densities and by heat treatment (for more details on such inhomogeneous distribution of defects in SI GaAs, see e.g. a review by Yacobi and Holt 1990, pp. 170–8, and references therein). From CL studies of LEC SI GaAs, Warwick and Brown (1985) also determined that the dominant intrinsic defect, EL2, segregated from the cell interior to the dislocation cell walls, too.

5.6.3 *Localized radiative recombination and dislocation bright contrast*

Some defects, including dislocations, appear bright at longer wavelengths (lower photon energies) than band-gap-CL. Such CL emission can arise from radiative recombination involving energy levels in the gap. The spectra of the CL emission then gives information on the luminescence mechanisms (centres) responsible. The luminescent centres can be intrinsic to the dislocation core structure or related to impurities at the dislocation.

Dislocation emission in diamond

Early in the study of 'cathode rays' at the end of the nineteenth century, natural diamonds were found by Crookes in 1879 to emit cathodoluminescence in the first report of the phenomenon. Hence the name 'cathodoluminescence', which means cathode-ray-induced light emission. Diamond CL was and is much studied due to the many emission bands in the diamond CL spectrum and the economic importance of natural and synthetic diamonds. Now there is additional technological interest in synthetic diamonds and especially in epitaxial diamond films both for hard surface coatings and electronic applications. Diamond CL bandgap or 'edge' emission occurs in the ultraviolet but there are many forms of relatively strong emission in the visible. Dislocations emitting yellow-green and others blue CL occur in natural diamonds (Hanley *et al.* 1977 give an extensive account with references to the earlier literature). Lang and his co-workers employed x-ray topography combined with direct photography using visible CL to identify the dislocations responsible. Kiflawi and Lang (1974) found that the CL from the blue-emitting dislocations was linearly polarized with the electric vector always parallel to the defect. The degree of polarization was greatest, 90% or more, when the dislocations were straight and ran in $\langle 110 \rangle$ directions suggesting that the core structure was important. CL from the yellow-green emitting dislocations was ascribed to the H3 centre (Hanley *et al.* 1977). Further spectral information on the two colours of dislocation CL emission was reported by Lang (1977 and 1980). This was difficult to explain in detail, however, due to the relatively large amount of impurities in and the complex growth morphologies of natural diamonds. Sumida and Lang (1981) reported dislocation CL in a natural semiconducting (type IIb) diamond showing anisotropic polarization in sheets parallel to {111} planes. This they interpreted as due to jogs in glide dislocations, which had arisen through cutting dislocations threading the slip planes. Woods and Lang (1975) studied CL from synthetic diamonds and observed different colours coming from separate growth sectors. They observed luminescent lines emitting polarized blue light and others appearing orange-pink and reported the CL spectrum, which they ascribed to decorated dislocations.

Pennycook *et al.* (1980) studied natural semiconducting type IIb diamonds in a STEM with a CL attachment. Almost all the CL was emitted from dislocations. Some screw and 60° dislocations, identified by STEM, were luminescent and some of each type were not. This variation was also observed from one part of a single dislocation to another. They suggested this could be due to some portions being extended and others constricted or to impurity decoration of some segments but not others.

Yamamoto *et al.* (1984) studied individual blue-emitting dislocations using TEM and TEM CL. They found that the dislocations emitted a relatively broad band at 2.85 eV and confirmed that the CL was polarized along the line of the defect. They considered several possible emission mechanisms and eliminated all except recombination via donor-acceptor pairs along the dislocation line.

Diamond films grown by CVD typically include a high density of dislocations, stacking faults, and grain boundaries. In these films, there is a broad luminescence band at around 435 nm (2.85 eV). This is referred to as band-A emission (in analogy with that observed in natural diamonds) and associated with dislocations (Robins *et al.* 1989) and donor-acceptor pairs trapped at dislocations

(Graham *et al.* 1991, Ruan *et al.* 1991). (For a review of luminescence centres in diamond, see e.g. Collins 1992). It is thought that in natural diamond this emission is associated with closely spaced donor-acceptor pairs (boron being an acceptor). Ruan *et al.* (1992), however, examined the CL spectra of undoped and boron-doped diamond films and demonstrated that the broad band centred around 2.83 eV in undoped samples is not related to boron or to donor-acceptor pairs, but it is associated with dislocations.

CL studies of CVD diamond films also demonstrated the connection of the distribution of various impurities and structural defects to some specific growth sectors (e.g. Vavilov *et al.* 1980, Yacobi *et al.* 1993, Kanda 2003).

The preferential incorporation of defects in monocrystalline diamond films was studied by Yacobi *et al.* (1993) by CL microscopy and spectroscopy. In these studies, the CL spectra typically included broad bands centred around 440 nm and 550 nm and a narrow band at about 740 nm. The films grown on the (111) faces displayed extensive non-uniformities, that is, the spatial distribution of the sources of the 550 nm and 740 nm bands correlated with topographic features in these films. In contrast, the films grown on (001) faces did not show those bands, and high-quality regions exhibited uniform monochromatic (440 nm) CL images.

The surface of a diamond film grown on a natural type Ia (001) diamond substrate is shown in Fig. 5.54. This reveals square growth features with ⟨110⟩ sides. CL monochromatic images and spectra recorded from one of the square regions in

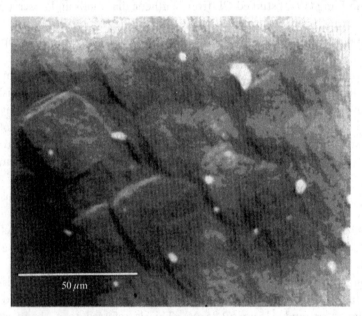

Figure 5.54   Secondary electron image of a diamond film (18 μm-thick) sample grown on natural type Ia (001) diamond substrate. The morphology shows squares with ⟨110⟩ sides. (After Yacobi *et al.* 1993.)

Fig. 5.54 are presented in Fig. 5.55. Strong non-uniformities in the distribution of impurities and defects are clearly observable. Fig. 5.55 shows CL monochromatic images (a) recorded at 440 nm, (b) recorded at 600 nm, and (c) CL spectra from corresponding bright regions in (a) (denoted 1) and (b) (denoted 2). These CL spectra consist of a broad band centred at about 440 nm and other features at longer wavelengths. The 440 and 600 nm monochromatic CL images exhibit strong variations in luminescence intensity (Figs. 5.55a and b). The non-uniform distribution of the impurities associated with the 440-nm band was related to the dislocations. The 600-nm band was associated with the ⟨110⟩ sides of the squares (see Fig. 5.54).

Cremades *et al.* (1993) used CL SEM to examine the top surface and cross-sectional samples of CVD diamond films. They concluded that the dislocation-related CL emission was mostly localized at the grain boundaries of the columnar grains and that the concentration of associated radiative centres was greater in boundaries parallel to the growth axis.

Graham *et al.* (1994) employed CL in a transmission electron microscope to study the distribution of dopants and defects in boron-doped (100) textured diamond films.

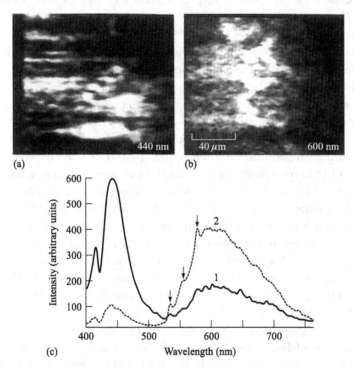

Figure 5.55 CL monochromatic images and spectra corresponding to Fig. 5.54. CL monochromatic images recorded at (a) 440 nm and (b) 600 nm. (c) CL spectra from corresponding bright regions in (a) (denoted 1) and (b) (denoted 2). (After Yacobi *et al.* 1993.)

They observed that the luminescence (related to dislocations and unrelated to boron) was generated (inhomogeneously) from dislocations within the grains and at grain boundaries (Graham *et al.* 1994). In contrast, luminescence due to donor-acceptor pairs (related to boron) was found to be uniformly distributed within these films. It was also found that the dislocation density in undoped films was much lower than in boron-doped samples, so boron doping during growth leads to higher dislocation densities. (Graham *et al.* 1994).

Dislocation exciton luminescence in germanium and dislocation impurity-activated CL in silicon

Recombination radiation, associated with dislocations was observed in germanium by Newman (1957). Although studies of this radiation have not used SEM or TEM CL, and so have not resolved individual dislocations, they have been important for the development of dislocation exciton emission theory. Benoit a la Guillaume (1959) studied electroluminescence (EL) produced by passing an electrical current through crystals containing more than 10^6 dislocations cm^{-2}. He showed that the less-than-bandgap infrared EL was emitted from dislocation sites identified by chemical etch pits. Dislocation EL in Ge was further studied by Gippius and Vavilov (1963, 1965a, b), Gippius *et al.* (1965) and Kolyubakin *et al.* (1984) and over 30 years by several other groups (see Lelikov *et al.* 1989 for references). The dislocation emission peak is at about 0.5 eV (i.e. 2.4 μm with a band width from 2 to 2.6 μm as reported by Kolyubakin *et al.* 1984). (The conduction band to valence band recombination radiation from Ge is at 0.7 eV.)

Lelikov *et al.* (1989) observed the PL (photoluminescence) from modern material with lower densities of dislocations ($10^3-10^5 cm^{-2}$). They reported a sharp (about 1 meV wide) line at 513 meV plus a background band in the range 500−600 meV. They studied the temperature variation of this spectral line and concluded that it is due to the radiative recombination of dislocation excitons (DEs) (see Section 5.1.6). These are formed by an electron and a hole in one-dimensional bands that are split off of the conduction and valence bands by the deformation potential. They found the experimental value of the DE binding energy, E_{DE}, to be 3 meV in good agreement with theory and discussed many other features of the observations and their interpretation. Possibly as a result of this extensive work on dislocation luminescence in Ge, Russian workers have published extensively on DEs: Rebane and Shreter (1993), Rebane *et al.* (1997), Shreter and Rebane (1996), Shreter *et al.* (1993, 1996a, 1996b, 1997).

The most conclusive experimental study of the mechanism of dislocation CL emission was that of Higgs, Lightowlers and their co-workers on Si as discussed in Section 5.6.4. This showed conclusively that light contamination with 'heavy' metals is essential for CL D-line emission from both dislocations and from the point defect debris on the slipped planes between the dislocations.

Dislocation luminescence in III-V materials

Complex dot-and-halo impurity-dependent CL contrast was intensively studied very early in GaAs (see Section 5.6.2). Since then CL contrast at dislocations and other defects has been studied in many other III-V compounds and alloys due to their importance for optoelectronic applications.

Davidson *et al.* (1975) used SEM CL to observe plastically bent GaP and found that dislocations appeared in dark contrast. Brantley *et al.* (1975) reported that the electroluminescence efficiency was reduced by 50% in green emitting GaP by $2-5 \times 10^5 \, \text{cm}^{-2}$ dislocations. Davidson and Dimitriadis (1980) put forward evidence that indicated that this reduction is essentially due to a Cottrell atmosphere of non-radiative recombination centres segregated to the dislocations.

The use of III-V alloys in laser heterostructures led to much study of the effects of dislocations on initial properties and their role in the dark line defect (DLD), rapid failure mechanism. Dislocations provide a competing non-radiative recombination path for injected carriers. By allowing the carriers intended to participate in the light amplification by stimulated emission of radiation to recombine non-radiatively, defects can locally delay the onset of laser action. It was observed in the earliest days of semiconductor injection lasers that misfit dislocations (Abrahams and Pankove 1966, Osvenskii *et al.* 1967) and other defects (Hatz 1968, Kressel *et al.* 1969 and Tretola and Irvin 1968) correlated with non-uniform emission in LEDs and lasers.

Defects were found to be involved in degradation (failure) mechanisms of the short-lived early lasers also (Unger 1968, Kressel *et al.* 1970, Mettler and Pawlik 1972, and Eliseev 1973). The relatively rapid degradation of some stripe-geometry, double-heterostructure $GaAs/Ga_xAl_{1-x}As$ lasers was found to be due to the development of dark lines visible in infrared microscopy (DeLoach *et al.* 1973). These large dark line defects (DLDs) were shown by transmission electron microscopy to be tangled dislocation dipole networks (Petroff and Hartman 1973). A large amount of work was carried out on this using a range of techniques, listed together with their findings in Table 5.5.

It became accepted that DLDs (i) were complex tangles of irregular dislocation dipoles and loops, as shown in Fig. 5.56, (ii) lie in the active layer of the laser structure and are of interstitial character, (iii) grow by climb during injection or photopumped laser operation, and (iv) originate at grown-in threading dislocations (i.e. those running in the growth direction and roughly perpendicular to the active layer).

The enhancement of defect reactions by recombination and other forms of electronic excitation is well established (Dean and Choyke 1977). Petroff *et al.* (1977) suggested that electronic excitation provides the activation energy for the rapid climb, which forms the dipole tangles, constituting DLDs, from threading dislocations. The point defects involved in the rapid climb they identify as the 'DX' centres that are known from deep level transient spectroscopy to be present in large densities in the alloys used in the lasers (Petroff 1979). This mechanism is

Table 5.5. *Studies of DLDs in degraded GaAs/Ga$_x$Al$_{1-x}$As double heterostructure lasers*

Technique	Main findings	References
Infrared photoluminescence and electroluminescence microscopy or visible light microscopy	DLDs are large enough to be seen in light microscopy and correlate with dislocation etch pits	DeLoach *et al.* (1973), Ito *et al.* (1974), Johnston *et al.* (1974), Nakashima *et al.* (1977), Hartman and Koszi (1978)
Transmission electron microscopy	DLDs are complex groups of interstitial type dislocation dipoles	Petroff and Hartman (1973, 1974), Hutchinson *et al.* (1975), Hutchinson and Dobson (1975, 1980), Ueda *et al.* (1977), Monemar and Woolhouse (1977), Dobson *et al.* (1977), O'Hara *et al.* (1977)
SEM CL and EBIC	DLDs appear in dark contrast, i.e. they are sites of enhanced non-radiative recombination	Petroff *et al.* (1978a), Chin *et al.* (1979), Petroff *et al.* (1980a,b), Petroff (1981)
High-resolution, field emission scanning CL, EBIC, STEM and SDLTS analyses	Additional structure and property information	Petroff *et al.* (1978b), Petroff (1979), Petroff (1981)

supported by SDLTS observations in a field emission gun scanning transmission electron microscope (FEGSTEM). This showed that there is a depletion of DX centres at DLDs whose position could be determined by EBIC as shown in Fig. 5.57.

Rapid point defect migration also occurs under laser operating conditions in response to the elastic stresses arising from the stripe structure of these early DH lasers. This results in a general darkening of the central stripe area and a brightening of the edges of the stripe as seen in either SEM CL or EBIC micrographs (Wakefield 1979, Stevenson *et al.* 1980, Robertson *et al.* 1981, Holt 1996). This effect has nothing to do with any extended defect and would occur in lasers free of any initial microstructural defects.

The recombination effects of DLD dislocations were studied by Petroff and his co-workers who built a unique, field-emission, ultra-high vacuum, scanning transmission electron microscope with low-temperature CL, EBIC and SDLTS facilities. They used it to study both DLDs and misfit dislocations in Ga$_x$Al$_{1-x}$As heterostructures (Petroff *et al.* 1978b). They used STEM to identify defects, SDLTS to find the defect energy level in the gap and CL to observe its radiative or non-radiative behaviour. Dislocation loops correlated with reduced CL emission

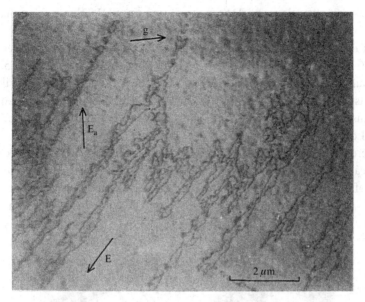

Figure 5.56 A dark line defect in a degraded (001) double heterostructure GaAs/Ga$_x$Al$_{1-x}$As laser seen by transmission electron microscopy. This area was part of a large dislocation network containing dipoles with an elongation parallel to [$\bar{1}$10]. The operative Bragg reflection had a reciprocal lattice vector **g** = $\bar{2}$20. (After Hutchinson and Dobson 1980. Reprinted with permission from *Philosophical Magazine*, **41** (1980), pp. 601–14; Taylor & Francis Ltd., the Journal's web site: http://www.tandf.co.uk/journals.)

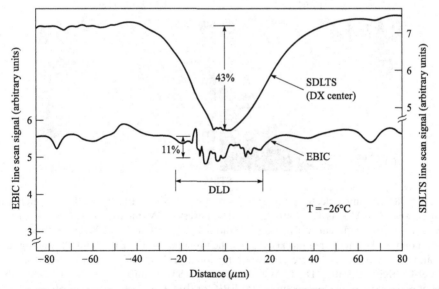

Figure 5.57 Scanning deep level transient spectroscopy and EBIC signal strengths for a line scan across a DLD in a Ga$_x$Al$_{1-x}$As DH structure. The specimen temperature T = 126° C was such that the SDLTS line scan displays the DX centre distribution. (After Petroff *et al.* 1978b.)

while low concentrations of DX centres seen in SDLTS micrographs did not correlate with dark lines in EBIC images.

Petroff *et al.* (1980a,b) examined a misfit dislocation network at a heterojunction between $Ga_xAl_{1-x}As_yP_{1-y}$ epitaxial layers with the results shown in Fig. 5.58. Their STEM analysis of dislocation, D_1, showed it to be a Lomer-Cottrell sessile. It gave no contrast in CL or EBIC, i.e. it was electrically inactive. They suggested that this was because core reconstruction eliminates dangling bonds and kink sites from the cores of Lomer-Cottrell sessiles.

A difference of about 10% in the EBIC signal reduction was found between the 60° dislocations of the two orthogonal grids in Fig. 5.58. The same difference in the strength of the non-radiative recombination between the two sets of dislocations was observed by CL measurements. The specimen and heterojunction interface

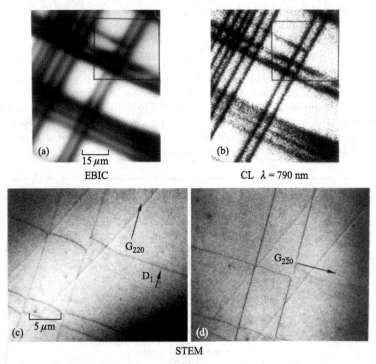

Figure 5.58   Scanning electron micrographs of a misfit dislocation network at a heterojunction between $Ga_xAl_{1-x}As_yP_{1-y}$ (001) layers. The lower STEM micrographs, (c) and (d), were taken with two different diffraction conditions G = 220 and = 2$\bar{2}$0 respectively. They correspond to the small area marked out by the fine lines on the EBIC and CL micrographs, (a) and (b), respectively. There are two orthogonal grids of 60° dislocations and a sessile Lomer-Cottrell dislocation D_1. D_1 is clearly visible in STEM diffraction contrast especially in (c) but it gives rise to no contrast in the (a) EBIC or (b) CL images. [After Petroff *et al.* 1980a. Reprinted with permission from *Physical Review Letters*, **44**, pp. 287–91, 1980 (http://publish.aps.org/linkfaq.html). Copyright 1980 by the American Physical Society.]

orientation was (001) so the dislocations of the two grids are of different polarity (see Section 4.1.3). The misfit dislocations aligned with the two ⟨110⟩ directions in epitaxial (001) interfaces in semiconducting compounds were already known to differ in mobility (Abrahams *et al.* 1972, Rozgonyi *et al.* 1974a) and Petroff *et al.* found that the less mobile type were the more effective in recombination. This was contrary to the early models, which associated both the electrical effects and the mobility with kinks in the dislocation lines. On this basis the more mobile type would be expected to have a higher density of kinks and hence be more electrically active.

Heterostructures of other materials were also found to develop DLDs during degradation. Mahajan *et al.* (1979) and Mahajan (1981) used TEM to observe the dislocation clusters in optically degraded regions in $In_xGa_{1-x}As_yP_{1-y}$ layers grown on (001) InP substrates. Temkin *et al.* (1981) used EBIC and transmission CL SEM to observe DLDs in rapidly degraded high-radiance DH structures in the same materials. Elliott *et al.* (1982) used transmission CL in an SEM and polarized infrared transmission (light) microscopy and electrochemical etching. They showed that dislocation clusters in the InP substrates gave rise to non-radiative regions in the epitaxially grown layers.

Dislocation luminescence in II-VI materials

In addition to diamond, in the earliest days of cathode ray studies, other materials were found in which bright CL was excited. ZnS was one that emitted brightly and with a wide range of colours, depending on the doping impurities present so it became one of the main phosphors. These are materials used in powder form to coat the CRTs (cathode ray tubes) long used in oscilloscopes, radar displays, computer monitors and TV sets. These materials were also of interest for possible use in electroluminescent (EL) displays. Much research and development was, therefore, carried out on phosphors. Fischer (1962) showed that efficient electroluminescent powder particles emitted from line sources. These he suggested were copper-sulphide-decorated conducting imperfection lines but no decorating precipitates were visible in light microscopy (Lendvay and Kovacs 1966). The importance of impurity decoration is also emphasized by the work of Daniels and Meadowcroft (1968). They found that twist boundaries in ZnS had no effect on electroluminescence. These boundaries consist of crossed grids of screw dislocations and were introduced by plastic deformation and so were cleaner than in the deliberately heavily doped and heat-treated phosphor material. Thus the cleaner dislocations introduced by deformation are not luminescent, just as in the case of Si (see Section 5.6.4).

Brümmer and Schreiber (1974) observed liquid nitrogen temperature dislocation CL in CdS. Dislocations, freshly introduced by deformation, gave bright contrast while lightly decorated dislocations gave dark contrast.

Myhajlenko *et al.* (1984) and Batstone and Steeds (1985) studied bright contrast at individual dislocations and dislocation tangles in ZnSe using both CL and TEM.

Both the so-called Y-band (2.60 eV) and S-band (2.52 eV) CL in ZnSe were related to dislocations. Donor-acceptor pair (DAP) emission at 2.68 eV was also observed, originating from some local areas in dislocation tangles. Steeds *et al.* (1991) proposed a model for the Y_o radiation from dislocations in ZnSe and some other II-VI semiconductors and previously explained as excitonic (Dean 1984). The model of Steeds *et al.* (1991) involved dislocation excitons formed from electrons and holes in the one-dimensional energy bands associated with the dislocations.

Negrii and Osipyan (1978, 1979 and 1982) studied dislocation recombination radiation in CdS crystals plastically deformed at temperatures from 4.2 K to 300° C. To establish the correspondence of the emission sites to the positions of the dislocations, Negrii and Osipiyan (1982b) recorded observations like those in Fig. 5.59. Fig. 5.59a shows the emission from a deformed crystal at 6 K, and for comparison, the surface etch pits in the same area are shown in Fig. 5.59b. They found, by comparison, that, surprisingly, the luminescent bands do not correspond to slip lines or dislocation pile-up at the ends of slip bands. Moreover, the subgrain boundaries shown by rows of closely spaced pits in Fig. 5.59b appeared dark on

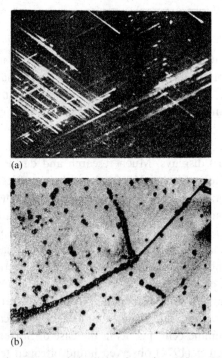

(a)

(b)

Figure 5.59   (a) Photomicrograph of the light, from $\lambda = 505$ to 510 nm, emitted by a CdS crystal. The area shown is 200 µm wide. (b) Light microscope image of the dislocation etch pits on the same area of the specimen. (After Negrii and Osipyan 1982b. Reprinted with permission from *Soviet Physics Solid State*, **24**, pp. 197–9. Copyright 1982, American Institute of Physics.)

emission photos like 5.59a. Geometrical analysis showed that the luminescent lines are tracks of moving isolated dislocations and terminate at those places where pits appear on etching. They concluded that, during slip at low temperatures, individual moving dislocations in CdS produce along the path behind them, specific crystal defects with a characteristic emission spectrum. To establish the character of this specific debris centre they applied elastic stresses to lift the orientational degeneracy of the centres. They observed splitting of the lines in the dislocation emission region of the luminescence spectrum. They concluded that the centres possessed C_S symmetry and so were anisotropic associations that could be composed of both intrinsic defects and impurity atoms. This spatial separation of passing dislocations and the point defect debris left behind them was only possible in low-temperature plastic deformation. In their earlier work on higher temperature indentation rosettes, Negrii and Osipiyan (1978, 1979) found that the luminescence emission and etch pit patterns did correspond. Hence, at these temperatures, a large fraction of the luminescent-centre debris left behind dislocations probably diffused to the dislocations.

Klassen and Osipiyan (1979) reviewed a substantial body of work carried out in Osipiyan's laboratories. These studies compared the effects of basal (0001) slip plane dislocations in wurtzite structure CdS and CdSe, with those of $(10\bar{1}0)$ prismatic slip plane dislocations. The basal dislocations are of opposite core polarities, α and β (e.g. A(s) and B(s) respectively), whereas the prismatic dislocations are non-polar, as discussed in Section 4.1. Thus the basal dislocations produce strong piezo-electric fields whereas the prismatic dislocations do not. It was found that the introduction of basal dislocations resulted in a strongly non-exponential decay of photoconductivity. Thus at high concentrations of injected carriers, the recombination rate greatly exceeded that in non-dislocated crystals whereas at low concentrations the recombination rate is reduced compared with that of non-dislocated crystals. Both types of dislocation shift the luminescence emission intensity between the bands of bound excitons to increase that of the excitons bound to neutral acceptors, and in this case the effect of the prismatic dislocations was the stronger. Basal dislocations produce a wide infrared luminescent emission, polarized parallel to the glide planes. Its frequency was about half that for recombination of carriers across the forbidden gap. The prismatic dislocations did not produce this effect.

A number of narrow, non-polarized emission bands were introduced into the edge region of dislocated CdS. The bands at 2.430 and 2.447 eV arose only at basal dislocations in high resistivity crystals as shown in Fig. 5.60.

Both types of dislocation produce additional optical absorption on the low energy side of the absorption edge. The basal dislocation absorption is stronger but the prismatic absorption tail contains a polarization-dependent structure shown in Fig. 5.61. At about 0.3 eV from the absorption edge, the vertical rise at the left for the undeformed specimen, there is a shoulder that is polarized parallel to the Burgers

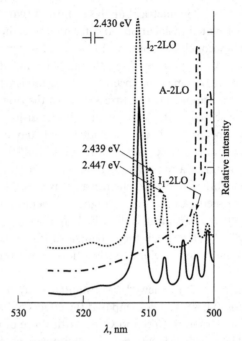

Figure 5.60 The local cathodoluminescence spectra produced at basal and prismatic dislocations in CdS at 10 K. The dot-dashed line is for a dislocation free region. The dotted line is the spectrum from a region containing basal plane dislocations. The solid curve is the spectrum from a region containing prismatic plane dislocations. (After Klaasen and Osipiyan 1979.)

vector of the dislocations. At shorter wavelengths the dichroism weakens. That is, below about 520 nm the relative absorption coefficient values for light polarized parallel and perpendicular to the slip plane are reversed.

Microscopic studies of light propagation through deformed crystals showed that the light propagation was strongly rearranged in the vicinity of the dislocation arrays in the glide bands. The light underwent local concentration and focusing under various circumstances. This Klassen and Osipiyan attributed to elastic and electrostatic field induced refractive index changes.

Klassen and Osipiyan point out that the absorption step for the prismatic dislocations of Fig. 5.61 and the wide infrared emission from basal dislocations show that there are bound electron states associated with both types of dislocations. Only the polar dislocations in the basal plane are likely to contain unpaired electrons forming a level in the mid-gap and giving rise to a wide emission band in the infrared at the half-gap photon energy.

The occurrence of the shallow level emission line doublet at 2.430 and 2.447 eV in Fig. 5.60, for both types of dislocation suggests a common origin, which could be the deformation potential.

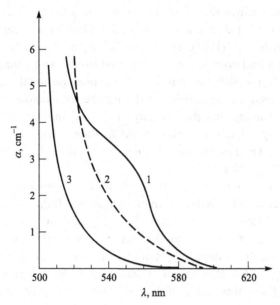

Figure 5.61 Absorption spectra of CdS containing a density of 10^8 prismatic-plane dislocations cm^{-2}. The solid curve on the left is the absorption edge spectrum of an undeformed reference specimen. The solid curve on the right is for the deformed specimen, with the polarization parallel to the slip planes and the dashed curve is for polarization perpendicular to the slip planes. (After Klassen and Osipiyan 1979.)

Thus a picture very similar to the three-band model for Ge dislocations appears likely to apply to II-VI dislocations. That is, the basal plane dislocations appear to have a deep level like the dangling-bond states and both types of dislocation have associated shallow states like those split off the energy bands by the elastic field in the case of Ge.

The differences between the effects of polar, basal-plane dislocations and non-polar prismatic plane dislocations were reported in Section 4.1. These effects were interpreted by Osipiyan and his co-workers in terms of the exciton scattering effects of the dislocations. For a recent summary of this work see Klassen and Osipiyan (1979).

For convenience, other luminescence properties are discussed in the following sections.

Dislocation excitons and CL emission

Steeds (1989) and his co-workers used TEM CL to study three cases of dislocation emission that could be thoroughly analysed. These included the bright CL contrast of dislocations in ZnSe epitaxially grown on GaAs, and of misfit dislocations in AlGaAs/GaAs quantum well structures. They also studied the suppression of exciton luminescence by dislocations in InP. Here we will discuss only the work on ZnSe.

The dislocation exciton model proposed by Emtage (1967) and developed by Bozhokin *et al.* (1982) (see Section 5.1.6) was applied to dislocation Y_o radiation from ZnSe by Steeds *et al.* (1991). The dislocation excitons again were formed by the pairing of electrons and holes in one-dimensional bands, with a binding energy E_{DE}. Using a classical Maxwell-Boltzmann statistics expression for the free energy of the system, they obtained an expression for the number of excitons per unit length in equilibrium, n. Assuming that the intensity of the Y_o emission as a function of the temperature is proportional to n, they could fit their theoretical expression to the experimental data. The best fit was obtained for a value of the dislocation exciton binding energy, $E_{DE} = 50$ meV.

The model of Bozhokin *et al.* (1982) led to an expression for the energy of a dislocation exciton. This is larger than that of the Bohr model of the exciton for ZnSe, $E_{BE} = 19$ meV, by a factor that depends on the elongation factor a_{par}/a_{perp}. Here a_{par} and a_{perp} are the dimensions of the exciton parallel and perpendicular to the dislocation line, respectively. For $a_{par}/a_{perp} = 3$, taking account of the anisotropy of the effective mass tensors for electrons and holes for edge dislocations, the theoretical value of E_{DE} is close to the experimental one (Steeds *et al.* 1991).

The band gap of ZnSe is $E_g = 2.8$ eV and the photon energy of the Y_o radiation, hv_y is 2.6 eV. Steeds *et al.* (1991) pointed out that since we can write

$$hv_y = E_g - E_e - E_h - E_{DE} \tag{5.64}$$

and the best fit value for the dislocation exciton energy $E_{DE} = 0.050$ eV (see Fig. 5.62), values for the energy depths of the electron (E_e) and hole (E_h) one-dimensional bands of 75 meV would correctly give $hv_y = 2.8 - 0.075 \times 2 - 0.050 = 2.6$ eV

There were no theoretical treatments leading to values for these band depths in ZnSe. However, calculations for CdTe, which also emits Y_o CL, indicate that these band depth values are 'perhaps not unreasonable' (Steeds *et al.* 1991).

Polarity dependence of dislocation CL in compounds

Mitsuhashi *et al.* (1967) studied the effects of dislocations on the emission spectra of CdS under X-ray excitation at 77 K. They bent wurtzite structure crystals in opposite polar senses to introduce known densities of Cd(s) or of S(s) basal plane dislocations after annealing. (The use of plastic bending for introducing known densities of parallel dislocations was discussed in Section 4.1.3). (They identified the type of dislocation in each case on the assumption that they were of the shuffle set form (see Section 4.1.2). If their cores are of the glide-set, the polarity would be reversed as usual.) They reported that the red emission at 635 and 740 nm from the crystals with S dislocations was much brighter than from the crystals with Cd dislocations. Bending about the *c*-axis, presumably introducing non-polar prismatic glide plane dislocations, resulted in green edge emission. The effect of low densities of grown-in

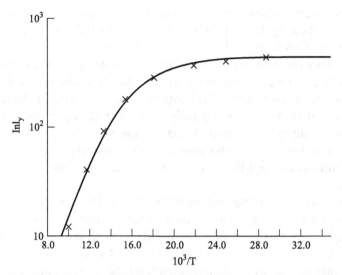

Figure 5.62 Plot of the logarithm of the intensity, I_y, of Y_o dislocation CL from ZnSe versus reciprocal temperature. The crosses are experimental points and the continuous curve shows the excellent fit of the temperature dependence of n, the number of dislocation excitons per unit length taking $E_{DE} = 50$ meV. [After Steeds *et al.* 1991. In *Polycrystalline Semiconductors II.* (Springer Proc. In Phys. Vol. **54**), ed. J. H. Werner and H. P. Strunk (Berlin: Springer-Verlag), pp. 45–49 (1991). With kind permission of Springer Science and Business Media.]

dislocations was also studied. In relatively dislocation-free regions containing only about 10^4 dislocations cm^{-2}, the blue edge emission was found to be weaker than in regions of higher densities.

Esquivel *et al.* (1973, 1976) deformed GaAs at 700° C in opposite polar senses in four point bending (see Section 4.1.3) to introduce either α (i.e. A(s)) or β (i.e. B(s)) dislocations. Both appeared in indistinguishable dark contrast.

Schreiber *et al.* (1997) were able to show that the Y emission (Dean 1984) in CdTe could be attributed to a particular type of dislocation. To do this it was necessary to study the SEM CL at low temperatures in indentation-deformed samples. There have been numerous studies of the detailed form of plastic deformation produced by indentations in sphalerite-structure semiconducting compounds (see Section 4.1.3 and e.g. Hirsch *et al.* 1985, George and Rabier 1987, Louchet and Thibault-Desseaux 1987). These authors studied the deformation in the plastic zone around indentations of the type made for hardness measurements on various crystallographic faces of the sphalerite structure. This led to a detailed interpretation in terms of the slip geometry, the sense and type of slip expected, dislocation interactions and the known differences in velocities of As(g) and Ga(g) dislocations. Only the results of these analyses will be reported here, to interpret the micrographic CL observations.

Y-band emission was found and ascribed to dislocations by Dean *et al.* (1984) in CdTe and ascribed by Dean (1984) to radiative recombination at dislocations. Schreiber *et al.* (1997) were able to show that in CdTe this emission could be attributed to only one polar type of dislocation. For this they studied the SEM CL at low temperatures in indented samples. Hirsch *et al.* (1985) analysed the anisotropy of hardness and the detailed form of the plastic deformation around indentations made on {111} faces of GaAs. Combining their results with the approach of George and Rabier (1987) and Louchet and Thibault-Desseaux (1987) led to a detailed interpretation in terms of the slip geometry, the sense and type of slip expected, dislocation interactions and the known differences in velocities of As(g) and Ga(g) dislocations.

Schreiber *et al.* (1997, 1999a) indented (111)Cd, ($\bar{1}\bar{1}\bar{1}$)Te and (110) faces of CdTe and used SEM CL to study configurations of dislocations that had been identified crystallographically. This revealed that CL emission originated at Te(g) dislocations only, while Cd(g) and screw segments gave dark CL contrast. CL spectroscopy determined that the radiative recombination from the Te(g) dislocations had a zero-phonon peak at 1.476 eV (5 K) and was the well-known Y luminescence due to extended defects in CdTe (Schreiber *et al.* 1997). Both the bright and dark CL contrast varied with temperature. They concluded the CL emitted from the Te(g) dislocations was due to radiative decay of excitons bound to these dislocations, while the Cd(g) dislocation dark CL contrast was due to non-radiative recombination. Schreiber *et al.* (1999a) studied the glide geometry and distribution of polar dislocations in the plastic deformation zones of microindentations. They employed CL contrast to distinguish Cd(g) from Te(g) dislocations in the deformed zones. The CL observations of defect arrangements were interpreted in the framework of a ⟨110⟩ glide prism model (Schreiber *et al.* 1999a). The results supported the three-dimensional ⟨110⟩ glide prism arrangement, the coexistence of Cd(g) and Te(g) dislocations in tangential and tetrahedral deformation zones and correlated propagation of these dislocations during indentation-induced plastic flow (Schreiber *et al.* 1999a).

Wurtzite-structure (hexagonal) CdS crystals deformed by prismatic slip (see Section 4.1.5) were found to emit luminescence bands, different from those seen in crystals deformed by basal plane slip (Negrii *et al.* 1991). Tarbaev (1998) examined the low-temperature (4.2−77 K) luminescence emission from wurtzite-structure CdSe crystals deformed by basal (0001) and prismatic {1$\bar{1}$00} plane slip, both with the same Burgers vector $\mathbf{b} = 1/3\langle11\bar{2}0\rangle$. Both slip systems operate in deformation at temperatures below half that of the melting point for CdSe. Indentation of the (0001) and {10$\bar{1}$0} planes was used to selectively produce essentially only prismatic or only basal plane dislocation half-loops respectively, as shown schematically in Fig. 5.63. Indentations in the (0001) plane produce rosettes with rows of dislocations gliding away in the ⟨1$\bar{2}$10⟩ directions to form the rays or petals of the rosette as in Fig. 5.63a and producing near-surface dislocation half loops in {1$\bar{1}$00} type planes of

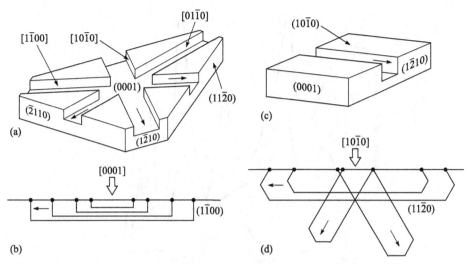

Figure 5.63 (a) The crystallographic alignment of the rosette around an indentation in the (0001) basal surface, and (b) the form of the near-surface dislocation loops on a prismatic $\{1\bar{1}00\}$ slip plane that result. The surface flow produced by indenting an $\{10\bar{1}0\}$ surface, as shown in (c) is crystallographically simpler and it results in (d) near-surface dislocation half-loops on the basal slip plane only. (After Tarbaev 1998. Reprinted with permission from *Physics of the Solid State*, **40**, pp. 1672–5. Copyright 1998, American Institute of Physics.)

the form shown in Fig. 5.63b. Indentation of $\{10\bar{1}0\}$ planes does not produce a rosette, only rays in the two opposite directions along $\langle 1\bar{2}10 \rangle$ as in Fig. 5.63c. This results in dislocation half loops in the (0001) basal plane like those represented in Fig. 5.63d. (For indentations on both types of surface, dislocations of the other kind are, in fact, also produced but only at depth in the CdSe. Due to strong self-absorption, luminescence from these does not escape from the crystal to contribute to the observed spectra.)

The emission spectra obtained in these alternative cases are shown in Fig. 5.64. These 4.2 K spectra from the high dislocation density, sub-surface volume around the indentations, contain three strong bands that do not appear in the as-grown crystals. These are at 1.7645, 1.7731 and 1.792 eV, labelled *a*, *b* and *c*. As the temperature rises, emission *a* becomes weaker relative to *b* until, at 77 K, they are of comparable intensity as shown by the dashed curve. Emission *c* predominates for basal dislocations (spectrum 1), whereas prismatic dislocations mainly emit bands *a* and *b* (spectrum 2) at 4.2 K. The ball-and-wire models of Osipiyan and Smirnova (1968 see Section 4.1.5) showed that basal plane dislocations in wurtzite-structure materials like CdSe are polar while prismatic dislocations are not so these spectral differences may be polarity related.

Later, systematic studies of CL emission from misfit dislocations of opposite polarity [i.e. B (g) versus A (g) type dislocations] were carried out in

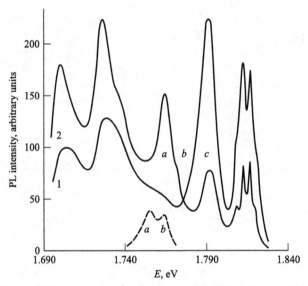

Figure 5.64 Photoluminescence spectra from the zone round indentations on the {10$\bar{1}$0} surface (spectrum 1, i.e. the lower curve at the left, due to basal plane dislocations) and on the (0001) face (spectrum 2, due to prismatic plane dislocations) of wurtzite-structure CdSe crystals at 4.2 K. The dashed curve represents bands *a* and *b* at 77 K. (After Tarbaev 1998. Reprinted with permission from *Physics of the Solid State*, **40**, pp. 1672–5. Copyright 1998, American Institute of Physics.)

ZnSe/GaAs (001) structures by Hilpert *et al.* (2000) and Schreiber *et al.* (1999b, 2000). In high quality pseudomorphic films, above the critical thickness, Se(g) misfit dislocations were found to nucleate first (Hilpert *et al.* 2000). Only the Se(g) misfit segments emitted Y luminescence, and this was found to be strongly polarized parallel to the dislocation line. These observations further supported attributing the defect Y luminescence to one-dimensional excitonic states at the dislocations (Hilpert *et al.* 2000).

Deformation triboluminescence in II-VI compounds

Osipiyan and Steinman (1973) plastically deformed CdS wurtzite-structure single crystals by basal plane slip. At low temperature, the radiation consisted mainly of lines due to bound excitons and the effect of introducing basal plane dislocations by strain was to reduce the quantum yield of luminescence, especially of the main, higher-photon-energy peak (see the peak at the right in Fig. 5.65). The dislocation density in the deformed crystals of Fig. 5.65 was about 10^7cm^{-2}. Their spectral studies showed that the reduction of the luminescence was due to an exciton-lifetime reduction produced by the dislocations, probably due to the energy of the excitons being consumed in a dislocation ionization process.

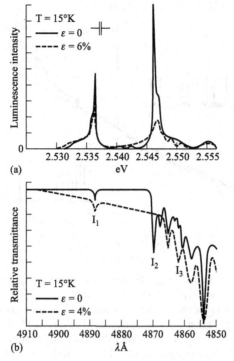

Figure 5.65 (a) Luminescence, and (b) absorption spectra of undeformed crystals (shown by solid curves) and samples plastically deformed to $\varepsilon = 6\%$ (shown by dashed curves). The CdS crystals had the wurtzite-structure and were deformed at 77 K while illuminated with $\lambda = 488$ nm radiation from an argon laser. [After Osipiyan and Steinman 1973. In *Luminescence of Crystals, Molecules and Solutions*, ed. F. Williams (New York: Plenum Press), pp. 467–72 (1973). © 1973 Plenum Press. With kind permission of Springer Science and Business Media.]

Triboluminescence has long been known in ZnS. It is the emission of light by materials under mechanical stress. Triboluminescence can result from plastic deformation or fracture. Bredikhin and Shmurak (1974) observed that the light emitted by ZnS during plastic deformation consisted of short pulses, as shown by the oscillograms A and B in Fig. 5.66.

Bredikhin and Shmurak (1975) studied the effect on this emission of applied electric fields. It was then known that an applied electric field, U, significantly increased the plastic flow stress, i.e. hardened ZnSe crystals. This is independent of the direction of the field and this they termed the (even) electroplastic effect (EPE, see Section 4.3.4). Bredikhin and Shmurak found, however, that depending on the sign of the field, ZnS could be either hardened or softened. This they termed the odd EPE. Their samples were melt-grown, wurtzite-structure ZnS crystals oriented for single slip on the basal plane and deformed in compression. The crystals were loaded to a stress above the elastic limit for a certain time which produced a series of flashes

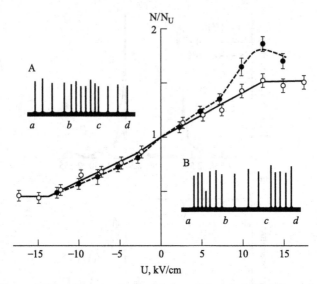

Figure 5.66   The variation with magnitude and sign of electric field of the ratio of the number of flashes without (N) and with (N_U) the field applied to the plates of a capacitor (white dots) or soldered specimen contacts (black dots). A and B are oscillograms showing the flashes emitted between the application (*a*) and removal (*d*) of the field. The external voltage was turned on at (*b*) and off at (*c*). T = 340 K. (After Bredikhin and Shmurak 1975. Reprinted with permission from *Soviet Physics Solid State*, **17**, pp. 1628–9. Copyright 1975, American Institute of Physics.)

of light as shown by the oscillograms A and B in Fig. 5.66. Each flash was produced by of the order of 10^3 dislocations sweeping over a slip band. The number of flashes per unit time was proportional to the deformation rate, $\dot\varepsilon$.

It can be seen that a 'positive' field (in one direction) decreased the number of flashes per unit time (from *b* to *c* in B) thus increasing N/N_U while a field in the opposite direction increased it (from *b* to *c* in A) thus decreasing $N/N\text{-}_U$. Thus this, like the EPE, is an odd order, field-direction-dependent effect.

The EPE can be explained by the field acting on the charged dislocations to give a driving force that, depending on the sign of the field, adds to or opposes the resolved shear stress under which they move during plastic deformation. It was found that the movement of the about 10^3 charged dislocations of a slip band, corresponding to a light pulse, produced a simultaneous electric pulse lasting about 40 ns at the $(10\bar10)$ side faces of the specimen where the slip dislocations escape.

Plastic deformation triboluminescence was studied systematically by Negrii and Osipiyan (1978, 1979) who used light microscopy to observe PL and CL from hexagonal CdS, undergoing slip on the basal and prismatic systems, at room temperature and below. The dislocation emission (DE, i.e. bright contrast) spectrum, consists of lines, in the near edge and exciton range, with wavelengths from $\lambda = 505$ to 510 nm, as shown in Fig. 5.67.

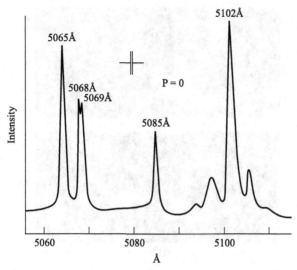

Figure 5.67   The dislocation emission (DE) spectrum of CdS. [After Osipiyan 1989. In *International Symposium on Structural Properties of Dislocations in Semiconductors*, Oxford, Conference Series No. 104, 1989 (Bristol: Institute of Physics), pp. 109–18.]

Using single crystals of CdS, ZnS and ZnSe of greater purity, as well as careful sample treatment and an improved deformation procedure, Negrii and Osipiyan (1982a, 1982b, Negrii *et al.* 1991, Negrii 1992) were able to reduce the brittle-ductile transition temperature sufficiently to allow these materials to be plastically deformed down to 4.2 K. (Plastic deformation is, in fact, possible down to 1.8 K, Tarbaev and Shepel'skii 1998.) This resulted in greater luminescence quantum efficiency, so moving dislocations could be observed by the light they emitted. The lower temperature also reduced impurity diffusion rates to keep the disloca-tions cleaner for longer. Moreover, at this low temperature, the annealing out of the defect debris introduced by deformation, that had been found to occur at room temperature and above, was avoided. They found that the generation and motion of individual dislocations could then be observed. The moving dislocations appear as small comet-like blobs of light moving along the slip plane as shown in Fig. 5.68.

Negrii and Osipiyan studied the dependence of the emission spectra of CdS crystals on the optical excitation intensity, dislocation types (basal or prismatic) and temperature as well as the time, excitation spectrum and polarization charac-teristics of this emission (Negrii and Osipiyan 1979, 1982a, 1982b). This photo-emission was due to debris generated by dislocations during glide. This debris contained effective radiative recombination centres with a characteristic emission spectrum and was observed as short light-emitting, tear-drop- or comet-like tracks trailing behind the moving dislocations as can be seen in Fig. 5.68. They found that

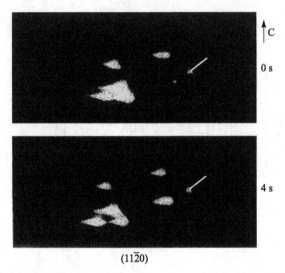

(11$\bar{2}$0)

Figure 5.68   Light photograph showing small comet- or droplet-shaped luminescent regions associated with dislocations, moving over slip planes in a CdS crystal oriented as marked, during deformation at 77 K. The arrows mark the appearance of a luminescent region due to a dislocation at one end of a loop at time t = 0 and after moving for 4 seconds. (After Negrii 1992. Reprinted from *Journal of Crystal Growth*, **117**, Dynamic and optical properties of screw dislocations introduced by plastic deformation of CdS crystals at 77–4.2 K, pp. 672–6, Copyright 1992, with permission from Elsevier.)

the polarization of the emitted radiation had two orientations which occurred in different domains. This structure was unstable with the result that the emission varied spatially and with time, as shown in Fig. 5.69.

Osipiyan and Negrii (1987) discussed the results of Figs. 5.67 and 5.69 in terms of ball-and-wire models of the atomic cores of 'configuration defects' arising in the low-temperature plastic-deformation of CdS crystals that they proposed to account for their observations. These defects are closely spaced dipoles that interact to become orientationally ordered so that the emission from them all has the same polarization. Osipiyan and Negrii (1989) discussed their evidence and suggest that the defects responsible for the emission from tracks behind the moving dislocations were narrow dislocation loops with extended screw components. The time varying emission (e.g. in Fig. 5.69) they suggested was associated with dislocation dipoles. For a brief review of this work see Osipiyan and Negrii (1989).

The nature of the sites responsible for the emission from behind the moving dislocations, i.e. the nature of the dipoles or strings of electronically interacting vacancies generated by jogs on moving screw dislocations, is still controversial. For the recent viewpoints of the two groups mainly involved see Negrii (1992) as well as Tarbaev *et al.* (1988), Tarbaev (1998) and Tarbaev and Shepel'skii (1998).

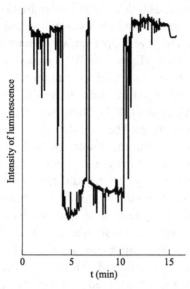

Figure 5.69 The variation with time of the intensity of dislocation luminescence in CdS under deformation at 4.2 K. [After Osipiyan 1989. In *International Symposium on Structural Properties of Dislocations in Semiconductors*, Oxford, Conference Series No. 104, 1989 (Bristol: Institute of Physics), pp. 109–18.]

5.6.4 CL D-line emission and the nearly universal role of traces of transition metals in Si and $Si_{1-x}Ge_x$

Most observed electronic and luminescent effects of dislocations in semiconductors are probably strongly affected, if not totally determined, by native point defects and especially impurity atoms in the defect core. The possible role of impurities is pointed out in our accounts of many dislocation-related phenomena. Usually, however, it was possible neither to be certain impurities were present nor how much. The extensive, systematic studies of the infrared D line emission from dislocations in Si and SiGe by Lightowlers, Higgs and their co-workers, however, was conclusive, and will now be discussed.

Four infrared dislocation emission lines were first identified in silicon photo-luminescence spectra by Sauer *et al.* (1985). They numbered these and found that D1 and D2 behaved alike as did D3 and D4. Sauer *et al.* ascribed D3 and D4 to dislocation core transitions while D1 and D2 they thought to be associated with point defects in the strain fields of the dislocations.

D line emission was unexpectedly missing from MBE-grown epitaxial Si, although etching showed it to contain 10^3 to 10^5 dislocations cm^{-2} (Higgs *et al.* 1989). However, D-band luminescence was seen after slight contamination with one of a number of transition metal elements.

Higgs *et al.* (1990a) proved, as follows, that contamination was indeed essential for D line emission. A heat treatment and PL spectral test had been shown to be able to detect transition metals and to have copper surface contamination sensitivity down to 10^{10} cm^{-2} (Canham *et al.* 1989). Higgs *et al.* (1990a) used it to examine several ingots of floating zone (FZ) Si. One was found to contain no trace of transition metals. Specimens were cut from it for deformation, using the high-low temperature strain technique of Wessel and Alexander (1977). These were deformed by quartz equipment to avoid metallic contamination. The high-low temperature strain technique is designed to deform specimens at the lowest possible temperature after starting dislocation generation at a necessarily higher temperature. The low temperature dislocation glide allows little if any impurity diffusion so keeping the dislocations as clean as possible. It was found that dislocations so produced in such uncontaminated silicon produced very little hole-electron pair recombination. That is, they neither emitted D lines in the PL spectrum nor showed dark line EBIC contrast (Higgs *et al.* 1990a,b). However, these effects appeared strongly when the silicon was lightly brushed with a copper wire and heated. This shows the minimal contamination required and how easily it can be acquired. No impurity precipitation at the dislocations, at this level of contamination, could be detected by TEM examination.

Oxidation-induced stacking faults in CVD (chemically vapour deposited) Si as well as dislocations in FZ Si were studied by Higgs *et al.* (1991). Contamination at the atomic level as described above or at a higher level to produce TEM-visible precipitates at the defects produced dark EBIC defect contrast. D line PL emission, however, disappeared as the contamination increased to the precipitation level. D line PL and EBIC contrast were made to appear when clean dislocations were lightly contaminated with not only Cu, but also Fe, Ni, Ag or Au (Higgs and Lightowlers 1992a and Higgs *et al.* 1992).

Higgs and Lightowlers (1992a) and Higgs *et al.* (1992) used the spatial resolution of SEM CL to show, by both spectroscopy (Fig. 5.70) and CL microscopy (Fig. 5.71), that the D3 and D4 line emission came from the dislocation lines but the D1 and D2 emission came from the slip plane between dislocation lines. Apparently D1 and D2 thus come from transition metal contaminated point defect debris left behind the dislocations.

Dislocations in Si$_{1-x}$Ge$_x$/Si behaved similarly. That is, Higgs and Lightowlers (1992b) and Higgs and Kittler (1994) showed that they produced neither EBIC or CL contrast nor D-line PL emission until they were lightly contaminated with a transition metal.

Fell and Wilshaw (1991), Wilshaw and Fell (1991) and Fell *et al.* (1993) applied the Wilshaw recombination theory to analyse the significant differences in EBIC contrast of clean dislocations produced by low (420°C) and higher (650°C) temperature deformation of Si and of dislocations lightly contaminated with Cu and Fe. The effects of copper decoration were found to differ for dislocations produced by low and higher temperature deformation. The weakest electrical activity was

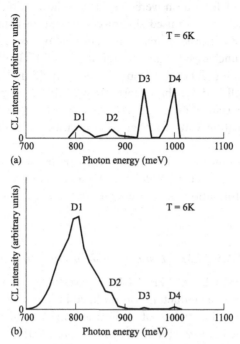

(a)

(b)

Figure 5.70 CL spectra of FZ Si plastically deformed at 800° C and lightly Cu contaminated (a) on a slip plane, and (b) between the slip planes (after Higgs *et al.* 1992).

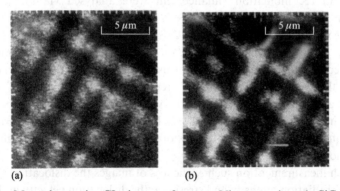

(a)   (b)

Figure 5.71 Monochromatic CL images from a Ni contaminated SiGe/Si epilayer taken through (a) a D2-transmitting filter, and (b) a D4 filter. (After Higgs and Lightowlers 1992a.)

associated with the uncontaminated 650° C deformed case. The energy states associated with these dislocations were determined to be shallow, and Cu decoration introduced a deeper state, pinned to the Fermi level. This raised the shallow state above the Fermi level and made it inactive as a recombination centre (Fell and Wilshaw 1991). In contrast, the states associated with dislocations produced by low

(420° C) temperature deformation were very deep. The Coulombic potential due to the occupation of these states raised shallower (dislocation) states (including those due to Cu) above the Fermi level, making them inactive as recombination centres. This explained the unchanged character of these (420° C) dislocations, whether 'clean' or lightly or more heavily impurity decorated, in terms of the charge-controlled model (Fell and Wilshaw 1991). It was also determined that a dislocation with a lower concentration of deep states can exhibit stronger recombination activity than a dislocation with a higher concentration of shallow states (Wilshaw and Fell 1991).

Fell *et al.* (1993) demonstrated the value of EBIC in the study of impurity gettering to dislocations and other extended defects. An analysis using the Wilshaw model, gave quantitative information on the energy and concentration of the electronic states concerned.

5.6.5 *SEM CL studies of dislocation motion*

The use of both SEM-CL and TEM-CL to investigate recombination-enhanced dislocation motion in semiconductors was outlined in Section 5.3.3. To reiterate, the usefulness of CL-SEM in studies of dislocation motion in semiconductors was demonstrated by Maeda *et al.* (1983), Hoering *et al.* (2001), and more recently by Schreiber and Vasnyov (2004) and Vasnyov *et al.* (2004), who employed computer-aided acquisition and fast processing of images in their observations. Dislocation dynamics and recombination enhancement were analysed via the resulting CL videotape movies.

Such video recording was applied to *in situ* observations that used an SEM CL stage with a bending attachment. Then, on sequentially recorded CL micrographs, the measured distance travelled by the dislocation during the application of the load gave the velocity. This allowed repeatable measurements of dislocation velocity as a function of independently varied applied stress, temperature and irradiation intensity (Maeda *et al.* 1983, Maeda and Takeuchi 1983). Fig. 5.72 presents sequential CL-SEM TV images of a dislocation, in *n*-GaAs, gliding under an applied load.

From such measurement on such sequences of images the dislocation velocity as a function of irradiation intensity and stress, i.e. the dislocation velocity in the presence of excitation, can be evaluated analytically (for details, see Section 5.3.3).

5.7 Scanning probe microscopy of extended defects

Ebert *et al.* (2001) used cross-sectional scanning tunnelling microscopy to simultaneously determine the type and electrical charge state of a dissociated perfect dislocation in highly doped GaAs. This revealed that both partial dislocation cores, and the stacking fault between them, were negatively charged.

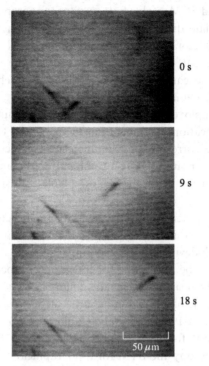

0 s

9 s

18 s

50 µm

Figure 5.72   CL-SEM TV images of a moving dislocation in *n*-GaAs. (After Maeda and Takeuchi 1983.)

Scanning Kelvin probe microscopy, through electrostatic interaction, makes possible the direct measurement of potential variations around dislocations. Such studies of dislocations in *n*-GaN and AlGaN/GaN heterostructures revealed that the dislocations are negatively charged (Koley and Spencer 2001). Krtschil *et al.* (2003) used scanning surface potential microscopy together with atomic force microscopy to study the electrical charge state of threading dislocations in doped GaN. They found the dislocations were either negatively charged or neutral depending on the type of doping atoms in the layers. This was explained as being due to the decoration of the dislocations with other defects and consequent partial compensation of the core charge (Krtschil *et al.* 2003). Scanning Kelvin probe microscopy and conductive atomic force microscopy studies of GaN surfaces were used to image and compare surface potential variations due to negatively charged threading dislocations and localized current leakage paths associated with these dislocations (Simpkins *et al.* 2003). From these and complementary studies it was deduced that dislocations with an edge component are negatively charged, whereas the pure screw dislocations, responsible for the observed leakage paths, are neutral (Simpkins *et al.* 2003).

Combining AFM and SCM, yields both surface topography and capacitance data and can directly correlate dislocations, terminating as surface pits, with electrical properties of semiconductors. Such observations on GaN films on sapphire, plus capacitance-voltage measurements on regions around the surface terminations of threading dislocations revealed the presence of negative charge in the vicinity of dislocations (e.g. Hansen *et al.* 1998).

Evoy *et al.* (1999) employed STL (scanning tunnelling luminescence) microscopy to study threading dislocations in GaN at low temperature. They observed near band edge emission with the corresponding luminescence images showing a correlation between dislocations and non-radiative recombination. From the observations of the luminescence profiles at the dislocation sites, a hole diffusion length of $30-55\,\text{nm}$ was estimated.

Porous Si can be prepared by electrochemical anodic etching and the material aroused great interest when it was found to be luminescent in the visible (Canham 1990). (Light emitting Si devices, fabricated in this material during the production of integrated circuits, could be invaluable for integrated optoelectronics.) The surface layers on such material consist of a partially columnar structure of nanometer-sized strands. It appears that the observed luminescence arises from quantum confinement in these columns (Canham 1990, Cullis and Canham 1991) although other explanations have also been put forward. Dumas *et al.* (1993, 1994a and b) used STL spectroscopy and microscopy and STM for topographical studies of 85% porous Si. The STL luminescence spectra recorded were similar to the PL spectra of like samples. Panchromatic STL images contained sharp contrast on the nanometer scale that corresponded to similar detail in the STM topographic images. They interpreted their results as consistent with the models involving quantum size effects which increase the band gap and shift it to a direct form at $k = 0$ (the origin of momentum space).

Hsu *et al.* (1994, 1996) used near-field scanning optical microscopy (NSOM, see Section 3.10.4) to investigate threading dislocations in $Ge_x Si_{1-x}$ films. Their electrical activity was determined by measurements of both spatially resolved photoresponse and surface topography near individual dislocations. This demonstrated the value of this high-resolution optical method in investigating the effect of defects on carrier recombination.

It was demonstrated by von Kanel and Meyer (1998, 2000) that ballistic-electron-emission microscopy (BEEM) allows the imaging of interfacial point defects and dislocations, in $CoSi_2/Si$ interfaces. BEEM was also used for the characterization of dislocations in silicon carbide by Reddy and Narayanamurti (2001) and in GaN by Im *et al.* (2001).

5.8 Effect of dislocations on transport properties of epitaxial heterostructures

Heterostructures are used in semiconductor devices, such as high-electron-mobility transistors (HEMT) and quantum cascade lasers (see Section 1.7.4). Such structures allow the control of carrier transport by energy band engineering (see Section 1.7). Heterostructures can also confine optical radiation.

5.8.1 *Introduction*

To obtain high-quality epitaxial heterostructures, it is important that the crystal structures and the lattice parameters of the constituent materials are matched as closely as possible, to avoid the formation of misfit dislocations (see Section 2.3.11 and Section 4.6). Nearly perfect matching is realized in $Al_xGa_{1-x}As/GaAs$ heterojunctions. In lattice mismatched systems, however, such as Ge_xSi_{1-x}/Si, high densities of dislocations may be present, to accommodate the misfit, if the layer thickness exceeds the critical value (see Section 4.6.2). In such cases, the effects of misfit dislocations on transport properties can be important.

In quantum wells, charge carrier motion is confined to two dimensions so the propagating electrons form a two-dimensional electron gas (2DEG). Their mobility in a 2DEG is limited by scattering at ionized impurities and dislocations and by phonons. This section is concerned with the role of dislocations in the electron transport in 2DEGs.

5.8.2 *Effect of dislocations on carrier mobility*

Dislocations strongly influence the transport properties of bulk semiconductors (see Section 5.2 and Section 5.3) and act as effective scatterers, thus reducing carrier mobilities. This was reviewed by Jaszek (2001), who offered a critical analysis of carrier scattering by dislocations, including scattering by misfit dislocations in heterostructures. Jaszek outlined issues such as incomplete understanding of dislocation core structures and the nature of dislocation energy levels, as well as the ambiguities in the determination of some parameters (e.g. the occupation probability of a dislocation core, or the charge per unit length).

Dislocation scattering is known to be anisotropic, i.e. the carrier mobility is significantly reduced for transport perpendicular rather than parallel to the dislocations and this leads to carrier mobility anisotropy in heterostructures.

The effect of dislocations, perpendicular to the direction of current flow, on electron transport in GaN was studied by Weimann *et al.* (1998). They modelled the dislocations as charged lines and compared the calculated and measured mobilities as a function of free electron concentration. The observed initial increase of mobility with increasing free electron concentration (with mobility maximum at an electron concentration of about $10^{18}\,cm^{-3}$ was attributed to increased screening of the charged dislocation core. A subsequent mobility decrease was ascribed to ionized impurity scattering (Weimann *et al.* 1998). They suggested that transport in vertical devices (e.g. lasers and LEDs) will be unaffected by scattering at charged dislocations due to the repulsive band bending around them and the directional dependence of the scattering (Weimann *et al.* 1998).

The influence of dislocation scattering on transport in 2DEGs is important in HEMTs. In principle, dislocations may affect the conductivity of a 2DEG through their (i) core-charge-related Coulomb potential and (ii) strain-field-induced deformation and piezoelectric potentials. Bougrioua *et al.* (1996a) presented a theoretical study of the effect of dislocations on the transport properties of 2DEGs in GaAlAs/GaInAs/GaAs. They showed the main influence to be the Coulomb potential of the line charge, rather than the strain field potentials. Two experimental cases, in GaAlAs/GaInAs/GaAs, were considered: (i) plastic-deformation-induced parallel dislocation segments, and (ii) long parallel misfit dislocations introduced during epitaxial growth. In both cases Bougrioua *et al.* demonstrated that the presence of dislocations results in conductivity anisotropy of the 2DEG. Bougrioua *et al.* (1996b) also studied the effect of dislocations on the conductivity of 2DEGs measured at helium temperature and high magnetic field, and showed that dislocation potentials account for the extra broadening of the Landau levels.

High densities, of the order of $10^{10}\,\mathrm{cm}^{-2}$, of dislocations may result in overlap of the electrical fields (the screening cylinders) around the charged lines. This may in principle lead to an abrupt change of the transport properties of the heterostructure. As noted by Jaszek (2001, and references therein), such a change may also be a result of the occupation statistics.

Ohori *et al.* (1994) investigated the effect of threading dislocations on the mobility of a two-dimensional electron gas (2DEG) in AlGaAs/GaAs heterostructures. They calculated the potential variation and scattering rate due dislocations as scattering line charges running normal to the interface and obtained the mobility of the 2DEG as a function of the dislocation density. Agreement between theory and experiment, for the variation with dislocation density, at 77 K was good. However, the agreement between room-temperature experimental and theoretical results was not as good. The model appears to overestimate the scattering rate. Ohori *et al.* (1994) also showed that, below densities of about $10^{8}\,\mathrm{cm}^{-3}$, the dislocation-induced reduction of room-temperature mobility does not influence significantly high-electron-mobility transistor performance.

Jena *et al.* (2000) analysed scattering by charged dislocation lines in 2DEGs, and applied their theory to elucidate transport in AlGaN/GaN HEMTs. It was assumed that the 2DEG is perfect (no carrier spread along the growth direction) and that the dislocations were perpendicular to the quantum well plane. Jena *et al.* (2000) showed that high densities of dislocations significantly influence carrier mobility, but not as greatly as in the 3D bulk case (Jena *et al.* 2000). They compared their theory with the 3D bulk scattering theory for dislocations and pointed out that the strong screening effect of carriers in a 2DEG is responsible for greater mobilities than in the 3D bulk case, as we shall now discuss.

The 3D bulk case was treated by Pödör (1966) and Look and Sizelove (1999). The Coulombic carrier scattering rate (for f ractionally filled states) due to the negatively charged dislocations can be expressed as (Look and Sizelove 1999):

$$\tau_{dis}^{3D} = \frac{\hbar^3 \varepsilon^2 c^2}{N_{dis} m^* q^4 f^2} \frac{(1 + 4\lambda_D^2 k_\perp^2)^{3/2}}{\lambda_D^4} \tag{5.65}$$

where $k_\perp$ is the wave vector for electron motion perpendicular to the dislocations, and λ_D is the Debye screening parameter, $\lambda_D = (\varepsilon k T/q^2 n')^{1/2}$, where n' is the effective screening concentration that may involve both free and bound carriers (Look and Sizelove 1999). (Note that variations in the 2DEG density would result in changes in dislocation scattering due to modifications in screening.)

The 2D dislocation scattering rate derived by Jena *et al.* (2000) is:

$$\tau_{dis}^{2D} \approx \frac{\hbar^3 \varepsilon^2 c^2}{N_{dis} m^* q^4 f^2} \frac{16\pi k_F^4}{\left(\dfrac{1.84 k_F}{q_{TF}} - 0.25\right)} \tag{5.66}$$

where q is the electronic charge, f is the fraction of filled states, k_F is the Fermi wave vector ($k_F = \sqrt{2\pi n_s}$, where n_s is the 2DEG carrier concentration, q_{TF} is the 2D Thomas Fermi wavevector ($q_{TF} = 2/a_B^*$) and a_B^* is the effective Bohr radius in the material.

Jena *et al.* (2000) argued that the 2DEG screening length depends on k_F and q_{TF}. The latter is constant so with increasing free carrier density, k_F increases and the Fermi wavelength (λ_F) gets shorter, resulting in better screening (Jena *et al.* 2000). Also, in contrast to 3D bulk case, the 2DEG carrier density does not freeze out at low temperatures. Thus, as noted by Jena *et al.* (2000), these characteristics play a role in the higher mobilities in a 2DEG. In the 3D bulk case, the Debye screening parameter q_D controls screening and scattering ($q_D = 1/\lambda_D = (q^2 n'/\varepsilon k T)^{1/2}$), n' being the effective screening concentration (i.e. both free and bound carriers). In a semiconductor, free carriers freeze out exponentially at low temperatures. Thus, the Debye screening length $\lambda_D = 1/q_D$ increases, resulting in a weaker screening. Also, since the carriers are less energetic, scattering is stronger, resulting in lower mobilities (Jena *et al.* 2000).

The effect of dislocation scattering on 2DEG mobility, derived from $\mu_{dis}^{2D} = q\tau_{dis}^{2D}/m^*$ (assuming that all the acceptor states in the dislocation are filled, i.e. $f = 1$) is presented in Fig. 5.73. The theory by Jena *et al.* (2000) predicts $\mu_{dis}^{2D} \propto n_s^{3/2}/N_{dis}$ that is shown in the plot.

Jena and Mishra (2002) presented a theory of deformation potential scattering of carriers in 2DEGs due to the strain fields of edge dislocations. They concluded that in the AlGaN/GaN system, deformation potential scattering has a significant effect on low-temperature mobilities for high dislocation densities, of the order of $10^{10}\,cm^{-2}$, and a low carrier density of about $10^{12}\,cm^{-2}$ (Jena and Mishra 2002).

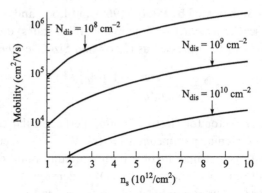

Figure 5.73   The effect of dislocation scattering on 2DEG mobility. The theory of Jena *et al.* (2000) leads to $\mu_{dis}^{2D} \propto n_s^{3/2}/N_{dis}$ which is plotted here. (After Jena *et al.* 2000. Reprinted with permission from *Applied Physics Letters*, **76**, pp. 1707–9. Copyright 2000, American Institute of Physics.)

However, Mishima *et al.* (2004) concluded from their theoretical calculations that in InSb/Al$_x$In$_{1-x}$Sb QW samples grown on GaAs (001) substrates, dislocation-related deformation potential scattering does not contribute significantly to the measured electron mobilities at low temperatures.

Gökden (2004) considered the effects of scattering on 2DEGs in AlGaN/GaN heterostructures. The conclusion of his calculation of the dependence of mobility on temperature was that the electron mobility at high temperatures is dominated by optical phonon interactions, whereas piezoelectric scattering and deformation-potential acoustic phonon scattering become dominant at intermediate temperatures (Gökden 2004). (Dislocations with their axes perpendicular to the quantum well plane were considered.) Experimental Hall mobilities of a 2DEG were compared with theoretical calculations. The mobility, predicted by the theory, was much greater than the experimental value, at low temperatures, which was explained by invoking dislocation scattering (Gökden 2004). The relative importance of different scattering mechanisms can be evaluated by measuring the ratio of the transport and quantum relaxation times. The former is a measure of the extent of time that a carrier remains moving in a specific direction, whereas the latter is a measure of the mean time that a carrier remains in a particular state prior to being scattered to a different state (see, e.g. Gökden 2004). It was also shown that the transport to quantum scattering ratio is greater for charged dislocation scattering than for impurity scattering, since the former is more anisotropic (Gökden 2004).

Naidenkova *et al.* (2002) demonstrated experimentally that misfit dislocations affect the surface morphology and transport properties of In$_{0.52}$Al$_{0.48}$As/In$_{0.75}$Ga$_{0.25}$As/InP pHEMT structures. This they attributed to the fact that misfit dislocations introduce directional surface and interface steps that were assumed to be responsible for carrier scattering and mobility variations between different crystallographic directions (Naidenkova *et al.* 2002).

Joshi *et al.* (2003) carried out a Monte Carlo analysis of 2DEG transport in order to elucidate the effects of the interface polarization charges due to edge dislocation strains, as well as relevant carrier–phonon scattering processes, on carrier mobilities in wurtzite GaN HEMTs. Electron mobilities were calculated as a function of temperature for various dislocation densities, and the results were in good agreement with the reported data (Joshi *et al.* 2003). They found that for the highest dislocation density of $10^{10}\,cm^{-2}$, reductions in mobility were 16.8% and 8.6% for 77 and 300 K, respectively.

The spatial variations of transconductance in heterostructures can be mapped by employing an AFM conducting tip acting as the gate, i.e. scanning gate microscopy (SGM). This offers a means, e.g. of assessing the effect of dislocations on the performance of AlGaN/GaN high electron mobility transistors (Hsu *et al.* 2003). The tip induces a local modulation of the 2DEG and the drain current variation is monitored as a function of tip position. Thus the influence of threading dislocations on device performance can be examined. Specifically, for near-depletion bias conditions the transconductance map reveals low signal regions associated with the positions of threading dislocations (Hsu *et al.* 2003).

5.8.3 Magneto-conductance oscillations in heterostructures

As mentioned in Section 5.1.10, in low-temperature magneto-conductance measurements, dislocations in semiconductors show signs of a low-dimensional quantum effect (Figielski *et al.* 1998, Figielski *et al.* 2000, Wosinski *et al.* 2002). In these measurements, a strong magnetic field was applied parallel to the axes of heterointerface misfit dislocations and magnetoconductance oscillations were observed. The period depended on the angle between the magnetic field direction and the dislocation axes, in GaAsSb/GaAs (Figielski *et al.* 1998, 2000) and InGaAs/GaAs heterojunctions (Wosinski *et al.* 2002). Figielski, Wosinski and their co-workers attributed the magnetoconductance oscillations to the trapping of charge carriers on localized quasi-stationary orbits surrounding the charged dislocations.

5.8.4 Dislocation-induced superconductivity?

Superconductivity is the property of zero-electrical-resistance that some materials exhibit when cooled below a critical temperature, T_c. Bardeen, Cooper and Schrieffer (1957) put forward a theory of superconductivity, now known by their initials (BCS) for which they received the Physics Nobel Prize in 1972. According to this, superconductivity is due to the formation of pairs of electrons (Cooper pairs) at temperatures below T_c. These pairs are bound by the exchange of phonons so they are not scattered by the vibrating atoms of the lattice and pass through without resistance. This accounted for the superconductivity of all the then-known metal, alloy and intermetallic compound superconductors.

Bardeen and co-workers (Allender *et al.* 1973, see also Ginzburg 1971) then suggested, on theoretical grounds, that, near certain semiconductor interfaces, electron pairs, bound by the exchange of excitons, could be formed. Such exciton binding should be stronger than phonon binding in the original class of superconductors. Thus, such new semiconductor heterostructure superconductors should have higher values of T_c. This idea led to much research.

Later, a new group of high T_c superconducting materials, copper-oxide-based ceramics, also known as the cuprate superconductors, were discovered with properties not accounted for by the BCS theory. This suggested that other possible mechanisms of superconductivity should also be considered.

Of interest in the context of this book are epitaxial superlattices of lead chalcogenides (relatively narrow gap IV-VI semiconducting compounds with the NaCl structure, see Section 4.1.6), grown on cleavage faces of KCl. Some of these superlattices are superconducting (Murase *et al.* 1986, Mironov *et al.* 1988). Periodic structures of alternating layers (superlattices) of PbTe and PbS, for example, have a critical temperature of 6 K although PbTe and PbS are not superconductors. It was suggested that superconductivity in these superlattices is related to the regular grids of misfit dislocations in the interfaces between these lattice parameter mismatched materials (Mironov *et al.* 1988, Dmitrenko *et al.* 1993).

Fogel *et al.* (2001), who present further evidence for this, even subtitle their paper 'Dislocation-induced superconductivity?' They reported that the superconducting transitional drop leads to zero resistance only in those superlattices containing continuous misfit dislocation networks, i.e. in those samples with relatively large thicknesses d_1 and d_2 of layers of PbTe and PbS. It will be recalled (see Section 4.6.2) that for the lowest thicknesses of epitaxial film, the lattice parameter misfit with the substrate is taken up by long-range elastic strain in the film. As the thickness increases above some critical thickness, misfit dislocations begin to appear and beyond some larger critical thickness, all the mismatch is accommodated by misfit dislocations. The variation of T_c with the PbS layer thickness is shown in Fig. 5.74.

In further experiments they investigated lead chalcogenide superlattices in which the thicknesses of the layers of the two compounds were equal, $d_1 = d_2$. They also found T_c to vary with the separation of the dislocations in the grids in the interfaces, D_g, as shown in Fig. 5.75. Moreover, they point out, for the most closely spaced dislocation grids, $D_g = 3.3$ to 8.6 nm there is virtually no variation of T_c. In a later paper Fogel *et al.* (2002) discuss possible mechanisms for dislocation-induced superconductivity and report the discovery of an additional superconducting lead chalcogenide superlattice, PbTe/PbSe.

Finally, Fogel *et al.* (2001) reported finding superconductivity in four more types of lead chalcogenide superlattices, PbS/PbSe, PbS/YbS, PbTe/YbS and PbSe/EuS all

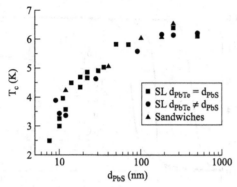

Figure 5.74   The superconducting transition temperature of PbTe/PbS superlattices (SLs) as a function of the thickness, d_{PbS}, of the PbS layers. [After Fogel *et al.* 2001. Reprinted with permission from *Physical Review Letters*, **86**, pp. 512–5, 2001 (http://publish.aps.org/linkfaq.html). Copyright 2001 by the American Physical Society.]

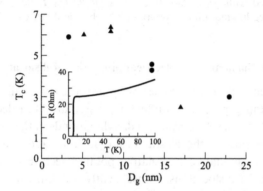

Figure 5.75   The variation of T_c with the dislocation grid period D_g (spacing). The inset shows resistance as a function of temperature for a PbS/PbSe superlattice. [After Fogel *et al.* 2001. Reprinted with permission from *Physical Review Letters*, **86**, pp. 512–15, 2001 (http://publish.aps.org/linkfaq.html). Copyright 2001 by the American Physical Society.]

grown on KCl substrates. The variation of relative resistance with temperature is shown for three of them in Fig. 5.76.

Erenburg *et al.* (2001) used precision x-ray diffractometry and TEM and concluded that for superconductivity in lead chalcogenide superlattices (SLs) it is essential that regular grids of misfit dislocations (MDs) cover the interfaces. When MDs cover only isolated patches (islands) on the interface, only partial superconducting transitions occur like that in the insert in Fig. 5.76. They reported that evidence from measurements of critical magnetic fields showed that the superconducting layers in the SLs are confined to the interfaces between pairs of compounds. This is convincing evidence that the superconductivity is exclusively associated with the MDs.

Fogel *et al.* (2002) give a more complete account of the experimental data and propose a model to account for their observations.

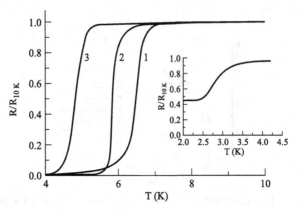

Figure 5.76 Resistance transitions for semiconducting superlattices. Curve 1 is for PbS/YbS, 2 is for PbTe/YbS and 3 is for PbS/PbSe. The insert shows the incomplete disappearance of resistance of PbTe/PbS superlattices with layer thicknesses of $d_1 = d_2 = 7$ nm. [After Fogel *et al.* 2001. Reprinted with permission from *Physical Review Letters*, **86**, pp. 512−5, 2001 (http://publish.aps.org/linkfaq.html). Copyright 2001 by the American Physical Society.]

5.9 Summary: the electrical properties of dislocations

Dislocations influence electrical properties by introducing localized levels or bands into the energy band gap of semiconductors. These are both deep (near mid-gap) recombination and trap levels ascribed to dangling bonds in the dislocation core and shallow trap states due to the elastic strain field. In addition, impurity atoms, especially metal contaminants, and native point defects can strongly affect the electrical properties of dislocations. Consequently, the intrinsic electrical effects of dislocations can only be seen in materials in which the segregation of impurity atoms to the dislocations is prevented or at least minimized. Otherwise, dislocations are generally either electrically neutralized, or activated, by impurity decoration. For low concentrations, impurities appear to be tightly bonded to the dislocation core, whereas at higher concentrations they form a cloud in the dislocation strain field.

The dislocation core states are closely spaced so the occupied sites interact electrostatically, affecting the occupation statistics and the energy (see Section 5.2.1, Section 5.2.6 and Jaszek 2001).

The earliest studies concerned with changes in conductivity and Hall mobility due to the introduction of high densities of dislocations, dipoles and point defect debris by plastic deformation (Section 5.1 and Section 5.2). This led to the Read model (1954a,b, see Section 5.2) of dislocations trapping electrons to produce a negative line charge, Q, per unit length, screened by an equal positive space charge in a surrounding volume of radius r_d. This model gave order of magnitude agreement with experiment, and was the starting point for later work (Section 5.2.6, Labusch and Schröter 1978, Figielski 2002 and Jaszek 2001). It is now accepted that, as originally suggested by Shockley (1953), overlap of dangling bond states broadens them into partly filled bands. The dislocation line charge then depends on the Fermi level so dangling-bonds can act as donors as well as acceptors. Thus dislocations can be neutral, negatively or positively charged and line charges of both signs are screened by space charges (Jaszek 2001).

Dislocation lines were expected to conduct along their partly filled energy bands but this was not found in d.c. measurements, because, it is believed, the lengths along which there are continuous bands are short. Supporting evidence of ultra-high frequency a.c. conduction, along short segments of dislocations only, at low temperatures was found in Si by Grazhulis *et al.* (1977a and b), Osipiyan (1981, 1983) and Kveder *et al.* (1985). The small number of electrically active deep levels per unit length of dislocation found by EPR studies suggests that the Hornstra rearrangement (Section 4.1.1) eliminates dangling bonds except at certain 'special sites'. Wilshaw and co-workers provided some evidence that kinks are not these electrically active sites (Section 5.5).

EPR studies also suggest that dangling bond states several times as numerous as those on dislocations are introduced by point defect debris during plastic deformation (Section 5.1.4). The importance of this point defect debris for optoelectronic effects is discussed in Section 5.9.1.

Carrier mobilities are reduced by dislocation scatterering (Section 5.2.6 and Jaszek 2001). This is strongest when the carriers move perpendicular to dislocations, which can lead to carrier mobility anisotropy. This occurs in heterostructures, due to interfacial misfit dislocations and can result in conductivity anisotropy of two-dimensional electron gases in quantum wells.

The non-centrosymmetric sphalerite and wurtzite structure compounds are piezoelectric. Consequently, in these materials, the elastic strain fields of edge and 60° dislocations produce electric fields (Section 5.1.7) and these were shown by Miyazawa and Hyuga (1986) to affect the threshold voltage of GaAs MESFETs (metal semiconductor field-effect transistors).

Dislocations capture and emit electrons and holes to act as generation/recombination centres with a dynamic equilibrium charge (Section 5.3). Thus, dislocations contribute to generation-recombination noise in devices (Section 5.1.2 and Section 5.11.4). Dislocation recombination is responsible for many significant electrical effects including more rapid decay of photoconductivity (Section 5.3.1), localized luminescence (Section 5.3.2) and enhanced dislocation motion (Section 5.3.3). Non-radiative recombination at dislocations competes with the desired radiative recombination in light emitting diodes and lasers and so reduces their efficiency (Section 5.11). It also produces dark contrast in both SEM CL and SEM EBIC images. Theories of charge-controlled and of Shockley-Read-Hall recombination at dislocations have been developed to interpret the results of the extensive studies of, especially, SEM EBIC dislocation contrast with considerable success (Section 5.5 and Section 5.6).

Quantized magnetoconductance oscillations at dislocations were observed in low-temperature measurements on GaAsSb/GaAs heterostructures (Figielski *et al.* 1998, Figielski *et al.* 2000, Wosinski *et al.* 2002). These oscillations were attributed to charge carrier trapping by localized quasi-stationary orbits around charged dislocations, and had a period dependent on the angle between the magnetic field direction and the dislocation axes.

Although PbTe and PbS are not superconductors, epitaxial superlattices of the two have critical temperatures up to 6 K (Section 5.8.4, Murase *et al.* 1986, Mironov *et al.* 1988). The superconductivity in these superlattices was ascribed to the regular grids of interfacial misfit dislocations between these lattice-parameter-mismatched materials (Mironov *et al.* 1988, Dmitrenko *et al.* 1993).

Much work is still required to extend our understanding of these electrical effects, especially of misfit dislocations in epitaxial heterostructures. These are confined to the interfaces between lattice parameter mismatched materials and as noted by Jaszek (2001), conduction perpendicular to such dislocations may have to be interpreted in terms of tunnelling through the defect cores. Carrier scattering is often mainly due to the Coulomb potential rather than the strain field potential of dislocations and the relative importance of the two needs to be established for each material or combination of heterostructure materials (Jaszek 2001). For other recent reviews see Mil'shtein (1999) and Schröter and Cerva (2002).

5.9.1 The electrical effects of debris left behind moving dislocations

As noted repeatedly throughout this book, the presence of impurities strongly influences the electrical and luminescence effects of dislocations. Also ESR and DLTS studies (see Section 5.1.4 and Section 5.5.7) provide ample evidence that the point-defect-associated debris left behind by jogs in moving screw dislocations during plastic deformation is electrically active.

Evidence from recent etching studies on Si and SiGe (see Section 4.7) revealed a new type of needle-shaped debris, so-called trails, left behind moving 60° dislocations. Feklisova *et al.* (2003, 2005) observed (using EBIC) in Si that the concentration of recombination centres in these trails was greater than near the dislocations and determined by DLTS that the concentration of these centres was related to the concentration of defects in the trails. Moreover, such point defect debris affects optoelectronic properties as shown in the SEM CL imaging of the D1 and D2 luminescence emitted from plastically deformed Si which is seen to originate from the slip plane between the dislocations. This was shown to be due to transition metal contaminated debris left behind the dislocations (see Section 5.6.4). The overwhelming evidence of many types of studies thus shows that the dislocations contribute little to the defect electrical activity in plastically deformed Si. Much of that is due to point defects near to or on the dislocation line.

The vital role of impurities in determining the electrical activity of both dislocations and debris was shown by the SEM EBIC and CL studies of Higgs *et al.* (1990a) and Higgs and Lightowlers (1992a,b) as discussed in Section 5.5.8 and Section 5.6.4. They found that the dislocations in pure and low-temperature deformed Si gave no EBIC dark contrast until they were deliberately lightly contaminated with transition metal atoms. The importance of impurities for the luminescence of both dislocations and debris was shown by their observations using the D1 to D4 lines of the CL emission from plastically deformed Si. They imaged these lines separately and found that D3 and D4 came from the dislocations while D1 and D2 came from between them, i.e. from the debris, but neither was emitted until the material had been lightly contaminated with a transition metal.

The present evidence thus suggests that the physical effects ascribed to dislocations in the earlier literature generally must be ascribed to debris and impurity contamination. It will be necessary, first, to obtain atomically clean samples of all materials other than Si, before the roles of the various types of defects can be similarly sorted out in these cases.

Observations on II-VI compound semiconductors (see Section 5.6.3) showed that moving triboluminescent tracks observed during low temperature plastic deformation were associated with emission from debris close behind moving dislocations.

5.10 The electrical and luminescent effects of grain boundaries in semiconductors

Grain boundaries (GBs) are classed into small- or large-angle types. The grains on either side of small-angle GBs are misoriented by tilts of less than 10 to 15 degrees and can be resolved into walls of dislocations. Large-angle GBs are best discussed in terms of the coincidence lattice and related concepts and of Hornstra's (1959, 1960) ball-and wire models (see Section 4.5.2).

Grain boundaries, like dislocations, can contain dangling bonds and have strain fields that produce localized states via the deformation potential. Thus we must expect them to have electrical and luminescent effects like dislocations. Again, separating intrinsic GB effects from extrinsic ones due to point defects, especially impurities, segregated to GBs is difficult and progress in this field has been limited. Twin boundaries are a special case, however. Models show that coherent first order twin boundaries are free of dangling bonds and bond strain (see Section 4.5.3). They also do not attract impurities. Consequently they are to be expected to have, and were early found in fact to have, very little effect on electrical and luminescent properties (Billig and Ridout 1954, Queisser 1963).

There was a good deal of interest in the properties of GBs in the 1950s and 1960s. This was partly because they were then still a diminishing problem in the immature Ge, Si and III-V technology of that period and partly because it was at one time thought that grain boundary devices might have some practical value.

5.10.1 Early studies of the electrical and luminescent effects of grain boundaries

Much of the early work was done on GBs in the then dominant material Ge. These were found to provide acceptor levels so the GB region was p-type in n-type material. Moreover, there was a negative surface charge on the boundary and a potential barrier on each side, like that in Fig. 5.77, as first pointed out by Taylor *et al.* (1952). Such GBs act like two Schottky barriers back-to-back and this is how they are modelled. Tweet (1954, 1955), Matare (1956a,b,c), Matare and Wegener (1957) and Mueller (1959a,b) showed that the resistance across the GBs was high but the resistance along them was low.

Systematic studies of the properties of a variety of GBs in Ge and Si were carried out by Sosnowski (1959) by several methods. He found that their properties divided GBs into two types. Figielski (1960) studied such boundaries in detail in n-type Ge. A theoretical analysis identified the two types as $n\text{-}p^+\text{-}n$ and $n\text{-}p\text{-}n$ barriers. In the $n\text{-}p^+\text{-}n$ type, p_b/n_o was greater than 100, where p_b is the hole density at the boundary and n_o is the electron density in the grains. Such boundaries generated significant photovoltages when illuminated at one side only, but did not locally enhance recombination. They were large-angle GBs and presented large electrical barriers. The $n\text{-}p\text{-}n$ boundaries were low-angle GBs. They were characterized by enhanced hole recombination and the absence of photoelectric phenomena.

GB studies were originally carried out on specimens cut from large-grain-size polycrystalline samples. However, Matare and Wegener (1957) grew Ge containing well-defined GBs. To do this they used a holder for two seed crystals with carefully chosen orientations. These seeds were dipped together into molten Ge and slowly raised while rotating. Thus a bicrystal boule could be produced in a Czochralski

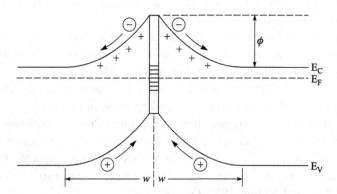

Figure 5.77 Energy band diagram of a grain boundary, containing acceptor levels, which becomes negatively charged in n-type material. This results in band bending by an amount ϕ, as shown, and the formation of compensating positive space charge regions of width w, on either side. Charge carriers, formed by electron bombardment, are collected in the directions shown by the arrows to give EBIC signals.

growth apparatus. Such bicrystal boules contained a single central GB of the desired structure. The method was later used to grow InSb bicrystals of defined misorientation and polar character, α or β, by Mueller and Jacobson (1962).

Since Ge *n*-type bicrystals constitute *n-p-n* structures, contacts applied to the two grains and to the boundary enable them to act as transistors (Matare 1955, 1956b,c). However, because of the high carrier recombination rate in the GB region, GB transistors were inferior to conventional Ge *n-p-n* transistors (Amelinckx and Dekeyser 1959).

Large-angle GBs in InSb, then the most highly developed III-V material, were studied by Mackintosh (1956). He found a high resistivity across the boundaries and used temperature dependence measurements to deduce the energy depths of donor levels. Mueller and Jacobson (1962) used the two-seed Czochralski method of Matare and Wegener (1957) to grow 6° tilt boundaries in InSb. Mueller and Jacobson (1959) grew different samples in such a way as to produce these boundaries with opposite polarities. They interpreted the GB structures on the early Hornstra (shuffle-set) model (see Section 4.1.2). Hence, they labelled them in terms of the types of edge dislocations that they contained as α and β GBs (containing dangling bonds from In and from Sb atoms, respectively). If the dislocation cores are really of the Hirth and Lothe (glide-set) core structure, of course, the types of dangling bonds in the two kinds of boundary would be reversed. Current flow across these boundaries in both *n*- and *p*-type material was observed. They found that the α boundaries acted as barriers to current flow in both the *n*- and *p*-type InSb samples but the β boundaries only in *p*-type material. Mueller and Maffitt (1964) reported that the conductance along their β boundaries was higher than that along their α GBs.

Similar early studies of GBs were reported for CdTe (De Nobel 1959 and Lang *et al.* 1963).

It was early found that GBs in GaP emit electroluminescence and it was suggested that this was because they constituted *n-p-n* structures in *n*-type polycrystalline material (Holt *et al.* 1958). The existence of *p-n* junctions on either side of the grain boundaries was confirmed by early voltage contrast observations in an experimental SEM. Coherent twin boundaries were not electroluminescent, however (Alfrey and Wiggins 1960). Shaw *et al.* (1966) reported early EBIC observations of decorated grain boundaries (Fig. 5.78). The bright line in Fig. 5.78a is a good, flat *p-n* junction in an early GaAs homojunction laser diode. Impurities can diffuse faster along grain boundaries than through defect-free material. In Fig. 5.78b the dopant diffusing down to the junction has thus penetrated more deeply along a network of subgrain boundaries (i.e. GBs with misorientations of less than a degree). These, therefore, also appear bright.

Brümmer and Schreiber (1972) published an SEM CL image of a small angle grain boundary in a CdS crystal appearing in dark contrast and (1974) of some in bright contrast. Brümmer and Schrieber (1974) then showed that the GB contrast, as shown

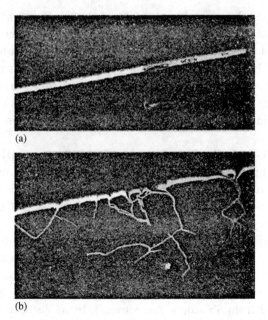

(a)

(b)

Figure 5.78   EBIC images of sections through GaAs laser diodes. (a) In monocrystalline material the bright line is the planar p-n junction. (b) The in-diffused dopant has penetrated deeply along subgrain boundaries below the junction so they also appear as bright p-n junction lines. (After Shaw *et al.* 1966. Reprinted from *Solid State Electronics*, **9**, Crystal mosaic structures and the lasing properties of GaAs laser diodes, pp. 664−5. Copyright 1966, with permission from Elsevier.)

by the series of line scans in Fig. 5.79 could be reversed, from dark as in Fig. 5.79a to bright in Fig. 5.79d by annealing for 40 hours at 350° C. This altered the impurity decoration of the boundary.

5.10.2 Carrier transport across grain boundaries

Electrically active grain boundaries (GBs) in polycrystalline semiconductors introduce deep levels in the band gap. These deep levels, which are due to the dangling bonds or impurity contamination, may act as strong recombination centres for minority carriers.

As discussed above, the presence of trapped charge in GBs results in band bending and energy barriers to electrical transport. The barrier heights of individual GBs are basically determined by their structures. Impurity segregation at grain boundaries affects many of their properties. (For reviews see Grovenor 1985, Seager 1985, Greuter and Blatter 1990, Möller 1993 and Kamins 1998). In some cases, this can be beneficial. An example, discussed in Section 5.12.1, is the use of grain boundaries as sinks to gettering unwanted impurity atoms, rendering them electrically inactive.

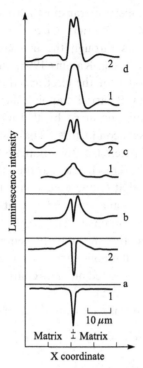

Luminescence intensity

X coordinate

Figure 5.79 The heat-treatment dependence of CL line profiles across a small-angle grain boundary in CdS. (a) Unannealed, curve 1 is the room temperature and 2 the liquid nitrogen temperature integral CL signal. (b) After 6 hours annealing at 350° C in argon, room temperature integral CL signal. (c) After 16 hours annealing at 350° C in argon, room temperature. Curve 1 is the profile for 630 nm, curve 2 for 510 nm. (d) After 40 hours annealing at 350° C, room temperature. Curve 1 is the profile for 630 nm, curve 2 for 510 nm. (After Brümmer and Schreiber 1974.)

Another case of 'defect engineering' uses interfacially trapped grain boundary charges to produce potential barriers that can be controlled by the applied bias (Section 5.12.4). Such GBs are employed in thermistors, varistors, and boundary layer capacitors. In these devices, impurity-dependent properties (like the critical voltage for switching) are controlled by doping (see Section 5.12.4).

Charged GBs are modelled by two Schottky barriers back-to-back (see also Section 5.12.4). Traps present at GBs capture carriers and the GB charge depletes the adjacent grains. The energy band bending at the GB forms double Schottky barriers (see Figs. 5.77 and 5.94).

It is now well accepted that both impurity segregation and carrier trapping at GBs affect the transport properties of polycrystalline semiconductors (Seto 1975, Baccarani *et al.* 1978, Kazmerski 1980, Orton and Powell 1980, Grovenor 1985, Blatter and Greuter 1986a, b, Greuter and Blatter 1990, Kamins 1998). The model of diffusive charge carrier transport across GB barriers was presented by

Taylor *et al.* (1952). Later models of transport across GB barriers employ thermionic emission. At the top of the barrier, where it is narrow, quantum tunnelling of sufficiently energetic carriers through the barrier can be included to form a thermionic-field-emission model (e.g. Kamins 1998, and references therein).

Seto (1975) presented a model for the electrical transport properties of a *p*-type semiconductor, controlled by GB carrier trapping due to dangling bonds, in which current flow between the grains occurred by thermionic emission. (The model is equally applicable to *n*-type semiconductors.) The main effects of this trapping are to (i) reduce the number of carriers free to conduct and (ii) reduce carrier mobility due to scattering by potential energy barriers at the GBs, as illustrated in Fig. 5.80. For simplicity, Seto assumed that (i) the grains have the same size L, (ii) only one impurity is present, fully ionized and uniformly distributed, (iii) the GB thickness is negligible compared to L and contains Q_t/cm^2 traps located at energy E_t relative to the intrinsic Fermi level, and (iv) the initially neutral traps become charged by trapping. This model assumes all mobile carriers in a region $(L/2-l)$ at the GB to be trapped so a depletion region is formed and any contribution from mobile carriers in this region is disregarded (Seto 1975).

The 1D Poisson equation in this case can be expressed as:

$$d^2V/dx^2 = qN/\varepsilon \qquad l < |x| < L/2 \qquad (5.67)$$

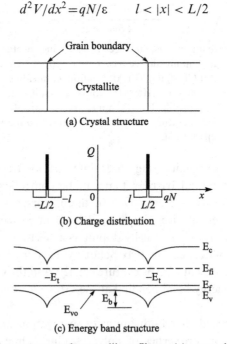

(a) Crystal structure

(b) Charge distribution

(c) Energy band structure

Figure 5.80 Model for p-type polycrystalline films: (a) crystal structure, (b) charge distribution, and (c) energy band structure. (After Seto 1975. Reprinted with permission from *Journal of Applied Physics*, **46**, pp. 5247–54. Copyright 1975, American Institute of Physics.)

For a given L, the trap densities relative to the doping concentrations, N, are such that $LN < Q_t$ or $LN > Q_t$. If $LN < Q_t$, the grain is completely depleted of carriers, the traps are partially filled and the potential barrier can be expressed as:

$$V_B = \frac{qL^2N}{8\varepsilon} \tag{5.68}$$

i.e. the barrier height increases linearly with N. When $LN > Q_t$, only part of the grain is depleted of carriers and the potential barrier can be expressed as (Seto 1975):

$$V_B = \frac{qQ_t^2}{8\varepsilon N} \tag{5.69}$$

Conduction in the grain is assumed to be much higher than that through the GB so it is sufficient to consider the current limiting transport across the GB. As mentioned above, thermionic emission and tunnelling are assumed to be the important contributions to the carrier transport across the GB. If the tunnelling current is neglected the GB thermionic emission current density J is related to the applied voltage V as (Seto 1975):

$$J = qp_a(kT/2\pi m^*)^{1/2}\exp(-qV_B/kT)\,[\exp(qV/kT)-1] \tag{5.70}$$

where p_a is the average carrier concentration, m^* is the effective mass of the carrier, and V is the voltage applied across the GB.

For a small applied-bias V (i.e. $V \ll kT/q$), the conductivity is:

$$\sigma = \sigma_0\exp(-qV_B/kT) \tag{5.71}$$

where $\sigma_0 = Lq^2p_a(2\pi m^* kT)^{-1/2}$.

By inserting the expressions derived for the average carrier concentration, p_a, for the two possible cases of doping concentration, the solutions for these two cases are found to be (Seto 1975):

$$\sigma \propto \exp[-(E_g/2 - E_F)/kT] \quad \text{if } LN < Q_t \tag{5.72}$$

$$\sigma \propto T^{-1/2}\exp(-E_B/kT) \quad \text{if } LN > Q_t \tag{5.73}$$

where $E_B = qV_B$.

The grain boundary controlled (effective) mobility is thermally activated and can be expressed as (e.g. Seto 1975):

$$\mu^* = \mu_0^*\exp(-E_B/kT) \tag{5.74}$$

where $\mu_0^* = Lq\,(2\pi m^* kT)^{-1/2}$.

Good agreement between experiment and theory was reported by Seto, who performed Hall and resistivity measurements over the range from -50 to $250°\,C$ and calculated the (i) carrier concentration and mobility as a function of doping concentration and (ii) mobility and resistivity as a function of temperature.

Nevertheless, Seto (1975) pointed out some limitations of the model. These include (i) neglecting the contribution of grain resistivities, which may become comparable to the resistivity of GB regions for large grain sizes and high doping levels, (ii) the assumption of discrete energy states (the trapping states at GBs may in fact be distributed over an energy range), (iii) using the depletion approximation that may result in inaccurate values of the barrier height if the carrier concentration in the depletion layer is considerable (since the mobility depends exponentially on the barrier height, this inaccuracy strongly influences the calculated mobility values). Seto estimated that the depletion approximation is valid up to a grain size of about 60 nm. Also, as noted by Baccarani *et al.* (1978), the Seto model did not include the possibility of GB traps being only partially filled, i.e. it is limited to the case of the energy of the trapping states being lower than that of the Fermi level.

Baccarani *et al.* (1978) extended the Seto grain-boundary trapping model to include both the case of a single trap level and of a continuous energy distribution of trap states within the band gap (for details, see Baccarani *et al.* 1978, and Kazmerski 1980). Baccarani *et al.* (1978) performed resistivity and Hall measurements and found very good agreement between the experimental data and the single trap level model. However, the model based on a continuous distribution of trapping states did not fit the experimental data (Baccarani *et al.* 1978).

Orton and Powell (1980) reviewed the early work and showed that idealized two-phase geometrical models for the resistivity and Hall coefficient can be used to interpret transport measurements in polycrystalline films. The two phases are the bulk grain characterized with the conductivity σ_1, carrier concentration n_1, and mobility μ_1, surrounded by a thin intergrain material having σ_2, n_2, and μ_2. The most common case is that of high resistance intergrain regions, i.e. $\sigma_1 \gg \sigma_2$. Then, as assumed by Seto, the resistivity of polycrystalline semiconductors is determined by transport across the GBs which results in thermally activated Hall mobility. The grain size, or more precisely, the ratio of the grain size to the depletion width is also important. Orton and Powell (1980) also emphasized differences depending on whether the depletion layers extend fully or partially through the grains, and whether the Debye length and mean free path are greater or less than the grain size.

Blatter and Greuter (1986a,b) and Greuter and Blatter (1990) discussed carrier transport across an electrically active grain boundary in semiconductors. They concluded that the non-linearity of the steady-state current-voltage characteristic in this case relates to the relationship between the applied bias and the occupation of the defect states at the interface and in the depletion regions (Greuter and Blatter 1990). They also discussed hot-electron phenomena at GBs. For high barriers, doping levels and elevated bias, large electric field build-up in the depletion regions causes minority carrier generation due to impact ionization by hot majority carriers. This results in significant increase in the non-linear characteristics (Greuter and Blatter 1990).

The effects of photoexcitation on carrier transport and photoconductivity across GBs were summarized by Bube (1992). These phenomena were discussed by Bube in terms of the increase of free carrier density, the intergrain barrier height decrease as the result of the charge variations in the intergrain states, and enhanced tunnelling through the barriers as the result of reduced GB depletion layer widths.

Tringe and Plummer (2000) examined the effect of a distribution of grain structures and orientations on the electrical properties of polycrystalline silicon. They demonstrated that electrical properties represent the average value of substantially varied individual GB properties. They also concluded that the GB structure has the primary effect on device resistance and that the distribution of single boundary barrier heights controls the resistance variations in individual boundaries, since GB resistance depends exponentially on the barrier height (Tringe and Plummer 2000).

As discussed in Section 4.5.3, reproducible GB properties can be obtained by employing oxidation and annealing to control the GB potential barriers (Kamiya *et al.* 2002). In practice, this is accomplished by selective GB oxidation at around 700° C followed by annealing at about 1000° C that results in an increased potential barrier height and resistance (Kamiya *et al.* 2002).

5.10.3 Recombination at grain boundaries

The reduction of conversion efficiency in polycrystalline, compared to single crystal, solar cells is usually ascribed, at least in part, to enhanced carrier recombination via GB states. These states can be due to dangling bonds, impurities, or dislocations.

Early models of GB recombination were outlined by Edmiston *et al.* (1996, and references therein). These include a model by Oualid *et al.* (1984), based on Shockley-Read-Hall (SRH) recombination theory (see below). To model the electrical and device properties of polycrystalline semiconductors and to quantify carrier recombination mechanisms at GBs, the concepts of the effective minority carrier diffusion length and lifetime, as well as the effective surface recombination velocity are introduced. Based on these early models, Edmiston *et al.* (1996) favoured a model of a GB with a high density of defect states. They presented a 2D numerical model of GB recombination, and compared it with other models related to both bulk and p-n junction regions of Si solar cells. They concluded that, under certain conditions, those models overestimate the effective surface recombination velocity at GBs if the region around the GB is not fully depleted of majority carriers. Their model for GB recombination within the p-n junction depletion region, which takes into account the effect of the GB charge on the electric field within that depletion region, was found to be in agreement with the numerical results over a wide range of grain boundary recombination rates (Edmiston *et al.* 1996). (In this case, the GB charge reduces the *p-n* junction electric field at the GB, resulting in enhanced recombination in this region.)

Non-radiative recombination, by minority carrier capture at defects having levels in the energy gap of the semiconductor, can be described by the Shockley-Read-Hall (SRH) recombination. For a single GB trap level in the energy gap, the SRH recombination rate, under steady-state non-equilibrium conditions, can be expressed as:

$$R_{SRH} = \frac{N_t \sigma_n \sigma_p v_{th}(np - n_i^2)}{\sigma_n(n + n_1) + \sigma_p(p + p_1)} \qquad (5.75)$$

where N_t is the density of recombination centres; σ_n and σ_p are the electron and hole capture cross-sections, respectively; v_{th} is the thermal velocity, E_t is the energy level of the trap and:

$$n_1 = n_i \exp[(E_t - E_i)/kT] \qquad (5.76)$$

$$p_1 = n_i \exp[(E_i - E_t)/kT] \qquad (5.77)$$

The term $(np - n_i^2)$ is the deviation from the thermal equilibrium. For $np = n_i^2$, in thermal equilibrium, $R_{SRH} = 0$. For electron traps $\sigma_n \gg \sigma_p$, and for hole traps $\sigma_p \gg \sigma_n$, and if both the electron and hole traps act as non-radiative recombination centres (i.e. $\sigma_n = \sigma_p = \sigma$), the equation for the recombination rate can be written as:

$$R_{SRH} = \frac{N_t \sigma v_{th}(np - n_i^2)}{\{n + p + 2n_i \cosh[(E_t - E_i)/kT]\}} \qquad (5.78)$$

From these equations, it follows that the non-radiative recombination rate increases as E_t approaches E_i, the mid-gap energy, so the maximum recombination rate occurs at centres with levels at mid-gap. For defect levels near the band edges (i.e. shallow levels), thermal emission of the trapped carrier back to the nearby band is more likely than recombination. Thus, the defect levels can be either a recombination (deep) centre or a (shallow) trap depending on E_i.

The recombination rate at the GB can also be expressed in terms of the surface recombination velocity ($S = N_{st} \sigma_s v_{th}$) as:

$$R_{SRH} = \frac{S_n S_p(np - n_i^2)}{S_n(n + n_1) + S_p(p + p_1)} \qquad (5.79)$$

Edmiston *et al.* (1996) concluded from their analysis that the definition of the effective surface recombination velocity must be related to device geometry and depend on whether the GB is horizontal or vertical, or dark or illuminated. They also concluded that the effective surface recombination velocity is only useful for device modelling if it is independent of carrier concentration (Edmiston *et al.* 1996).

An effective method for rendering dangling bonds (and associated states) electrically inactive is passivation (see Section 5.5.8).

5.10.4 More recent SEM CL and REBIC analyses of grain boundaries in semiconducting materials

The properties of grain boundaries were outlined in Section 5.10.1 and their role in devices will be discussed in Section 5.12.4. In this section we discuss SEM CL, EBIC and REBIC characterization of grain boundaries. These techniques are especially useful in characterizing the electrical properties of GBs in thin film transistor arrays, terrestrial solar cells, as well as varistors, thermistors, and boundary layer capacitors (see Section 5.12.4).

To reiterate briefly, grain boundaries are modelled by two Schottky barriers back-to-back (see Fig. 5.94). This represents the effect of grain boundary trapped charge, leading to energy band bending at the GB (Section 5.12.4).

Leach (2001), using SEM-REBIC, demonstrated that there is a correlation between the electrical properties of individual GBs in a ZnO varistor and the orientations of the crystal faces on either side of the boundary. In these studies (Leach 2001), two main types of contrast were observed. Some grain boundaries exhibited 'bright and dark' contrast (termed type I contrast, or symmetric) indicative of symmetrical and opposed electric fields on either side of a charged GB. However, most of the interfaces were observed having electrically asymmetric contrast, i.e. either bright or dark contrast (termed type II contrast, or asymmetric). In the latter cases, the symmetrical contrast could be obtained by applying a small external voltage bias across an electrically active GB (Leach 2001). Thus, the contrast type can be changed, revealing electrical activity in the space-charge regions on both sides of the GB plane and a common source for contrast types (Leach 2001). Such observations also help ascertain the relation of the GB crystallographic structure to the EBIC contrast (i.e. electrical properties) of a varistor (Leach 2001). Fig. 5.81 shows GB-EBIC images of GB contrast within the varistor. Figs. 5.81a and 5.81b correspond to EBIC images of bright type II GB with (a) zero and (b) −90 mV applied. The application of a bias of −90 mV results in the symmetrical barrier structure and type I contrast. Fig. 5.81c presents a zero bias GB-EBIC image corresponding to a dark type II contrast. The application of a bias of +50 mV in this case results in type I contrast (see Fig. 5.81d).

The orientations of the grains on either side of the boundaries were determined by EBSD analysis and depth resolved EBIC (DREBIC) microscopy was employed to ascertain the GB plane orientations (Leach 2000). The latter were found to control the asymmetry in the electrical characteristics of the barrier (Leach 2001). Thus, it was concluded that, while specific dopants have a major influence on charged GB structures and their electrical properties, there is also some crystallographic influence on the electrical behaviour of the barrier (Leach 2001). In DREBIC microscopy, various electron beam energies are employed to excite the GB at different depths

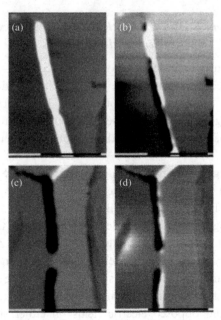

Figure 5.81 GB-EBIC images of grain boundaries within a varistor. (a) and (b) are EBIC images of a bright type II GB with (a) zero and (b) −90 mV applied. (c) and (d) are EBIC images of a dark type II GB with (c) zero and (d) +50 mV applied. The scale bars are 10 μm wide. (After Leach 2001. Reprinted from *Journal of the European Ceramic Society*, **21**, Crystal plane influence of the EBIC contrast in zinc oxide varistors, pp. 2127–30. Copyright 2001, with permission from Elsevier.)

below the surface, and the EBIC signal offset obtained for the inclined GB allows the determination of the GB plane slope (Leach 2000).

SEM CL and EBIC were extensively employed to characterize grain boundaries and interfaces in II-VI (CdTe/CdS) (Edwards *et al.* 1997, Galloway *et al.* 1997, 1999, Durose *et al.* 1999) and I-III-VI$_2$ [Cu(In,Ga)Se$_2$] thin-film polycrystalline solar cells (e.g. Romero *et al.* 2002a, Romero *et al.* 2002b). Interestingly, some thin-film polycrystalline solar cells (e.g. CdTe/CdS) perform better than their single-crystal counterparts. This was attributed to the GB band bending (due to the depletion space charge regions) facilitating separation of electron-hole pairs and thus enhancing the device photoresponse at the GBs (e.g. Visoly-Fisher *et al.* 2004). This was supported by earlier work (Visoly-Fisher *et al.* 2003), using scanning capacitance microscopy (SCM) and scanning Kelvin-probe microscopy (SKPM), which confirmed GB depletion in polycrystalline CdTe. Smith *et al.* (2004) employed near-field scanning optical microscopy (NSOM) and OBIC to study local photocurrent generation in polycrystalline CdTe/CdS solar cells. They observed enhancement, compared to the intra-grain volume, in photocurrent collection at GBs. This confirmed the previous suggestion that GBs assist in electron-hole pair separation in these solar cells (Smith *et al.* 2004) and explained the GB contribution to the improved performance.

As outlined in Section 5.5.8, the EBIC technique was also found valuable in studying the effect of impurity contamination on grain boundaries in polycrystalline materials. Chen *et al.* (2004) employed EBIC together with EBSD to examine the influence of GB type and impurity contamination on the recombination activity of GBs in multicrystalline Si. These studies showed that the recombination activity of uncontaminated GBs was weak, and the GB type had no considerable influence on recombination activity; whereas with increasing Fe contamination the recombination activity of the GBs was significantly increased. Also, for the same degree of contamination, the recombination activity was greater for the high-Σ GBs than for the low-Σ ones (Chen *et al.* 2004).

As discussed in Section 4.8.1, Herrera Zaldivar *et al.* (2001) observed REBIC dark-bright contrast at the perimeter of growth hillocks in Si-doped GaN films (see Fig. 4.146). They concluded, therefore, that the hillock edges had associated space-charge regions with corresponding energy band bending, acting like back-to-back Schottky barriers. This was explained on the basis of non-uniform distributions of charged defects and impurities at the hillocks (Herrera Zaldivar *et al.* 2001).

Gaevski *et al.* (2002) performed 3D studies of electrically active grain boundaries in ZnO varistor materials, by using SEM EBIC together with examination of subsurface geometry using focused ion beam (FIB) erosion. The EBIC profiles across GBs were compared with calculations using theoretical models for the electron-hole generation function (Gaevski *et al.* 2002). They concluded that the EBIC contrast is strongly influenced by both the tilt of the GB relative to the sample surface and the depletion region asymmetries at the boundary (Gaevski *et al.* 2002). They also found that the barrier asymmetry under zero applied voltage could be attributed to small variations of deep donor concentrations between grains.

Maestre *et al.* (2004) employed SEM REBIC and CL to study the electrical activity of grain boundaries in SnO_2 semiconducting oxide ceramic samples. They observed characteristic black-white PAT REBIC contrast at GBs in samples annealed in oxygen at 600° C. CL measurements showed that such oxygen treatment also results in reduced radiative recombination at the GBs (Maestre *et al.* 2004).

Urbieta *et al.* (2004) employed REBIC in the STM (STM-REBIC) for nanoscale characterization (i.e. with a spatial resolution in the few nanometer range) of electrically active boundaries in ZnO with grain sizes between about 200 and 500 nm. Electrically active grain boundaries were imaged, and in some of the small grains it was revealed that the space charge region extended over the whole grain.

Microcathodoluminescence spectroscopy studies of defects in Bi_2O_3-doped ZnO revealed much greater donor defect densities within the ZnO grains (suggesting higher *n*-type conductivity) than at their GBs (Sun *et al.* 2002). The GL (green luminescence) emission was found to be much stronger in the grain bulk than at the GBs. The dominant GL emission is usually attributed to donor–acceptor recombination associated with oxygen vacancies or ionized Zn interstitials (as electron traps) and zinc vacancies (as hole traps) (e.g. Sun *et al.* 2002 and references therein).

5.11 The role of defects in devices

To reiterate briefly, dislocations in semiconductors may:

(i) Introduce states deep in the band gap that (a) act as non-radiative recombination centres for carriers and reduce luminescence efficiency for light-emitting devices or, (b) in other cases, act as radiative recombination centres, producing luminescent emission at a photon energy of a fraction of the band gap value (producing bright dislocation CL contrast (see Section 5.6.3) or induce shallow states that (c) act as generation-recombination centres that produce electrical noise.
(ii) Cause impurity segregation into Cottrell atmospheres and precipitates.
(iii) Induce leakage current by acting as electrical shorts in devices.
(iv) Get charged by capturing electrons and act as scattering centres for electrons.
(v) Reduce reliability in devices.

To understand these properties and the role of defects in devices we must consider defect generation in materials. An important distinction is that between bulk and interface defects generated during crystal growth. Defects in the bulk include precipitates, dislocations and dislocation loops. Interface defects include dislocations, dislocation clusters, stacking faults, microtwins, and misfit dislocations. As discussed in Section 4.6, defects such as dislocations present in the substrate can also propagate into the epitaxial layer.

5.11.1 Introduction

Long-term reliability of semiconductor devices is essential. (Integrated circuits, ICs, eliminated large numbers of unreliable contacts beween the discrete devices used previously in electronics. The resulting increase in reliability provided a major motivation for developing ICs originally.) Thus, the role of defects in semiconductor materials and devices and in device degradation is most important (for reviews see Holt 1996, Ueda 1999, Mahajan 2000). Generally defects affect devices either (i) immediately, or (ii) by changing the microstructure during device operation, resulting in eventual device failure.

Device reliability does not always correlate inversely with the defect density. An important reason for this is that defects in the material outside the active volume of the device have no immediate effects. However, they can have indirect or delayed effects. A beneficial indirect effect is impurity gettering by deliberately introduced defects (see Section 5.12.1). On the other hand, some defects present in the substrate can gradually slip or climb up into the device and participate in device degradation and failure.

5.11.2 Immediate effects: 'Fatal' defects, device yields and good housekeeping

The reliability and yield (the percentage of usable devices per slice processed) of semiconductor devices are among the most important measures of the success of their manufacture. Device yield is also the probability of obtaining a part with no fatal defects. (Fatal defects are those that result in immediate device failure.) Defects can be introduced during any processing step of device fabrication. Fatal defects are typically large and affect a substantial fraction of the device. Examples of such defects are: large precipitates producing premature reverse breakdown, diffusion pipes shorting through base layers, and regions of slip at the edge of the wafer.

Fatal defects are a major factor determining the yield and, therefore, price competitiveness. Hence, eliminating fatal defects is a major goal of process development and quality control. In the development of ever larger scale integration, knowledge of defect characteristics and the ways they are introduced in growth and processing, is essential to eliminate these defects.

Mechanisms of semiconductor device failure are often categorized as (i) intrinsic failures, (ii) extrinsic failures, and (iii) electrical stress failures. Intrinsic failure mechanisms arise from the wafer fabrication process and are often due to crystal defects, like dislocations or processing defects like piping, gate charge effects and oxide breakdown. Extrinsic failures are related to, for example, metallization, die attachment, bonding, particulate contamination, and device packaging. Electrical stress failures include those due to electrostatic discharge (ESD) and electrical overstress (EOS). They are due to inadequate design and handling of components.

In the industrial fabrication of electronic devices, problems are often solved by 'good housekeeping'. That is, the use of improved substrates, chemicals, and process gases plus greater cleanliness of the production setting and more accurate control of process parameters commonly improve yields.

5.11.3 Defects and accelerated device failure mechanisms

Device degradation is divided into (i) gradual, (ii) rapid or (iii) catastrophic failure (see Fig. 5.82).

Gradual degradation (over thousands hours of operation) of light-emitting devices appears to be caused by recombination-enhanced point defect processes. It is characterized by a slow reduction in output power and the emergence of dark-spot defects (DSDs) or uniform darkening in the active device region. Ueda (1999) reviewed the results of TEM studies, and concluded that gradual degradation involves (i) non-radiative recombination at defects causing point defect generation and reactions, and (ii) the new defects then also act as non-radiative recombination centres, (iii) the migration and condensation of the point defects generated at nucleation centres, and (iv) the formation of microloops and/or defect clusters.

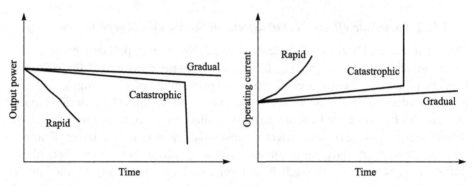

Figure 5.82   Schematic diagrams of the degradation modes of light-emitting devices (LEDs and laser diodes). (Adapted from Ueda 1996.)

Gradual degradation can be accelerated by stress. To minimize gradual degradation would require stoichiometry control (see Section 6.3.2), lattice matching in the heterostructure, and prevention of stress introduction during fabrication (Ueda 1999).

Rapid degradation occurs over several minutes to a few hundred hours of operation. It is characterized by a steep reduction in output power and the emergence of characteristic dark-line defects (DLDs) and/or dark-spot defects (DSDs) in the active device region. As mentioned in Section 5.3.3, recombination-enhanced dislocation glide (REDG) and recombination-enhanced dislocation climb (REDC) cause rapid degradation of optoelectronic devices (e.g. Ueda 1999). It can be reduced by inhibiting the formation of dislocation clusters and loops.

Catastrophic failure in GaAlAs/GaAs double-heterostructure (DH) lasers (e.g. Ueda 1999 and references therein) is typically caused by a current surge or by strong optical excitation. They result in catastrophic optical damage (COD) at the mirror surface. TEM studies show that the DLDs (generated at the mirror surface), seen in luminescence images of catastrophically degraded lasers, are arrays of dislocation networks or dislocation loops connected by dark knots (see Fig. 5.83).

Henry *et al.* (1979) described COD as follows. Above a critical output power density, strong optical absorption at the mirror surface and the consequent local heating of the crystal cause the energy band gap to narrow in the surface region, which results in more optical absorption. This leads to rapid thermal runaway and the crystal melts, followed by rapid cooling causing defects to form in the degraded area. No reports of COD in InGaAsP/InP DH lasers have appeared. This may be because the critical output power density to cause COD is very high in this case (Ueda 1999). According to Ueda (1999), dissimilar surface recombination velocities can explain the variation in COD level in different materials. In some cases, COD was also found to be related to a grown-in inclusion. This indicates that the mirror protection using dielectric films (e.g. SiO$_2$) and reduction of grown-in defects are required to eliminate catastrophic failure in lasers (Ueda 1999).

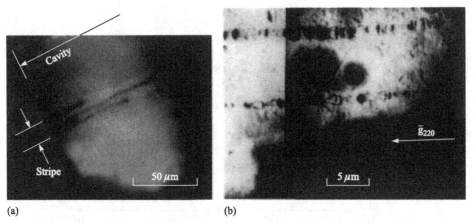

Figure 5.83 (a) An electroluminescence image of ⟨110⟩ DLDs in a catastrophically degraded GaAlAs DH laser. (b) A TEM image of dislocation networks corresponding to the ⟨110⟩ DLDs. (After Ueda *et al.* 1979. Reprinted with permission from *Journal of Applied Physics*, **50**, pp. 6643–7. Copyright 1979, American Institute of Physics.)

The relatively long working life of GaN-based devices containing high dislocation densities is mainly related to the very low dislocation mobility due to the large shear stress for dislocation motion and the small radiation enhancement effect (see Section 5.3.3 and Sugiura 1997). Thus, degradation by dislocation glide in GaN is slow. In contrast, GaAlAs/GaAs light-emitting devices exhibit a strong recombination enhancement effect producing relatively fast dislocation motion and rapid degradation through dislocation glide (Sugiura 1997).

The generation of defects in 4H-SiC devices and their degradation due to carrier injection under forward bias was investigated by Skowronski *et al.* (2002) and Zhang *et al.* (2003). The degradation of 4H-SiC *p-n* diodes during prolonged operation was related to an increasing stacking fault density in the active device region (Bergman *et al.* 2001). This was attributed to motion of the bounding partial dislocations through a recombination-enhanced dislocation glide mechanism (Skowronski *et al.* 2002).

The effect of dislocations on the radiative recombination efficiency, or the internal quantum efficiency of light emitting devices, as shown in Fig. 5.84, can be expressed (e.g. Yamaguchi *et al.* 1986) as:

$$\frac{\eta}{\eta_0} = \frac{1}{1 + L_0^2 \, \pi^3 N_{\mathrm{d}}/4} \tag{5.80}$$

where L_0 is the minority carrier diffusion length, N_{d} is the dislocation density, η and η_0 are the efficiencies with and without dislocations, and η is defined as the ratio of the radiative recombination rate to the total recombination rate, i.e. $\eta_{\mathrm{rr}}/\eta_{\mathrm{tot}}$.

Numerous studies also showed that the presence of dislocations results in reduced solar cell efficiency due to their recombination activity (e.g. Yamaguchi *et al.* 1986,

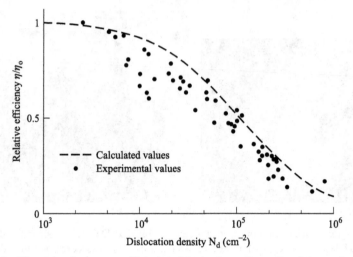

Figure 5.84   Electroluminescence efficiency (black dots, from Roedel *et al.* 1977) as a function of dislocation density for GaAs light-emitting diodes. The dashed line represents the efficiency calculated using $\eta/\eta_0 = (1 + L_0^2 \pi^3 N_d/4)^{-1}$ (see Yamaguchi *et al.* 1986. Reprinted with permission from *Journal of Applied Physics*, **59**, pp. 1751−3. Copyright 1986, American Institute of Physics).

Jain and Flood 1993, Möller 1996, Kittler *et al.* 2002). This is relatively well understood. The recombination strength varies strongly with the amount of defect contamination (decoration) by impurities (e.g. Kittler *et al.* 1995, 2003). It is also well documented that dislocations cause a reduction of the diffusion length and minority-carrier lifetime (e.g. Yamaguchi *et al.* 1985, 1986, 1989, Möller 1993). In thin-film GaAs solar cells, the dominant loss mechanism is due to recombination at dislocations decreasing the short-circuit current and increasing the excess leakage current (Yamaguchi *et al.* 1985, 1986, 1989). The minority carrier diffusion length due to recombination at dislocations is obtained from the one-dimensional continuity equation and can be expressed as (Yamaguchi *et al.* 1985):

$$\frac{1}{L_d^2} = \frac{\pi^3 N_d}{4} \tag{5.81}$$

The effective minority carrier diffusion length can thus be written as:

$$\frac{1}{L^2} = \frac{1}{L_0^2} + \frac{\pi^3 N_d}{4} \tag{5.82}$$

where L_0 is the intrinsic diffusion length. Note that the diffusion length L is related to the minority carrier lifetime τ by $L = (D\tau)^{1/2}$, where D is the minority carrier diffusion coefficient.

Solar cell efficiency depends on the short circuit current, the open circuit voltage and the fill factor. The open circuit voltage and the fill factor are decreased with an increased dark forward current, i.e. a shunt, attributed to dislocation tangles (Langenkamp and Breitenstein 2002).

5.11.4 Device performance limiting effects

There are a number of effects that reduce the performance of a device so it cannot be used as intended but do not prevent it being used in less demanding applications. These are less deleterious than fatal defects that prevent the device functioning at all. We will deal with those that are due to extended defects.

Excess noise

Typical sources of excess noise in semiconductor devices are defects, including dislocations that are sites of strong carrier scattering and trapping (e.g. Brophy 1956, 1959, Morrison 1956, 1992, Mil'shtein 2002). This was discussed in detail by Morrison (1992), who considered the fluctuations in electron trapping at dislocations in semiconductors. (The cylindrical space charge regions around the dislocations dominate the trapping.) According to that study, these fluctuations are related to mobility changes and are in reasonable agreement with the Hooge expression for $1/f$ noise (Morrison 1992). According to Hooge (1969), the empirical relationship for the noise spectrum for the current S_{current} and mobility S_{mobility} can be expressed as:

$$S_{\text{current}} = \alpha_{\text{H}} \frac{I^2}{N} \frac{1}{f} \tag{5.83}$$

$$S_{\text{mobility}} = \alpha_{\text{H}} \frac{\mu^2}{N} \frac{1}{f} \tag{5.84}$$

In these equations, I and μ are the current and mobility, respectively; α_{H} is the Hooge parameter, and N is the total number of free carriers. Thus, according to Hooge, $1/f$ noise is related to mobility fluctuations.

According to Morrison's calculations (1992), the noise power is proportional to the dislocation density and depends inversely on the doping level at high values. It was also concluded that the frequency range of $1/f$ noise due to dislocations will be wide only for a random distribution of dislocations. For aligned and impurity free dislocations the spectrum would be analogous to that due to generation-recombination noise (Morrison 1992).

Excess noise in semiconductor photodiodes is typically of $1/f$ form, which has been often linked to generation-recombination at junction dislocations. Kuksenkov et al. (1998) investigated the low-frequency noise of GaN p-n junction photodetectors. They attributed the photodiode dark conductivity to carrier hopping via dislocation states in the depletion region. The dark current noise under reverse bias followed the Hooge relationship for $1/f$ noise (Kuksenkov et al. 1998).

The $1/f$ frequency noise in AlGaN/GaN heterostructure field-effect transistors was studied by Garrido *et al.* (2000). The origin of this noise they related to mobility fluctuations, resulting from the changing carrier scattering rate of charged dislocations. Fang *et al.* (1990) measured low-frequency electrical current noise in double-heterojunction AlGaAs/GaAs laser diodes fabricated on GaAs and on Si substrates. They also observed $1/f$ frequency dependencies in the noise spectra. Diodes fabricated on Si substrates exhibited 50 times more noise than those fabricated on GaAs substrates. This was attributed to the much greater dislocation densities at the Si interfaces than at GaAs substrate/device interfaces (Fang *et al.* 1990).

Zimin *et al.* (2002) studied photovoltaic $p-n^+$ PbTe-on-Si infrared sensors and reported that their sensitivity is limited by generation-recombination. This they ascribe to the effect of dislocations crossing the active areas of the devices. They suggested that these dislocations cause local fluctuations of the built-in electric field.

Low-frequency noise can in principle be employed for semiconductor device characterization (Claeys and Simoen 1998). The degradation of Zener diodes was correlated with the $1/f$ noise by Zhuang and Du (2002), (diodes with greater initial noise degraded earlier). They attributed both the $1/f$ noise and the device degradation to dislocations in the space-charge region of the $p-n$ junction, and proposed to use a $1/f$-noise as a screening tool. Jones (2002), however, pointed out the additive nature of various noise contributions that makes any detailed understanding of various noise sources difficult and the value of using the noise as a diagnostic tool questionable.

Excess leakage currents

Excess leakage currents in semiconductors devices are often attributed to the presence of dislocations. However, the effect of any defect on device characteristics is likely to depend on the decoration of the defect by impurities (Section 4.7, Section 5.1.8, Section 5.5.8 and Section 5.6.4).

Beam *et al.* (1992) studied the influence of dislocation density on the current-voltage $(I-V)$ characteristics of InP photodiodes under reverse bias. They observed an increase in the leakage current with increasing dislocation density (see Fig. 5.85) and a reduction in the breakdown voltage. They attributed the increases in leakage current to the overlap of the space charge cylinders around dislocations, threading the p-n junction, at higher dislocation densities (Beam *et al.* 1992).

Ross *et al.* (1993) found the reverse leakage current increased linearly with dislocation density in GeSi/Si diodes. Hence they concluded that the current generated is about 10^{-5} to 10^{-4} A/m of misfit dislocation line. This was in good agreement with the value derived by Bull *et al.* (1979) for densities of dislocations threading through base-emitter junctions of a Si transistor. Ross *et al.*'s (1993) calculations showed that the derived leakage current per unit length would require much greater numbers of generation-recombination sites than the total number of

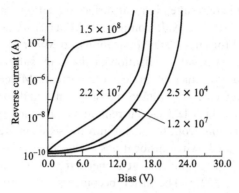

Figure 5.85 Reverse bias $I-V$ characteristics for InP photodiodes (fabricated on Fe-doped substrates) containing different dislocation densities. (After Beam *et al.* 1992. *Semiconductor Science and Technology*, 7, pp. A229–A232, 1992.)

dislocation core sites per unit length. Thus, they proposed that point defects around the dislocations or on the planes swept by the threading dislocations in extending the interfacial misfit dislocation segments must be involved.

The bulk leakage current depended on the threading dislocation density in GeSi *p-i-n* diodes according to Giovane *et al.* (2001), who distinguished the surface leakage current from the bulk leakage current. They treated the leakage current using a recombination-generation model supported by DLTS data for the capture cross-section and density of dislocation traps.

Reverse-bias leakage current in GaN devices was also extensively investigated. Kozodoy *et al.* (1998) demonstrated that threading dislocations in GaN *p-n* junction diodes result in significant increases in leakage current. They compared *p-n* diodes fabricated in low dislocation density areas and in adjacent high dislocation density areas, all produced by lateral epitaxial overgrowth (LEO) (see Section 4.6.7). In areas of low dislocation density, reverse-bias leakage current was lower by three orders of magnitude.

Parish *et al.* (1999) also reported improved performance (fast response times and low leakage currents) in AlGaN-based photodiodes on LEO GaN, compared to diodes produced on dislocated GaN. This improved performance was again ascribed to the absence of leakage paths (via the depletion region) due to dislocations, which are otherwise present in large numbers in GaN (Parish *et al.* 1999).

In a series of articles, Miller *et al.* (2002, 2003a, 2004) presented an analysis of reverse-bias leakage current in GaN-based devices, as well as some methods for its reduction. Currents due to dislocation leakage can be blocked by using an atomic force microscope (AFM) to form self-terminating insulating layers over the leakage paths (Miller *et al.* 2002). As the layer thickness reached 2 to 3 nm, the leakage current was blocked and further layer growth discontinued. Auger electron spectroscopy measurements on the modified surface showed that probe scanning in an ambient atmosphere resulted in local surface oxidation, and that the insulating layer is most likely a gallium oxide compound (Miller *et al.* 2003a). However, this method is impractical for large-scale applications, so Miller *et al.* (2003b) presented an alternative. Electrochemical anodization of the GaN surface in NaOH solution produced a comparable reduction in leakage current. Miller *et al.* (2004) used temperature-dependent current−voltage measurements and identified two distinct mechanisms for reverse-bias leakage current in GaN Schottky diodes. These are (i) field emission tunnelling from the metal into the semiconductor, and (ii) leakage current paths associated with dislocations. The former is dominant at low temperatures and becomes significant at higher temperatures for large reverse-bias voltages (Miller *et al.* 2004). The latter becomes significant above about 275 K with an exponential temperature dependence due to either trap-assisted tunnelling or one-dimensional variable-range-hopping along the threading dislocations (Miller *et al.* 2004).

Hsu *et al.* (2001, 2002) employed scanning current−voltage microscopy (using a modified AFM equipped with a voltage bias between a conducting tip and the sample to detect current). Together with TEM this demonstrated that reverse bias leakage in GaN Schottky diodes takes place predominantly at dislocations with a screw component. They also found that the reverse bias current distribution was very non-uniform. Moreover, the electrical activity of dislocations depends strongly on the growth conditions (e.g. V/III ratio during the epitaxial growth). This indicated that dislocation properties depend on local structural and/or chemical changes (Hsu *et al.* 2001). Hsu *et al.* (2002) also found that pure screw dislocations produce more reverse-bias leakage than edge or mixed dislocations.

McCarthy *et al.* (2001) investigated the influence of threading dislocations on collector−emitter leakage in an AlGaN/GaN heterojunction bipolar transistor (HBT) grown on a substrate using the LEO method. Emitter−collector leakage is one of the main problems in the applications of these devices. McCarthy *et al.* (2001) reported a correlation between threading dislocations and collector−emitter leakage. This leakage is the result of base layer compensation near the dislocations, producing under bias a punch-through from the collector to emitter. McCarthy *et al.* (2001) showed that the LEO wing regions (see Fig. 5.86) had four orders of magnitude lower emitter−collector leakage than adjacent window regions having a dislocation density of about $10^8 \, cm^{-2}$ (see Fig. 5.87). In this case, the width of window regions was 5 μm with a period of 40 μm. The threading dislocations continue above window regions, but not above laterally grown GaN wing regions that are nearly dislocation free.

5.11.5 *Mechanisms of defect-induced failure in devices*

Some defect-induced failure mechanisms in semiconductor devices are better understood than others. As mentioned above, on the basis of their effects on

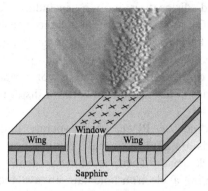

Figure 5.86 Above: AFM image of a LEO substrate. Spiral growth mode in the window region is related to the screw component of threading dislocations. (After McCarthy *et al.* 2001. Reprinted with permission from *Applied Physics Letters*, **78**, pp. 2235–7. Copyright 2001, American Institute of Physics.)

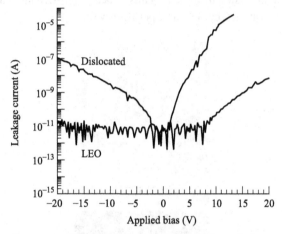

Figure 5.87 The collector–emitter leakage current of the wing relative to the window region (see Fig. 5.86), showing a reduction of leakage by four orders of magnitude for the wing region as compared to the window region. (After McCarthy *et al.* 2001. Reprinted with permission from *Applied Physics Letters*, **78**, pp. 2235–7. Copyright 2001, American Institute of Physics.)

devices, defects can be distinguished as those having immediate effects and those affecting microstructure (during device operation) resulting in device failure.

As mentioned in Section 5.11.4, plausible mechanisms for excess leakage current in devices have been proposed but none is universally accepted yet.

Mechanisms of defect-induced failure in optoelectronic devices were outlined in Section 5.11.3. Device degradation can be gradual, rapid or catastrophic. These are relatively well understood. Gradual degradation of light-emitting devices involves recombination-enhanced point defect processes. Rapid degradation is attributed to recombination-enhanced dislocation motion, including recombination-enhanced dislocation glide (REDG) and recombination-enhanced dislocation climb (REDC) resulting in dark-line defects (DLDs). Catastrophic failure observed in double-heterostructure lasers is manifested by catastrophic optical damage (COD) at the mirror surface. In this case the DLDs (at the mirror surface) observable in luminescence images are arrays of dislocation networks or dislocation loops connected by dark knots.

In avalanche photodiodes (APDs), which operate under reverse bias, near-breakdown electric fields are applied to make the junction depletion region 'reach through' the whole thick device. Then, at whatever depth an incoming photon is absorbed to produce an initial electron-hole pair, an 'avalanche' of carrier pairs will be created by subsequent impact ionization. Hence an easily detectable current pulse is obtained for each incoming photon. Due to the high electrical field, defects like dislocations, stacking faults and precipitates that act as field concentrators can produce premature electrical breakdown as shown in Fig. 5.88b. That is, for example, metal precipitates at dislocations can cause local high-density carrier generation (microplasmas) at voltages below the operating value.

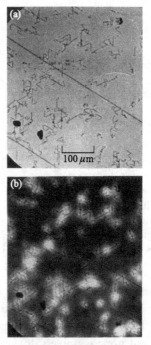

Figure 5.88   EBIC micrographs of a Si avalanche photodiode (APD) under (a) zero and (b) 266 V reverse bias. The fine zigzag dark lines in (a) are bits of a network of diffusion-induced misfit dislocations. Microplasma sites appear as bright spots in (b) and, as can be seen, these are associated with the dislocations and especially with nodal points in the network. (After Holt and Lesniak 1985.)

Rose (1957), Champlin (1959), and McIntyre (1961) reported cases of microplasma formation and microplasma breakdown in power diodes was found to be associated with dislocations by Chynoweth and Pearson (1958). In uniform (defect-free) reverse biased *p-n* junctions, at the breakdown voltage non-destructive electrical current flow occurs over the whole area, e.g. by Zener tunnelling. (So-called Zener diodes, operating in this way, are used to provide a fixed voltage, their reproducible reverse breakdown voltage.) However, as we saw above, defects can result in localized avalanche breakdown at voltages below that for uniform break-down (see Fig. 5.88). The breakdown voltage at microplasma sites is substantially less than in adjacent areas (Lesniak and Holt 1987). Microplasmas also increase electrical noise in these devices (McKay 1954, Champlin 1959).

Shockley (1961) suggested that microplasmas were not due to dislocations but to precipitates formed at dislocations that acted as electric field concentrators. Magnea *et al.* (1985) showed that microplasmas occur at local enhancements of the electric field due to doping fluctuations resulting from impurity segregation to crystal defects such as dislocations. These local electric field increases appear as local reductions in the breakdown voltage. Lesniak and Holt (1983, 1987) and Holt and Lesniak (1985) concluded on the basis of their EBIC studies (see Fig. 5.88) that neither dislocations nor precipitates were responsible for microplasmas in their Si avalanche photodiodes. Their microplasmas were due to dislocation-retarded diffusion resulting in the *p-n* junction being much less deep below the dislocations (see Fig. 5.89). The resultant sharp cusps in the junction produced field concentrations facilitating localized premature breakdown.

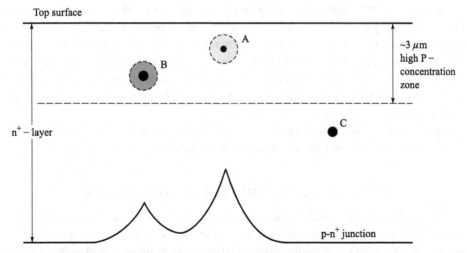

Figure 5.89 Schematic diagram illustrating how diffusion induced dislocations (i.e., A and B), by absorbing the diffusing impurity, retard the diffusion front and produce cusps in the p-n junction. It is thought that microplasma breakdown occurs at the field-concentrating cusps. (After Lesniak and Holt 1987.)

Some workers reported finding similar densities of microplasma sites and dislocations in reverse-biased diodes (e.g. Lee and Burrus 1980, Magnea *et al.* 1985), whereas others (e.g. Susa *et al.* 1982, Lesniak and Holt 1983) found far fewer microplasmas than dislocations. This is not surprising in the light of the mechanisms suggested above for premature breakdown. Impurity segregation or precipitation at the dislocations may be fairly uniform in which case they will all breakdown at about the same voltage as sometimes found. However, segregation may be non-uniform and/or the forms of the cusps in the junction may differ widely. In such cases, as the reverse bias increases, first a few and then more and more microplasmas will appear at the most favourable sites.

In strongly reverse-biased *p-n* junctions, avalanche breakdown also leads to light emission in the visible range (from about 1.5 to 3.0 eV) (Newman 1955, Chynoweth and McKay 1956). Several mechanisms were proposed for this emission. These include (i) direct, free electron-hole recombination in the microplasma emitting light on the high-energy side of the spectrum, and intraband transitions of energetic carriers contributing to the low-energy side (Chynoweth and McKay 1956), (ii) direct interband transitions (Wolff 1960), (iii) bremsstrahlung i.e. the emission of photons by energetic electrons that are strongly decelerated (Figielski and Torum 1962, Toriumi *et al.* 1987), (iv) indirect interband recombination of electrons and holes (Gautam *et al.* 1988), (v) a combination of direct and phonon-assisted intraband transitions (between levels in the conduction-band) (Bude *et al.* 1992). None of these models, however, can account for the whole of the observed spectra, the majority of which peak at around 2 eV.

Akil *et al.* (1998, 1999) proposed a multimechanism model for silicon avalanche light emission spectra. This model ascribes emission below about 2 eV to indirect interband transitions between high-field generated carriers. For emission from about 2 eV to about 2.3 eV, bremsstrahlung (intraband) is a primary process. Above 2.3 eV, direct interband transitions dominate. Variations between the observed spectra can be ascribed to differences in electric field intensity and ionization length for carriers in different samples (Akil *et al.* 1998, 1999).

Light emission by devices operated in the avalanche breakdown mode enables them to be employed as light-emitting diodes (e.g. Snyman *et al.* 1998).

5.12 Device benefits of dislocations and grain boundaries

In recent years the term 'dislocation engineering' has been adopted for the deliberate use of dislocations in semiconductor device technology. The aim is to introduce specific dislocations in particular concentrations and groupings at chosen locations inside the material to improve device operation. Methods for generating dislocations in such a controllable manner include ion-implantation and annealing (to eliminate the point defect 'radiation damage'), and the generation of misfit dislocations at epitaxial interfaces. Such dislocation engineering has the advantage of compatability with the existing microelectronics technology.

Such defect engineering involves primarily gettering (or removing) of unwanted impurities (see Section 5.12.1), as well as 'carrier lifetime engineering' using controlled introduction of recombination-active defects in distinct regions of the device (e.g. Claeys and Vanhellemont 1993).

Interfacially trapped charges form at grain boundaries producing potential barriers, which can be controlled with the applied bias. So, if the grain boundary structure and chemistry and thus electrical properties can be controlled, they can be used with benefit in a number of electronic devices. These devices include thermistors, varistors, and boundary layer capacitors.

5.12.1 Gettering

Gettering renders unwanted impurity atoms electrically inactive. It uses large densities of defects introduced, e.g. by abrading the back of a slice of the semiconductor, as sinks. The slice is heated so the impurities diffuse to the defects where they precipitate, becoming electrically inactive. Gettering is extensively employed for improving microelectronic and photovoltaic devices characteristics. It is used particularly to eliminate detrimental transition-metal contaminants, introduced during wafer growth and processing, from the device region (see a review by Myers *et al.* 2000). In addition to dislocations, grain boundaries and stacking faults can act as sinks for gettering (see Myers *et al.* 2000), as well as micropores (Chelyadinskii and Komarov 2003).

Gettering is either (i) intrinsic or (ii) extrinsic. In the intrinsic case, the dislocations are generated as a result of thermal precipitation of dissolved oxygen in silicon. In extrinsic gettering, dislocations are generated at the backside of the wafer by ion implantation, high-concentration diffusion, mechanical abrasion or laser-induced damage. Intrinsic gettering has the advantage of inserting the gettering sites closer to the active device regions.

Impurity gettering was reviewed by Sumino (2003), who distinguished *reaction-induced* from *interaction-induced* gettering. In the former, reaction products are preferentially formed at structural defects. This is often referred to as preferential precipitation of impurities at structural defects. The interaction between impurity atoms and structural defects at elevated temperatures occurs through their strain fields, and is treated by elasticity theory (see Section 4.7). Elastic strain fields extend to the surfaces of the crystal, but diminish rapidly to negligible levels over microscopic distances. In Si, gettering does not occur through (elastic) interactions for typical impurity concentrations. Gettering in Si is principally reaction induced (e.g. Sumino 2003).

Gettering is affected by various factors. These include (i) the type of structural defect acting as a gettering centre, (ii) the gettered impurities and their concentration, (iii) the crystal thermal history, and (iv) the gettering temperature range (e.g. Sumino 2003).

5.12.2 Devices based on dislocation luminescence

Silicon-based light-emitting devices, compatible with silicon microelectronics, are being sought for applications including optical on-chip interconnects. Bulk silicon has an indirect energy gap so it is not an efficient light source (Section 1.5.6). Nevertheless, several methods for obtaining visible light emission from Si-based devices have been proposed. These include devices based on porous silicon, on silicon/silicon dioxide superlattices, and on silicon nanoprecipitates (quantum dots, Section 1.7.2) in silicon dioxide. Infrared-emitting devices were also proposed, based on erbium-doped silicon, on iron disilicide and employing a silicon/germanium quantum cascade laser (see Section 1.7.4 and for a review, see Pavesi 2003).

Implantation-induced dislocations were employed by Ng *et al.* (2001) to develop an efficient silicon LED operating at room temperature. They implanted boron into silicon to form a *p-n* junction and introduce dislocation loops and associated local strain fields. The dislocation loops were about 100 nm from the junction. The loops were of the order of 100 nm across and about 20 nm apart (Ng *et al.* 2001). The strain field of this dislocation loop array alters the band structure of silicon to provide spatial confinement of the charge carriers. This charge carrier spatial confinement stops diffusion to point defects or the surface preventing non-radiative recombination and facilitates room-temperature electroluminescence (Ng *et al.* 2001). Electroluminescence spectra recorded at various temperatures are shown in Fig. 5.90.

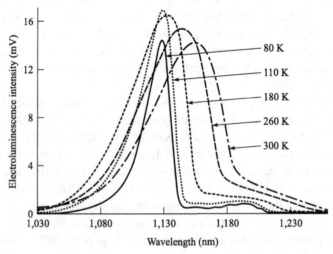

Figure 5.90 Dislocation facilitated silicon electroluminescence spectra at the temperatures marked (After Ng *et al.* 2001. Reprinted by permission from Macmillan Publishers Ltd: *Nature*, **410**, pp. 192–4. Copyright 2001).

However, the interpretation, of Ng *et al.* (2001), i.e. that extended defects provide spatial confinement of the charge carriers, preventing them diffusing to non-radiative recombination centres was disputed by Sobolev *et al.* (2004). They suggested that extended defects probably affect luminescence by gettering non-radiative recombination centres and/or by forming new radiative recombination centres.

Dislocation engineering has been employed for other silicon-based room temperature light emitting diodes also, in materials such as β-FeSi$_2$ and Er-implanted Si producing electroluminescence at about 1.5 μm (Lourenço *et al.* 2003). The β-FeSi$_2$ is employed as a radiative recombination path in a conventional Si device to produce efficient light emission. But due to significant thermal quenching, the room-temperature luminescence for such devices is not sufficiently high. Therefore, the devices were fabricated by Fe implantation into pre-grown silicon *p-n* junctions followed by boron implantation (Lourenço *et al.* 2003). As in the work of Ng *et al.* (2001) immediately above, the boron implantation was employed to form dislocation loops and their local strain fields. Again the dislocation-induced strain field raises the Si energy gap sufficiently to spatially confine the injected carriers and suppress their diffusion to non-radiative recombination centres. Thus, the controlled introduction of the dislocation loops minimizes thermal quenching. This results in relatively strong room-temperature luminescence. In the absence of boron implantation to form dislocation loops the room-temperature luminescence was much weaker than from the devices given those dislocation loops (Lourenço *et al.* 2003).

The mechanism of ion-implantation-induced dislocation luminescence in Si was further studied by Stowe *et al.* (2003). They used cathodoluminescence (CL) and observed relatively efficient room-temperature luminescence at 1.07 eV (i.e., 50 meV lower than the energy-gap value). This was associated with the presence of ion-implantation-induced dislocations and was ascribed to electron-hole recombination at the dislocations. Their model (Stowe *et al.* 2003) related to the shallow one-dimensional (1D) energy bands due to the strain field of dislocations. [The energy-gap dependence on strain (the deformation potential) means that the dislocation strain field causes local modification of the band structure (see Section 5.1.5) in Si forming potential wells (for both types of carrier).] Shallow 1D dislocation energy bands split off from both the valence band and conduction band edges. Theory predicts that the energy difference between the 1D-bands and the band edges is of the order of 50 meV. Hence, Stowe *et al.* (2003) suggested that carrier confinement due to the deformation potential results in relatively efficient room-temperature radiative electron-hole recombination via these 1D-energy bands.

Dislocation electroluminescence (lines D1 through D4, see Section 5.3.2) with greater than 0.1% external efficiency at room temperature was reported by Kveder *et al.* (2004). Impurity gettering and hydrogen passivation allowed this enhanced D1-related luminescence efficiency. They assumed that some radiative transitions involved dislocation-induced defects, whereas the non-radiative transitions are associated with impurities and core defects at dislocations (Kveder *et al.* 2004). Fig. 5.91 shows that although all the dislocation-related lines (D1 through D4) are observable at low temperatures, the D1 line makes the principal contribution to the luminescence intensity at elevated temperatures.

5.12.3 Dislocations and grain boundaries as microelectronic components

Imperfections and inhomogeneities are not always undesirable. Doping impurities and interfaces like *p-n* and hetero-junctions, metal-semiconductor contacts and semiconductor-oxide interfaces are the components of semiconductor device structures. As we shall now see, many attempts were made down the years, to use defects to produce device functions or to enhance device performance. None have had any commercial success.

Proposals to utilize electronic properties of dislocations and grain boundaries in electronic devices were first made in the 1950s and early 1960s. One of these was a proposed grain-boundary transistor (Matare 1955). Layers were found at grain boundaries of type opposite to that of the material (e.g. *p*-type boundary layers between *n*-type grains). Matare developed a method for growing specimens of Ge containing tilt boundaries of defined orientation and tilt angle (Matare and

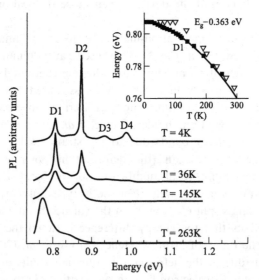

Figure 5.91   Normalized photoluminescence spectra of hydrogen passivated *n*-type Si at four temperatures. The inset shows the temperature variation of the energy position of the D1 line with closed symbols corresponding to PL and open symbols to EL; the solid curve represents calculated values according to $E_g(T) - 0.363\,\text{eV}$, where $E_g(T)$ is the band gap of Si. (After Kveder *et al.* 2004. Reprinted with permission from *Applied Physics Letters*, **84**, pp. 2106–8. Copyright 2004, American Institute of Physics.)

Wegener 1957). He showed that when the two *n*-type grains and the *p*-type grain boundary region were contacted, these specimens would function as *n*-*p*-*n* transistors (Matare 1955).

Mueller and Jacobson (1959) also used a two-seed Czochralski method to grow 6° tilt boundaries in InSb (the most studied III-V compound at that time). Interpreting the structures on the early Hornstra (shuffle-set) model, they labelled them in terms of the types of edge dislocations that they contained as α and β grain boundaries (assumed to contain dangling bonds only from In and only from Sb atoms, respectively). (If the dislocation cores are really of the Hirth and Lothe (glide-set) core structure, of course, the types of dangling bonds in the two types of boundary would be reversed). Current flow across these boundaries in both *n*- and *p*-type material was observed. They found that the α boundaries acted as barriers to current flow in both the *n*- and *p*-type samples but the β boundaries only in *p*-type material. Mueller (1959a) studied the transient photovoltaic response of grain boundaries in Si and Ge and proposed that it be used to detect light. Mueller and Jacobson (1959) made a similar proposal after a study of the photovoltaic response of grain boundaries in InSb.

The reduction in minority carrier lifetime due to recombination at dislocations was proposed to be used to reduce diode switching times in plastically deformed germanium (Pearson and Riesz 1959). They fabricated diffused diodes in plastically deformed Ge. The high densities of dislocations reduced the lifetime of the minority carriers and greatly reduced the minority carrier storage effect. This resulted in diodes with turn-off times of the order of 10^{-9} sec, advantageous in high-speed switching applications. Similarly, Schumann and Rideout (1964) showed that plastic deformation of the material could reduce the turn-off delay of Ge *n*-*p*-*n* mesa transistors. Unfortunately, while deformation results in a fast recombination time it also produces a slow trapping time as well as a reduction in the carrier density in the diode (Neubert *et al.* 1973).

Subsequent work (for a review, see e.g. Mil'shtein 1999) demonstrated further device effects related to dislocations. However, several problems limit practical applications. Two main reasons that none of these proposals were adopted commercially are that (i) it is difficult, if not impossible, to control sufficiently well the number and type of dislocations introduced and to do so without additional (electrically active) point defect debris being produced, and (ii) technologies have always become available to achieve the same results economically without using defects, in fact by rigorously eliminating them. Detailed problems include (i) non-uniform distributions of dislocations, point defects and impurities, (ii) decoration of dislocations with impurities, (iii) the presence of dislocation-related intrinsic defects, such as kinks and jogs, (iv) variable stress distributions, and (v) difficulties with placement control. Hence, in practical applications, the difficulty of controlling various device parameters would be highly complex and intractable.

Although the use of dislocations as microelectronic components appears unlikely, there still might be some electronic applications related to the presence of dislocations at specific sites, e.g. interfaces. Rozgonyi *et al.* (1987) pointed out that in the case of interfacial misfit dislocations the depth, type and density of the dislocations could be controlled. Rozgonyi and Kola (1989) discussed defect engineering (control simultaneously of dopants, impurities and defects) for ULSI silicon. They were able to obtain controlled minority carrier lifetime reduction by the careful placement of misfit dislocations generated by a multi-layer Si(Ge)/Si structure. Radzimski *et al.* (1991) showed that the electrical activity of misfit dislocations in Si/SiGe/Si interfaces was strongly dependent on the amount of Au decorating them. They intentionally placed such gold-decorated dislocations within the space charge layer of a power diode to enhance its switching properties. Whether practical device applications will follow for even such especially favourable cases, however, remains to be seen.

Gopal *et al.* (1998) demonstrated the presence, at the InAs/GaP heterointerface, of a high-density sheet of electrons, due to the network of interfacial misfit dislocations. According to their model, these dislocations pin the Fermi level at the interface in the conduction band, giving rise to a sheet of electronic charge (Gopal *et al.* 1998). They suggested this may have device applications, such as the possibility for the misfit dislocations to act as conducting channels. Fig. 5.92 is a cross-sectional HRTEM image of the InAs/GaP interface including a regular array of misfit dislocations with a spacing of about 4 nm. Fig. 5.93 is a carrier-concentration depth profile obtained employing an electrochemical capacitance voltage profiler (Gopal *et al.* 1998). The sharp carrier concentration peak at the InAs/GaP heterointerface demonstrates the presence of a high-density sheet of charge localized at that interface.

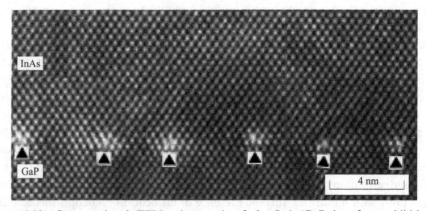

Figure 5.92 Cross-sectional TEM micrograph of the InAs/GaP interface exhibiting a regularly spaced array of 90° misfit dislocations indicated by arrows. (After Gopal *et al.* 1998. Reprinted with permission from *Applied Physics Letters*, **72**, pp. 2319–21. Copyright 1998, American Institute of Physics.)

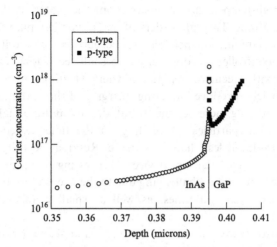

Figure 5.93 Carrier-concentration depth profile (obtained using an electrochemical capacitance voltage profiler) of the InAs/GaP heterostructure. (After Gopal *et al.* 1998. Reprinted with permission from *Applied Physics Letters*, **72**, pp. 2319–21. Copyright 1998, American Institute of Physics.)

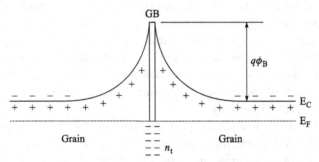

Figure 5.94 A simplified schematic diagram of the energy band structure and spatial charge distribution around a negatively charged grain boundary, showing the formation of a double Schottky barrier.

5.12.4 Devices controlled by grain boundaries

Some electronic devices are made of polycrystalline semiconductors. These are devices that must be large in area and relatively inexpensive like (i) thin film transistor arrays used to address large displays, and (ii) terrestrial solar cells. In addition, there are important electrical devices (varistors, thermistors and boundary layer capacitors), which depend for their operation on the electrical properties of the grain boundaries that they contain. These are not often thought of as semiconductor devices as they are made of oxides and their technology is typical of electroceramics. However, the principles underlying the role of grain boundaries (GBs) are the same in both cases.

Ceramics are high-temperature, non-stoichiometric materials that are processed at elevated temperatures. They are orders of magnitude less pure than semiconductors. Nevertheless, impurity-dependent properties like the critical voltage for switching can be controlled by doping, as we shall see below. Typically ceramics fabrication starts with precursor material powders. Heating and compression of the mixed powder particles produces sintering (merging of the particles starting at their contact points). Sintering increases the overall density and gradually eliminates the voids (pores) between particles. Depending on the time heated under pressure, material can be produced less than fully dense. Residual porosity can be valuable for applications like gas sensing devices. Processing to provide reproducible properties in these materials is challenging due to their sensitivity to small variations in sintering temperatures and times, as well as small changes in stoichiometry (see Section 6.3.2).

Ceramics microstructure can vary in grain size and shape. (In device modelling, a regular lattice of cubical grains often represents such microstructures.) The technological applications of semiconducting oxides are not limited to electroceramic materials, but also include thin films. Ceramics thin-film synthesis techniques include sputter-deposition, laser ablation, MOCVD, and sol-gel techniques (e.g. Auciello *et al.* 1998). Electrical and dielectric properties, critical for applications of these materials, depend strongly on small concentrations of dopants and defects at grain boundaries. Semiconducting oxides typically have greater band gaps and higher temperature stability than the classical semiconductors (e.g. Si and GaAs). The main advantages of devices based on oxide ceramics are stability in harsh environments and the low cost of processing. The chief problems concern the control of microstructure and homogeneity.

Control of the structure and chemistry of grain boundaries is vital for these devices since they control GB electrical properties. As noted in Section 5.10.2, these properties of grain boundaries are generally modelled by two Schottky barriers back-to-back. This model is based on the fact that traps are typically present at grain boundaries, these trap carriers resulting in a GB charge. This depletes the adjacent grains of free carriers to form shielding space charge barriers. Thus, depending on the doping type of the grains, this results in the upward (for an *n*-charged GB) or downward (for a *p*-charged GB) energy band bending at the GB. The effect is equivalent to the formation of double Schottky barriers.

The double Schottky barriers produce non-linear $I-V$ behaviour employed in electroceramic device applications. These GB barriers and their applications in devices have been extensively investigated, especially for *n*-type polycrystalline oxide semiconductors (e.g. Greuter and Blatter 1990, Clarke 1999, Waser and Hagenbeck 2000, Van de Krol and Tuller 2002, and references cited therein). However, these grain-boundary Schottky barriers have no reality. They are just the standard way of modelling the effect of a negatively (or positively) charged grain boundary (see Fig. 5.94).

Grain boundary controlled devices include varistors, positive temperature coefficient (PTC) thermistors, and boundary layer capacitors. A varistor (variable resistor) switches from a low to an orders-of-magnitude higher resistance at a certain voltage. A PTC thermistor is a thermally sensitive resistor, whose electrical resistance rises rapidly by orders-of-magnitude after a certain (reference) temperature is surpassed.

The materials used in these devices are polycrystalline oxide semiconductors like ZnO, $BaTiO_3$ and $SrTiO_3$, although other types of polycrystalline materials are also used. The unique characteristics of such oxide semiconductors are related to their defect chemistry (see, e.g. reviews by Waser and Hagenbeck 2000, and Van de Krol and Tuller 2002). These materials may exhibit unintentional *n*- or *p*-type doping behaviour due to intrinsic point defects. For example, in ZnO unintentional *n*-type doping is typically attributed to zinc interstitials and oxygen vacancies (see Section 6.3). This is, however, still debatable. Since the oxygen vacancy is a deep donor and the zinc interstitial has high formation energy, Van de Walle (2001) proposed that unintentionally incorporated hydrogen is a likely candidate as shallow donor.

Important applications are as varistors, devices made of appropriately doped ZnO (standard material), $SrTiO_3$, TiO_2, and SiC, and as positive temperature coefficient (PTC) thermistors using $BaTiO_3$. In practical applications, these are typically donor-doped ceramic materials. Depletion space-charge layers are formed next to the grain boundaries as the positively charged donor centres in the bulk compensate the negatively charged grain-boundary states. Then the electronic behaviour of the GBs can be represented by back-to-back double Schottky barriers. Optimal and reliable operation of devices, such as ZnO varistors, necessitates incorporation of specific additives (e.g. Bi, O, and transition elements at around the grain boundaries) and processing steps (e.g. heat treatments) to produce optimal GB properties (see Steele 1991).

The current flow across the boundaries in varistors (variable high resistors) is Ohmic (*I* varies linearly with *V*) at low fields, whereas at and above a critical applied device voltage the fields are sufficiently high to tilt the energy levels until GB barrier breakdown (or switching) occurs resulting in a steep resistance drop. Varistors are used to protect electrical equipment. Large high-switching-voltage ones are connected between power-lines and earth. If lightning strikes, the varistor switches to a low resistance and shunts the lightning current harmlessly to earth. Smaller, lower breakdown voltage varistors are used similarly to protect electrical machinery or electronic circuitry from voltage spikes.

The possible breakdown mechanisms include tunnelling, thermionic emission, and impact ionization (e.g. Pike and Seager 1979, Pike 1982, Blatter and Greuter 1986b, Greuter and Blatter 1990, Van de Krol and Tuller 2002), with the most likely being the impact ionization mechanism (Blatter and Greuter 1986b, Greuter and Blatter 1990). In this electron-hole pairs are generated by impact ionization, followed by the separation of carriers by the space charge field and the breakdown of the barrier due to the recombination of the minority carriers with the electrons trapped at the interface states (e.g. Blatter and Greuter 1986b, Greuter and Blatter 1990, Van de Krol and Tuller 2002).

PTC thermistors are typically based on polycrystalline *n*-type $BaTiO_3$. This ferroelectric material exhibits a dramatic reproducible increase in resistance (by several orders of magnitude) just above its characteristic Curie temperature (see Fig. 5.95). This is widely employed in, for example, surge protection devices, temperature sensors, and self-regulating heaters. The sharp resistance increase starts at the Curie temperature T_C. (T_C can be increased or reduced for different applications by additions of suitable dopants.) The positive temperature coefficient of resistance (PTCR) effect is attributed to grain boundary-controlled properties of the material. The first model of the PTCR effect was proposed by Heywang (1961, 1964). According to Heywang the PTCR effect is related to the increase in the grain-boundary barrier height above the Curie temperature. This is due to a decrease in the temperature-dependent dielectric constant (obeying the Curie-Weiss law). The latter occurs above T_C as the material experiences a ferroelectric to paraelectric transition, associated with a tetragonal to cubic crystallographic transformation. Indeed, the resistivity in the grain-boundary region can be expressed as:

$$\rho_{GB} = A \exp\left(\frac{q\phi_B}{kT}\right) \tag{5.85}$$

where A is a constant. The barrier height, according to this model, can be expressed as:

$$\phi_B = \frac{qn_t^2}{8\varepsilon\varepsilon_0 N_d} \tag{5.86}$$

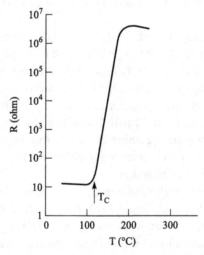

Figure 5.95 Typical dependence of electrical resistance on temperature of positive temperature coefficient of resistance (PTCR) ceramic. T_C denotes the Curie temperature.

where n_t is the trapped electron concentration at the grain boundary states, N_d is the donor concentration in the grains, ε is the static dielectric constant, and ε_0 is the permittivity of the vacuum. Thus, the barrier height ϕ_B, and hence the resistivity vary at T_C due to the inverse dependence of ϕ_B on the dielectric constant.

The Heywang model was further modified by Jonker (1964) to account for the effect of spontaneous polarization below the Curie temperature, and numerous further modifications have been advanced, e.g. to account for anomalous $I-V$ characteristics, as well as to elucidate the origin of the interface states at grain boundaries. Some investigations also related differences in PTCR characteristics at individual grain boundaries to variations in grain boundary misorientation at the interface (Hayashi *et al.* 1996). Greater PTCR effects were found for random grain boundaries, than for coherent boundaries (Hayashi *et al.* 1999).

The PTCR effect is greatly influenced by the types and concentrations of impurities and intrinsic defects in $BaTiO_3$ ceramics, as well as by the microstructure of the material. The importance of such factors as the amount of grain-to-grain contact area (i.e. density), number of boundaries accessible to conduction, and grain boundary domain orientation and coherence was reviewed by Roseman and Mukherjee (2003). They concluded that the presence of a variety of grain boundary structures, potential barriers and depletion widths (due to structural heterogeneities) results in preferential conduction through aligned domain pathways and low potential barrier boundaries (Roseman and Mukherjee 2003). These should be taken into consideration for both the refinement of modelling studies and interpretation of various observations and their discrepancies.

The electrical characterization of individual grain boundaries has demonstrated the heterogeneity of the PTCR effect. Some boundaries were found to be highly resistive, and others exhibited small or no PTCR effect. This is likely due to the differences (as the result of variations in grain boundary structure and/or non-uniform processing) in the dopant concentrations segregating at various grain boundaries. The complexity of the combination and interdependence of all these factors does not allow the earlier models based on a relatively limited number of assumptions to provide any universal description or to be equally valid for observations on materials prepared and processed in different ways.

The formation of grain boundary barriers in $SrTiO_3$ or $BaTiO_3$ ceramics can be also employed to produce boundary layer capacitors. These are based on narrow depletion regions around grain boundaries acting like parallel plate capacitors. The capacitance C (per unit cross section) for a symmetric junction is $C = \varepsilon\varepsilon_0/2d$, where d is the width of the depletion region. This capacitance is expected to be especially large for high ε and small d. The network of these boundary layer capacitors determines the macroscopic capacitance of the material (for more application details and references, see Greuter and Blatter 1990).

The ease of diffusion of gases (e.g. oxygen) in porous polycrystalline material can be used in gas sensors. In this case, oxygen adsorbed between two grains causes the barrier (and hence the resistance) to increase, facilitating the gas sensing capability.

There are two valuable ways to use SEM-EBIC to study grain-boundaries in electroceramics. They differ in the separation of the electrodes through which the signal current flows. If the electrodes are mm or cm apart the method is called remote electron beam-induced current (REBIC, see Section 5.10.4), whereas if the contacts are on either side of a single grain boundary (micrometers apart) it is called GB-EBIC. Using these techniques it was possible to detect negatively charged GB barriers in positive temperature coefficient (PTC) thermistors of polycrystalline n-type BaTiO$_3$ (Seaton and Leach 2004). By hot-stage microscopy, Seaton and Leach also showed that only some of the GBs were electrically active, i.e. showed EBIC contrast above the Curie temperature. Electron backscattering pattern analysis showed that it was mainly high-angle grain boundaries that were EBIC active (Seaton and Leach 2004).

Fig. 5.96 shows the occurrence (at 180° C) of EBIC contrast (the paired dark-bright lines). These run along those GBs that are electrically active. This contrast arises from the separation of electron-hole pairs (generated by the primary electron beam) by the oppositely directed space-charge fields on either side of charged GBs (Seaton and Leach 2004). As shown in Fig. 5.96b, at 100° C (i.e. below the Curie temperature) no contrast is observed, indicating the absence of electrostatic barriers at the GBs. This is consistent with the model predicting the suppression (due to the interaction with the spontaneous polarization of the ferroelectric phase) of GB barriers below the Curie temperature. But above that temperature, some grain boundaries exhibit EBIC contrast (Fig. 5.96c).

Fig. 5.97 reveals the distinct dissimilarity in contrast of two GBs labelled A and B. This is due to the differences in their structure (Seaton and Leach 2004). Pronounced

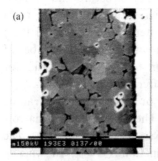

Figure 5.96 SEM images of the same area of a Barium-Titanate-based PTC thermistor: (a) secondary electron image of the inter-electrode area showing the grain structure in the material, (b) conductive mode image at 100° C and (c) conductive mode image at 180° C showing EBIC contrast. The black and white scale markers along the bottoms of the images each equal 10 μm. (After Seaton and Leach 2004. Reprinted from *Journal of the European Ceramic Society*, **24**, Conductive mode imaging of thermistor grain boundaries, pp. 1191–4. Copyright 2004, with permission from Elsevier.)

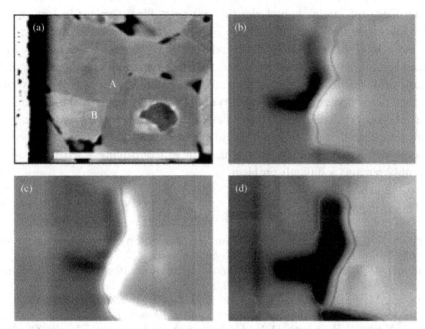

Figure 5.97   SEM images of a Barium-Titanate-based PTC thermistor: (a) secondary electron image showing two electrically active grain boundaries labelled A and B, (b)–(d) conductive mode images of these grain boundaries at 200° C under (b) zero applied bias, (c) +0.5 V bias, and (d) −0.5 V bias. The white scale marker in (a) equals 5 µm. (After Seaton and Leach 2004; Reprinted from *Journal of the European Ceramic Society*, **24**, Conductive mode imaging of thermistor grain boundaries, pp. 1191–4. Copyright 2004, with permission from Elsevier.)

contrast is observed only at the GB labelled A. The effects of applied bias voltages show that this is EBIC contrast. With a positive or negative applied bias across GB-A, the space-charge field is compensated on one side of the GB plane. There the signal is suppressed. This EBIC contrast at GB-A demonstrates the presence of a charge-separating electrostatic barrier. Under similar conditions, GB-B, however, shows no measurable EBIC at zero applied bias and exhibits single-line contrast under applied bias. This is characteristic of another effect (β-conductivity) and implies the presence of a resistive layer only. Thus, conductive mode microscopy can demonstrate distinct variations in GB contrast, indicating structural variations between individual GBs (Seaton and Leach 2004).

To summarize, polycrystalline oxide semiconductors such as ZnO, BaTiO$_3$ and SrTiO$_3$ are widely employed in grain boundary controlled devices. The double Schottky barrier model provides a standard quantitative description of these materials. The practical applications and performance of the devices based on grain boundary engineering depend on the microstructure (grain size distribution) and the presence and distribution of small amounts of dopants and defects at grain boundaries. These are understood empirically.

Knowledge of grain boundaries in polycrystalline semiconductors remains somewhat incomplete with the details of the mechanisms concerned unresolved (e.g. the role of native point defects, relative crystallographic orientation of the adjacent grains, interfacial segregation, and the formation of new inter-granular phases). Uniformity of microstructure and distribution of various species is vital for consistent performance of grain boundary controlled devices. Since the essential non-linear electrical behaviour of these devices is related to the grain boundary electrostatic charge, it is also important to elucidate how that charge is compensated by point defects to preserve charge neutrality. Controlling microstructure and uniformity remains a difficult problem.

References

Aberle, A. G. (2001). Overview on SiN surface passivation of crystalline silicon solar cells. *Solar Energy Materials and Solar Cells*, **65**, 239–48.

Abrahams, M. S. and Pankove, J. I. (1966). Orientation effect in GaAs injection lasers. *Journal of Applied Physics*, **37**, 2596–7.

Abrahams, M. S., Blanc, J. and Buiocchi, C. J. (1972). Like-sign asymmetric dislocations in zinc-blende structure. *Applied Physics Letters*, **21**, 185–6.

Akil, N., Kerns, S. E., Kerns, D. V., Hoffmann, A. and Charles, J. P. (1998). Photon generation by silicon diodes in avalanche breakdown. *Applied Physics Letters*, **73**, 871–2.

Akil, N., Kerns, S. E., Kerns, D. V., Hoffmann, A. and Charles, J. P. (1999). A multimechanism model for photon generation by silicon junctions in avalanche breakdown. *IEEE Transactions on Electron Devices*, **46**, 1022–8.

Alexander, H. (1986). Dislocations in covalent crystals. In *Dislocations in Solids*, 7, ed. F. R. N. Nabarro (Amsterdam: North-Holland), pp. 113–234.

Alexander, H. (1989). Changes of electrical properties of silicon caused by plastic deformation. In *Point and Extended Defects in Semiconductors*, eds. G. Benedek, A. Cavallini and W. Schröter. New York: Plenum.

Alexander, H. (1991). Chapter 6 Dislocations. In *Materials Science and Technology, Vol. 4 Electronic Structure and Properties of Semiconductors*, ed. W. Schröter (Basel: VCH), pp. 249–319.

Alexander, H. (1994). What information on extended defects do we obtain from beam-injection methods? *Materials Science and Engineering*, **B24**, 1–7.

Alexander, H., Labusch, R. and Sander, W. (1965). Electron spin resonance in deformed silicon crystals. *Solid State Communications*, **3**, 357–60.

Alexander, H., Dietrich, S., Hüne, M., Kolbe, M. and Weber, G. (1990). EBIC microscopy applied to glide dislocations. *Physica Status Solidi*, **A117**, 417–28.

Alfrey, G. F. and Wiggins, C. S. (1960). Electroluminescence at grain boundaries in gallium phosphide. In *Solid State Physics in Electronics and Telecommunications*, **2** (London: Academic Press), pp. 747–50.

Allender, D., Bray, J. and Bardeen, J. (1973). Model for an exciton mechanism of superconductivity. *Physical Review*, **B7**, 1020–9.

Amelinckx, S. and Dekeyser, W. (1959). The structure and properties of grain boundaries. *Solid State Physics*, **8**, 325–499.

Auciello, O., Foster, C. M. and Ramesh, R. (1998). Processing technologies for ferroelectric thin films and heterostructures. *Annual Review of Materials Science*, **28**, 501−31.

Baccarani, G., Ricco, B. and Spadini, G. (1978). Transport properties of polycrystalline silicon films. *Journal of Applied Physics*, **49**, 5565−70.

Balk, L. J., Kubalek, E. and Menzel, E. (1976). Investigations of as-grown dislocations in GaAs single crystals in the SEM. *Scanning Electron Microscopy*, **1**, 257−64.

Ballutaud, D., Riviere, A., Rusu, M., Bourdais, S. and Slaoui, A. (2002). EBIC technique applied to polycrystalline silicon thin films: minority carrier diffusion length improvement by hydrogenation. *Thin Solid Films*, **403**, 549−52.

Bardeen, J., Cooper, L. N. and Schrieffer, J. R. (1957). Theory of superconductivity. *Physical Review*, **108**, 1175−204.

Bardsley, W. (1960). The electrical effects of dislocations in semiconductors. *Progress in Semiconductors*, **4**, 155−203.

Barth, W. and Güth, W. (1970). Absorptionsmessungen am Plastisch Deformiertem Germanium. *Physica Status Solidi*, **38**, K141−K144.

Barth, W. and Elsässer, K. (1971). Polarization of the infrared absorption of dislocations in germanium. *Physica Status Solidi*, **B48**, K147−K149.

Barth, W., Elsässer, K. and Güth, W. (1976). The optical absorption of 60° dislocations in germanium. *Physica Status Solidi*, **A34**, 153−63.

Batstone, J. L. and Steeds, J. W. (1985). TEM and CL characterization of dislocations in OMCVD ZnSe. In *Microscopy of Semiconducting Materials 1985*. Conf. Series No. 76 (Bristol: Institute of Physics), pp. 383−8.

Beam, E. A., Temkin, H. and Mahajan, S. (1992). Influence of dislocation density on I−V characteristics of InP photodiodes. *Semiconductor Science and Technology*, **7**, A229−A232.

Bell, R. L. and Willoughby, A. R. F. (1966). Etch-pit studies of dislocations in InSb. *Journal of Materials Science*, **1**, 219−28.

Bell, R. L. and Willoughby, A. R. F. (1970). The effect of plastic bending on the electrical properties of indium antimonide 2. Four-point bending of n-type material. *Journal of Materials Science*, **5**, 198−217.

Bell, R. L., Latkowski, R. and Willoughby, A. R. F. (1966). The effect of plastic bending on the electrical properties of indium antimonide. *Journal of Materials Science*, **1**, 66−78.

Benoit a la Guillaume, C. (1959). Recombinaison radiative par l'intermediare des dislocations dans le rermanium. *Physics and Chemistry of Solids*, **8**, 150−3.

Bensahel, D. and Dupuy, M. (1979). SEM and TEM. Diffusion of lithium in ZnTe. *Physica Status Solidi*, **A55**, 203−10.

Bergman, J. P., Lendenmann, H., Nilsson, P. A., Lindefelt, U. and Skytt, P. (2001). Crystal defects as source of anomalous forward voltage increase of 4H-SiC diodes. *Materials Science Forum*, **353−356**, 299−302.

Billig, E. and Ridout, M. S. (1954). Transmission of electrons and holes across a twin boundary in germanium. *Nature*, **173**, 496−7.

Blatter, G. and Greuter, F. (1986a). Carrier transport through grain-boundaries in semiconductors. *Physical Review*, **B33**, 3952−66.

Blatter, G. and Greuter, F. (1986b). Electrical breakdown at semiconductor grain boundaries. *Physical Review*, **B34**, 8555−72.

Blumtritt, H., Kittler, M. and Seifert, W. (1989). On the formation of bright EBIC contrasts at crystal defects. In *International Symposium on Structural Properties of Dislocations in Semiconductors*, Institute of Physics Conference Series 104 (Bristol: Institute of Physics), pp. 233–8.

Bode, M., Jakubowicz, A. and Habermeier, H. U. (1987). Characterization of dislocations in GaAs by simultaneous EBIC/CL measurements. In *Proceedings of the Second International Symposium on Defect Recognition and Image Processing in III–V Compounds* (DRIP II) (New York: Elsevier), pp. 155–62.

Bonch-Bruevich, V. L. and Glasko, V. B. (1961). The theory of electron states connected with dislocations. I. Linear dislocations. *Soviet Physics Solid State*, **3**, 26–33.

Bondarenko, I. E., Eremenko, V. G., Farber, B. Ya., Nikitenko, V. I. and Yakimov, E. B. (1981). On the real structure of monocrystalline silicon near dislocation. *Physica Status Solidi*, **A68**, 53–60.

Bondarenko, I. E., Blumtritt, H., Heydreich, J., Kazmirruk, V. V. and Yakimov, E. B. (1986). Recombination properties of dislocation slip planes. *Physica Status Solidi*, **A95**, 173–7.

Booker, G. R., Ourmazd, A. and Darby, D. B. (1979). Electrical recombination behavior at dislocations in gallium-phosphide and silicon. *Journal de Physique*, **40**, Suppl. 6, 19–21.

Booyens, H., Vermaak, J. S. and Proto, G. R. (1977). Dislocations and the piezoelectric effect in III–V crystals. *Journal of Applied Physics*, **48**, 3008–13.

Booyens, H., Vermaak, J. S. and Proto, G. R. (1978a). The piezoresistance effect and dislocations in III–V compounds. *Journal of Applied Physics*, **49**, 1149–55.

Booyens, H., Vermaak, J. S. and Proto, G. R. (1978b). The anisotropic carrier mobility due to dislocations in III–V compounds. *Journal of Applied Physics*, **49**, 1173–6.

Bougrioua, Z., Farvacque, J. L. and Ferre, D. (1996a). Effects of dislocations on transport properties of two dimensional electron gas. 1. Transport at zero magnetic field. *Journal of Applied Physics*, **79**, 1536–45.

Bougrioua, Z., Farvacque, J. L. and Ferre, D. (1996b). Effects of dislocations on transport properties of two dimensional electron gas. 2. The quantum regime. *Journal of Applied Physics*, **79**, 1546–55.

Bozhokin, S. V., Parshin, D. A. and Karchenko, V. A. (1982). Dislocation Mott exciton. *Soviet Physics Solid State*, **24**, 800–3.

Brantley, W. A., Lorimor, O. C., Dapkus, P. D., Haszko, S. E. and Saul, R. H. (1975). Effect of dislocations on green electroluminescence efficiency in GaP grown by liquid phase epitaxy. *Journal of Applied Physics*, **46**, 2629–37.

Bredikhin, S. I. and Shmurak, S. Z. (1974). Deformation-stimulated emission of ZnS crystals. *JETP Letters*, **19**, 367–8.

Bredikhin, S. I. and Shmurak, S. Z. (1975). Effect of electric field on deformation induced light emission of ZnS crystals. *JETP Letters*, **21**, 156–7.

Breitenstein, O. (1989). Scanning DLTS. *Review de Physique Applique*, **C6**, 101–10.

Breitenstein, O. and Heydenreich, J. (1985). Review: Scanning deep level spectroscopy. *Scanning*, **7**, 273–89.

Brohl, M. and Alexander, H. (1989). Microwave conductivity in plastically deformed silicon. In *International Symposium on Structural Properties of Dislocations in Semiconductors*, Oxford. Conference Series No. 104, pp. 163–8.

Brophy, J. J. (1956). Excess noise in deformed germanium. *Journal of Applied Physics*, **27**, 1383–4.

Brophy, J. J. (1959). Crystalline imperfections and 1/f noise. *Physical Review*, **115**, 1122–5.

Broudy, R. M. (1963). The electrical properties of dislocations in semiconductors. *Advances in Physics*, **12**, 135–84.

Brümmer, O. and Schreiber, J. (1972). Microskopische untersuchungen des katodolumineszenz-verhaltens von kristallbaufehlern in CdS-einkristallen mit der elektronenstrahlmikrosonde. *Annalen der Physik*, **28**, 105–17.

Brümmer, O. and Schreiber, J. (1974). Zum lumineszenzverhalten von versetzungen in CdS-einkristallen. *Kristall und Technik*, **9**, 817–29.

Brümmer, O. and Schreiber, J. (1975). Mikroskopische kathodolumineszenz-untersuchungen an halbleitern mit der elektronenstrahlmikrosonde. *Microchimica Acta (Supplement)*, **6**, 331–44.

Bube, R. H. (1992). *Photoelectronic Properties of Semiconductors*. Cambridge: Cambridge University Press.

Bude, J., Sano, N. and Yoshii, A. (1992). Hot-carrier luminescence in Si. *Physical Review*, **B45**, 5848–56.

Bull, C., Ashburn, P., Booker, G. R. and Nicholas, K. H. (1979). Effects of dislocations in silicon transistors with implanted emitters. *Solid State Electronics*, **22**, 95–104.

Buonassisi, T., Heuer, M., Vyvenko, O. F. *et al.* (2003). Applications of synchrotron radiation x-ray techniques on the analysis of the behavior of transition metals in solar cells and single-crystalline silicon with extended defects. *Physica B: Condensed Matter*, **340–342**, 1137–41.

Cai, W., Bulatov, V. V., chang, J., Li, J. and Yip, S. (2004). Dislocation core effects on mobility. In *Dislocations in Solids*, eds. F. R. N. Nabarro and J. P. Hirth (North-Holland Publishers), vol. **12**, p. 1.

Canham, L. T. (1990). Silicon quantum wire array fabrication by electrochemical and chemical dissolution of wafers. *Applied Physics Letters*, **57**, 1046–8.

Canham, L. T., Dyball, M. R. and Barraclough, K. G. (1989). Surface copper contamination of as-received float-zone silicon-wafers. *Journal of Applied Physics*, **66**, 920–7.

Casey, H. C. (1967). Investigation of inhomogeneities in GaAs by electron beam excitation. *Journal of the Electrochemical Society*, **114**, 153–8.

Castaldini, A. and Cavallini, A. (1989). Imaging of extended defects by quenched infra-red beam induced currents (Q-IRBIC). In *Point and Extended Defects in Semiconductors*, eds. G. Benedek, A. Cavallini and W. Schröter (New York: Plenum Press), pp. 257–68.

Castaldini, A., Cavallini, A. and Gondi, P. (1987). IRBIC semiconductor defect pictures. *Bulletin of the Academy of Sciences of the USSR Division of Physical Science*, **51**, 77–80.

Castaldini, A., Cavallini, A. and Cavalcoli, D. (1989). Electrical activity associated with dislocations in silicon. In *International Symposium on Structural Properties of Dislocations in Semiconductors*, Oxford. Conference Series 104 (Bristol: Institute of Physics), pp. 169–74.

Cavalcoli, D., Cavallini, A. and Gombia, E. (1997). Defect states in plastically deformed n-type silicon. *Physical Review*, **B56**, 10208–14.

Cavallini, A. and Castaldini, A. (1991). Developments of IRBIC and QIRBIC in defect studies: A review. *Journal de Physique*, **C6**, 89−99.

Chakrabarti, U. K., Pearton, S. J., Hobson, W. S., Lopata, J. and Swaminathan, V. (1990). Hydrogenation of GaAs-on-InP. *Applied Physics Letters*, **57**, 887−9.

Chamonal, J. P., Molva, E., Dupuy, M., Accomo, R. and Pautrat, J. L. (1983). Spectral resolution in low temperature cathodoluminescence. Application to CdTe. *Physica*, **116B**, 519−26.

Champlin, K. S. (1959). Microplasma fluctuations in silicon. *Journal of Applied Physics*, **30**, 1039−50.

Chan, D. S. H., Pey, K. L. and Phang, J. C. H. (1993). Semiconductor parameters extraction using cathodoluminescence in the scanning electron microscope. *IEEE Transactions on Electron Devices*, **40**, 1417−25.

Chatterjee, B., Ringel, S. A., Sieg, R., Hoffman, R. and Weinberg, I. (1994). Hydrogen passivation of dislocations in InP on GaAs heterostructures. *Applied Physics Letters*, **65**, 58−60.

Chatterjee, B. and Ringel, S. A. (1995). Hydrogen passivation and its effects on carrier trapping by dislocations in InP/GaAs heterostructures. *Journal of Applied Physics*, **77**, 3885−98.

Chelyadinskii, A. R. and Komarov, F. F. (2003). Defect-impurity engineering in implanted silicon. *Physics Uspechi*, **46**, 789−820.

Chen, J., Sekiguchi, T., Yang, D. *et al.* (2004). Electron-beam-induced current study of grain boundaries in multicrystalline silicon. *Journal of Applied Physics*, **96**, 5490−5.

Chim, W. K., Chan, D. S. H., Low, T. S. *et al.* (1992). Modelling techniques for the quantification of some electron beam induced phenomena. *Scanning Microscopy*, **6**, 961−78.

Chin, A. K., Temkin, H., Mahajan, S. *et al.* (1979). Evaluation of defects in InP and InGaAsP by transmission cathodoluminescence. *Journal of Applied Physics*, **50**, 5707−9.

Chynoweth, A. G. and McKay, K. G. (1956). Photon emission from avalanche breakdown in silicon. *Physical Review*, **102**, 369−76.

Chynoweth, A. G. and Pearson, G. L. (1958). Effect of dislocations on breakdown in silicon p-n junctions. *Journal of Applied Physics*, **29**, 1103−10.

Claesson, A. (1979). Effect of disorder and long range strain field on the electron states. *Journal de Physique*, **C6**, 39−41.

Claeys, C. and Vanhellemont, J. (1993). Recent progress in the understanding of crystallographic defects in silicon. *Journal of Crystal Growth*, **126**, 41−62.

Claeys, C. and Simoen, E. (1998). Noise as a diagnostic tool for semiconductor material and device characterization. *Journal of the Electrochemical Society*, **145**, 2058−67.

Clarke, D. R. (1999). Varistor ceramics. *Journal of the American Ceramic Society*, **82**, 485−502.

Collins, A. T. (1992). The characterization of point defects in diamond by luminescence spectroscopy. *Diamond and Related Materials*, **1**, 457−69.

Cook, J. W. and Schetzina, J. F. (1995). Blue-green light-emitting diodes promise full-color displays. *Laser Focus World* (March), pp. 101−4.

Cremades, A., Dominguez-Adame, F. and Piqueras, J. (1993). Study of defects in chemical-vapor-deposited diamond films by cross-sectional cathodoluminescence. *Journal of Applied Physics*, **74**, 5726−8.

Crookes, W. W. (1979). Contributions to molecular physics in high vacua. *Philosophical Transactions of the Royal Society*, **170**, 641–62.

Cullis, A. G. and Canham, L. T. (1991). Visible light emission due to quantum size effects in highly porous crystalline silicon. *Nature*, **353**, 335–8.

Czyzewski, Z. and Joy, D. C. (1990). Monte Carlo simulation of CL and EBIC contrasts for isolated dislocations. *Scanning*, **12**, 5–12.

Daniels, B. K. and Meadowcroft, D. B. (1968). Twist boundaries and electroluminescence. *Physica Status Solidi*, **27**, 535–9.

Darby, D. B. and Booker, G. R. (1977). Scanning electron microscope EBIC and CL micrographs of dislocations in GaP. *Journal of Materials Science*, **12**, 1827–33.

Davidson, S. M. and Rasul, A. (1977). Applications of high performance SEM-based CL analysis system to compound semiconductor devices. In *Scanning Electron Microscopy 1977/I*, ed. O. Johari (Chicago: SEM Inc), pp. 225–31.

Davidson, S. M. and Dimitriadis, C. A. (1980). Advances in the electrical assessment of semiconductors using the scanning electron microscope. *Journal of Microscopy*, **118**, 275–90.

Davidson, S. M., Iqbal, M. Z. and Northrop, D. C. (1975). SEM cathode-luminescent studies of plastically deformed gallium phosphide. *Physica Status Solidi*, **A29**, 571–8.

Dean, P. J. (1984). Comparison of MOCVD-grown with conventional II–VI materials parameters for EL thin films. *Physica Status Solidi*, **A81**, 625–46.

Dean, P. J. and Choyke, W. J. (1977). Recombination-enhanced defect reactions, strong new evidence for an old concept in semiconductors. *Advances in Physics*, **26**, 1–30.

Dean, P. J., Williams, G. M. and Blackmore, G. (1984). Novel type of optical transition observed in MBE grown CdTe. *Journal of Physics D: Applied Physics*, **17**, 2291–300.

DeLoach, B. C., Hakki, B. W., Hartman, R. L. and D'Asarg, L. A. (1973). Degradation of CW GaAs double-heterojunction lasers at 300K. *Proceedings of IEEE*, **61**, 1042–4.

De Nobel, D. (1959). Phase equilibria and semiconducting properties of CdTe. *Philips Research Reports*, **14**, 361–99.

Dmitrenko, I. M., Fogel, N. Ya., Cherkasova, V. G., Fedorenko, A. I. and Sipatov, A. Yu. (1993). Dimension crossover and the nature of the superconducting layers in PbTe/PbS semiconductor superlattices. *Low Temperature Physics*, **19**, 533–8.

Dobson, P. S., Hutchinson, P. W., O'Hara, S. and Newman, D. H. (1977). TEM observation of dark defects in degraded gallium arsenide heterojunction lasers. In *Gallium Arsenide and Related Compounds 1976*. Conf. Series No. 33A (Inst. Phys.: Bristol and London), pp. 419–26.

Donolato, C. (1978/79). On the theory of SEM charge-collection imaging of localized defects in semiconductors. *Optik*, **52**, 19–36.

Donolato, C. (1983). Quantitative evaluation of the EBIC contrast of dislocations. *Journal de Physique*, **44**, Colloque C4, 269–75.

Donolato, C. (1985). Beam induced current characterization in polycrystalline semiconductors. In *Polycrystalline Semiconductors. Physical Properties and Applications*, ed. G. Harbeke (Berlin: Springer-Verlag), pp. 138–54.

Donolato, C. and Klann, H. (1980). Computer simulation of SEM electron beam induced current images of dislocations and stacking faults. *Journal of Applied Physics*, **51**, 1624–33.

Dow, J. D. and Redfield, D. (1972). Toward a unified theory of Urbach's rule and exponential absorption edges. *Physical Review*, **B5**, 594–610.

Drozdov, N. A., Patrin, A. A. and Tkachev, V. D. (1976). Recombination radiation on dislocations in silicon. *JETP Letters*, **23**, 597–9.

Duerinckx, F. and Szlufcik, J. (2002). Defect passivation of industrial multicrystalline solar cells based on PECVD silicon nitride. *Solar Energy Materials and Solar Cells*, **72**, 231–46.

Dumas, P., Gu, M., Syrykh, C. *et al.* (1993). Direct observation of individual nanometer-sized light-emitting structures on porous silicon surfaces. *Europhysics Letter*, **23**, 197–202.

Dumas, P., Gu, M., Syrykh, C. *et al.* (1994a). Photon spectroscopy, mapping and topography of 85% porous silicon. *Journal of Vacuum Science and Technology*, **B12**, 2064–6.

Dumas, P., Gu, M., Syrykh, C. *et al.* (1994b). Nanostructuring of porous silicon using scanning tunneling microscopy. *Journal of Vacuum Science and Technology*, **B12**, 2067–9.

Durose, K., Edwards, P. R. and Halliday, D. P. (1999). Materials aspects of CdTe/CdS solar cells. *Journal of Crystal Growth*, **197**, 733–42.

Ebert, P., Domke, C. and Urban, K. (2001). Direct observation of electrical charges at dislocation in GaAs by cross-sectional scanning tunneling microscopy. *Applied Physics Letters*, **78**, 480–2.

Edmiston, S. A., Heiser, G., Sproul, A. B. and Green, M. A. (1996). Improved modeling of grain boundary recombination in bulk and p-n junction regions of polycrystalline silicon solar cells. *Journal of Applied Physics*, **80**, 6783–95.

Edwards, P. R., Halliday, D. P. and Durose, K. (1997). The influence of $CdCl_2$ treatment and interdiffusion on grain boundary passivation in CdTe/CdS solar cells. In *Proceedings of 14th Photovoltaic Solar Energy Conference* (Barcelona: WIP), pp. 2083–6.

Eliseev, P. G. (1973). Degradation of injection lasers. *Journal of Luminescence*, **7**, 338–56.

Elliott, C. R., Regnault, J. C. and Wakefield, B. (1982). Nonradiative regions in GaInAsP/InP double heterostructure material: Correlation with dislocation clusters in the substrates. *Electronics Letters*, **18**, 7–8.

Elsner, J., Jones, R., Heggie, M. I. *et al.* (1998). Deep acceptors trapped at threading-edge dislocations in GaN. *Physical Review*, **B58**, 12571–4.

Emtage, P. R. (1967). Binding of electrons, holes and excitons to dislocations in insulators. *Physical Review*, **163**, 865–72.

Eremenko, V. G. and Fedorov, A. V. (1995). New effect of interaction between moving dislocation and point defects in silicon. *Materials Science Forum*, **196**, 1219–23.

Eremenko, V. G. and Yakimov, E. B. (2004). Anomalous electrical properties of dislocation slip plane in Si. *European Physical Journal – Applied Physics*, **27**, 349–51.

Eremenko, V., Jimenez, J., Fedorov, A. *et al.* (1997). Characterization of the new type of structural defects in Si by the scanning optical and electron beam techniques. Institute of Physics Conference Series No. 160, pp. 269–72.

Eremenko, V., Abrosimov, N. and Fedorov, A. (1999). The origin and properties of new extended defects revealed by etching in plastically deformed Si and SiGe. *Physica Status Solidi*, **A171**, 383–8.

Erenburg, A. I., Bomze, Y. V., Fogel, N. Y. *et al.* (2001). Structural investigations of superconducting multilayers consisting of semiconducting materials. *Low Temperature Physics*, **27**, 93–5.

Esquivel, A. L., Lin, W. N. and Wittry, D. B. (1973). Cathodoluminescence study of plastically deformed GaAs. *Applied Physics Letters*, **22**, 414–16.

Esquivel, A. L., Sen, S. and Lin, W. N. (1976). Cathodoluminescence and electrical anisotropy from α and β dislocations in plastically deformed gallium arsenide. *Journal of Applied Physics*, **47**, 2598–603.

Ettenberg, M. (1974). Effects of dislocation density on the properties of liquid phase epitaxial GaAs. *Journal of Applied Physics*, **45**, 901–6.

Evoy, S., Craighead, H. G., Keller, S., Mishra, U. K. and DenBarrs, S. P. (1999). Scanning tunneling microscope-induced luminescence of GaN at threading dislocations. *Journal of Vacuum Science and Technology*, **B17**, 29–32.

Fang, R. Z., Van Rheenen, A. D., Van der Ziel, A., Young, A. C. and Van der Ziel, J. P. (1990). $1/f$ noise in double-heterojunction AlGaAs/GaAs laser diodes on GaAs and on Si substrates. *Journal of Applied Physics*, **68**, 4087–90.

Feklisova, O. V., Yakimov, E. B. and Yarykin, N. (2003). Contribution of the disturbed dislocation slip planes to the electrical properties of plastically deformed silicon. *Physica*, **B340**, 1005–8.

Feklisova, O. V., Pichaud, B. and Yakimov, E. B. (2005). Annealing effect on the electrical activity of extended defects in plastically deformed p-Si with low dislocation density. *Physica Status Solidi*, **A202**, 896–900.

Fell, T. S. and Wilshaw, P. R. (1989). Recombination at dislocations in the depletion region in silicon. In *International Symposium on Structural Properties of Dislocations in Semiconductors*, Institute of Physics Conference Series 104, pp. 227–32.

Fell, T. S. and Wilshaw, P. R. (1991). Quantitative EBIC Investigation of Deformation-Induced and Copper Decorated Dislocations in Silicon. In *Microscopy of Semiconducting Materials 1991*. Conf. Series No. 117 (Bristol: Institute of Physics), pp. 733–6.

Fell, T. S., Wilshaw, P. R. and de Coteau, M. D. (1993). EBIC investigations of dislocations and their interactions with impurities in silicon. *Physica Status Solidi*, **A138**, 695–704.

Figielski, T. (1960). Electronic processes at intercrystalline barriers in germanium. *Acta Physica Polonica*, **19**, 607–30.

Figielski, T. (1978). Recombination at dislocations. *Solid State Electronics*, **21**, 1403–12.

Figielski, T. (2002). Dislocations as electrically active centers in semiconductors, half a century from the discovery. *Journal of Physics: Condensed Matter*, **14**, 12665–72.

Figielski, T. and Torum, A. (1962). On the origin of light emitted from reverse biased p-n junctions. In *Proceedings of 6th International Conference on Physics of Semiconductors*, Exeter (London: Pergamon), pp. 863–8.

Figielski, T., Wosinski, T., Makosa, A., Dobrowolski, W. and Raczynska, J. (1998). Solid-state Aharonov-Bohm effect at dislocations in semiconductors. *Philosophical Magazine Letters*, **77**, 221–7.

Figielski, T., Wosinski, T. and Makosa, A. (2000). Mesoscopic conductance oscillations associated with dislocations in semiconductors. *Physica Status Solidi*, **B222**, 151−8.

Fischer, A. G. (1962). Electroluminescent lines in ZnS powder particles. I Embedding media and basic observations. *Journal of the Electrochemical Society*, **109**, 1043−9.

Fogel, N. Y., Pokhila, A. S., Bomze, Y. V. *et al.* (2001). Novel superconducting semiconducting superlattices: Dislocation-induced superconductivity? *Physical Review Letters*, **86**, 512−15.

Fogel, N. Y., Buchstab, E. I., Bomze, Y. V. *et al.* (2002). Interfacial superconductivity in semiconducting monochalcogenide superlattices. *Physical Review*, **B66**, 174513−1−174513−11.

Fornari, R., Franzosi, P., Salviati, G., Ferrari, C. and Ghezzi, C. (1985). A study of microdefects in n-type doped GaAs crystals using cathodoluminescence and x-ray techniques. *Journal of Crystal Growth*, **72**, 717−25.

Gaevski, M., Elfwing, M., Olsson, E. and Kvist, A. (2002). Three-dimensional investigations of electrical barriers using electron beam induced current measurements. *Journal of Applied Physics*, **91**, 2713−24.

Gallagher, C. J. (1952). Plastic deformation of germanium and silicon. *Physical Review*, **88**, 721.

Galloway, S. A., Wilshaw, P. R. and Fell, T. S. (1993). An EBIC investigation of alpha, beta and screw dislocations in gallium arsenide. In *Microscopy of Semiconducting Materials 1993*. Conference Series No. 134 (Bristol: Institute of Physics), pp. 71−6.

Galloway, S. A., Edwards, P. R. and Durose, K. (1997). EBIC and cathodoluminescence studies of grain boundary and interface phenomena in CdTe/CdS solar cells. In *Microscopy of Semiconducting Materials 1997*, Conference Series. No. 157 (Bristol: Institute of Physics), pp. 579−82.

Galloway, S. A., Edwards, P. R. and Durose, K. (1999). Characterization of thin film CdS/CdTe solar cells using electron and optical beam induced current. *Solar Energy Materials and Solar Cells*, **57**, 61−74.

Garrido, J. A., Foutz, B. E., Smart, J. A. *et al.* (2000). Low-frequency noise and mobility fluctuations in AlGaN/GaN heterostructure field-effect transistors. *Applied Physics Letters*, **76**, 3442−4.

Gautam, D. K., Khokle, W. S. and Garg, K. B. (1988). Effect of absorption on photon-emission from reverse-biased silicon p-n-junctions. *Solid State Electronics*, **31**, 1119−21.

George, A. and Rabier, J. (1987). Dislocations and plasticity in semiconductors. I Dislocation structures and dynamics. *Review de Physique Applique*, **22**, 941−66.

Ginzburg, V. L. (1971). Manifestation of exciton mechanism in case of granulated superconductors. *JETP Letter*, **14**, 396.

Giovane, L. M., Luan, H. C., Agarwal, A. M. and Kimerling, L. C. (2001). Correlation between leakage current density and threading dislocation density in SiGe *p-i-n* diodes grown on relaxed graded buffer layers. *Applied Physics Letters*, **78**, 541−3.

Gippius, A. A. and Vavilov, V. S. (1963). Radiative recombination at dislocations in germanium. *Soviet Physics Solid State*, **4**, 1777−82.

Gippius, A. A. and Vavilov, V. S. (1965a). Mechanism for radiative recombination at dislocations in germanium. *Soviet Physics Solid State*, **6**, 1873–9.

Gippius, A. A. and Vavilov, V. S. (1965b). On the mechanism of radiative recombination on dislocations in germanium. In *Proceedings of 7th International Congress on Physics of Semiconductors* (Paris: Dunod), pp. 137–42.

Gippius, A. A., Vavilov, V. S. and Konoplev, V. S. (1965). Determination of the yield of luminescence associated with dislocations in germanium. *Soviet Physics Solid State*, **6**, 1741–2.

Gleichmann, R., Blumtritt, H. and Heydenreich, J. (1983). New morphological types of CuSi precipitates in silicon and their electrical effects. *Physica Status Solidi*, **A78**, 527–38.

Gökden, S. (2004). Dislocation scattering effect on two-dimensional electron gas transport in GaN/AlGaN modulation-doped heterostructures. *Physica*, **E23**, 19–25.

Gopal, V., Kvam, E. P., Chin, T. P. and Woodall, J. M. (1998). Evidence for misfit dislocation-related carrier accumulation at the InAs/GaP heterointerface. *Applied Physics Letters*, **72**, 2319–21.

Graham, R. J., Moustakas, T. D. and Disko, M. M. (1991). Cathodoluminescence imaging of defects and impurities in diamond films grown by chemical vapor deposition. *Journal of Applied Physics*, **69**, 3212–18.

Graham, R. J., Shaapur, F., Kato, Y. and Stoner, B. R. (1994). Imaging of boron dopant in highly oriented diamond films by cathodoluminescence in a transmission electron microscope. *Applied Physics Letters*, **65**, 292–4.

Grazhulis, V. A. (1979). Application of EPR and electric measurements to study dislocation energy spectrum in silicon. *Journal de Physique*, **C6**, 59–61.

Grazhulis, V. A. and Osipyan, Yu. A. (1970). Electron paramagnetic resonance in plastically deformed silicon. *Soviet Physics JETP*, **31**, 677.

Grazhulis, V. A., Kveder, V. V. and Mukhina, V. Yu. (1977a). Investigation of the energy spectrum and kinetic phenomena in dislocated Si crystals. *Physica Status Solidi*, **43**, 407–15.

Grazhulis, V. A., Kveder, V. V. and Mukhina, V. Yu. (1977b). Investigation of energy-spectrum and kinetic phenomena in dislocated Si crystals. 2. Microwave conductivity. *Physica Status Solidi*, **A44**, 107–15.

Greuter, F. and Blatter, G. (1990). Electrical properties of grain boundaries in polycrystalline compound semiconductors. *Semiconductor Science and Technology*, **5**, 111–37.

Grovenor, C. R. M. (1985). Grain boundaries in semiconductors. *Journal of Physics*, **C18**, 4079–119.

Güth, W. (1972). Electronic states of dislocations in germanium. *Physica Status Solidi*, **B51**, 143–7.

Hanley, P. L., Kiflawi, I. and Lang, A. R. (1977). On topographically identifiable sources of cathodoluminescence in natural diamonds. *Philosophical Transactions of the Royal Society*, **A284**, 329–68.

Hanoka, J. I., Seager, C. H., Sharp, D. J. and Panitz, J. K. G. (1983). Hydrogen passivation of defects in silicon ribbon grown by the edge-defined film-fed growth process. *Applied Physics Letters*, **42**, pp. 618–20.

Hansen, P. J., Strausser, Y. E., Erickson, A. N. *et al.* (1998). Scanning capacitance microscopy imaging of threading dislocations in GaN films grown on (0001) sapphire by metalorganic chemical vapor deposition. *Applied Physics Letters,* **72**, 2247–9.

Hartman, R. L. and Koszi, L. A. (1978). Characterization of (Al, Ga) As injection lasers using the luminescence emitted from the substrate. *Journal of Applied Physics,* **49**, 5731–44.

Hastenrath, M., Lohnert, K. and Kubalek, E. (1979). Zeitaufgeloste kathodolumineszenz im raster-rlektronenmikroskop mit hilf einer streak kamera. *BEDO* **12/1**, 163–76, see also the account in Pfefferkorn, G., Brocker, W. and Hastenrath, M. (1980). The cathodoluminescence method in the scanning electron microscope. *Scanning Electron Microscopy,* **1**, 250–8.

Hatz, J. (1968). Some effects of material inhomogeneities on the near-field pattern of GaAs diode lasers. *Physica Status Solidi,* **28**, 233–45.

Hayashi, K., Yamamoto, T. and Sakuma, T. (1996). Grain orientation dependence of the PTCR effect in niobium-doped barium titanate. *Journal of the American Ceramic Society,* **79**, 1669–72.

Hayashi, K., Yamamoto, T., Ikuhara, Y. and Sakuma, T. (1999). Grain boundary character dependence of potential barrier in barium titanate. *Materials Science Forum,* **294–2**, 711–14.

Henry, C. H., Petroff, P. M., Logan, R. A. and Merritt, F. R. (1979). Catastrophic damage of $Al_xGa_{1-x}As$ double-heterostructure laser material. *Journal of Applied Physics,* **50**, 3721–32.

Hergert, W. and Hildebrandt, S. (1988). Unified theoretical description of EBIC, LBIC, CL and PL experiments. Transient analysis. *Physica Status Solidi,* **A109**, 625–33.

Hergert, W., Hildebrandt, S. and Pasemann, L. (1987). Theoretical investigations of combined EBIC, LBIC, CL, and PL experiments. *Physica Status Solidi,* **A102**, 819–28.

Hergert, W. and Pasemann, L. (1984). Theoretical study of the information depth of the cathodoluminescence signal in semiconductor materials. *Physica Status Solidi,* **A85**, 641–8.

Herrera Zaldivar, M., Fernández, P. and Piqueras, J. (2001). Study of growth hillocks in GaN:Si films by electron beam induced current imaging. *Journal of Applied Physics,* **90**, 1058–60.

Heywang, W. (1961). Bariumtitanat als sperrschichthalbleiter. *Solid State Electronics,* **3**, 51–8.

Heywang, W. (1964). Resistivity anomaly in doped. *Journal of the American Ceramic Society,* **47**, 484–90.

Higgs, V. and Kittler, M. (1994). Influence of hydrogen on the electrical and optical activity of misfit dislocations in Si/SiGe epilayers. *Applied Physics Letters,* **65**, 2804–6.

Higgs, V., Lightowlers, E. C., Davies, G., Schaffler, E. and Kasper, E. (1989). Photoluminescence from MBE Si grown at low-temperatures – donor bound excitons and decorated dislocations. *Semiconductor Science and Technology,* **4**, 593–8.

Higgs, V., Lightowlers, E. C. and Kightley, P. (1990a). Dislocation related D-band luminescence; The effects of transition metal contamination. In *Materials Research Society Symposia Proceedings,* **163**, pp. 57–62.

Higgs, V., Norman, C. E., Lightowlers, E. C. and Kightley, P. (1990b). Characterization of clean dislocations and the influence of transition metal contamination. In *Proceedings of 20th International Conference on Physics of Semiconductors*, eds. E. M. Anastassakis and J. D. Joannopoulos (Singapore: World Scientific), pp. 706–9.

Higgs, V., Norman, C. E., Lightowlers, E. C. and Kightley, P. (1991). Characterization of dislocations in the presence of transition-metal contamination. *Institute of Physics Conference Series* (117), pp. 737–42.

Higgs, V. and Lightowlers, E. C. (1992a). Characterization of extended defects in Si and $Si_{1-x}Ge_x$ alloys: The influence of transition metal contamination. In *Mechanisms of Heteroepitaxial Growth*, eds. M. F. Chisholm, R. Hull, L. J. Schowalter and B. J. Garrison (Materials Research Society Symposia Proceedings No. 263, Pittsburgh, PA, 1992), pp. 305–16.

Higgs, V. and Lightowlers, E. C. (1992b). Characterization of extended defects in Si and $Si_{1-x}Ge_x$ alloys: The influence of transition metal contamination. In *Materials Research Society Symposia Proceedings*, **263**, 305–16.

Higgs, V., Lightowlers, E. C., Norman, C. E. and Kightley, P. (1992). Characterization of dislocations in the presence of transition metal contamination. *Materials Science Forum*, **83–87**, 1309–14.

Higgs, V., Chin, F., Wang, X., Mosalski, J. and Beanland, R. (2000). Photoluminescence characterization of defects in Si and SiGe structures. *Journal of Physics: Condensed Matter*, **12**, 10105–21.

Hildebrandt, S. and Hergert, W. (1990). Unified theoretical description of the CL, EBIC, PL, and LBIC contrast profile area of an individual surface-parallel dislocation. *Physica Status Solidi*, **A119**, 689–99.

Hildebrandt, S., Schreiber, J. and Hergert, W. (1991). Recent results in the theoretical description of CL and EBIC defect contrasts. *Journal de Physique*, **IV 1**(C6), 39–44.

Hildebrandt, S., Schreiber, J., Hergert, W., Uniewski, H. and Leipner, H. S. (1998). Theoretical fundamentals and experimental materials and defect studies using quantitative scanning electron microscopy-cathodoluminescence/electron beam induced current on compound semiconductors. *Scanning Microscopy International*, **12**, 535–52.

Hilpert, U., Schreiber, J., Worschech, L. *et al.* (2000). Optical characterization of isolated Se(g)-type misfit dislocations and their influence on strain relief in thin ZnSe films. *Journal of Physics: Condensed Matter*, **12**, 10169–74.

Hirsch, P. B. (1981). Electronic and mechanical properties of dislocations in semiconductors. In *Defects in Semiconductors* (New York: North-Holland), pp. 257–71.

Hirsch, P. B. (1985). Dislocations in semiconductors. *Materials Science and Technology*, **1**, 666–77.

Hirsch, P. B., Pirouz, P., Roberts, S. G. and Warren, P. D. (1985). Indentation plasticity and polarity of hardness on {111} faces of GaAs. *Philosophical Magazine*, **52**, 759–84.

Hoering, L., Schreiber, J. and Hilpert, U. (2001). SEM CL in-situ observation during dislocation motion in GaAs and CdTe. *Solid State Phenomena*, **78–79**, 139–48.

Holt, D. B. (1989). The conductive mode. In *SEM Microcharacterization of Semiconductors*, eds. D. B. Holt and D. C. Joy (London: Academic Press), pp. 241–338.

Holt, D. B. (1996). The role of defects in semiconductor materials and devices. *Scanning Microscopy*, **10**, 1047–78.

Holt, D. B. and Lesniak, M. (1985). Recent developments in electrical microcharacterization using the charge collection mode of the scanning electron-microscope. *Scanning Electron Microscopy*, Part 1, pp. 67–86.

Holt, D. B. and Napchan, E. (1994). Quantitation of SEM EBIC and CL signals using Monte Carlo electron-trajectory simulations. *Scanning*, **16**, 78–86.

Holt, D. B., Alfrey, G. F. and Wiggins, C. S. (1958). Grain boundaries and electroluminescence in gallium phosphide. *Nature*, **181**, 109.

Hooge, F. N. (1969). 1/f is no surface effect. *Physics Letters*, **A29**, 139–40.

Hornstra, J. (1958). Dislocations in the diamond lattice. *Journal of Physics and Chemistry of Solids*, **5**, 129–41.

Hornstra, J. (1959). Models of grain boundaries in the diamond lattice. I. Tilt about ⟨110⟩. *Physica*, **25**, 409–22.

Hornstra, J. (1960). Models of grain boundaries in the diamond lattice. II. Tilt about ⟨001⟩ and theory. *Physica*, **26**, 198–208.

Hsu, J. W. P., Fitzgerald, E. A., Xie, Y. H. and Silverman, P. J. (1994). Near-field scanning optical microscopy imaging of indiviual threading dislocations on relaxed Ge_xSi_{1-x} films. *Applied Physics Letters*, **65**, 344–6.

Hsu, J. W. P., Fitzgerald, E. A., Xie, Y. H. and Silverman, P. J. (1996). Studies of electrically active defects in relaxed GeSi films using a near-field scanning optical microscope. *Journal of Applied Physics*, **79**, 7743–50.

Hsu, J. W. P., Manfra, M. J., Lang, D. V. et al. (2001). Inhomogeneous spatial distribution of reverse bias leakage in GaN Schottky diodes. *Applied Physics Letters*, **78**, 1685–7.

Hsu, J. W. P., Manfra, M. J., Molnar, R. J., Heying, B. and Speck, J. S. (2002). Direct imaging of reverse-bias leakage through pure screw dislocations in GaN films grown by molecular beam epitaxy on GaN templates. *Applied Physics Letters*, **81**, 79–81.

Hsu, J. W. P., Weimann, N. G., Manfra, M. J. et al. (2003). Effect of dislocations on local transconductance in AlGaN/GaN heterostructures as imaged by scanning gate microscopy. *Applied Physics Letters*, **83**, 4559–61.

Huang, Y., Chen, X. D., Fung, S., Beling, C. D. and Ling, C. C. (2003). Experimental study and modeling of the influence of screw dislocations on the performance of Au/n-GaN Schottky diodes. *Journal of Applied Physics*, **94**, 5771–5.

Hunter, D. R., Paxman, D. H., Burgess, M. and Booker, G. R. (1973). Use of the SEM for measuring minority carrier lifetimes and diffusion lengths in semiconductor devices. In *Scanning Electron Microscopy: Systems and Applications*, Conference Series No. 18 (London: Institute of Physics), pp. 208–13.

Hutchinson, P. W., Dobson, P. S., O'Hara, S. and Newman, D. H. (1975). Defect structure of degraded heterojunction GaAsAs–GaAs lasers. *Applied Physics Letters*, **26**, 250–2.

Hutchinson, P. W. and Dobson, P. S. (1975). Defect structure of degraded GaAsAs–GaAs double heterostructure lasers. *Philosophical Magazine*, **32**, 745–54.

Hutchinson, P. W. and Dobson, P. S. (1980). Climb assymmetry in degraded gallium arsenide lasers. *Philosophical Magazine*, **41**, 601–14.

Iber, H., Peiner, E. and Schlachetzki, A. (1996). The effect of dislocations on the optical absorption of heteroepitaxial InP and GaAs on Si. *Journal of Applied Physics*, **79**, 9273–7.

Im, H. J., Ding, Y., Pelz, J. P., Heying, B. and Speck, J. S. (2001). Characterization of individual threading dislocations in GaN using ballistic electron emission microscopy. *Physical Review Letters*, **87**, 106802-1–106802-4.

Ito, R., Nakashima, H. and Nakada, O. (1974). Growth of dark lines from crystal defects in GaAs–GaAlAs double heterostructure crystals. *Japanese Journal of Applied Physics*, **13**, 1321–2.

Jain, R. K. and Flood, D. J. (1993). Influence of the dislocation density on the performance of heteroepitaxial indium-phosphide solar-cells. *IEEE Transactions on Electron Devices*, **40**, 1928–34.

Jain, S. C., Willander, M., Narayan, J. and Van Overstraeten, R. (2000). III–nitrides: Growth, characterization, and properties. *Journal of Applied Physics*, **87**, 965–1006.

Jakubowicz, A. (1986). Theory of cathodoluminescence contrast from localized defects in semiconductors. *Journal of Applied Physics*, **59**, 2205–9.

Jakubowicz, A. and Habermeier, H.-U. (1985). Electron-beam-induced current investigations of oxygen precipitates in silicon. *Journal of Applied Physics*, **58**, 1407–9.

Jakubowicz, A., Bode, M. and Habermeier, H.-U. (1987). Simultaneous EBIC/CL investigations of dislocations in GaAs. In *Microscopy of Semiconducting Materials 1987*, Conference Series No. 87 (Bristol: Institute of Physics), pp. 763–8.

Jaszek, R. (2001). Carrier scattering by dislocations in semiconductors. *Journal of Materials Science: Materials Electronics*, **12**, 1–9.

Jena, D., Gossard, A. C. and Mishra, U. K. (2000). Dislocation scattering in a two-dimensional electron gas. *Applied Physics Letters*, **76**, 1707–9.

Jena, D. and Mishra, U. K. (2002). Effect of scattering by strain fields surrounding edge dislocations on electron transport in two-dimensional electron gases. *Applied Physics Letters*, **80**, 64–6.

Jiang, F., Stavola, M., Rohatgi, A. *et al.* (2003). Hydrogenation of Si from SiNx(H) films: Characterization of H introduced into the Si. *Applied Physics Letters*, **83**, 931–3.

John, H. F. (1967). Silicon power device material problems. *Proceedings of IEEE*, **55**, 1249–71.

Johnston, W. D., Callahan, W. M. and Miller, B. T. (1974). Observation of dark-line degradation sites in a GaAs/GaAsAs DH laser material by etching and phase contrast microscopy. *Journal of Applied Physics*, **45**, 505–7.

Jones, R. (1979). Theoretical calculations of electron states associated with dislocations. *Journal de Physique*, **C6**, 33–8.

Jones, R. (2000). Do we really understand dislocations in semiconductors? *Materials Science and Engineering*, **B71**, 24–9.

Jones, B. K. (2002). Electrical noise as a reliability indicator in electronic devices and components. *IEE Proceedings – Circuit and Device Systems*, **149**, 13–22.

Jones, R., Coomer, B. J., Goss, J. P., Oberg, S. and Briddon, P. R. (2000). Intrinsic defects and the D1 to D4 optical bands detected in plastically deformed Si. *Physica Status Solidi*, **B222**, 133–40.

Jonker, G. H. (1964). Some aspects of semiconducting barium titanate. *Solid-State Electronics*, **7**, 895–903.

Joshi, R. P., Viswanadha, S., Jogai, B., Shah, P. and del Rosario, R. D. (2003). Analysis of dislocation scattering on electron mobility in GaN high electron mobility transistors. *Journal of Applied Physics*, **93**, 10046–52.

Joy, D. C. (1988). An introduction to Monte Carlo simulations. In *Eurem 88*, Conference Series No. 93, eds. P. J. Goodhew and H. G. Dickinson (Bristol: Institute of Physics), pp. 23–32.

Joy, D. C. (1995). Monte Carlo Modeling for Electron Microscopy and Microanalysis, Oxford: Oxford University Press.

Kamieniecki, E. (1979). Photoconductivity produced by polarized light in plastically deformed Ge. *Journal de Physique*, **40**, Colloque **C6**, 87–9.

Kamins, T. (1998). *Polycrystalline Silicon for Integrated Circuits and Displays* (Boston: Kluwer), Chap. 5, pp. 195–243.

Kamiya, T., Durrani, Z. A. K. and Ahmed, H. (2002). Control of grain-boundary tunneling barriers in polycrystalline silicon. *Applied Physics Letters*, **81**, 2388–90.

Kanda, H., Watanabe, K., Koizumi, S. and Teraji, T. (2003). Characterization of phosphorus doped CVD diamond films by cathodoluminescence spectroscopy and topography. *Diamond and Related Materials*, **12**, 20–5.

Kazmerski, L. L. (1980). *Polycrystalline and Amorphous Thin Films and Devices* (New York: Academic Press), pp. 59–133.

Kazmerski, L. L. (1991). Specific atom imaging, nanoprocessing and electrical nanoanalysis with scanning tunneling microscopy. *Journal of Vacuum Science and Technology*, **B9**, 1549–56.

Kiflawi, I. and Lang, A. R. (1974). Linearly polarised luminescence from linear defects in natural and synthetic diamond. *Philosophical Magazine*, **30**, 219–23.

Kiflawi, I. and Lang, A. R. (1976). On the correspondence between cathodoluminesence images and x-ray diffraction contrast images of individual dislocations in diamond. *Philosophical Magazine*, **33**, 697–701.

Kisielowski-Kemmerich, C. (1989). LCAO analysis of dislocation-related EPR spectra in deformed silicon. In *International Symposium on Structural Properties of Dislocations in Semiconductors*, Oxford. Conf. Series No. 104 (Bristol: Institute of Physics), pp. 187–92.

Kisielowski-Kemmerich, C., Weber, G. and Alexander, H. (1985). In *Proceedings of the Thirteenth International Conference on Defects in Semiconductors* (Metallurgical Society of AIME, Warrendale, PA, 1985), p. 387.

Kisielowski, C., Plam, J., Bollig, B. and Alexander, H. (1991). Inhomogeneities in plastically deformed silicon single-crystals. I. ESR and photo-ESR investigations of p-doped and n-doped silicon. *Physical Review*, **B44**, 1588–99.

Kittler, M. and Seifert, W. (1981). On the sensitivity of the EBIC technique as applied to defect investigations in silicon. *Physica Status Solidi*, **A66**, 573–83.

Kittler, M. and Seifert, W. (1993a). On the origin of EBIC defect contrast in silicon: A reflection on injection- and temperature-dependent investigations. *Physica Status Solidi*, **A138**, 687–93.

Kittler, M. and Seifert, W. (1993b). Two classes of defect recombination behaviour in silicon as studied by SEM-EBIC. *Scanning*, **15**, 316–21.

Kittler, M. and Seifert, W. (1993c). On the sensitivity of the EBIC technique as applied to defect investigations in silicon. *Physica Status Solidi*, **A66**, 573–83.

Kittler, M. and Seifert, W. (1994). Two types of electron-beam-induced current behaviour of misfit dislocations in Si(Ge): Experimental observations and modelling. *Materials Science and Engineering*, **B24**, 78–81.

Kittler, M., Ulhaq-Bouillet, C. and Higgs, V. (1994). Recombination activity of 'clean' and contaminated misfit dislocations in Si(Ge) structures. *Materials Science and Engineering*, **B24**, 52–5.

Kittler, M., Ulhaq-Bouillet, C. and Higgs, V. (1995). Influence of copper contamination on recombination activity of misfit dislocations in SiGe/Si epilayers: temperature dependence of activity as a marker characterizing the contamination level. *Journal of Applied Physics*, **78**, 4573–83.

Kittler, M., Seifert, W. and Krüger, O. (2001). Electrical behaviour of crystal defects in silicon solar cells. *Solid State Phenomena*, **78–79**, 39–48.

Kittler, M., Seifert, W., Arguirov, T., Tarassov, I. and Ostapenko, S. (2002). Room-temperature luminescence and electron-beam-induced current (EBIC) recombination behaviour of crystal defects in multicrystalline silicon. *Solar Energy Materials and Solar Cells*, **72**, 465–72.

Kittler, M., Seifert, W. and Knobloch, K. (2003). Influence of contamination on the electrical activity of crystal defects in silicon. *Microelectronic Engineering*, **66**, 281–28.

Klassen, N. V. and Osipiyan, Yu. A. (1979). Optical properties of II–VI compounds with dislocations. *Journal de Physique*, **C6**, 91–4.

Knobloch, K., Kittler, M. and Winfried Seifert, W. (2003). Influence of contamination on the dislocation-related deep level $C1$ line observed in deep-level-transient spectroscopy of n-type silicon: A comparison with the technique of electron-beam-induced current. *Journal of Applied Physics*, **93**, 1069–74.

Koley, G. and Spencer, M. G. (2001). Scanning Kelvin probe microscopy characterization of dislocations in III-nitrides grown by metalorganic chemical vapor deposition. *Applied Physics Letters*, **78**, 2873–5.

Kolyubakin, A. I., Osipiyan, Yu. A., Shevchenko, S. A. and Steinman, E. A. (1984). Dislocation luminescence in germanium. *Soviet Physics Solid State*, **26**, 407–11.

Kozodoy, P., Ibbetson, J. P., Marchand, H. *et al.* (1998). Electrical characterization of GaN p-n junctions with and without threading dislocations. *Applied Physics Letters*, **73**, 975–7.

Kressel, H., Nelson, H., McFarlane, S. H. *et al.* (1969). Effect of substrate imperfections on GaAs injection lasers prepared by liquid-phase epitaxy. *Journal of Applied Physics*, **40**, 3587–97.

Kressel, H., Byer, N. E., Lockwood, H. *et al.* (1970). Evidence for role of certain metallurgical flaws in accelerating electroluminescent diode degradation. *Metallurgical Transactions*, **1**, 635–8.

Krtschil, A., Dadgar, A. and Krost, A. (2003). Decoration effects as origin of dislocation-related charges in gallium nitride layers investigated by scanning surface potential microscopy. *Applied Physics Letters*, **82**, 2263–5.

Krüger, O., Seifert, W., Kittler, M. and Vyvenko, O. F. (2000). Extension of hydrogen passivation of intragrain defects and grain boundaries in cast multicrystalline silicon. *Physica Status Solidi*, **222**, 367–78.

Kuksenkov, D. V., Temkin, H., Osinsky, A., Gaska, R. and Khan, M. A. (1998). Low-frequency noise and performance of GaN *p-n* junction photodetectors. *Journal of Applied Physics*, **83**, 2142–6.

Kurtz, A. D., Kulin, S. A. and Averbach, B. L. (1956). Effects of growth rate on crystal perfection and lifetime in germanium. *Journal of Applied Physics*, **27**, 1287–90.

Kusanagi, S., Sekiguchi, T. and Sumino, K. (1992). Difference of the electrical properties of screw and 60° dislocations in silicon as detected with temperature-dependent electron beam induced current technique. *Applied Physics Letters*, **61**, 792–4.

Kusanagi, S., Sekiguchi, T., Shen, B. and Sumino, K. (1995). Electrical activity of extended defects and gettering of metallic impurities in silicon. *Materials Science and Technology*, **11**, 685–90.

Kveder, V. V., Labusch, R. and Osipiyan, Y. A. (1985). Frequency-dependence of the dislocation conduction in Ge and Si. *Physica Status Solidi*, A**92**, 293–302.

Kveder, V. V., Osipiyan, Yu. A., Schröter, W. and Zoth, G. (1982). On the energy spectrum of dislocations in silicon. *Physica Status Solidi*, A**72**, 701–13.

Kveder, V., Kittler, M. and Schröter, W. (2001). Recombination activity of contaminated dislocations in silicon: A electron-beam-induced current contrast behaviour. *Physical Review*, **B63**, 115208–1 to 115208–11.

Kveder, V., Badylevich, M., Steinman, E. *et al.* (2004). Room-temperature silicon light-emitting diodes based on dislocation luminescence. *Applied Physics Letters*, **84**, 2106–8.

Kyser, D. F. and Wittry, D. B. (1964). Cathodoluminescence in gallium arsenide. In *The Electron Microprobe*, eds. T. D. McKinley, K. F. J. Heinrich and D. B. Wittry (New York: Wiley), pp. 691–714.

Labusch, R. (1997). Conductivity and photoconductivity at dislocations. *Journal de Physique III*, **7**, 1411–24.

Labusch, R. and Schröter, W. (1978). Electrical properties of dislocations in semiconductors. In *Dislocations in Solids*, **5**, ed. F. R. N. Nabarro (Amsterdam: North-Holland), pp. 127–91.

Landauer, R. (1954). Bound states in dislocations. *Physical Review*, **94**, 1386–8.

Lang, A. R. (1977). Defects in natural diamonds – recent observations by new methods. *Journal of Crystal Growth*, **42**, 625–31.

Lang, A. R. (1980). Polarized infrared cathodoluminescence from synthetic diamonds. *Philosophical Magazine*, **B41**, 689–98.

Lang, R. G., Kren, J. G. and Patrick, W. J. (1963). Vacuum evaporation of cadmium telluride. *Journal of the Electrochemical Society*, **110**, 407–12.

Langenkamp, M. and Breitenstein, O. (2002). Classification of shunting mechanisms in crystalline silicon solar cells. *Solar Energy Materials Solar Cells*, **72**, 433–40.

Leach, C. (2000). SEM based estimation of the grain boundary plane orientation in zinc oxide varistors using conductive mode microscopy. *Scripta Materialia*, **43**, 529–34.

Leach, C. (2001). Crystal plane influence of the EBIC contrast in zinc oxide varistors. *Journal of the European Ceramic Society*, **21**, 2127–30.

Leamy, H. J. (1982). Charge collection scanning electron microscopy. *Journal of Applied Physics*, **53**, R51–R80.

Lee, T. P. and Burrus, C. A. (1980). Dark current and breakdown characteristics of dislocation-free InP photodiodes. *Applied Physics Letters*, **36**, 587–9.

Lelikov, Yu. S., Rebane, Yu. T. and Shreter, Yu. G. (1989). Optical properties of dislocations in germanium crystals. In *Structure and Properties of Dislocations in Semiconductors*, Conference Series No. 104 (Bristol: Institute of Physics), pp. 119–29.

Lendvay, E. and Kovacs, P. (1966). Luminescence and impurity precipitation in ZnS single crystals with high Cu concentrations. In *Proceedings of International Conference on Luminescence* (Budapest Academiai Kiado), pp. 1098–101.

Lesniak, M. and Holt, D. B. (1983). Electrically active defects in Si photodetector devices. In *Institute of Physics Conference Series 67*, pp. 439–44.

Lesniak, M. and Holt, D. B. (1987). Defect microstructure and microplasmas in silicon avalanche photodiodes. *Journal of Materials Science*, **22**, 3547–55.

Lester, S. D., Ponce, F. A., Craford, M. G. and Steigerwald, D. A. (1995). High dislocation densities in high efficiency GaN-based light-emitting diodes. *Applied Physics Letters*, **66**, 1249–51.

Levade, C., Faress, A. and Vanderschaeve, G. (1994). A TEM *in situ* investigation of dislocation mobility in the II–VI semiconductor compound ZnS. A quantitative study of the cathodoplastic effect. *Philosophical Magazine*, **A69**, 855–70.

Lipson, H. G., Burstein, E. and Smith, P. L. (1955). Optical properties of plastically deformed germanium. *Physical Review*, **99**, 444–5.

Logan, R. A., Pearson, G. L. and Kleinman, D. A. (1959). Anisotropic mobilities in plastically deformed germanium. *Journal of Applied Physics*, **30**, 885–95.

Lohnert, K. and Kubalek, E. (1983). Characterization of semiconducting materials and devices by EBIC and CL techniques. In *Microscopy of Semiconducting Materials 1983*, Conference Series No. 67 (Bristol: Institute of Physics), pp. 303–14.

Lohnert, K. and Kubalek, E. (1984). The cathodoluminescence contrast formation of localized non-radiative defects in semiconductors. *Physica Status Solidi*, **A83**, 307–14.

Lohnert, K., Hastenrath, M. and Kubalek, E. (1979). Spatially resolved cathodoluminescence studies of GaP LEDs in the scanning electron microscope using optical multichannel analysis. *Scanning Electron Microscopy*, **I**, 229–36.

Look, D. C. and Sizelove, J. R. (1999). Dislocation scattering in GaN. *Physical Review Letters*, **82**, 1237–40.

Louchet, F. and Thibault-Dessaux, J. (1987). Dislocation cores in semiconductors. From the 'shuffle or glide' dispute to the 'glide and shuffle' partnership. *Review de Physique Appliquee*, **22**, 207–19.

Lourenço, M. A., Siddiqui, M. S. A., Gwilliam, R. M., Shao, G. and Homewood, K. P. (2003). Efficient silicon light emitting diodes made by dislocation engineering. *Physica*, **E16**, 376–81.

Lüdemann, R. (1999). Hydrogen passivation of multicrystalline silicon solar cells. *Materials Science and Engineering*, **B58**, 86–90.

McCarthy, L., Smorchkova, I., Xing, H. *et al.* (2001). Effect of threading dislocations on AlGaN/GaN heterojunction bipolar transistors. *Applied Physics Letters*, **78**, 2235–7.

McIntyre, R. J. (1961). Theory of microplasma instability in silicon. *Journal of Applied Physics*, **32**, 983–95.

McKay, K. G. (1954). Avalanche breakdown in silicon. *Physical Review*, **94**, 877–84.

Mackintosh, I. M. (1956). Effects at high-angle grain boundaries in indium antimonide. *Journal of Electronics*, **1**, 554–8.

Maeda, K. and Takeuchi, S. (1983). Recombination enhanced mobility of dislocations in III–V-compounds. *Journal de Physique*, **44** (NC-4), 375–85.

Maeda, K., Sato, M., Kubo, A. and Takeuchi, S. (1983). Quantitative measurements of recombination enhanced dislocation glide in gallium arsenide. *Journal of Applied Physics*, **54**, 161–8.

Maeda, K. and Takeuchi, S. (1996). Enhancement of dislocation mobility in semiconducting crystals by electronic excitation. In *Dislocations in Solids*, **10**, eds. F. R. N. Nabarro and M. S. Duesbery (Amsterdam: Elsevier), pp. 443–504.

Maeda, K., Suzuki, K., Yamashita, Y. and Mera, Y. (2000). Dislocation motion in semiconducting crystals under the influence of electronic perturbations. *Journal of Physics: Condensed Matter*, **12**, 10079–91.

Maestre, D., Cremades, A. and Piqueras, J. (2004). Direct observation of potential barrier formation at grain boundaries of SnO_2 ceramics. *Semiconductor Science and Technology*, **19**, 1236–9.

Magnea, N., Petroff, P. M., Capasso, F., Logan, R. A. and Foy, W. (1985). Microplasma characteristics in InP-$In_{0.53}Ga_{0.47}As$ long wavelength avalanche photodiodes. *Applied Physics Letters*, **46**, 66–8.

Mahajan, S. (1981). The interrelationship between structure and properties in InP and InGaAsP materials. In *Defects in Semiconductors*. Proceedings of Materials Research Society Annual Meeting, eds. J. Narayan and T. Y. Tan (New York: North-Holland), pp. 465–79.

Mahajan, S. (2000). Defects in semiconductors and their effects on devices. *Acta Materialia*, **48**, 137–49.

Mahajan, S., Johnston, W. D., Pollack, M. A. and Nahorny, R. E. (1979). The mechanism of optically induced degradation in InP/$In_{1-x}Ga_xAs_yP_{1-y}$ heterostructures. *Applied Physics Letters*, **34**, 717–19.

Masut, R., Penchina, C. M. and Farvaque, J. L. (1982). Occupation statistics of dislocation deep levels in III–V compounds. *Journal of Applied Physics*, **53**, 4964–9.

Matare, H. F. (1955). Grain boundaries and transistor action. *Proceedings of the Institute of Radio Engineers*, **43**, 375–8.

Matare, H. F. (1956a). Zum elektrischen verhalten von bikristallzwischenschichten. *Zeitschrift fur Physik*, **145**, 206–34.

Matare, H. F. (1956b). Korngrenzen-transistoren. *Elektronische Rundschau*, **8**, 209–11.

Matare, H. F. (1956c). Korngrenzen-transistoren. *Elektronische Rundschau*, **9**, 253–5.

Matare, H. F. and Wegener, H. A. R. (1957). Oriented growth and definition of medium angle semiconductor bicrystals. *Zeitschrift fur Physik*, **148**, 631–45.

Matragrano, M. J., Watson, G. P., Ast, D. G., Anderson, T. J. and Pathangey, B. (1993). Passivation of deep level states caused by misfit dislocations in InGaAs on patterned GaAs. *Applied Physics Letters*, **62**, 1417–19.

McNally, P. J., McCaffrey, J. K. and Baric, A. (1995). Piezoelectrically-active defects and their impact on the performance of GaAs MESFETs. *Journal of Materials Processing Technology*, **55**, 303–10.

McNally, P. J., Cooper, L. S., Rosenburg, J. J. and Jackson, T. N. (1988). Investigation of stress effects on the direct current characteristics of GaAs metal semiconductor field effect transistors through the use of externally applied loads. *Applied Physics Letters*, **52**, 1800–2.

Melliar-Smith, C. M. (1977). Crystal defects in integrated circuits. In *Treatise on Materials Science and Technology*, ed. H. Herman, **11**. Properties and Microstructure, ed. R. K. MacCrone (New York: Academic Press).

Merten, L. (1964a). Modell einer schraubenversetzung in piezoelektrischen kristallen I and II (Model of a screw dislocation in piezoelectric crystals I and II). *Physik der Kondensiterten Materie*, **2**, 53–79.

Merten, L. (1964b). Piezoelektrische potentialfelder um stufenversetzungen belibiger richtung in piezoelektrischen kristallen mit elastischer isotropie. *Zeitschrift fur Naturforschung*, **19a**, 1161–9.

Mettler, K. and Pawlik, D. (1972). Effect of dislocations on the degradation of silicon-doped GaAs luminescent diodes. *Siemens Forschungs und Entwicklungsberichte*, **1**, 274–8.

Meyer, M., Miles, M. H. and Ninomiya, T. (1967). Some electrical and optical effects of dislocations on semiconductors. *Journal of Applied Physics*, **38**, 4481–6.

Miller, E. J., Schaadt, D. M., Yu, E. T. *et al.* (2002). Reduction of reverse-bias leakage current in Schottky diodes on GaN grown by molecular-beam epitaxy using surface modification with an atomic force microscope. *Journal of Applied Physics*, **91**, 9821–6.

Miller, E. J., Schaadt, D. M., Yu, E. T. *et al.* (2003a). Origin and microscopic mechanism for suppression of leakage currents in Schottky contacts to GaN grown by molecular-beam epitaxy. *Journal of Applied Physics*, **94**, 7611–15.

Miller, E. J., Schaadt, D. M., Yu, E. T. *et al.* (2003b). Reverse-bias leakage current reduction in GaN Schottky diodes by electrochemical surface treatment. *Applied Physics Letters*, **82**, 1293–5.

Miller, E. J., Yu, E. T., Waltereit, P. and Speck, J. S. (2004). Analysis of reverse-bias leakage current mechanisms in GaN grown by molecular-beam epitaxy. *Applied Physics Letters*, **84**, 535–7.

Mil'shtein, S. (1999). Dislocations in microelectronics. *Physica Status Solidi*, **A171**, 371–6.

Mil'shtein, S. (2002). Dislocation-induced noise in semiconductors. *Journal of Physics: Condensed Matter*, **14**, 13387–95.

Mironov, O. A., Savitskii, B. A., Sipatov, A. Y. *et al.* (1988). Superconductivity of semiconductor superlattices based on lead chalcogenides. *JETP Letters*, **48**, 106–9.

Mishima, T. D., Keay, J. C., Goel, N. *et al.* (2004). Structural defects in InSb/$Al_x In_{1-x}Sb$ quantum wells grown on GaAs (0 0 1) substrates. *Physica*, **E21**, 770–3.

Mitsuhashi, H., Komura, H. and Chikawa, J. (1967). Dislocation effects on the luminescence of CdS crystals. In *II-VI Semiconducting Compounds*, International Conference, ed. D. G. Thomas (New York: Benjamin), pp. 179–89.

Miyazawa, S. and Hyuga, F. (1986). Proximity effect of dislocations on GAAs MESFET threshold voltage. *IEEE Transactions on Electron Devices*, **ED-33**, 227–33.

Möller, H. J. (1993). *Semiconductors for Solar Cells*. Boston: Artech House.

Möller, H. J. (1996). Multicrystalline silicon for solar cells. *Solid State Phenomena*, **47–48**, 127–42.

Monemar, B. A. and Woolhouse, G. R. (1977). Optical studies of defects and degradation in GaAs–GaAlAs double-heterostructure laser material. In *Gallium Arsendie and Related Compounds 1976*. Conference Series No. 33A (Bristol: Institute of Physics), pp. 400–10.

Montelius, L., Owman, F., Pistol, M.-E. and Samuelson, L. (1991). Low temperature injection luminescence using a scanning tunneling microscope. In *Microscopy of Semiconducting Materials 1991*. Conference Series No. 117 (Bristol: Institute of Physics), pp. 719–22.

Montelius, L., Pistol, M.-E. and Samuelson, L. (1992). Low-temperature luminescence due to minority carrier injection from the scanning tunneling microscope tip. *Ultramicroscopy*, **42–44**, 210–14.

Morrison, S. R. (1956). Recombination of electrons and holes at dislocations. *Physical Review*, **104**, 619–23.

Morrison, S. R. (1992). $1/f$ noise from levels in a linear or planar array. III. Trapped carrier fluctuations at dislocations; IV. The origin of the Hooge parameter. *Journal of Applied Physics*, **72**, 4104–12 and 4113–17.

Mueller, R. K. (1959a). Transient response of grain boundaries and its application for a novel light sensor. *Journal of Applied Physics*, **30**, 1004–10.

Mueller, R. K. (1959b). Capture diameter of dislocations in low-angle grain boundaries in germanium. *Journal of Physics and Chemistry of Solids*, **8**, 157–61.

Mueller, R. K. and Jacobson, R. L. (1959). Grain boundary photovoltaic cell. *Journal of Applied Physics*, **30**, 121–2.

Mueller, R. K. and Jacobson, R. L. (1962). Alpha and beta grain boundaries in indium antimonide. *Journal of Applied Physics*, **33**, 2341–5.

Mueller, R. K. and Maffitt, K. N. (1964). Grain Boundary Conductance in InSb. *Journal of Applied Physics*, **33**, 734–5.

Murase, K., Ishida, S., Takaoka, S. *et al.* (1986). Superconducting behavior in PbTe-SnTe superlattices. *Surface Science*, **170**, 486–90.

Myers, S. M., Seibt, M. and Schröter, W. (2000). Mechanisms of transition-metal gettering in silicon. *Journal of Applied Physics*, **88**, 3795–819.

Myhajlenko, S. S., Batstone, J. L., Hutchinson, H. J. and Steeds, J. W. (1984). Luminescence studies of individual dislocations in II–VI (ZnSe) and III–V (InP) semiconductors. *Journal of Physics C: Solid State Physics*, **17**, 6477–92.

Naidenkova, M., Goorsky, M. S., Sandhu, R. *et al.* (2002). Interfacial roughness and carrier scattering due to misfit dislocations in $In_{0.52}Al_{0.48}As/In_{0.75}Ga_{0.25}As/InP$ structures. *Journal of Vacuum Science and Technology*, **B20**, 1205–8.

Nakashima, H., Kishino, S., Chinone, N. and Ito, R. (1977). Growth and propagation mechanism of ⟨110⟩-oriented dark-line defects in GaAs–$Ga_{1-x}Al_xAs$ double hererostructure crystals. *Journal of Applied Physics*, **48**, 2771–5.

Negrii, V. D. (1992). Dynamic and optical properties of screw dislocations introduced by plastic deformation of CdS crystals at 77–4.2 K. *Journal of Crystal Growth*, **117**, 672–6.

Negrii, V. D. and Osipyan (1978). Y. A. Influence of dislocations on radiative recombination processes in cadmium-sulfide. *Soviet Physics Solid State*, **20**, 432–6.

Negrii, V. D. and Osipiyan, Yu. A. (1979). Dislocation emission in CdS. *Physica Status Solidi*, **A55**, 583–8.

Negrii, V. D. and Osipyan, Y. A. (1982a). Cooperative behavior of defects introduced by plastic-deformation in cadmium-sulfide crystals. *JETP Letters*, **35**, 598–601.

Negrii, V. D. and Osipyan, Yu. A. (1982b). Distinctive features of the luminescence of cadmium sulfide deformed at low temperatures. *Soviet Physics Solid State*, **24**, 197–9.

Negrii, V. D., Osipiyan, Yu. A. and Lomak, N. V. (1991). Dislocation structure and motion in CdS crystals. *Physica Status Solidi*, **A126**, 49–61.

Neubert, D., Kos, J. and Hahn, D. (1973). Problems of lifetime doping by dislocation in silicon. In *Solid State Devices 1972*. Conference Series No. 15. (London: Institute of Physics), p. 220 (abstract only).

Newman, R. (1955). Visible light from a silicon *p-n* junction. *Physical Review*, **100**, 700–3.

Newman, R. (1957). Recombination radiation from deformed and alloyed germanium p-n junctions at 80° K. *Physical Review*, **105**, 1715–20.

Ng, W. L., Lourenço, M. A., Gwilliam, R. M. *et al.* (2001). An efficient room-temperature silicon-based light-emitting diode. *Nature*, **410**, 192–4.

Nickel, N. H. (1999). Hydrogen in semiconductors II. In *Semiconductors and Semimetals*, Volume **61**, eds. R. K. Willardson, A. C. Beer and E. R. Weber (San Diego: Academic Press).

Nikitenko, V. I., Farber, B. Ya. and Yakimov, E. B. (1981). Asymmetry of dislocation mobility in semiconductors. *JETP Letters*, **34**, 233–6.

O'Hara, S., Hutchinson, P. W., Davis, R. and Dobson, P. S. (1977). Defect-induced degradation in high radiance lamps. In *Gallium Arsenide and Related Compounds 1976*. Conference Series No. 33A (Bristol: Institute of Physics), pp. 379–87.

Ohori, T., Ohkubo, S., Kasai, K. and Komeno, J. (1994). Effect of threading dislocations on mobility in selectively doped heterostructures grown on Si substrates. *Journal of Applied Physics*, **75**, 3681–3.

Okada, J. (1955). Effects of dislocations on minority carrier lifetime in germanium. *Journal of the Physical Society of Japan*, **10**, 1110–11.

Omling, P., Weber, E. R., Montelius, L, Alexander, H. and Michel, J. (1985). Electrical properties and point defects in plastically deformed silicon. *Physical Review*, **B22**, 6571–81.

Ono, H. and Sumino, K. (1985). Defect states in p-type silicon crystals induced by plastic deformation. *Journal of Applied Physics*, **57**, 287–92.

Orton, J. W. and Powell, M. J. (1980). The Hall effect in polycrystalline and powdered semiconductors. *Reports on Progress in Physics*, **43**, 1263–307.

Osipyan, Yu. A. (1981). Dislocation microwave electrical conductivity of semiconductors and electron-dislocation spectrum. *Crystal Research and Technology*, **16**, 239–46.

Osipyan, Yu. A. (1983). Dislocation electron spectrum and the mechanism of dislocation microwave conduction in semiconductors. *Journal de Physique*, **C4**, 103–11.

Osipiyan, Yu. A. (1989). Electrical and optical phenomena of II–VI semiconductors associated with dislocations. In *International Symposium on Structural Properties of Dislocations in Semiconductors*, Oxford, Conference Series No. 104 (Bristol: Institute of Physics), pp. 109–18.

Osipyan, Yu. A. and Steinman, E. A. (1973). The effect of dislocations on the luminescence spectra of CdS and CdSe single crystals. In *Luminescence of Crystals, Molecules and Solutions*, ed. F. Williams (New York: Plenum Press), pp. 467–72.

Osipyan, Yu. A. and Negrii, V. D. (1987). Optical properties of configuration defects arising under low-temperature plastic deformation of CdS crystals. In *Microscopy of Semiconducting Materials 1987*, Conference Series No. 87 (Bristol: Institute of Physics), pp. 333–8.

Osipyan, Yu. A. and Negrii, V. D. (1989). Optical studies of cadmium sulphide crystals plastically deformed at low temperatures. In *International Symposium on Structural Properties of Dislocations in Semiconductors*, Oxford, Conference Series No. 104 (Bristol: Institute of Physics), pp. 217–20.

Osipyan, Y. A. and Petrenko, V. F. (1975). Short-circuit effect in plastic-deformation of ZnS and motion of charged dislocations. *Zhurnal Eksperimentalnoi i Teoreticheskoi Fiziki*, **69**, 1362–71.

Osipiyan, Yu. A. and Smirnova, I. S. (1968). Perfect dislocations in the wurtzite lattice. *Physica Status Solidi*, **30**, 19–29.

Osipyan, Yu. A., Timofeev, V. B. and Shteinman, E. A. (1972). Exciton scattering by dislocations in the CdSe crystal. *Soviet Physics JETP*, **35**, 146–9.

Osipyan, Yu. A., Petrenko, V. F., Zaretskii, A. V. and Whitworth, R. W. (1986). Properties of II–VI semiconductors associated with moving dislocations. *Advances in Physics*, **35**, 115–88.

Osvenskii, V. B., Proshko, G. P. and Milvidskii, M. G. (1967). Effect of dislocations on the structure of diffused p-n junctions in GaAs and on recombination radiation parameters. *Soviet Physics Semiconductors*, **1**, 755–60.

Oualid, J., Singal, C. M., Dugas, J., Crest, J. P. and Amzil, H. (1984). Influence of illumination on the grain-boundary recombination velocity in silicon. *Journal of Applied Physics*, **55**, 1197–205.

Pankove, J. I. and Johnson, N. M. (1991). Hydrogen in semiconductors. In *Semiconductors and Semimetals*, Volume **34**, eds. R. K. Willardson and A. C. Beer (San Diego: Academic Press).

Parish, G., Keller, S., Kozodoy, P. *et al.* (1999). High-performance (Al,Ga)N-based solar-blind ultraviolet *p–i–n* detectors on laterally epitaxially overgrown GaN. *Applied Physics Letters*, **75**, 247–9.

Pasemann, L. (1981). A contribution to the theory of the EBIC contrast of lattice defects in semiconductors. *Ultramicroscopy*, **6**, 237–50.

Pasemann, L. and Hergert, W. (1986). A theoretical study of the determination of the depth of a dislocation by combined use of EBIC and CL technique. *Ultramicroscopy*, **19**, 15–22.

Pasemann, L., Blumtritt, H. and Gleichmann, R. (1982). Interpretation of the EBIC contrast of dislocations in silicon. *Physica Status Solidi*, **A70**, 197–209.

Pavesi, L. (2003). Will silicon be the photonic material of the third millennium. *Journal of Physics: Condensed Matter*, **15**, R1169—R1196.

Pearson, G. L. and Riesz, R. P. (1959). High-speed switching diodes from plastically deformed germanium. *Journal of Applied Physics*, **30**, 311—12.

Pearton, S. J., Wu, C. S., Stavola, M. *et al.* (1987). Hydrogenation of GaAs on Si. Effects on diode reverse leakage current. *Applied Physics Letters*, **51**, 496—8.

Pearton, S. J., Corbett, J. W. and Stavola, M. (1992). *Hydrogen in Crystalline Semiconductors*. Berlin: Springer.

Peiner, E., Guttzeit, A. and Wehmann, H. H. (2002). The effect of threading dislocations on optical absorption and electron scattering in strongly mismatched heteroepitaxial III-V compound semiconductors on silicon. *Journal of Physics: Condensed Matter*, **14**, 13195—201.

Pennycook, S. J., Brown, L. M. and Craven, A. J. (1980). Observation of cathodoluminescence at single dislocations by STEM. *Philosophical Magazine*, **A41**, 589—600.

Petrenko, V. F. (1982). *Doctor of Science Thesis*, Institute of Solid State Physics, Chernogolovka as quoted by Osipiyan *et al.* (1986).

Petrenko, V. F. and Whitworth, R. W. (1980). Charged dislocations and the plastic-deformation of II—VI compounds. *Philosophical Magazine*, **A41**, 681—99.

Petroff, P. (1979). Point defects and dislocation climb in III—V compounds. *Journal de Physique*, **C6**, 201—5.

Petroff, P. (1981). Luminescence properties of GaAs epitaxial layers grown by liquid phase epitaxy and molecular beam epitaxy. In *Defects in Semiconductors*. Proceedings of the Materials Research Society (New York: North-Holland), pp. 457—64.

Petroff, P. and Hartman, R. L. (1973). Defect structure introduced during operation of heterojunction GaAs lasers. *Applied Physics Letters*, **23**, 469—71.

Petroff, P. and Hartman, R. L. (1974). Rapid degradation phenomenon in heterojunction GaAsAs—GaAs lasers. *Journal of Applied Physics*, **45**, 3899—903.

Petroff, P. M. and Lang, D. V. (1977). New spectroscopic technique for imaging spatial-distribution of nonradiative defects in a scanning-transmission electron-microscope. *Applied Physics Letters*, **31**, 60—2.

Petroff, P. M., Kimerling, L. C. and Johnston, W. D. (1977). Electronic excitation effects on the mobility of point defects and dislocations in GaAlAs-GaAs heterostructures. In *Radiation Effects in Semiconductors 1976*, Conference Series No. 31 (Bristol: Institute of Physics), pp. 362—7.

Petroff, P., Lang, D. V., Logan, R. A. and Johnston, W. D. (1978a). Deep level—dislocation interactions in $Ga_{1-x}Al_xAs$ (DH) structures. In *Defects and Radiation Effects in Semiconductors 1978*. Conference Series No. 46 (Bristol: Institute of Physics), pp. 427—32.

Petroff, P., Lang, D. V., Strudel, J. L. and Logan, R. A. (1978b). Scanning transmission electron microscopy techniques for simultaneous electronic analysis and observation of defects in semiconductors. In *SEM 1978*, **I** (Chicago: SEM Inc), pp. 325—32.

Petroff, P. M., Lang, D. V., Strudel, D. L. and Savage, A. (1978c). New STEM spectroscopic techniques for simultaneous electronic analysis and observation of defects in semiconductor materials and devices. In *Proceedings of Ninth*

International Congress on Electron Microscopy, Toronto 1978, **1** (Toronto: Microscopical Society of Canada), pp. 130–1.

Petroff, P. M., Logan, R. A. and Savage, A. (1980a). Nonradiative recombination at dislocations in III–V compound semiconductors. *Physical Review Letters*, **44**, 287–91.

Petroff, P. M., Logan, R. A. and Savage, A. (1980b). Nonradiative recombination at dislocations in III–V compound semiconductors. *Journal of Microscopy*, **118**, 255–61.

Pey, K. L., Chan, D. S. H. and Phang, J. C. H. (1993a). A numerical method for simulating cathodoluminescence contrast from localized defects. In *Microscopy of Semiconducting Materials 1993*, Conference Series No. 134 (Bristol: Institute of Physics), pp. 687–92.

Pey, K. L., Phang, J. C. H. and Chan, D. S. H. (1993b). Investigation of dislocations in GaAs using cathodoluminescence in the scanning electron microscope. *Scanning Microscopy*, **7**, 1195–206.

Pey, K. L., Chan, D. S. H. and Phang, J. C. H. (1995a). Cathodoluminescence contrast of localized defects Part I. Numerical model for simulation. *Scanning Microscopy*, **9**, 355–66.

Pey, K. L., Chan, D. S. H. and Phang, J. C. H. (1995b). Cathodoluminescence contrast of localized defects Part II. Defect investigation. *Scanning Microscopy*, **9**, 367–80.

Pfann, W. G. (1961). Improvement of semiconducting devices by elastic strain. *Solid State Electronics*, **3**, 261–7.

Pike, G. E. (1982). Electronic properties of ZnO varistors: a new model. In *Grain Boundaries in Semiconductors*, Materials Research Society Symposium Proceedings, **5**, eds. H. J. Leamy, G. E. Pike and C. H. Seager (Amsterdam: North-Holland), pp. 369–80.

Pike, G. E. and Seager, C. H. (1979). The DC voltage dependence of semiconductor grain-boundary resistance. *Journal of Applied Physics*, **50**, 3414–22.

Pödör, B. (1966). Electron mobility in plastically deformed germanium. *Physica Status Solidi*, **16**, K167.

Queisser, H. J. (1963). Properties of twin boundaries in silicon. *Journal of the Electrochemical Society*, **110**, 52–6.

Queisser, H. J. (1969). Observations and properties of lattice defects in silicon. In *Semiconductor Silicon*, eds. R. R. Haberecht and E. L. Kern (New York: Electrochemical Society), pp. 585–95.

Queisser, H. J. and Haller, E. E. (1998). Defects in semiconductors: Some fatal, some vital. *Science*, **281**, 945–50.

Radzimski, Z. J., Zhou, T. Q., Buczkowski, A. B. and Rozgonyi, G. A. (1991). Electrical activity of dislocations: Prospects for practical utilization. *Applied Physics*, **A53**, 189–93.

Rasul, A. and Davidson, S. M. (1977). SEM measurements of minority carrier lifetimes at dislocations in GaP, employing photon counting. In *Scanning Electron Microscopy 1977/I*, ed. O. Johari (Chicago: SEM Inc.), pp. 233–9.

Read, W. T. (1954a). Theory of dislocations in germanium. *Philosophical Magazine*, **45**, 775–96.

Read, W. T. (1954b). Statistics of the occupation of dislocation acceptor centres. *Philosophical Magazine*, **45**, 1119–28.

Read, W. T. (1955). Scattering of electrons by charged dislocations in semiconductors. *Philosophical Magazine*, **46**, 111–31.

Rebane, Y. T. and Shreter, Y. G. (1993). g-tensors of electrons bound to 60°-dislocations in Ge and Si. *Physics and Technology*, p. 35.

Rebane, Y. T., Shreter, Y. G. and Albrecht, M. (1997). Excitons bound to stacking faults in wurtzite GaN. In *Materials Research Society Symposium – Proceedings*, **468**, Gallium Nitride and Related Materials II, 1997, pp. 179–82.

Reddy, C. V. and Narayanamurti, V. (2001). Characterization of nanopipes/ dislocations in silicon carbide using ballistic electron emission microscopy. *Journal of Applied Physics*, **89**, 5797–9.

Ringel, S. A. (1997). Hydrogen-extended defect interactions in heteroepitaxial InP materials and devices. *Solid-State Electronics*, **41**, 359–80.

Robertson, M. J., Wakefield, B. and Hutchinson, P. (1981). Strain-related degradation phenomena in long-lived GaAlAs stripe lasers. *Journal of Applied Physics*, **52**, 4462–6.

Robins, L. H., Cook, L. P., Farabaugh, E. N. and Feldman, A. (1989). Cathodoluminescence of defects in diamond films and particles grown by hot-filament chemical-vapor deposition. *Physical Review*, **B39**, 13367–77.

Romero, M. J., Al-Jassim, M. M., Dhere, R. G. *et al.* (2002a). Beam injection methods for characterizing thin-film solar cells. *Progress in Photovoltaics*, **10**, 445–55.

Romero, M. J., Albin, D. S., Al-Jassim, M. M. *et al.* (2002b). Cathodoluminescence of Cu diffusion in CdTe thin films for CdTe/CdS solar cells. *Applied Physics Letters*, **81**, 2962–4.

Rose, D. J. (1957). Microplasmas in silicon. *Physical Review*, **105**, 413–18.

Roseman, R. D. and Mukherjee, N. (2003). PTCR effect in BaTiO$_3$: Structural aspects and grain boundary potentials. *Journal of Electroceramics*, **10**, 117–35.

Ross, F. M., Hull, R., Bahnck, D. *et al.* (1993). Changes in electrical device characteristics during the *in situ* formation of dislocations. *Applied Physics Letters*, **62**, 1426–8.

Rozgonyi, G. A., Petroff, P. M. and Panish, M. B. (1974a). Elimination of dislocations in heteroepitaxial layers by the controlled introduction of interfacial misfit dislocations. *Applied Physics Letters*, **24**, 251–4.

Rozgonyi, G. A., Petroff, P. M. and Panish, M. B. (1974b). Control of lattice parameters and dislocations in the system (Ga$_{1-x}$Al$_x$As$_{1-y}$P$_y$/GaAs). *Journal of Crystal Growth*, **27**, 106–17.

Rozgonyi, G. A., Salih, A. S. M., Radzimski, Z. J. *et al.* (1987). Defect engineering for VLSI epitaxial silicon. *Journal of Crystal Growth*, **85**, 300–7.

Rozgonyi, G. A. and Kola, R. R. (1989). Defect engineering for ULSI epitaxial silicon. *Solid State Phenomena*, **6–7**, 143–58.

Ruan, J., Choyke, W. J. and Partlow, W. D. (1991). Cathodoluminescence and annealing study of plasma-deposited polycrystalline diamond films. *Journal of Applied Physics*, **69**, 6632–6.

Ruan, J., Kobashi, K. and Choyke, W. J. (1992). On the 'band-A' emission and boron related luminescence in diamond. *Applied Physics Letters*, **60**, 3138–40.

Salerno, J. P., Gale, R. P., Fan, J. C. C. and Vaughan, J. (1981). Scanning cathodoluminescence microscopy of polycrystalline GaAs. In *Defects in*

Semiconductors. Proceedings of Materials Research Society Annual Meeting, eds. J. Narayan and T. Y. Tan (New York: North-Holland), pp. 509–14.

Samuelson, L., Gustafsson, A., Lindahl, J. *et al.* (1994a). Scanning tunneling microscope and electron beam induced luminescence in quantum wires. *Journal of Vacuum Science and Technology*, **B12**, 2521–6.

Sauer, R., Weber, J., Stolz, J., Weber, E. R., Kusters, K. H. and Alexander, H. (1985). Dislocation-related photoluminescence in silicon. *Applied Physics*, **A36**, 1–13.

Seto, J. Y. W. (1975). The electrical properties of polycrystalline silicon films. *Journal of Applied Physics*, **46**, 5247–54.

Schmidt, T. M., Justo, J. F. and Fazzio, A. (2000). The effect of a stacking fault on the electronic properties of dopants in gallium arsenide. *Journal of Physics: Condensed Matter*, **12**, 10235–9.

Schreiber, J. and Hergert, W. (1989). Combined application of SEM-CL and SEM-EBIC for the investigation of compound semiconductors. In *International Symposium on the Structure and Properties of Dislocations in Semiconductors*, 1989. Conf. Series No. 104 (Bristol: Inst. Phys.), pp. 97–107.

Schreiber, J. and Hildebrandt, S. (1991). Quantitative evaluation of recombination activity of dislocations by combined SEM-CL/EBIC. *Journal de Physique*, **C6**, 15–19.

Schreiber, J. and Vasnyov, S. (2004). The dynamic mode of high-resolution cathodoluminescence microscopy. *Journal of Physics: Condensed Matter*, **16**, S75–S84.

Schreiber, J., Hergert, W. and Hildebrandt, S. (1991). Combined application of SEM-CL and SEM-EBIC for the investigation of compound semiconductors. *Applied Surface Science*, **50**, 181–5.

Schreiber, J., Uniewski, H., Hildebrandt, S., Hoering, L. and Leipner, H. S. (1997). Distinction of the recombination properties and identification of Y luminescence at glide dislocations in CdTe. In *Microscopy of Semiconducting Materials 1997*. Conference Series No. 157 (Bristol: Institute of Physics), pp. 651–4.

Schreiber, J., Hoering, L., Uniewski, H., Hildebrandt, S. and Leipner, H. S. (1999a). Recognition and distribution of A(g) and B(g) dislocations in indentation deformation zones on {111} and {110} surfaces of CdTe. *Physica Status Solidi*, **A171**, 89–97.

Schreiber, J., Hilpert, U., Hoering, L. *et al.* (1999b). Study of plastic relaxation of layer stress in ZnSe/GaAs (001) heterostructures. In *Microscopy of Semiconducting Materials 1999*. Conference Series No. 164 (Bristol: Institute of Physics), pp. 299–304.

Schreiber, J., Hilpert, U., Hoering, L. *et al.* (2000). Luminescence studies on plastic stress relaxation in ZnSe/GaAs (001). *Physica Status Solidi*, **A222**, 169–77.

Schreiber, J. and Vasnyov, S. (2004). The dynamic mode of high-resolution cathodoluminescence microscopy. *Journal of Physics: Condensed Matter*, **16**, S75–S84.

Schröter, W. and Cerva, H. (2002). Interaction of point defects with dislocations in silicon and germanium: Electrical and optical effects. *Solid State Phenomena*, **85–86**, 67–143.

Schröter, W., Scheibe, E. and Schoen, H. (1980). Energy spectra of dislocations in silicon and germanium. *Journal of Microscopy*, **118**, 23–34.

Schröter, W., Queisser, I. and Kronewitz, J. (1989). Capacitance transient spectroscopy of dislocations in semiconductors. In *Structure and Properties of Dislocations in Semiconductors 1989*. Conference Series No. 104 (Bristol: Institute of Physics), pp. 75–84.

Schröter, W., Kronewitz, J., Gnauert, U., Riedel, F. and Seibt, M. (1995). Bandlike and localized states at extended defects in silicon. *Physical Review*, **B52**, 13726–9.

Schröter, W., Kveder, V. and Hedemann, H. (2002a). Electrical effects of point defect clouds at dislocations in silicon, studied by deep level transient spectroscopy. *Solid State Phenomena*, **82–84**, 213–18.

Schröter, W., Hedemann, H., Kveder, V. and Riedel, F. (2002b). Measurements of energy spectra of extended defects. *Journal of Physics: Condensed Matter*, **14**, 13047–59.

Schumann, P. A. and Rideout, A. J. (1964). Reduction of the turn-off delay of a germanium NPN mesa by plastic deformation. *Solid State Electronics*, **7**, 849–51.

Seager, C. H. (1985). Grain boundaries in polycrystalline silicon. *Annual Review of Materials Science*, **15**, 271–302.

Seager, C. H. and Ginley, D. S. (1979). Passivation of grain boundaries in polycrystalline silicon. *Applied Physics Letters*, **34**, 337–40.

Seager, C. H., Ginley, D. S. and Zook J. D. (1980). Improvement of polycrystalline silicon solar cells with grain-boundary hydrogenation techniques. *Applied Physics Letters*, **36**, 831–3.

Seager, C. H. and Ginley, D. S. (1981). Studies of the hydrogen passivation of silicon grain boundaries. *Journal of Applied Physics*, **52**, 1050–5.

Seaton, J. and Leach, C. (2004). Conductive mode imaging of thermistor grain boundaries. *Journal of the European Ceramic Society*, **24**, 1191–4.

Seifert, W. and Kittler, M. (1987). Negative (bright) EBIC contrast at oxygen induced defects in silicon. *Physica Status Solidi*, **A99**, K11–K14.

Seifert, W., Knobloch, K. and Kittler, M. (1997). Modification of the recombination activity of dislocations in silicon by hydrogenation, phosphorous diffusion and heat treatments. *Solid State Phenomena*, **57–8**, 287–92.

Shaw, D. A. and Thornton, P. R. (1968). Cathodoluminescent studies of laser quality GaAs. *Journal of Materials Science*, **3**, 507–18.

Shaw, D. A., Hughes, K. A., Neve, N. F. B., Sulway, D. V., Thornton, P. R. and Gooch, C. (1966). Crystal mosaic structures and the lasing properties of GaAs laser diodes. *Solid State Electronics*, **9**, 664–5.

Shockley, W. (1953). Dislocations and edge states in the diamond crystal structure. *Physical Review*, **91**, 228.

Shockley, W. (1961). Problems related to p-n junctions in silicon. *Solid State Electronics*, **2**, 35–67.

Shreter, Y. G. and Rebane, Y. T. (1996). Dislocation-related luminescence in GaN. In *23rd International Conference on the Physics of Semiconductors*, 1996, pt. 4, pp. 2937–40.

Shreter, Yu. G., Rebane, Yu. T. and Peaker, A. R. (1993). Optical properties of dislocations in silicon crystals. *Physica Status Solidi*, **A138**, 681–6.

Shreter, Y. G., Rebane, Y. T., Klyavin, O. V. *et al.* (1996a). Dislocation-related absorption and photoluminescence in deformed n-ZnSe crystals. *Journal of Crystal Growth*, **159**, 883–8.

Shreter, Yu. G., Rebane, Y. T., Klyavin, O. V. *et al.* (1996b). Dislocation-related absorption, photoluminescence and birefringence in deformed n-ZnSe crystals. *Diffusion and Defect Data Part B (Solid State Phenomena)*, **51–52**, 93–8.

Shreter, Y. G., Rebane, Y. T., Davis, T. J. *et al.* (1997). Dislocation luminescence in wurtzite GaN. In *III-V Nitrides*, eds. F. A. Ponce, T. D. Moustakas, I. Akasaki and B. A. Monemar, Materials Research Society Symposium Proceedings **449** (Pittsburgh: Materials Research Society), pp. 683–8.

Simpkins, B. S., Yu, E. T., Waltereit, P. and Speck, J. S. (2003). Correlated scanning Kelvin probe and conductive atomic force microscopy studies of dislocations in gallium nitride. *Journal of Applied Physics*, **94**, 1448–53.

Skowronski, M., Liu, J. Q., Vetter, W. M. *et al.* (2002). Recombination-enhanced defect motion in forward-biased 4H−SiC *p-n* diodes. *Journal of Applied Physics*, **92**, 4699–704.

Smith, S., Zhang, P., Gessert, T. and Mascarenhas, A. (2004). Near-field optical beam-induced currents in CdTe/CdS solar cells: Direct measurement of enhanced photoresponse at grain boundaries. *Applied Physics Letters*, **85**, 3854–6.

Snyman, L. W., Aharoni, H., du Plessis, M. and Gouws, R. B. J. (1998). Increased efficiency of silicon light-emitting diodes in a standard 1.2 μm silicon complementary metal oxide semiconductor technology. *Optical Engineering*, **37**, 2133–41.

Sobolev, N. A., Emel'yanov, A. M., Shek, E. I. and Vdovin, V. I. (2004). Influence of extended structural defects on the characteristics of electroluminescence in efficient silicon light-emitting diodes. *Solid State Phenomena*, **95–96**, 283–8.

Sosnowski, L. (1959). Electronic properties at grain boundaries. *Journal of Physics and Chemistry of Solids*, **8**, 142–6.

Steckenborn, A., Munzel, H. and Bimberg, D. (1981). Cathodoluminescence lifetime pattern of GaAs surfaces around dislocations. *Journal of Luminescence*, **24/25**, 351–4.

Steeds, J. W. (1989). High spatial resolution cathodoluminescence from dislocations in semiconductors studied in a TEM. In *International Symposium on Structural Properties of Dislocations in Semiconductors*, Oxford. Conf. Series No. 104 (Bristol: Institute of Physics), pp. 199–202.

Steeds, J. W., Batstone, J. L., Rebane, Yu. T. and Schreter, Yu. G. (1991). Dislocation luminescence in zinc selenide. In *Polycrystalline Semiconductors II*. Springer Proc. *In Phys.* **54**, eds. J. H. Werner and H. P. Strunk (Berlin: Springer-Verlag), pp. 45–9.

Steele, B. C. H. (ed.) (1991). *Electronic Ceramics*. London: Elsevier Applied Science.

Stevenson, J. L., Skeats, A. P. and Heckingbottom, R. (1980). EBIC microscopy of double heterostructure laser materials and devices. *Journal of Microscopy*, **118**, 321–7.

Stowe, D. J., Galloway, S. A., Senkader, S. *et al.* (2003). Near-band gap luminescence at room temperature from dislocations in silicon. *Physica*, **B340–342**, 710–13.

Stringfellow, G. B., Lindquist, P. F., Cass, T. R. and Burmeister, R. A. (1974). Dislocations in vapour phase epitaxial gaP. *Journal of Electronic Materials*, **3**, 497–515.

Suezawa, M. and Sumino, K. (1989). Electron spin resonance study of deformation-induced Si-Kl centers in silicon. *Journal of the Physical Society of Japan*, **58**, 2463–71.

Sugiura, L. (1997). Comparison of degradation caused by dislocation motion in compound semiconductor light-emitting devices. *Applied Physics Letters*, **70**, 1317–19.

Sumida, N. and Lang, A. R. (1981). Cathodoluminescence evidence of dislocation interactions in diamond. *Philosophical Magazine*, **A43**, 1277–87.

Sumino, K. (2003). Basic aspects of impurity gettering. *Microelectronics Engineering*, **66**, 268–80.

Sumino, K. and Yonenaga, I. (2002). Interactions of impurities with dislocations: Mechanical effects. *Solid State Phenomena*, **85–86**, 145–76.

Sun, X. L., Brillson, L. J., Chiang, Y. M. and Luo, J. (2002). Microcathodoluminescence spectroscopy of defects in Bi_2O_3-doped ZnO grains. *Journal of Applied Physics*, **92**, 5072–6.

Susa, N., Yamauchi, Y. and Ando, H. (1982). Effects of imperfections in InP avalanche photodiodes with vapour phase epitaxially grown p^+-n junctions. *Journal of Applied Physics*, **53**, 7044–50.

Sutton, A. P. and Balluffi, R. W. (1995). *Interfaces in Crystalline Materials*. Oxford: Oxford University Press.

Sze, S. M. (1985). *Semiconductor Devices. Physics and Technology*. New York: Wiley.

Tarbaev, N. I. (1998). Low-temperature photoluminescence determination of dislocation slip systems in CdSe single crystals. *Physics of the Solid State*, **40**, 1672–5.

Tarbaev, N. I. and Shepelskii, G. A. (1998). One-dimensional structures formed by low-temperature slip of dislocations that act as sources of dislocation absorption and emission in II–VI semiconductor crystals. *Semiconductors*, **32**, 580–6.

Tarbaev, N. I., Schreiber, J. and Shepelskii, G. A. (1988). Physical properties of $A^{II}B^{VI}$ semiconductor crystals after plastic deformation at low temperature. *Physica Status Solidi*, **A110**, 97–106.

Taylor, W. E., Odell, N. H. and Fan, H. Y. (1952). Grain boundary barriers in germanium. *Physical Review*, **88**, 867–75.

Temkin, H., Zipfel, C. L. and Keramidas, V. G. (1981). High-temperature degradation of InGaAsP/InP light emitting diodes. *Journal of Applied Physics*, **52**, 5377–80.

Thornton, P. R. (1963). Electrical effects of dislocations in high resistivity GaAs. *Solid State Electronics*, **6**, 677–8.

Toriumi, A., Yoshimi, M., Iwase, M., Akiyama, Y. and Taniguchi, K. (1987). A study of photon-emission from n-channel MOSFETS. *IEEE Transactions on Electron Devices*, **34**, 1501–8.

Toth, A. L. (1981). Measurement of EBIC Contrast and Resolution of Dislocations in Silicon. Microscopy of Semiconducting Materials 1981. Conf. Series no. 60 (Bristol: Institute of Physics), pp. 221–2.

Tretola, A. R. and Irvin, J. C. (1968). Correlation of the physical location of crystal defects with electrical imperfections in GaAs p-n junctions. *Journal of Applied Physics*, **39**, 3563–8.

Tringe, J. W. and Plummer, J. D. (2000). Electrical and structural properties of polycrystalline silicon. *Journal of Applied Physics*, **87**, 7913–26.

Tweet, A. G. (1954). Grain boundary conduction in gold-doped Ge. *Physical Review*, **96**, 828.

Tweet, A. G. (1955). Properties of grain boundaries in gold-doped germanium. *Physical Review*, **99**, 1182–9.

Ueda, O. (1996). *Reliability and Degradation of III–V Optical Devices*. Boston: Artech House.

Ueda, O. (1999). Reliability issues in III–V compound semiconductor devices: optical devices and GaAs-based HBTs. *Microelectronics Reliability*, **39**, 1839–55.

Ueda, O., Isozumi, S., Kotani, T. and Yaoki, T. (1977). Defect structure of ⟨100⟩ dark lines in the active region of a rapidly degraded $Ga_{1-x}Al_xAs$ LED. *Journal of Applied Physics*, **48**, 3950–2.

Ueda, O., Imai, H., Kotani, T., Wakita, K. and Saito, H. (1979). TEM observation of catastrophically degraded $Ga_{1-x}Al_xAs$ double-heterostructure lasers. *Journal of Applied Physics*, **50**, 6643–7.

Unger, K. (1968). Theoretical study of the filamentary radiation and the surface damage of junction lasers. In *Proceedings of the Ninth International Conference on Physica of Semiconductors*, Moscow (Leningrad: Nauka), **I**, pp. 537–9.

Urbieta, S., Fernandez, P., Piqueras, J., Vasco, E. and Zaldo, C. (2004). Nanoscopic study of ZnO films by electron beam induced current in the scanning tunneling microscope. *Journal of Optoelectronics and Advanced Materials*, **6**, 183–8.

Van de Krol, R. and Tuller, H. L. (2002). Electroceramics – the role of interfaces. *Solid State Ionics*, **150**, 167–79.

Vanderschaeve, G., Levade, C. and Caillard, D. (2001). Dislocation mobility and electronic effects in semiconductor compounds. *Journal of Microscopy*, **203**, 72–83.

Van de Walle, C. G. (2001). Defect analysis and engineering in ZnO. *Physica*, **B308–310**, 899–903.

Vasnyov, S., Schreiber, J. and Hoering, L. (2004). A quantitative evaluation of the dynamic cathodoluminescence contrast of gliding dislocations in semiconductor crystals. *Journal of Physics: Condensed Matter*, **16**, S269–S277.

Vavilov, V. S., Gippius, A. A., Zaitsev, A. M. *et al.* (1980). Investigation of the cathodoluminescence of epitaxial diamond films. *Soviet Physics Semiconductors*, **14**, 1078–9.

Visoly-Fisher, I., Cohen, S. R. and Cahen, D. (2003). Direct evidence for grain-boundary depletion in polycrystalline CdTe from nanoscale-resolved measurements. *Applied Physics Letters*, **82**, 556–8.

Visoly-Fisher, I., Cohen, S. R., Ruzin, A. and Cahen, D. (2004). How polycrystalline devices can outperform single-crystal ones: Thin film CdTe/CdS solar cells. *Advanced Materials*, **16**, 879–83.

von Kanel, H. and Meyer, T. (1998). Recent progress on BEEM. *Ultramicroscopy*, **73**, 175–83.

von Kanel, H. and Meyer, T. (2000). Nano-scale defect analysis by BEEM. *Journal of Crystal Growth*, **210**, 401–7.

Vyvenko, O. F., Krüger, O. and Kittler, M. (2000). Cross-sectional electron-beam-induced current analysis of the passivation of extended defects in cast multicrystalline silicon by remote hydrogen plasma treatment. *Applied Physics Letters*, **76**, 697–9.

Vyvenko, O. F., Buonassisi, T., Istratov, A. A. *et al.* (2002a). X-ray beam induced current – a synchrotron radiation based technique for the *in situ* analysis of recombination properties and chemical nature of metal clusters in silicon. *Journal of Applied Physics*, **91**, 3614–17.

Vyvenko, O. F., Buonassisi, T., Istratov, A. A. *et al.* (2002b). Application of synchrotron-radiation-based x-ray microprobe techniques for the analysis of recombination activity of metals precipitated at Si/SiGe misfit dislocations. *Journal of Physics: Condensed Matter*, **14**, 13079–86.

Vyvenko, O. F., Buonassisi, T., Istratov, A. A. and Weber, E. R. (2004). X-ray beam induced current/microprobe x-ray fluorescence: synchrotron radiation based x-ray microprobe techniques for analysis of the recombination activity and chemical nature of metal impurities in silicon. *Journal of Physics: Condensed Matter*, **16**, S141–S151.

Wakefield, B. (1979). Strain-enhanced luminescence degradation in GaAs-GaAlAs double-heterostructure lasers revealed by photoluminescence. *Journal of Applied Physics*, **50**, 7914–16.

Wakefield, B., Leigh, P. A., Lyons, M. H. and Elliott, C. R. (1984). Characterization of semi-insulating liquid encapsulated Czochralski GaAs by cathodoluminescence. *Applied Physics Letters*, **45**, 66–8.

Wang, G., Ogawa, T., Soga, T., Jimbo, T. and Umeno, M. (2001). Passivation of dislocations in GaAs grown on Si substrates by phosphine (PH₃) plasma exposure. *Applied Physics Letters*, **78**, 3463–5.

Warwick, C. A. and Brown, G. T. (1985). Spatial distribution of 0.68-eV emission from undoped semi-insulating gallium arsenide revealed by high resolution luminescence imaging. *Applied Physics Letters*, **46**, 574–6.

Warwick, C. A., Gill, S. S., Wright, P. J. and Cullis, A. G. (1985). Spatial variation of dopant concentration in Si implanted Czochralski and metal organic vapour phase epitaxial GaAs. In *Microscopy of Semiconducting Materials 1985*. Conf. Series No 76 (Bristol: Institute of Physics), pp. 365–72.

Waser, R. and Hagenbeck, R. (2000). Grain boundaries in dielectric and mixed-conducting ceramics. *Acta Materialia*, **48**, 797–825.

Watson, C. C. R. and Durose, K. (1993). Cathodoluminescence microscopy of bulk CdTe crystals. *Journal of Crystal Growth*, **126**, 325–9.

Weber, J. (1994). Correlation of structural and electronic properties from dislocations in semiconductors. *Solid State Phenomena*, **37–38**, 13–24.

Weber, E. R. and Alexander, H. (1983). EPR of dislocations in silicon. *Journal de Physique*, **C4**, 319–28.

Wederoth, M., Gregor, M. J. and Ulbrich, R. G. (1992). Luminescence from gold-passivated gallium arsenide surfaces excited with a scanning tunneling microscope. *Solid State Communications*, **83**, 535–7.

Weimann, N. G., Eastman, L. F., Doppalapudi, D., Ng, H. M. and Moustakas, T. D. (1998). Scattering of electrons at threading dislocations in GaN. *Journal of Applied Physics*, **83**, 3656–9.

Werkhoven, C., van Opdorp, C. and Vink, A. T. (1977). Non-radiative recombination in n-Type LPE GaP. In *GaAs and Related Compounds 1976*. Conf. Series No. 33A (Bristol: Institute of Physics), pp. 317–25.

Werkhoven, C., van Opdorp, C. and Vink, A. T. (1978/79). Influence of crystal defects on the luminescence. *Philips Technical Review*, **38**, 41–50.

Werner, M., Weber, E. R., Bartsch, M. and Messerschmidt, U. (1995). Carrier injection enhanced dislocation glide in silicon. *Physica Status Solidi*, **A150**, 337–41.

Wertheim, G. K. and Pearson, G. L. (1957). Recombination in plastically deformed germanium. *Physical Review*, **107**, 694–8.

Wessel, K. and Alexander, H. (1977). On the mobility of partial dislocations in silicon. *Philosophical Magazine*, **35**, 1523–36.

Wilshaw, P. R. and Booker, G. R. (1985). New results and an interpretation for SEM EBIC contrast arising from individual dislocations in silicon. In *Microscopy of Semiconducting Materials*, 1985. Conf. Series No. 76 (Bristol: Institute of Physics), pp. 329–36.

Wilshaw, P. R. and Booker, G. R. (1987). The theory of recombination at dislocations in silicon and an interpretation of EBIC results in terms of fundamental dislocation parameters. *Bulletin of the Academy of Sciences of the USSR Division of Physical Science*, **51**, 109–13.

Wilshaw, P. R. and Fell, T. S. (1989). The electronic properties of dislocations in silicon. In *Structure and Properties of Dislocations in Semiconductors*, 1989. Conf. Series No. 104 (Bristol: Institute of Physics), pp. 85–96.

Wilshaw, P. R. and Fell, T. S. (1991). The electrical activity of dislocations in the presence of transition metal contaminants. In *Polycrystalline Semiconductors* II, eds. J. H. Werner and H. P. Strunk (Berlin: Springer-Verlag), pp. 77–83.

Wilshaw, P. R. and Fell, T. S. (1995). Electron beam induced current investigations of transition metal impurities at extended defects in silicon. *Journal of the Electrochemical Society*, **142**, 4298–304.

Wilshaw, P. R., Fell, T. S. and Booker, G. R. (1989). Recombination at dislocations in silicon and gallium arsenide. In *Point and Extended Defects in Semiconductors*, eds. G. Benedek, A. Cavallini and W. Schroter, Nato ASI Series B Physics, **202** (New York: Plenum), pp. 243–56.

Wilshaw, P. R., Fell, T. S. and de Coteau, M. D. (1991). EBIC contrast of defects in semiconductors. *Journal de Physique*, **C6**, 3–14.

Wilshaw, P. R., Blood, A. M. and Braban, C. F. (1997). Carrier recombination at defects in silicon: the effect of transition metals and hydrogen passivation. In *Microscopy of Semiconducting Materials*, Conf. Ser. No. 157 (Bristol: Inst. Phys.), pp. 623–8.

Wolff, P. A. (1960). Theory of optical radiation from breakdown avalanches in germanium. *Journal of Physics and Chemistry of Solids*, **16**, 184–90.

Woods, G. S. and Lang, A. R. (1975). Cathodoluminescence, optical absorption and x-ray topographic studies of synthetic diamonds. *Journal of Crystal Growth*, **28**, 215–26.

Wosinski, T. and Figielski, T. (1989). Electronic properties of dislocations and associated point-defects in GaAs. *Institute of Physics Conference Series* (104), pp. 151–62.

Wosinski, T., Figielski, T., Makosa, A. *et al.* (2002). Quantum effects associated with misfit dislocations in GaAs-based heterostructures. *Materials Science and Engineering*, **B91–92**, 367–70.

Yacobi, B. G. and Holt, D. B. (1990). *Cathodoluminescence Microscopy of Inorganic Solids*. New York: Plenum Press.

Yacobi, B. G., Lebens, J., Vahala, K. J., Badzian, A. R. and Badzian, T. (1993). Preferential incorporation of defects in monocrystalline diamond films. *Diamond and Related*. Materials **2**, 92–9.

Yakimov, E. B., Eremenko, V. G. and Nikitenko, V. I. (1976). Photoconductivity of silicon with dislocations. *Soviet Physics Semiconductors*, **10**, 231–2.

Yamaguchi, M. and Amano, C. (1985). Efficiency calculations of thin-film GaAs solar cells on Si substrates. *Journal of Applied Physics*, **58**, 3601–6.

Yamaguchi, M., Yamamoto, A. and Itoh, Y. (1986). Effect of dislocations on the efficiency of thin-film GaAs solar cells on Si substrates. *Journal of Applied Physics*, **59**, 1751–3.

Yamaguchi, M., Yamamoto, A. and Itoh, Y. (1986). Effect of dislocations on the efficiency of thin-film GaAs solar cells on Si substrates. *Journal of Applied Physics*, **59**, 1751–3.

Yamaguchi, M., Amano, C. and Itoh, Y. (1989). Numerical analysis for high-efficiency GaAs solar cells fabricated on Si substrates. *Journal of Applied Physics*, **66**, 915–19.

Yamamoto, N., Spence, J. C. H. and Fathy, D. (1984). Cathodoluminescence and polarization studies from individual dislocations in diamond. *Philosophical Magazine*, **49**, 609–29.

Yarykin, N. and Steinman, E. (2003). Comparative study of the plastic deformation- and implantation-induced centres in silicon. *Physica*, **B340–342**, 756–9.

Yastrubchak, O., Wosinski, T., Makosa, A., Figielski, T. and Toth, A. L. (2001). Capture kinetics at deep-level defects in lattice-mismatched GaAs-based heterostructures. *Physica*, **B308**, 757–60.

Yonenaga, I., Werner, M., Bartsch, M., Messerschmidt, U. and Weber, E. R. (1999). Recombination-enhanced dislocation motion in SiGe and Ge. *Physica Status Solidi*, **A171**, 35–40.

Zaretskii, A. V., Osipiyan, Yu. A., Petrenko, V. F. and Strukova, G. K. (1977). Experimental determination of dislocation charges in CdS. *Fizika Tverdogo Tela*, **19**, 418–23.

Zhang, M., Pirouz, P. and Lendenmann, H. (2003). Transmission electron microscopy investigation of dislocations in forward-biased 4H-SiC $p-i-n$ diodes. *Applied Physics Letters*, **83**, 3320–2.

Zhuang, Y. Q. and Du, L. (2002). 1/f noise as a reliability indicator for subsurface Zener diodes. *Microelectronics Reliability*, **42**, 355–60.

Zimin, D., Alchalabi, K. and Zogg, H. (2002). Heteroepitaxial PbTe-on-Si pn-juction IR-sensors: correlations between material and device preperties. *Physica*, **E13**, 1220–3.

Zozime, A. and Castaing, J. (1996). Effect of hydrogenation on the properties of extended defects in semiconductors. *Materials Science and Engineering*, **B42**, 57–62.

6

Point defect materials problems

6.1 Introduction

One final category of extended defect remains, although it is not generally so described. This consists of undesired, non-uniform distributions of native point defects, impurities and alloy composition variations. These point defect maldistributions render the initial material properties non-uniform and interfere with the controlled introduction of the variations required for devices.

Semiconductor processing starts with material that is uniformly sufficiently pure and perfect to exhibit the intrinsic properties of the semiconductor. Controlled concentrations of selected impurities are then introduced into chosen volumes to achieve the desired extrinsic properties. These include, for example, p- or n-type conductivity of the necessary value or luminescent emission of a certain wavelength and efficiency. For this, the impurity must occur as a uniform, random distribution of single, isolated impurity atoms of the desired element on substitutional sites in the required concentration. This chapter is concerned with crystal growth phenomena affecting point defect distributions and so materials uniformity and, therefore, capable of leading to failure to achieve successful device fabrication.

Because of their importance and relative simplicity, point defects have been studied intensively throughout the history of semiconductor physics and chemistry. The properties of point defects are therefore well treated in many review articles (Queisser and Haller 1998) and books such as Stoneham (2000) as well as series of conferences. We shall, therefore, give only the necessary minimum background on a number of points required for the present purpose.

This chapter is concerned with the ways in which actual point defect and impurity atom distributions can differ from the ideal uniform doping described above. Each of these phenomena gives rise to some form of maldistribution of point defects constituting a particular type of volume defect.

6.2 Impurity precipitation

Precipitates in semiconductors are volumes of a second phase, e.g. of an impurity. They are less common and smaller than in metal alloys, mainly because the concentrations of dopants in semiconductors are low (usually 10^{15} to $10^{18}\,\mathrm{cm}^{-3}$). Precipitation is not infrequent for the heaviest 'degenerate' doping. (Degenerate doping is so called because at such concentrations, i.e. greater than 10^{18} or $10^{19}\,\mathrm{cm}^{-3}$, the wave functions of the outermost donor electrons on neighbouring dopant atoms overlap significantly. The donor level becomes degenerate and broadens into an impurity energy band.)

Precipitation will tend to occur if the concentration that can be held in solid solution falls, as the temperature decreases, to a level below that actually present in the material. This is known as retrograde solubility. Consider a semiconductor and dopant, introduced at a high temperature, e.g. by diffusion. Suppose that the solubility limit falls during cooling. Then any excess of the impurity must be precipitated from solution with two effects. Firstly, precipitates form. These may be metallic particles which can be harmful if they occur at *p-n* junctions as will be discussed next. Secondly, the concentration of the electrically active donor or acceptor atoms and hence the charge carrier concentration will be less than the number of dopant atoms.

It was early observed that Si diodes that exhibited 'soft' reverse-bias characteristics broke down at a number of specific light-emitting sites called microplasmas that often correlated with dislocation etch pit sites (for further details see Section 5.11.5). A number of TEM and SEM studies were carried out but only one was conclusive. Katz (1974) used etching to show that a rod defect had a one-to-one correspondence with light-emitting spots, soft characteristics and low breakdown voltages in silicon diodes and transistors. He found that a single rod defect in a complete device window was sometimes sufficient to produce virtual short circuiting. Cullis and Katz (1974) used TEM and SEM to study these rod defects which were sometimes associated with stacking faults. The rods penetrated (001) Si foils along inclined ⟨101⟩ directions. Electron diffraction and x-ray microanalytical data indicated that the rods, one of which is shown in Fig. 6.1, had a precipitate structure consistent with that of ζ_a-FeSi$_2$ (α-lebolite) which exhibits metallic conduction. Phosphorous-diffusion gettering was shown to lead to the dissolution of these rod precipitates, leaving a residue of dislocations and smaller precipitates.

6.3 Point defect interactions

Failure to obtain the desired results may be due to other point defects interacting with and altering the properties of the dopant atoms. Such point defect phenomena prevent the dopant impurities, even if present in the right place and right concentration from producing the intended properties.

6.3.1 Point defect thermodynamics

The formation energies of point defects are of the same order as kT at room and elevated temperatures. Therefore they form in thermodynamic equilibrium, unlike

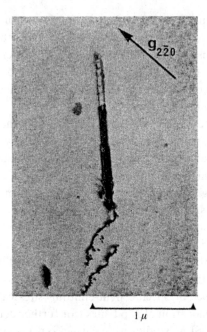

$\mathbf{g}_{2\bar{2}0}$

$1\,\mu$

Figure 6.1 Bright field TEM image of a rod shaped precipitate in silicon, with associated dislocations. These needle-like precipitates were identified as ζ_a-FeSi$_2$. [After Cullis and Katz 1974. Reprinted with permission from *Philosophical Magazine*, **30** (1974), pp. 1419–42; Taylor & Francis Ltd., the Journal's web site: http://www.tandf.co.uk/journals.]

the much higher-energy extended defects with which this book is mainly concerned. In fact, in equilibrium, the concentrations both of point defects and of bound and free charges can be found by a modified thermodynamic approach. This subject, founded by Kroger (Kroger and Vink 1956, Kroger 1964), is known as defect chemistry or defect thermodynamics. Defect thermodynamics is well dealt with at textbook level by Swalin (1972).

   This is important because of the occurrence of native point defects. These defects can occur in pure material and involve misplacement of atoms in covalent elements or compounds or ions in ionic compounds. In defect thermodynamics a letter or letters designate native point defects with a subscript giving the site occupied. They include vacancies, interstitials (i.e. interstitial atoms) (e.g. Si$_i$) and antisite or antistructure defects. Antisite defects are atoms or ions occupying the wrong sites in a compound like Ga$_{As}$ and As$_{Ga}$ in GaAs. In binary ionic AB compounds it has long been recognized that two types of pairs of point defects occur. Schottky defects are pairs of A and B vacancies. Frenkel defects are a vacancy and an interstitial of the same kind. These are important in ionically bonded crystals because the formation of either type of pair maintains the equality of charge in the crystal. The concentrations of each of the possible types of defect depend, in equilibrium, on the temperature and the formation energies of the defects. The number of possible defect types increases rapidly with the number of constituent elements in ternary and higher compounds and alloys (Section 1.6.1 and Section 1.6.2). This is a major reason these materials are so difficult and relatively little studied.

6.3.2 Non-stoichiometry

A possible source of large densities of native point defects is non-stoichiometry. Non-stoichiometry is deviation from the exact chemical combining ratio in a compound. For example, an AB compound may occur with the actual composition $A_{1-\delta}B$ or $AB_{1-\delta}$, where δ is small. Again the phenomenon can be considered thermodynamically and discussed in terms of the phase diagram of the alloy. For example, the existence region for GaAs in the Ga-As phase diagram in Fig. 6.2 can be seen to stretch from about 49.9 to 50.1% As i.e. δ can be up to 0.1%. That is, there can be up to one atom in a thousand or $\sim 10^{20}$ cm^{-3} of one element in excess of the stoichiometric ratio. If this limit were exceeded, however, this would cause a second phase to occur in the crystal. The excess of one element, δ, can result in a variety of native point defects. For example, $A_{1-\delta}B$ material could contain a fraction δ of V_A (vacancies on A sites) or of B_i (excess B atoms on interstitial sites). Similarly, $AB_{1-\delta}$ material could contain an atomic fraction δ of V_B or of A_i.

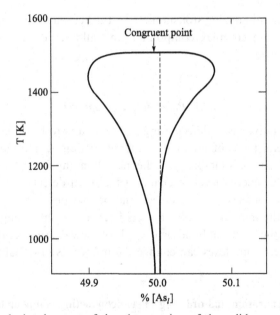

Figure 6.2   The calculated extent of the phase region of the solid compound GaAs in the Ga-As phase diagram. (After Hurle 1979. Reprinted from *Journal of Physics and Chemistry of Solids*, **40**, Revised calculation of point defect equilibria and non-stoichiometry in gallium arsenide, pp. 613–26. Copyright 1979, with permission from Elsevier.)

Any electrically active point defect present in a concentration comparable with that of the doping atoms can grossly alter many properties of the material. Doping densities range from 10^{15} or less to $10^{19}\,\text{cm}^{-3}$. These densities, assuming that there are roughly 10^{23} atomic sites cm^{-3}, correspond to fractional concentrations from 10^{-8} to 10^{-4}. The same values of δ would give native point defect concentrations equalling those of the dopants. Such fractional non-stoichiometries are difficult to measure by any techniques available at present but may well occur in many semiconducting compounds and are a major unsolved problem.

The non-stoichiometry can be altered by growth from a melt containing a surplus of one element and/or growth under a high overpressure of the gas of that element. A valuable application of non-stoichiometry is semi-insulating (S.I.) GaAs, which was first produced by Cr-doping. S.I. material can be grown from GaAs melts containing a surplus of As, which leads to the incorporation of numerous As_{Ga} antisites. These antisites give rise to a metastable deep donor level known as EL2 (Queisser and Haller 1998). These donors fully compensate the comparatively moderate concentrations of acceptors that occur in liquid-encapsulated Czochralski GaAs. This pins the Fermi level near the middle of the band gap giving very high resistivities ($>10^8$ ohm cm, hence the term 'semi-insulating'). The availability of semi-insulating wafers of GaAs and InP is a significant advantage over Si for microwave integrated circuits (MMICs) because it allows very effective electrical isolation of neighbouring devices.

The difficulty of minimizing stoichiometry is further reason for the limited interest in ternary and more complex compounds and alloys, despite the vast range of possibilities they offer.

6.3.3 Point defect complexes

Small groups of native point defects ranging from pairs to larger complexes tend to form due to several types of interaction energy, leading to attraction and binding. These include elastic, coulombic and chemical bonding interactions. Pairing and clustering alters the energy level spectrum of the point defects. This occurs because the native point defects are changed and/or because the symmetry of the coordination of the individual point defects is changed, the strength of the crystal field is altered or the chemical bonding orbitals of the valence electrons are modified, etc. All these defect complexes can change the properties of dopants.

6.4 Phase separation and ordering in semiconducting compounds and alloys

Semiconductor technology requires as starting material that which is uniformly highly perfect. Thermodynamics may make it difficult to nearly impossible to attain such uniformity or perfection. Depending on the enthalpy of mixing of the elements in a semiconductor alloy there may be a tendency to break down in one of two different ways. If the enthalpy is of such a sign that like atom neighbours are preferred, there will be a short-range tendency for like atoms to cluster. Phase separation into large-scale A-rich and B-rich regions may even occur. If the enthalpy is of the opposite sign, unlike neighbours are preferred. Consider a semiconducting alloy, e.g. of the form $A_xB_{1-x}C$. Such material is supposed to have the C atoms occupying one f.c.c. sublattice of the sphalerite structure while the A and B atoms are randomly distributed over the other f.c.c. sublattice. If unlike neighbours are preferred, however, the A and B atoms may be distributed in an ordered way. This may give a different crystal structure, e.g. chalcopyrite instead of sphalerite and significantly alter the properties of the material. Only if the enthalpy of mixing is zero will there be no tendency to either separation or ordering. Such a case is referred to as an ideal solution and is obviously a rarity. Phase separation and ordering in bulk crystals is dealt with by Swalin (1972) and a recent review of these phenomena in epitaxial films is given by Zunger and Mahajan (1994).

6.5 Large-scale, grown-in spatial maldistributions of point defects

We dealt very briefly above with phenomena that can introduce large concentrations of native point defects, or cause the precipitation of a fraction of the dopant atoms introduced into the semiconductor or cause separation or ordering in semiconducting alloys. These phenomena occur in thermodynamic equilibrium and affect the material uniformly throughout. We now turn to phenomena that result in large-scale non-uniform distributions of point defects. We shall refer to such distributions as maldistributions of point defects.

One group of such phenomena arise from dynamical, non-equilibrium phenomena in crystal growth. They involve heat and mass transport in the liquid and so are characteristic of melt-grown bulk crystals. Understanding these phenomena makes it possible to design crystal growth conditions to avoid their occurrence. Suppose, however, that the material grown, which then serves as the starting material for device production, is thermodynamically unstable or metastable, produced by carefully chosen growth kinetics. Then merely heating may result in undesired phase separation such as precipitation, ordering or more exotic processes such as spinodal decomposition (Swalin 1972).

6.5.1 Constitutional supercooling

This phenomenon received much attention in the early days of semiconductor technology. Chalmers (1964) and Delves (1975) reviewed the literature and treated mathematically the conditions for constitutional supercooling. Here we will emphasize the physical mechanism involved following the account given by

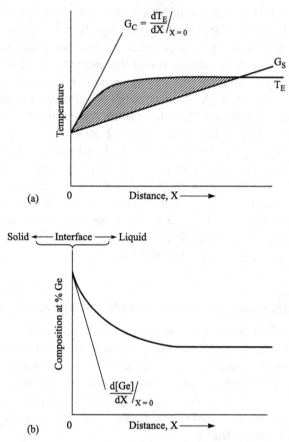

(a)   0   Distance, X ⟶

Solid ⟵ Interface ⟶ Liquid

(b)   0   Distance, X ⟶

Figure 6.3   The variation of (a) the temperature, and (b) composition in the melt ahead of the growing interface under conditions of constitutional supercooling. (After Dismukes and Ekstrom 1965.)

Dismukes and Ekstrom (1965) who grew bulk single crystals of Ge-Si alloys. They moved a molten zone of composition C_1 along a bar of starting composition C_s to produce zone levelling as in zone refining (see Section 1.1). Because there is a lower concentration C_s of Ge in the final solid alloy crystal than in the liquid C_1, excess Ge is continually rejected into the melt ahead of the growing solid-liquid interface. Consequently there will be an increased Ge concentration in some boundary layer in the melt as shown in Fig. 6.3. The equilibrium melting (liquidus) temperature T_1 thus varies as shown in Fig. 6.3. The line G_s represents a temperature gradient for which constitutional supercooling occurs. That is, throughout the cross-hatched region in Fig. 6.3 the actual temperature, G_s, is below the liquidus temperature T_1. For temperature gradients of the critical value, G_c, and above there is no constitutional supercooling. That is, high temperature gradients and fast growth are needed to avoid constitutional supercooling.

It can be shown (e.g. Dismukes and Ekstrom 1965) that the critical condition to avoid constitutional supercooling is given by:

$$\frac{G_c}{R_c} = \frac{m(C_1 - C_s)}{D} = \frac{mC_s(1 - k)}{kD} \tag{6.1}$$

where R_c is the critical growth rate, the critical temperature gradient G_c is defined in Fig. 6.3, m is the slope of the liquidus, i.e. $m = (dT_1/d[Ge])$ where [Ge] is the concentration of Ge, D is the diffusion coefficient for Ge and k is the distribution coefficient $k = C_s/C_1$.

The consequence of the phenomenon of constitutional supercooling is a tendency to grow-in forms of compositional inhomogeneity called cell structure and faceting. The consequence of setting up high temperature gradients to avoid constitutional supercooling is a tendency to form another type of maldistribution called impurity growth striations as will be discussed in Section 6.5.4.

This discussion has used Ge-Si alloys as the illustrative example not only because of the importance of this material but also because (1) the system is simple, and (2) these non-equilibrium problems are severe although Ge-Si alloys are regular and nearly ideal thermodynamically.

Constitutional supercooling can be encountered in any system of two or more components such as Ge or Si with a high doping concentration as well as doped compounds and semiconductor alloys. The problem tends to be especially severe in cases in which the liquidus and solidus are widely separated as in the lowest-lying alloy system of Fig. 6.4 (InSb-InAs) because in such cases k, the distribution coefficient, is small. Equation (6.1) shows this means that the critical growth rate R_c tends to be small and the critical temperature gradient G_c tends to be large. Such values are difficult to attain controllably without other problems such as impurity growth striations setting in.

6.5.2 Cell formation

When the melt boundary layer is constitutionally supercooled, it has a temperature below the liquidus for its enriched (e.g. with Ge in the case of Ge–Si) concentration. It is unlikely that solid will precipitate from the liquid ahead of the advancing growth interface due to the activation barrier for homogeneous nucleation. What usually happens is that any irregular protuberance in the otherwise flat growth face, arising from random fluctuations, grows more rapidly into the supercooled liquid than the rest of the interface. The fast-growing protuberance rejects excess Ge to the sides, and more protuberances develop as shown in Fig. 6.5. The material between the protrusions, having a higher Ge concentration, cannot freeze until a lower temperature is reached. Thus low and high Ge concentrations are frozen into the

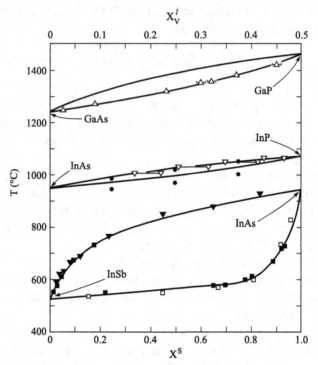

Figure 6.4   The liquidus and solidus curves for the InSb-InAs, InAs-InP, and GaAs-GaP pseudobinary alloy systems. The width of the liquidus-solidus gap varies greatly between these systems and with it the severity of the problem of constitutional supercooling. (After Panish and Ilgems 1972. Reprinted from *Progress in Solid State Chemistry*, **7**, Phase equilibria in ternary III-V systems, pp. 39−83. Copyright 1972, with permission from Elsevier.)

crystal. The cell structure, when examined in the growth interface plane normal to that of Fig. 6.5, is usually of hexagonal honeycomb form.

It was found in early work on the growth of bulk crystals from the melt that constitutional supercooling makes it difficult to produce homogeneous alloy (bulk) materials such as $GaAs_{1-x}P_x$ or $Pb_ySn_{1-y}Te$ (Hiscocks and West 1968). It was found e.g. that inhomogeneities of this type in the composition of $GaAs_{1-x}P_x$ material used to make early injection lasers affected the spontaneous luminescence spectrum which prevented the development of good lasers (Pankove *et al.* 1967). Other examples of inhomogeneities due to this cause were found in GaAs crystals doped with Te (Iizuka 1968) and in PbTe (Crocker 1966). This problem is one of the main reasons that so few III-V compounds have been produced as wafers of Czochralski crystals. The material, even in these few relatively good cases, is still inadequate for making devices. This is why all III-V devices must be made in material epitaxially grown on one of the few available types of Czochralski III-V wafers as starting substrates.

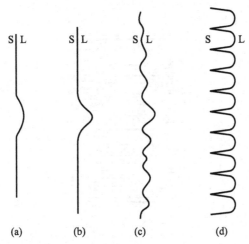

Figure 6.5 Schematic illustration of the development of a cellular interface from an initial protrusion in a solid liquid interface growing into a constitutionally supercooled melt. (After Delves 1975. Reprinted from *Theory of Interface Stability in Crystal Growth*. Copyright 1975, with permission from Elsevier.)

6.5.3 Crystallographic impurity facets

The distribution coefficient, k, and hence the critical, minimum growth rate to avoid constitutional supercooling and the concentration of an impurity or one component of an alloy, varies with the crystallographic orientation of the solid-liquid growth interface. There is therefore sometimes a tendency for crystallographic facets to form in the curved growth interfaces. This leads to anisotropic impurity segregation. The impurity distributions, whether planar, annular or high in a central core, are often referred to as facets. This phenomenon was extensively studied in the early days of crystal growth applied mainly to Ge, Si or InSb. Reviews of this work will be found in Mullin (1962) and Kane and Larrabee (1970).

6.5.4 Impurity growth striations

Growth striations often have strikingly regular stripe-like forms, hence the descriptive terms: striations or striae. They are characteristic of crystals grown from the melt, including by float-zoning, and have been studied mainly in relation to melt growth.

The first intensively studied striations were the spiral ramp impurity striations in Czochralski-grown crystal boules. In the Czochralski technique a seed crystal is dipped into the melt and then withdrawn ('pulled') while being rotated. The rotation eliminates the irregular crystal cross-sections that otherwise result from the non-uniform rates of heat flow radially outward from the axis of the crystal growth

Figure 6.6 Spiral ramp striations of oxide in a Czochralski-grown silicon crystal. The crystal was cross-sectioned down its length and imaged by Lang x-ray topography. (After Schwuttke 1962.)

apparatus. The impurity growth striations in crystals grown in this way were found to be spiral ramps with pitches given by dividing the pull velocity by the number of rotations per unit time (Fig. 6.6). The thermal asymmetry in the melt was early recognized as the cause of the helical impurity distributions in these cases. It was later shown that the principal effect is associated with the formation of planar facets, usually on {111} planes, at the solid liquid interface.

A more general cause of impurity growth striations in melt-grown crystals is convective effects in the melt. Under certain hydrodynamic conditions (low Prandtl number liquids under conditions of quite low Rayleigh number) the molten material exhibits convective motion with periodic components. These periodicities in the convection appear as temperature fluctuations, often with a markedly simple periodic character. It is well known from work on zone refining (Pfann 1957) that the concentration of many impurities that are incorporated during solidification depends strongly on the growth rate. The convective fluctuations mentioned above result in fluctuations in the growth rate. Consequently periodic variations in the grown-in impurity concentrations appear in the crystal. These are striations. A variety of methods such as magnetic quenching of the convection currents in electrically conducting melts have been developed and much detailed experimental work has been done in this field. Reviews by Bate (1968), Hurle (1972), Sangwal and Benz (1996), Wang *et al.* (2004), and Hurle and Rudolph (2004, see Section 6.6) cover defect formation, segregation, faceting, and impurity striations in melt-grown semiconductors.

Sangwal and Benz (1996) reviewed the formation of impurity striations in relation to the conditions and mechanisms of growth from melts and solutions. They outlined the mechanisms of formation of different types of impurity striations and facets, and described the effects of gravity, magnetic field and ultrasonic vibrations on them.

Wang *et al.* (2004) outlined the development of quantitative analytical tools for characterization of the crystal growth process. These proved invaluable in understanding the cause of segregation during crystal growth from the melt. The authors also emphasized the importance of the interdisciplinary approach in elucidating the relationships between crystal growth parameters and the properties of the grown crystals.

Striations are not commonly found in vapour-phase-grown material. Some work has been done on convective effects in vapour growth reactors which indicates that these effects should be of lesser importance than those in melt growth (Hurle 1972). In general, striations will result in a number of physical properties having anisotropies different from those to be expected on the basis of the symmetry of the crystal structure of the semiconducting material (Bate 1968).

Striations can be observed by a variety of microscopic techniques. The bulk electron voltaic effect and β-conductivity methods in the scanning electron microscope conductive mode were developed by Munakata (see Holt (1974) for a detailed account and the original Munakata references). These methods can give quantitative plots of resistivity variations with sufficiently high spatial resolution to reveal impurity striations. Optical microscopy can also be used in many cases. This may be either transmission microscopy, using an appropriate wavelength, e.g. in the infrared, or visible light reflection from etched surfaces. More recent techniques employed for the observation of impurity striations include, e.g. x-ray topography, x-ray double crystal diffraction, chemical etching, laser-light-scattering microtomography, and photoluminescence tomography (see Sangwal and Benz 1996 and references therein).

6.5.5 Swirl defects in silicon

Much silicon is grown as dislocation-free crystals. This is done by using Dash's (1958) method of 'necking' the seed crystal while growing from the melt either by the Czochralski method or the floating zone method (Section 1.1). In the absence of dislocation sinks, in regions away from the crystal surface, during cooling the point defects condense to produce microdefects. These are distributed in characteristic 'swirl' patterns seen in cross-sectional slices as shown in Fig. 6.7, hence the name swirl defects. De Kock (1973, 1979) and many others extensively studied these defects. The latter reference is a review of this work, to which reference should be made for an introduction to the original literature.

Floating zone (FZ) material can be of high purity but Czochralski (Cz) material contains up to about 3×10^{18} oxygen atoms cm^{-3} derived from the silica crucible holding the molten Si (see Fig. 1.1).

Figure 6.7   TV monitor picture obtained from a silicon vidicon camera with a photodiode array target made on a dislocation-free slice. The target was not illuminated. Leaky diode regions appear as bright spots and a background swirl pattern of leakage current is visible. Similar swirl patterns of precipitates were found in similar slices by means of x-ray topography. (After de Kock 1973.)

Two different forms of condensed swirl defects were found in the purer FZ material (de Kock 1973, Bernewitz *et al.* 1974). The larger type A defects occur in low concentrations around $10^6\,\mathrm{cm}^{-3}$ and the small type B defects in concentrations from 10^7 to $10^{11}\,\mathrm{cm}^{-3}$. When dislocations are present to act as sinks, no microdefects are present within a mm of the dislocation line. This indicates that the native point defects are extremely mobile. This is in agreement with expectations for both vacancies and interstitials.

Transmission electron microscope (TEM) analyses show the A type defects to be unit dislocation loops lying generally in {111} planes as shown in Fig. 6.8. The Burgers vector was found to be of $a/2\langle110\rangle$ form, inclined to the loop plane. The loops are of interstitial type. In less pure material impurity decoration of the loops is sometimes observed. The B swirl defects are not visible in the TEM and their structure is not known.

In crystals containing relatively high concentrations of oxygen and carbon a few faulted loops are found. In addition to simple loops, loop clusters and more complicated localized dislocation structures are also found (Bernewitz *et al.* 1974, Seeger *et al.* 1979).

Several early models of the formation of swirl defects were published. Some assumed the aggregation of interstitials on carbon atom nuclei, others the separate aggregation of vacancies and interstitials, local internal melting, or the formation of interstitial-type dislocation loops by the aggregation of vacancies (for details and references see de Kock 1979). Mahajan (2004) summarized the present consensus on the formation of swirl defects during crystal growth. It is based on mechanisms related to supersaturations of interstitials and vacancies during the cool-down from

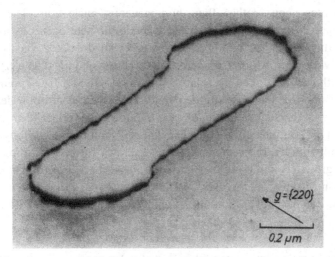

Figure 6.8 An A-type swirl defect consisting of a loop of unit dislocation seen by transmission electron microscopy. (After Foell *et al.* 1977. Reprinted from *Journal of Crystal Growth*, **40**, The formation of swirl defects in silicon by agglomeration of self-interstitials, pp. 90–108. Copyright 1977, with permission from Elsevier.)

the growth temperature. Thus, B swirls are related to the agglomeration of the interstitials due to supersaturation. Impurities (e.g. carbon) are sites of heterogeneous nucleation. With continuing cool-down, B swirls transform into extrinsic stacking faults that grow by the absorption of Si interstitials. At a specific size the defect is transformed into a dislocation loop, i.e. an A swirl (see Mahajan 2004, and references therein).

Swirl defects are harmful in electronic devices. This may be either in themselves (de Kock 1970, Ravi and Varker 1974) or through A type defects giving rise to larger stacking faults during oxidation treatments (de Kock 1973, Matsui and Kawamura 1972) or to stacking faults in the growth of epitaxial layers on FZ substrates (Ravi 1974). For these reasons methods have been developed for growing swirl-free FZ crystals. These are growth in a hydrogen vapour (de Kock 1971), or growth at high pull rates, $>5\,\mathrm{mm\,min^{-1}}$, in an argon atmosphere (de Kock *et al.* 1973) or growth at very low pull rates, $0.2\,\mathrm{mm\,min^{-1}}$ (Roksnoer *et al.* 1976).

In Cz material there are three types of microdefects that can form in numbers that depend on the doping. In crystals doped with $[B] > 10^{16}\,\mathrm{cm^{-3}}$ the largest type is predominant. In Sb and P doped crystals the two smaller types are found. TEM studies showed the largest defects to be interstitial type dislocation loops similar to the A type swirl defects in FZ silicon. Processing of Cz material at temperatures in the range from $1000°$ to $1200°\mathrm{C}$ leads to the precipitation of excess oxygen from solution, producing a variety of microdefects (Patel and Authier 1975, Maher *et al.* 1976). In dislocation-free crystals these defects are distributed in a striated pattern. There is a controversy concerning the mechanism of the precipitation of oxygen in silicon. This evidence suggests heterogeneous rather than homogeneous nucleation. The pre-existing condensed native defect aggregates could provide the heterogeneous nuclei and so account for the spatial distribution of the oxygen-related defects. The formation and evolution of various grown-in microdefects in Cz silicon and FZ silicon crystals during crystal growth were described by Talanin and Talanin (2004). They pointed out that the decomposition of the oversaturated solid-state solution of point defects during cooling below the crystallization temperature follows two independent mechanisms, i.e. vacancy-type and interstitial-type. The defect formation is caused by the primary oxygen-vacancy and carbon-interstitial agglomerates formed on impurity centres. For specific growth conditions during the cool-down, the aggregation of point defects may result in vacancy microvoids and interstitial-type dislocation loops around these oxygen-vacancy and carbon-interstitial aggregates (Talanin and Talanin 2004).

6.6 Major persisting issues

Hurle and Rudolph (2004) outlined the historical advances over the past several decades in understanding the formation of various defects in semiconductor crystals. They outlined major mechanisms of the formation of defects such as native point defects and dislocations, and discussed such issues as striations and the effects of constitutional supercooling, the relationship between facets and inhomogeneous dopant incorporation, and the problem of twinning in InP. They also point out that although dislocations can be eliminated in Si, this is not the case for III-V and II-VI semiconductor compounds. In such compounds, the lower thermal conductivity and yield stresses do not allow the reduction of the thermal stresses to such a level as to prevent multiplication of dislocations. Nevertheless, recent advances in vertical Bridgman and gradient freeze growth techniques resulted in obtaining undoped GaAs and InP with reduced dislocation densities (see Hurle and Rudolph 2004 and references quoted therein).

Hurle and Rudolph (2004) also discuss in their review some of the persisting problems. These include the need for improved understanding of the thermodynamics and kinetics of point defects and their interactions with dopant atoms, both during growth and in post-growth annealing employed to optimize the material. The issue of morphological instability related to constitutional supercooling is also of great importance; the resolution of this problem (for the growth of solid solution crystals of Si-Ge and of pseudo-binary semiconductor III-V and II-VI alloys) could result in new possibilities for device design. Scaling-up crystal diameters leads to greater thermal stresses during cooling, and reducing the density of dislocations in the larger crystals presently being grown and under development would necessitate continuous improvements of furnace design. The scaling-up also results in greater turbulence in the melt, and hence strong fluctuations in growth rate. The damping of this turbulence could be realized by employing magnetic fields (Hurle and Rudolph 2004).

Some of these issues were also discussed in a review of the melt growth of III-V compounds by Mullin (2004), who outlined advances in liquid encapsulation, vertical gradient freeze, vapour pressure controlled Czochralski and hot wall pulling techniques. Mullin (2004) reviewed the advantages and practical limitations of these techniques and such issues as twinning, cellular structure, dislocation formation, and constitutional supercooling. The review also addressed the potential of electro-magnetic stirring.

The formation of dislocation cellular structures (dislocation patterning) in melt-grown semiconductor compound crystals (III-V, II-VI and IV-VI) was discussed from a phenomenological perspective by Rudolph (2005). These cellular structures consist of high-dislocation-density walls separated by interior lower-density regions. Such cellular structures form in growing crystals due to the action of the thermo-mechanical stress field. Hence, the analysis of such structures is often directed at correlating the cell dimensions with the dislocation density and thermo-mechanical stress. Rudolph (2005) also outlined some possible methods for preventing dislocation patterning during growth. These include (i) high doping, (ii) minimization of the thermo-mechanical stress, (iii) minimization of the intrinsic point defect content by control of stoichiometry during growth, (iv) growth at temperatures substantially lower than the melting point, and (v) solid state recrystallization. The need for a theoretical framework for this phenomenon was also emphasized (Rudolph 2005).

References

Bate, R. T. (1968). Electrical properties of nonuniform crystals. In *Semiconductors and Semimetals, Vol 4. Physics of III-V Compounds*, eds. R. K. Willardson and A. C. Beer (New York: Academic Press), pp. 459–76.

Bernewitz, L. J., Kolbesen, B. O., Mayer, K. R. and Schuh, G. E. (1974). TEM observation of dislocation loops correlated with individual swirl defects in as-grown silicon. *Applied Physics Letters*, **25**, 277–9.

Chalmers, B. (1964). *Principles of Solidification*. New York: Wiley.

Cullis, A. G. and Katz, L. E. (1974). Electron microscope study of electrically active impurity precipitate defects in silicon. *Philosophical Magazine*, **30**, 1419–43.

Crocker, A. M. (1966). Mosaic-free PbTe crystals. *British Journal of Applied Physics*, **17**, 433.

Dash, W. C. (1958). The growth of silicon crystals free from dislocations. In *Growth and Perfection of Crystals*, eds. R. H. Doremus, B. W. Roberts and D. Turnbull (New York: Wiley), pp. 361–85.

de Kock, A. J. R. (1970). Vacancy clusters in dislocation-free silicon. *Applied Physics Letters*, **16**, 100–2.

de Kock, A. J. R. (1971). The elimination of vacancy-cluster formation in dislocation-free silicon crystals. *Journal of the Electrochemical Society*, **118**, 1851–6.

de Kock, A. J. R. (1973). Microdefects in dislocation-free silicon crystals. *Philips Research Reports Supplement*, **1**, 1–105.

de Kock, A. J. R. (1979). Introduction of defects into silicon during growth and processing. In *Defects and Radiation Effects in Semiconductors* 1978. Conf. Series No. 46 (Bristol: Institute of Physics), pp. 103–11.

de Kock, A. J. R., Roksnoer, P. J. and Boonen, P. G. T. (1973). Microdefects in swirl-free crystals. *Journal of the Electrochemical Society*, **120**, 94c.

Delves, R. T. (1975). *Theory of Interface Stability in Crystal Growth*, ed. B. R. Pamplin (Oxford: Pergamon), pp. 40–103.

Dismukes, J. P. and Ekstrom, L. (1965). Homogeneous solidification of Ge-Si alloys, *Transactions of the Metallurgical Society of AIME*, **233**, 672–80.

Foell, H., Goesele, U. and Kolbesen, B. O. (1977). The formation of swirl defects in silicon by agglomeration of self-interstitials. *Journal of Crystal Growth*, **40**, 90–108.

Hiscocks, S. E. R. and West, P. D. (1968). Crystal pulling and constitution in $Pb_{1-x}Sn_xTe$. *Journal of Materials Science*, **3**, 76–9.

Holt, D. B. (1974). The bulk electron voltaic effect. In *Quantitative Scanning Electron Microscopy*, eds. D. B. Holt, M. D. Muir, P. R. Grant and I. M. Boswarva (London: Academic Press), pp. 269–84.

Hurle, D. T. J. (1972). Hydrodynamics, convection and crystal growth. *Journal of Crystal Growth*, **13/14**, 39–43.

Hurle, D. T. J. (1979). Revised calculation of point defect equilibria and non-stoichiometry in gallium arsenide. *Journal of Physics and Chemistry of Solids*, **40**, 613–26.

Hurle, D. T. J. and Rudolph, P. (2004). A brief history of defect formation, segregation, faceting, and twinning in melt-grown semiconductors. *Journal of Crystal Growth*, **264**, 550–64.

Iizuka, T. (1968). Some observations of large imperfections in highly Te-doped GaAs crystals. Japan. *J. Appl. Phys.*, **7**, 485–9; and Investigation of microprecipitates in highly Te-doped GaAs Crystals. *Japanese Journal of Applied Physics*, **7**, 490–7.

Kane, P. F. and Larrabee, G. B. (1970). *Characterization of Semiconductor Materials*, New York: McGraw-Hill.

Katz, L. E. (1974). Relationship between process-induced defects and soft p-n junctions in silicon devices. *Journal of the Electrochemical Society*, **121**, 969–72.

Kroger, F. A. (1964). *The Chemistry of Imperfect Crystals*, Amsterdam: North-Holland.

Kroger, F. A. and Vink, H. J. (1956). Relations between the concentrations of imperfections in crystalline solids. *Solid State Physics*, **3**, 307–435.

Mahajan, S. (2004). The role of materials science in microelectronics: past, present and future. *Progress in Materials Science*, **49**, 487–509.

Maher, D. M., Staudinger, A. and Patel, J. R. (1976). Characterization of structural defects in annealed silicon containing oxygen. *Journal of Applied Physics*, **47**, 3813–25.

Matsui, J. and Kawamura, T. (1972). Spotty defects in oxidized floating-zoned dislocation-free silicon crystals. *Japanese Journal of Applied Physics*, **11**, 197–205.

Mullin, J. B. (1962). Segregation in indium antimonide. In *Compound Semiconductors*, **I**, Preparation of III-V compounds, eds. R. K. Willardson and H. L. Goering (New York: Reinhold), pp. 365–81 and subsequent papers in that volume.

Mullin, J. B. (2004). Progress in the melt growth of III-V compounds. *Journal of Crystal Growth*, **264**, 578–92.

Panish, M. B. and Ilgems, M. (1972). Phase equilibria in ternary III-V systems. *Progress in Solid State Chemistry*, **7**, 39–83.

Pankove, J. I., Nelson, H., Tietjen, J. J., Hegyi, I. J. and Maruska, H. P. (1967). $GaAs_{1-x}P_x$ lasers. *RCA. Review*, **28**, 560–8.

Patel, J. R. and Authier, A. (1975). X-ray topography of defects produced after heat-treatment of dislocation-free silicon containing oxygen. *Journal of Applied Physics*, **46**, 118–25.

Pfann, W. G. (1957). Techniques of zone melting and crystal growing. *Solid State Physics*, **4**, 428–521.

Queisser, H. J. and Haller, E. E. (1998). Defects in semiconductors: some fatal, some vital. *Science*, **281**, 945–50.

Ravi, K. V. (1974). The heterogeneous precipitation of silicon oxides in silicon. *Journal of the Electrochemical Society*, **121**, 1090–8.

Ravi, K. V. and Varker, C. J. (1974). Comments on the distinction between 'striations' and 'swirls' in silicon. *Applied Physics Letters*, **25**, 69–71.

Roksnoer, P. J., Bartels, W. J. and Bulle, C. W. T. (1976). Effect of low cooling rates on swirls and striations in dislocation free silicon crystals. *Journal of Crystal Growth*, **35**, 245–8.

Rudolph, P. (2005). Dislocation cell structures in melt-grown semiconductor compound crystals. *Crystal Research and Technology*, **40**, 7–20.

Sangwal, K. and Benz, K. W. (1996). Impurity striations in crystals. *Progress in Crystal Growth and Characterization of Materials*, **32**, 135–69.

Schwuttke, G. H. (1962). X-Ray diffraction microscopy of impurities in silicon single crystals. In *Direct Observation of Imperfections in Crystals*, eds. J. B. Newkirk and J. H. Warnick (New York: Interscience), pp. 497–508.

Seeger, A., Frank, W. and Gosele, U. (1979). Diffusion in elemental semiconductors: new developments. In *Defects and Radiation Effects in Semiconductors*, 1978. Conf. Series No. 46, ed. J. H. Albany (Bristol: Institute of Physics), pp. 148–67.

Stoneham, A. M. (2000). *Theory of Defects in Solids. Electronic Structure of Defects in Insulators and Semiconductors*. Oxford: Oxford University Press.

Swalin, R. A. (1972). *Thermodynamics of Solids*. New York: Wiley.

Talanin, V. I. and Talanin, I. E. (2004). Mechanism of formation and physical classification of the grown-in microdefects in semiconductor silicon. In *Defects and Diffusion in Semiconductors, Defect and Diffusion Forum*, **230**, 177–98.

Wang, C. A., Carlson, D., Motakef, S., Wiegel, M. and Wargo, M. J. (2004). Research on macro- and microsegregation in semiconductor crystals grown from the melt under the direction of August F. Witt at the Massachusetts Institute of Technology. *Journal of Crystal Growth*, **264**, 565–77.

Zunger, A. and Mahajan, S. (1994). Atomic ordering and phase separation in III-V alloys. In *Handbook on Semiconductors*, Vol. **3**. Amsterdam: Elsevier.

Index

625

Printed in the United States
By Bookmasters